QUANTITATIVE MICROBEAM

ANALYSIS

QUANTITATIVE MICROBEAM ANALYSIS

Proceedings of the Fortieth Scottish
Universities Summer School in Physics,
Dundee, August 1992.

A NATO Advanced Study Institute.

Edited by

A G Fitzgerald – University of Dundee

B E Storey – University of Dundee

D Fabian – University of Strathclyde

Series Editor

P Osborne – University of Edinburgh

Copublished by
Scottish Universities Summer School in Physics &
Institute of Physics Publishing, Bristol and Philadelphia

British Library Catalogue-in-Publication Data:

*A catalogue record for this book is available
from the British Library*

ISBN 0-7503-0256-9

Library of Congress Cataloguing-in-Publication Data are available.

Copublished by

SUSSP Publications
The Department of Physics, Edinburgh University,
The King's Buildings, Mayfield Road, Edinburgh EH9 3JZ, Scotland.

and

IOP Publishing Ltd, a company wholly owned by the
Institute of Physics, London.

IOP Publishing Ltd, Techno House, Redcliffe Way, Bristol BS1 6NX, UK.
US Editorial Office: IOP Publishing Inc., The Public Ledger Building,
Suite 1035, Independence Square, Philadelphia, PA 19106.

Printed in Great Britain by Galliard (Printers) Ltd, Great Yarmouth, Norfolk.

SUSSP Proceedings

1.	Hildegard Hammer	42.	Catherine Steukers
2.	Tony Tooke	43.	Steve Jenkins
3.	Florence Mercier	44.	Leon Seijbel
4.	Sandy Fitzgerald	45.	Michael Zemyan
5.	Valerie Michaud	46.	Paula Silva
6.	David Finlayson	47.	Eugenia Leitao
7.	Silvia Montoro	48.	Shengru Li
8.	Brian Storey	49.	Helmut Viefhaus
9.	Dima Shirokov	50.	Claes Olsson
10.	Leena Johansson	51.	Richard Watts
11.	Laurence Lanier	52.	Morten Raanes
12.	Eleni Pavlidou	53.	Michel Trudeau
13.	Laura Pereira	54.	Ricardo Berbara
14.	Sophie Rio	55.	Bart Kooi
15.	Edmond Redmard	56.	Francisco Yubero
16.	Teresa Pereira da Silva	57.	Janusz Lekki
17.	Anabela Rolo	58.	Bertil Stenbom
18.	Helmut Werner	59.	Maria Domenech Carbo
19.	Joanne Dumville	60.	Ray Egerton
20.	Nikolay Uzunov	61.	Rosario Maldonado
21.	Pierre Beccat	62.	Li Li Cao
22.	Randy Wood	63.	Jaques Cazaux
23.	Eleni Aloupi	64.	Colin Scott
24.	Jaqueline Smith	65.	Marek Faryna
25.	Claudia Bianchi	66.	Henrique Fonseco
26.	Igor Zotov	67.	Ian Borthwick
27.	Hong Chen	68.	Jan Solberg
28.	Stefan Baunack	69.	Gabor Horvath
29.	Aleksandras Plesanovas	70.	Alessandro Santucci
30.	Juan De Jesus	71.	Remzi Gürler
31.	Ramiro Perez-Campos	72.	Pierre Van Espen
32.	Mathias Bronold	73.	Panagiotis Aloupogiannis
33.	Elke Fakkeldij	74.	Ian Campbell
34.	Angelina Santos	75.	Pieter Kruit
35.	Arthur Berger	76.	John Walton
36.	Mark Farnworth	77.	John Titchmarsh
37.	Stefan Kuypers	78.	Jose Gimeno Adelantado
38.	Mimo Radmilovic	79.	Mark Baker
39.	Pierre Trebbia	80.	Juan Trigo
40.	Anna Belu	81.	Maria Caturla
41.	Teresa Bartlett		

40th SUSSP prizewinners (from left to right) – Arthur Berger (winner of the poster competition), Laurence Lanier (winner of Dr Ledingham's quiz), Eleni Aloupi (runner-up in the poster competition), and Igor Zotov (winner of the ten-pin bowling competition)

Executive Committee

Prof A G Fitzgerald	University of Dundee	Director and Co-Editor
Dr A O Tooke	University of Dundee	Secretary and Social Secretary
Dr D M Finlayson	University of St Andrews	Treasurer
Dr B E Storey	University of Dundee	Steward and Co-Editor
Dr D Fabian	University of Strathclyde	Co-Editor
Dr K M Ledingham	University of Glasgow	

International Organising Committee

Prof A G Fitzgerald	University of Dundee
Prof R F Egerton	University of Alberta
Prof J Cazaux	University of Reims
Prof P Karduck	RWTH Aachen
Prof D Briggs	ICI Wilton Research Centre

Lecturers

Martin Seah	National Physical Laboratory
Martin Prutton	University of York
Ray Egerton	University of Alberta
Pierre Trebbia	University of Reims
Jaques Cazaux	University of Reims
David Joy	University of Tennessee
John Titchmarsh	AEA Technology, Harwell
Pierre Van Espen	University of Antwerp
Iain Campbell	University of Guelph
Pieter Kruit	Delft University of Technology
Helmut Werner	University of Vienna
Klaus Wittmaack	Atomika Analysetechnik
David Briggs	ICI Wilton Research Centre
Ewa Maydell	University of Strathclyde
Kenneth Ledingham	University of Glasgow
Philip John	Heriot-Watt University

Seminars given by:

David Williams	Lehigh University
Sandy Fitzgerald	University of Dundee
Derek Fabian	University of Strathclyde

Preface

Recent developments in analytical electron microscopy and in surface analysis have resulted in a convergence of these artificially separated research activities. Microbeam analytical techniques have developed rapidly over the past thirty years. Surface analytical methods such as Auger electron spectroscopy have developed separately from analysis associated with high resolution electron microscopy. Scanning transmission electron microscopes and scanning electron microscopes are now being built which enable Auger electron spectra to be obtained from the surface of materials probed by electron energy loss spectroscopy or x-ray microanalysis. Instruments are now available which offer both high spatial resolution ion beam analytical methods and high spatial resolution electron spectroscopy. With the increasing use of different microbeam analytical techniques to provide complementary information it is important for the analyst in one area to understand the underlying principles and the potential of different microbeam analytical methods. The 40th Scottish Universities Summer School in Physics was organised with two objectives: first as an introduction to microbeam analysis and second as a forum for the discussion of the latest developments in a wide range of microbeam techniques. The aim of this Proceedings is to record the material discussed at the Summer School. This volume therefore provides both an introduction to and a survey of recent developments in microbeam analysis.

The 40th SUSSP was held between 16th August and 4th September 1992, in the Department of Applied Physics and Electronic and Manufacturing Engineering and in Airlie Hall in the University of Dundee. The seventy-eight participants came from twenty-six countries. The sixteen lecturers were all internationally known experts in their particular area of Microbeam Analysis. In addition to these lectures three speakers presented seminars on specialised topics. These Proceedings contain the material presented in the lectures and seminars in condensed form.

In addition to the lectures and seminars summarised in this volume, other activities contributed to the success of the 40th SUSSP. Discussion sessions were held on surface electron spectroscopies, electron energy loss spectroscopy, secondary ion mass spectrometry, accelerated ion spectroscopies, x-ray microanalysis and on the role of high resolution electron microscopy in microbeam analysis. The participants contributed actively in the programme, in extended question sessions at the end of each lecture and by contributing forty-five short papers which were given in three poster sessions. These papers were of high quality and were an important ingredient in the success of the School. The participants enjoyed an excellent Social Programme which led to the formation of firm friendships. The participants visited some of the major tourist attractions, the Edinburgh Tattoo, the Glasgow Burrell Collection, Glamis Castle, Falkland Palace, Pilochry Theatre and Scone

Palace as well as being introduced to malt whisky and the Scottish leisure pursuits of golf, hillwalking and country dancing.

We are grateful to numerous organisations and individuals for their help in making the School a success. Foremost among these, of course, is NATO, the School sponsor, the UK Science and Engineering Research Council, SUSSP, JEOL (UK), Philips Analytical, Oxford Instruments, Kratos, Gatan and Fisons Scientific Instruments. We wish also to thank our International Committee for their help in planning the Scientific Programme and the Local Committee who worked hard to produce a successful event. Thanks are due to Professor Arthur Cracknell and the staff of the Department of Applied Physics and Electronic & Manufacturing Engineering at the University of Dundee, for the facilities placed at our disposal.

The Editors are grateful to the lecturers for the considerable amount of work they have put into preparing an excellent set of contributions to this volume. We also thank Shengru Li and Allison Cook for their work during the School and their careful work in preparing publication-quality diagrams. We are grateful to Peggy Owens for her efforts in converting a large part of the text into the LATEX format. Thanks are due also to Heather Watton and Allison Cook for their preparatory work for the School.

Sandy Fitzgerald
Dundee, December 1992

Contents

Quantification in AES and XPS

M P Seah

National Physical Laboratory
Teddington, Middlesex

1   Introduction

This volume covers a broad range of microbeam analytical methods and so we shall consider, here, some very general background. We shall then consider quantification using different methods and different levels of complexity. Many problems are adequately solved using very simple relationships whilst others require much more detailed analysis. Many of the aspects discussed in this chapter are now being installed by manufacturers into commercial software and the basic structures of expert systems are already being countenanced. Thus, in a few years time it is to be hoped that analysts can concentrate on surface science and applied surface science rather than working through many of the details presented here. However, if they do not understand these basics they will not be able to take effective advantage of what the software systems have to offer.

1.1   The basic principles of AES and XPS

In both Auger electron spectroscopy (AES) and X-ray photoelectron spectroscopy (XPS) we use an incident radiation which strikes the surface under study. This surface emits low energy electrons, the energy spectrum of which exhibits lines characteristic of the atoms present at the solid surface. By measuring the spectrum with an electron spectrometer in ultra-high vacuum we obtain an analysis. In AES the incident radiation is usually an electron beam in the energy range 2 to 25 keV with a beam current in the range 1 na to 1μA. This flexibility allows both insulators and conductors to be studied with rather high sensitivity or, as discussed by Prutton in this volume, at high spatial resolution. At the current time the best spatial resolution observed is 5 nm (Janssen and Venables 1978). The incident electron ejects core electrons from atoms in the surface region and then these core holes are usually filled by electrons from higher levels in the same atom with the quantum of energy release being taken by a third electron

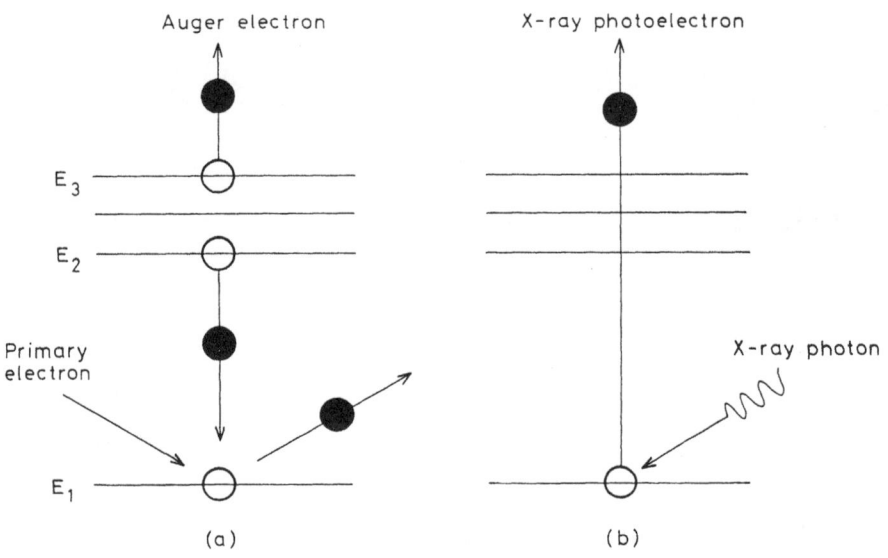

Figure 1. *Schematic representation of (a) the Auger process and (b) photoelectron creation showing the atomic core levels.*

which is ejected. In terms of the shell structure shown in Figure 1, the energy of the liberated electron, the Auger electron, is approximately E_A where

$$E_A \simeq E_1 - E_2 - E_3 \qquad (1)$$

Figure 2 shows an Auger electron spectrum with the lines associated with several elements marked. Figure 2a shows a spectrum in the direct mode and Figure 2b the traditional differential mode used to enhance the visibility of the peaks.

In XPS the radiation is usually of monochromatic X-rays from Al or Mg anodes. These give sharp lines with a width below 1 eV and in the useful energy range 1250 to 1500 eV. In some instruments the Bremsstrahlung background and weak satellite peaks are removed from the Al radiation by using a monochromator. In the past this has not been popular because considerable intensity was lost. However, in today's instruments, this is no longer the case and fully monochromated sources provide simpler spectra that are more fully resolved. The X-ray quantum of energy $h\nu$ directly ejects core electrons as shown in Figure 1 so that electrons are emitted with an energy

$$E_X = h\nu - E_1 \qquad (2)$$

to give an XPS spectrum where each core level is represented by an appropriate photoelectron line as shown in Figure 3. In this figure you will also notice Auger electron peaks produced in the subsequent filling of the initial core hole.

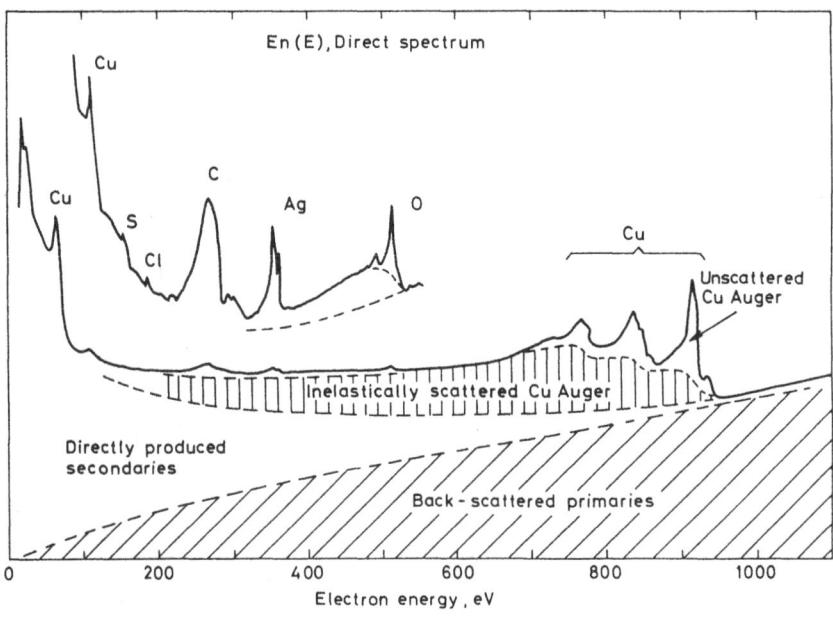

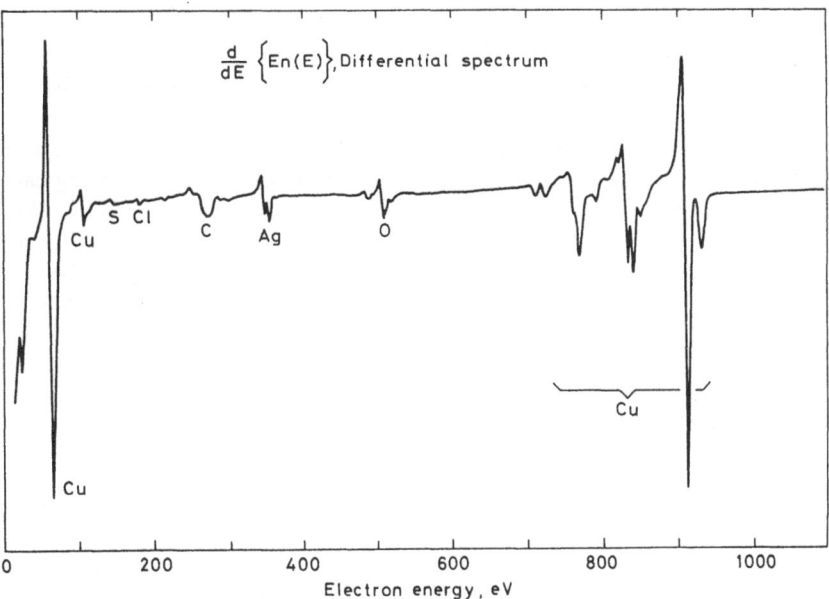

Figure 2. *Auger electron spectrum from contaminated copper, (a) direct mode, showing the separate process contributions, and (b) differential mode, after Seah (1984).*

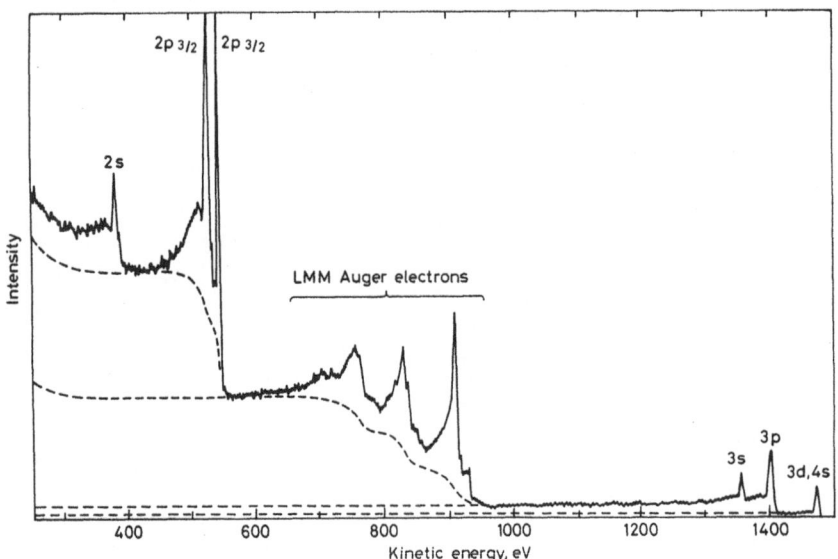

Figure 3. XPS *spectrum from copper using unmonochromated Al Kα radiation, showing a stepped background and the Auger electron peaks, after Seah (1979).*

1.2 General analytical strengths of the two techniques

Both methods analyse a zone of a few atom layers at solid surfaces with sensitivities of a few percent of an atom layer. The advantage of AES is in its easy imaging capability with submicron images readily available with signal to noise ratios of 50 for the major peaks in the spectrum. On the other hand, with XPS, resolutions below 100 μm and, more recently, 10 μm provide only moderate spatial resolution. The importance of XPS lies in the narrow and simple peak structures which permit the chemical binding state of the atoms to be identified. A second important feature is that the weak X-ray beam is less damaging than the electron beam used in AES so that organic and other delicate films may be studied. Many of these materials are insulators and these largely give little problem in XPS.

1.3 General experiments performed with AES and XPS

Many of the experiments using AES involve the spatial resolution attribute as in studies of highly inhomogeneous materials such as integrated circuits (Harris and Nowicki 1990), metal matrix composite materials and fracture samples (Seah 1990c), where elemental information is required with sub-micron spatial resolution. The remaining experiments mainly involve thin films where the surface is slowly removed with an inert gas ion beam and the exposed surface is analysed by AES (Hoffman 1990). In this way composition-depth profiles of high depth resolution may be measured to depths of 1 μm or so.

In XPS, on the other hand, depth profiling is much less popular since large areas of material must be sputtered and, in addition, the sputter-induced changes in the surface chemistry must be allowed for. Many studies concern the development of particular chemistry or functional groups in studies at polymer (Briggs 1990) or corrosion surfaces (McIntyre and Chan 1990) involved in practical situations. Some experiments traditionally performed with XPS, involving more robust materials, may, in the future, be studied with AES where chemical states may be identified at sub-micron spatial resolution and, conversely, experiments conducted by AES where high sensitivity was needed may, in the future be made by XPS where, with the lower backgrounds, higher sensitivities should be possible.

1.4   Overview of equipment, CMAs and SSAs

Electron spectrometers for AES and XPS developed historically in different ways. The first genuine AES system was a 127° electrostatic deflector analyser focusing only in the dispersion direction (Harris 1968). The concept of AES was then taken up by surface scientists with low energy electron diffraction equipment which could easily be converted to retarding field spectrometers (Weber and Peria 1967). These instruments were generally used for studying single crystal metal surfaces at poor energy resolution. To improve sensitivity the cylindrical mirror analyser (CMA) was introduced (Palmberg *et al.* 1969) and this instrument remains with us today as one of the main work-horses of AES.

For XPS, high energy resolution was required from the start (Siegbalm *et al.* 1967). These analysers were therefore based on the spherical sector analyser (SSA). To achieve the high energy resolution the spectrometer itself was operated at a moderate resolution and the electrons were retarded to a low energy, the pass energy E_p, before injection into the SSA. Transport of the electrons from the sample to the SSA and retardation were effected by an electron optical lens system as shown in Figure 4a. The resolution of the SSA is usually set by slits in the range of 0.25 to 2% of E_p so that, if E_p is 50 eV, the resolution is in the range 0.25 to 1.0 eV. The retardation of the SSA may be swept from zero to –1500 V so that the whole of the XPS spectrum is scanned at a constant resolution, the so called constant Δ E mode. These spectrometers may also be operated at a constant retardation ratio R such that $RE_p = E_A$ or E_x. In this case with R = 4 the above resolution $\Delta E/E_A$ would be 0.125 to 0.5%. This is called the constant $\Delta E/E$ mode and is often used for convenience for AES. In this mode the simpler but equally effective CMA, shown in Figure 4b is used. These instruments usually operate in the 0.25 to 1.0% resolution range and intercept 4 to 8% of the total emission from the sample (Seah and Hunt 1988). Both SSAs and CMAs are now available with multidetectors so that efficiencies reach an equivalent of 20 to 50% of the total emission current in any given energy interval.

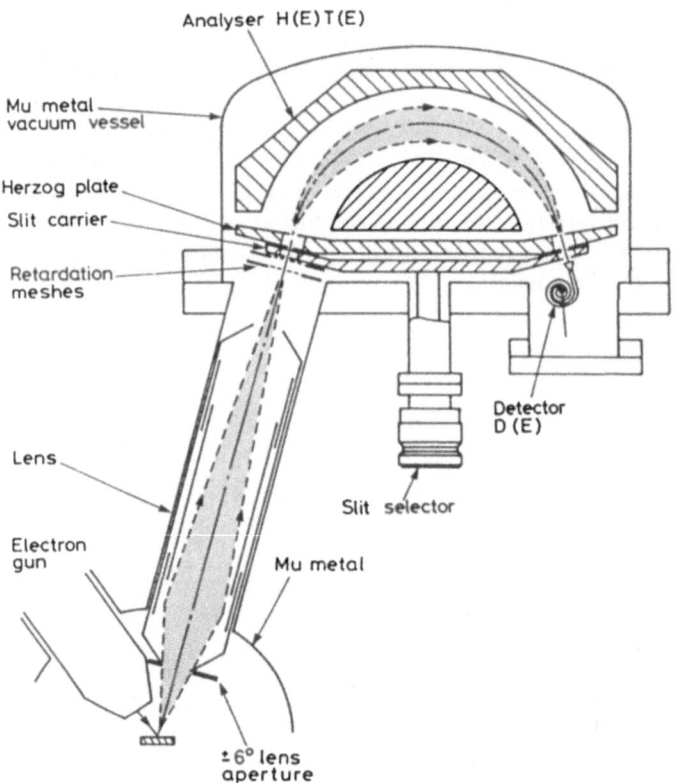

Figure 4. *(a) Schematic of a spherical sector analyser with input lens and an electron source for* AES *(for* XPS *this source would be replaced by an x-ray source).*

2   Quantification for homogenous samples

2.1   General equations for AES introducing sensitivity factors

The simplest equation for quantification which is available on most data systems is that for χ_i, the atomic fraction of element i:

$$\chi_i = \frac{I_i/I_i^\infty}{\sum_j (I_j/I_j^\infty)} \tag{3}$$

where I_i and I_j are the intensities for elements i and j in the solid and I_i^∞ and I_j^∞ are the intensities for pure bulk i and j. I_i^∞ and I_j^∞ are often known as relative sensitivity factors and are available in reference handbooks (Davis *et al.* 1976, McGuire 1979, Shiokawa *et al.* 1979 and Sekine *et al.* 1982).

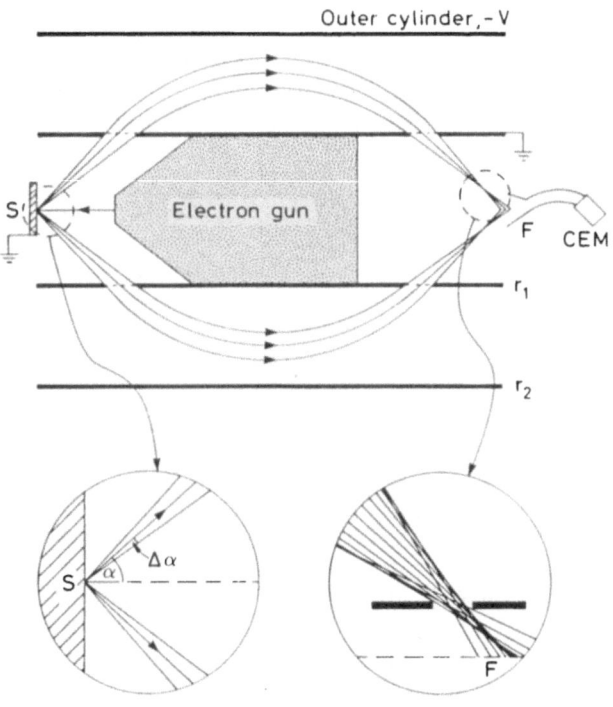

Figure 4. *(b) A cylindrical mirror analyser, after Seah (1989).*

To improve on Equation 1 we really need to know what constitutes I_i. This is often written (Seah 1990b)

$$I_i = I_0 \gamma \sigma_i(E_0) \sec \alpha \left[1 + r_m(E_i, \alpha)\right] T(E_i) D(E_i) \int_0^{+\infty} N_i(z) \exp\left(-\frac{z}{\lambda_m(E_i) \cos \theta}\right) dz \tag{4}$$

where I_0 is the primary electron beam current of energy E_0 and angle of incidence α, γ is the probability that the initially ionised core level at E_1 decays via the E_1-E_2-E_3 transition, $\sigma_i (E_0)$ is the cross section for ionising the initial core level, r_m is the additional ionisation of surface atoms due to the back reflected inelastically scattered primary electrons returning through the surface layers of the matrix m, $T(E_i)$ and $D(E_i)$ are the spectrometer transmission and detector efficiencies for the Auger electrons of energy E_i, $N_i(z)$ is the atomic density of i atoms as a function of depth z from the surface, and $\lambda_m (E_i)$ is the attenuation length of the matrix m for electrons of energy E_i. This equation is valid for a model in which the Auger electrons travel in a straight line between creation and subsequent removal from the peak intensity by inelastic scattering, *i.e.* elastic scattering is ignored. Additionally, the spectrometer is assumed to subtend a small solid angle at the angle θ to the surface normal and the surface is assumed to be atomically flat.

We may simplify Equation 4 by making some further assumptions. If the sample is homogeneous the integral over depth becomes $N_i\lambda_m(E_A)\cos\theta$. If we also normalise intensities to the pure bulk standards I_i^∞ many terms cancel so that

$$\frac{I_i/I_i^\infty}{I_j/I_j^\infty} = \frac{[1+r_m(E_i)]N_i\lambda_m(E_i)[1+r_j(E_j)]N_j^\infty\lambda_j(E_j)}{[1+r_m(E_j)]N_j\lambda_m(E_j)[1+r_i(E_i)]N_i^\infty\lambda_i(E_i)} \tag{5}$$

If a_i^3, a_j^3 and a_m^3 are the atomic volumes of i, j and the average matrix atoms, then

$$\frac{N_iN_j^\infty}{N_jN_i^\infty} = \frac{\chi_i}{\chi_j}\left(\frac{a_i}{a_j}\right)^3 \tag{6}$$

Combining Equations 5 and 6 we find a relation similar to Equation 3:

$$\chi_i = \frac{I_i/I_i^\infty}{\sum_j F_{ji}^A I_j/I_j^\infty} \tag{7}$$

where the sum j is over all elements and where F_{ji} is known as a matrix factor,

$$F_{ji}^A = \frac{[1+r_m(E_i)]\,[1+r_j(E_j)]\,\lambda_m(E_i)\lambda_j(E_j)a_i^3}{[1+r_m(E_j)]\,[1+r_i(E_i)]\,\lambda_m(E_j)\lambda_i(E_i)a_j^3} \tag{8}$$

Hall and Morabito (1979) show that F_{ji}^A depends on χ_i in any binary alloy system but that the dependence is weak. In general, therefore, it is assumed that F_{ji}^A does not depend on χ_i. Thus

$$F_{ji}^A = \left[\frac{1+r_j(E_j)}{1+r_i(E_j)}\right]\frac{\lambda_j(E_j)}{\lambda_i(E_j)}\frac{a_i^3}{a_j^3} \text{ or } \left[\frac{1+r_j(E_i)}{1+r_i(E_i)}\right]\frac{\lambda_j(E_i)}{\lambda_i(E_i)}\frac{a_i^3}{a_j^3} \tag{9}$$

Note that if I_i^∞ etc., are not from pure element standards but are derived from an alloy 'n', I_i^n, (*e.g.* a 50/50 binary alloy, 25/25/25/25 quaternary *etc.*), Equations 7 and 8 become:

$$\chi_i = \frac{I_i/I_i^n}{\Sigma_j F_{ji}^n I_j/I_j^n}$$

$$F_{ji}^n = \frac{[1+r_m(E_i)][1+r_n(E_j)]\lambda_m(E_i)\lambda_n(E_j)}{[1+r_m(E_j)][1+r_n(E_i)]\lambda_m(E_j)\lambda_n(E_i)} \tag{10}$$

If the dependence of both λ and r for both m and n have the same energy dependence then $F_{ji}^n = 1$. The matrix terms are thus sensitive to the source of the sensitivity factors I_i^∞ or I_i^n .

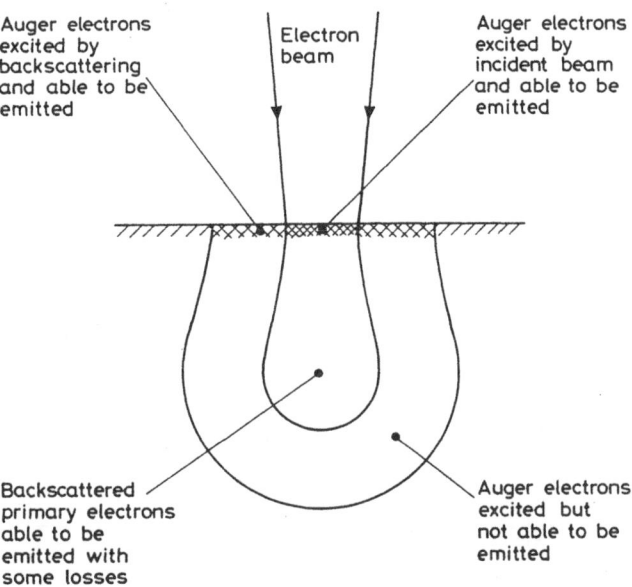

Auger electrons
excited by
backscattering
and able to be
emitted

Electron
beam

Auger electrons
excited by
incident beam
and able to be
emitted

Backscattered
primary electrons
able to be
emitted with
some losses

Auger electrons
excited but
not able to be
emitted

Figure 5. *Schematic presentation of electron scattering in the sample in* AES, *after Seah (1990b).*

2.2 Backscattering in AES

The backscattering term in Equation 8 may be very significant if the element i and the matrix m are of significantly different atomic number. Evaluations of the backscattering term, r_M, have been made by several authors by Monte Carlo calculations (Shimizu and Ichimura 1981, Ichimura and Shimizu 1981, Jablonski 1980, Shimizu 1983, El Gomati and Prutton 1978) and empirically by Reuter (1972). The calculations are roughly in agreement for deep core levels but Reuter's relations errs on the low side for core levels below 500 eV. Shimizu and Ichimura's calculations (Shimizu and Ichimura 1981, Ichimura and Shimizu 1981, Shimizu 1983) are most accessible and so are included here. Their Monte Carlo calculation assumes that the incident electron beam slows down according to the Bethe stopping-power equation with elastic scattering calculated numerically by partial wave expansion with a Thomas-Fermi-Dirac atomic potential. These energetic electrons ionise core levels according to the Gryzinski cross section (1965) in the volume shown in Figure 5 but Auger electrons generated from the subsequent decay only escape from the surface region. In Figure 5 we see two groups of Auger electrons, those generated by the primary electrons in the centre and those by the backscattered electrons over a larger area. These two groups are shown by the 1 and $r_m(E_i, \alpha)$ terms in Equation 4, respectively, and will be discussed further in their spatial context in section 5.1.

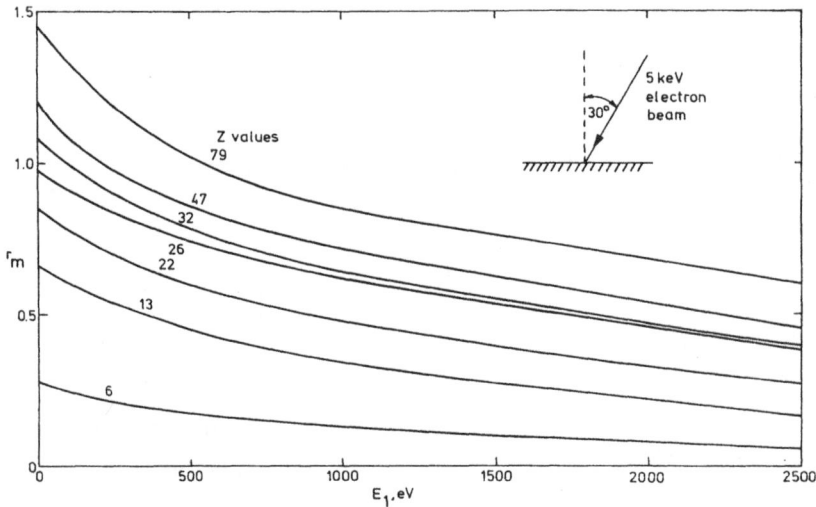

Figure 6. *The backscattering term r_m for a range of elements using 5 keV electrons at 30° from the surface normal, from the calculations of Shimizu and Ichimura (Shimizu and Ichimura 1981, Ichimura and Shimizu 1981, Shimizu 1983).*

Figure 6 shows how r_M increases with atomic number, Z, and reduces as the core level binding energy, E_1, increases for a 5 keV electron beam incident at $\alpha = 30°$ to the surface normal. Figure 7a shows how r_M varies with Z and Figure 7b the variation with α. Shimizu (1983) gives the following relations in summary:

$$r = (2.34 - 2.10Z^{0.14})\,U^{-0.35} + (2.58Z^{0.14} - 2.98); \qquad \alpha = 0°$$

$$r = (0.462 - 0.777Z^{0.20})\,U^{-0.32} + (1.15Z^{0.20} - 1.05); \qquad \alpha = 30° \qquad (11)$$

$$r = (1.21 - 1.39Z^{0.13})\,U^{-0.33} + (1.94Z^{0.13} - 1.88); \qquad \alpha = 45°$$

where U is E_0/E_1 and Z is the average atomic number of the sample in the near-surface region.

In the literature it is generally agreed that these equations describe experimental results accurately although this has been questioned (Palczynski 1991) for low values of the initial core level binding energy, E_1. Many calculations involve ratios of $(1+r)$ values and here any systematic errors are reduced. Other cross sections than Gryzinsky's have been used (Bethe 1930 and Lotz 1970) but generally the results seem worse.

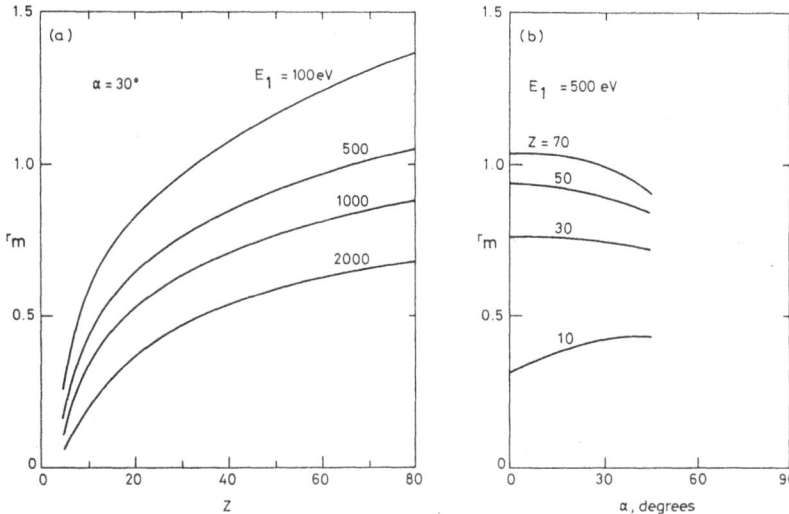

Figure 7. *The dependence of r_m on (a) the atomic number Z and (b) the angle of incidence of the electron beam α, from the data of Shimizu and Ichimura (Shimizu and Ichimura 1981, Ichimura and Shimizu 1981, Shimizu 1983).*

2.3 The inelastic mean free path and attenuation length

In Equation 4 it is assumed that the electron intensity decays exponentially with the thickness of material traversed. In this way, if an overlayer of one element is sequentially deposited on top of a second the signal of the latter decays exponentially whereas the signal of the former grows as a constant minus an exponential function. As a result of the analyses of the exponent in many experiments, Seah and Dench (1979) established the empirical expressions for the attenuation lengths, λ_a, for different classes of material over the energy range 1 eV to 6 keV.

For elements:

$$\lambda = 538E^{-2} + 0.41\,(aE)^{0.5} \text{ monolayers [1.36]} \tag{12}$$

as shown in Figure 8, whereas for inorganic compounds;

$$\lambda = 2170E^{-2} + 0.72\,(aE)^{0.5} \text{ monolayers [1.38]} \tag{13}$$

and for organic compounds;

$$\lambda = 49E^{-2} + 0.11E^{0.5} \text{ mg/m}^2 \text{ [2.10]} \tag{14}$$

where the number in square brackets represents one standard deviation uncertainty describing the scatter of the data. The inorganic compounds were composed largely of

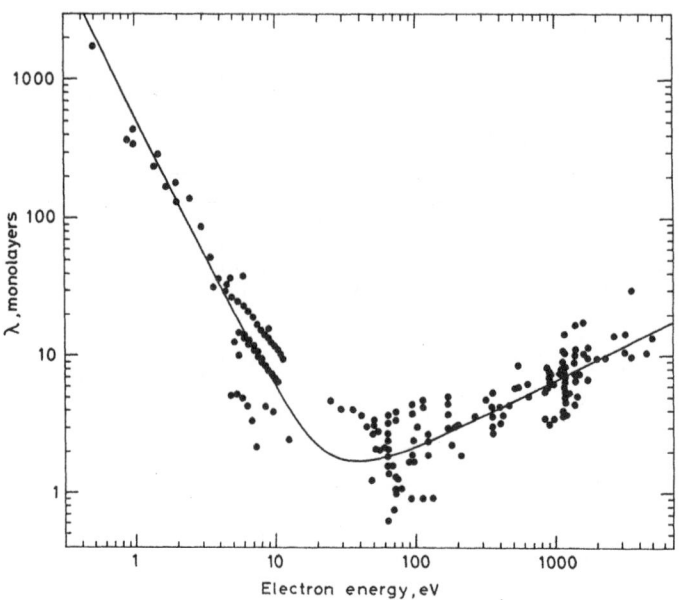

Figure 8. *The dependence of attenuation length, λ, on the emitted electron energy for elements, after Seah and Dench (1979).*

oxides or alkali halides and the factor 0.72 in Equation 13 could be replaced by 0.55 [1.37] and 0.89 [1.19] for each of these two separate groups, respectively (Seah 1986).

Analysing individual data sets, rather than the whole ensemble, Wagner *et al.* (1986) concluded that the energy dependence of the attenuation length (AL) was as E^m where m ranged from 0.54 to 0.81, rather than 0.5. The greater path lengths for the insulators compared to the elements arises as a result of the reduced inelastic scattering caused by the band gap, E_g, in insulators. Szajman *et al.* (1981) take this into account, formally, to show that both groups may be combined to form one universal curve. More recently Tanuma *et al.* (1990a,b and 1991a,b) have derived theoretical inelastic mean free paths (IMFPs) for 27 elements and 15 compounds and, for $50 \leq E \leq 2000$ eV, fit this data to the relation

$$\lambda_i = \frac{E}{E_p^2[\beta \log(\gamma E) - C/E + D/E^2]} \text{ Å} \qquad (15)$$

where E is the electron energy, eV, $E_p = 28.8(\rho N_v/A)^{\frac{1}{2}}$ eV, N_v is the total number of valence electrons per atom or molecule, ρ is in gm cm^{-3} and A is the atomic or molecular weight. In addition:

$$\beta = -0.0216 + \frac{0.944}{(E_p^2 + E_g^2)^{\frac{1}{2}}} + 0.000739\,\rho$$

$$\gamma = 0.191\,\rho^{-0.50}$$

$$C = 1.97 - 0.91U \qquad (16)$$

$$D = 53.4 - 20.8U$$

$$U = \rho N_v/A$$

Equation 15 gives the total distance travelled by an electron between inelastic scattering events. Because the electrons are also elastically scattered the net distance travelled is less. The average net distance is called the attenuation length and appears in Equations 12 and 14. The AL and IMFP would be equal in the absence of elastic scattering. An analysis of the theoretical predictions for λ/λ_i at 1000 eV shows that the ratio decreases as Z increases (Ebel *et al.* 1988). If this is used with the calculations at different energies by Jablonski (1987) we find that, to a standard deviation of 4%, the data in the latter reference may be fully described by:

$$\lambda_a/\lambda_i = \{1 - 0.028Z^{0.5}\}\,\{0.501 + 0.068\log E\} \qquad (17)$$

with an average value around 0.8. Equations 15 to 17 may be combined to give values of λ_a based on theoretical principles. However, at the present time the experimental data are too poor to test if that approach or Equations 12 to 14 are more generally effective. The latter remain popular for their simplicity. There has also been concern that relations of the type shown in Equation 4, which assume an exponential attenuation of the electrons, may be invalid because of the effects of elastic scattering. However, it is fortunate that, on the whole the angular emission of electrons is still predicted to follow the $\cos\theta$ dependence (Jablonski 1987 and Jablonski *et al.* 1988) and the attenuation is still near to an exponential function (Jablonski 1987 and Jablonski *et al.* 1988).

The effect of elastic scattering is clearly illustrated in the work of Gries and Werner (1990). They use an efficient Monte Carlo calculation to determine the intensity of electrons emitted in a given direction θ as a function of the depth of the emitter. The results of their analysis for 250 eV electrons emitted from Si are shown in Figure 9 where the elastic scattering is taken from cross sections calculated from relativistic Hartree-Fock-Slater scattering potentials and λ is assumed to be 0.91 nm. The straight lines in Figure 9a show the result expected in the absence of elastic scattering. In Figure 9b the abscissa is the reduced depth scale $z/\cos\theta$ so that the set of straight lines become the one line shown. It is seen that down to an intensity of about 7% the simple attenuation length concept is reasonably valid. At much greater depths the change in the intensity as a function of depth does not really depend on the angle of emission (the functions become parallel). This arises because the average elastic angular deflections are relatively small for trajectories of up to $2\lambda_i$ but have led to angular deflections of 90° at paths of 6 λ_i or greater. More detailed results are given by Werner (Werner 1991 and Werner *et al.* 1991).

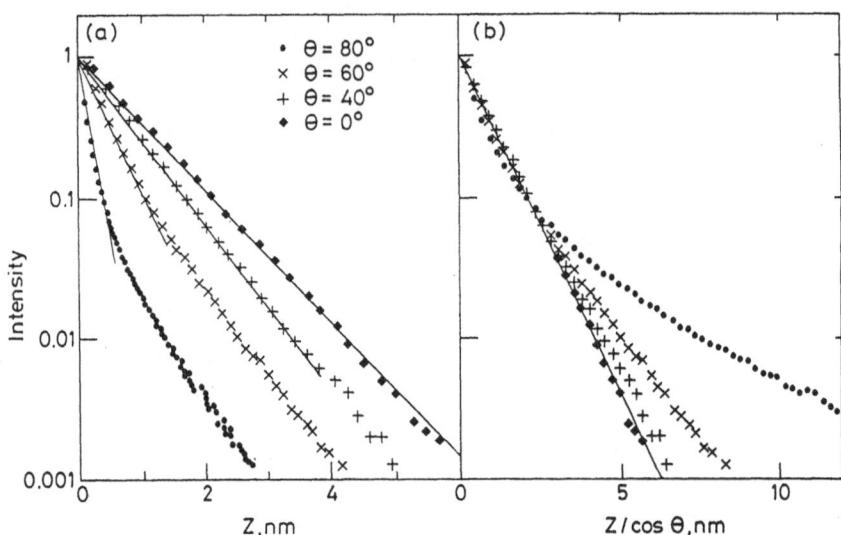

Figure 9. *The intensity emitted at an angle θ to the surface normal as at function of the depth, z, of the emitting atom. The calculations are for 250 eV electrons leaving a silicon substrate, after Gries and Werner (1990). The straight lines represent the simple exponential attenuation law for inelastic scattering in the absence of elastic scattering.*

2.4   Matrix factors for AES

As a numerical example of the calculation of matrix factors we now calculate the figure for carbon in iron with a 5 keV electron beam at 30° to the normal from Equation 9. We shall use the Seah and Dench relation (1979) for λ_a for elements and Shimizu's backscattering factor (1983).

$$F^A_{CFe} = \left[\frac{1 + r_C(E_C)}{1 + r_{Fe}(E_C)}\right] \left(\frac{a_{Fe}}{a_C}\right)^{1.5}$$

$$= \frac{1.34}{1.75} \left(\frac{0.22754}{0.20696}\right)^{1.5} \tag{18}$$

$$= 0.88$$

and, similarly, for aluminium in nickel $F^A_{AlNi} = 0.72$ for the high energy Al peak and 0.70 for the low energy one. A detailed analysis of calculations of this type, using Equations 11 and 12, has been made by Zagorenko and Zaporozchenko (1989). They find that, for five stoichiometric compounds with scribed clean surfaces, the inclusion of the above matrix factor, leads to a 7 times improvement in accuracy compared with the calculation from Equation 3. On average the matrix factors differ from unity by a factor of 1.5 (Hall and Morabito 1979).

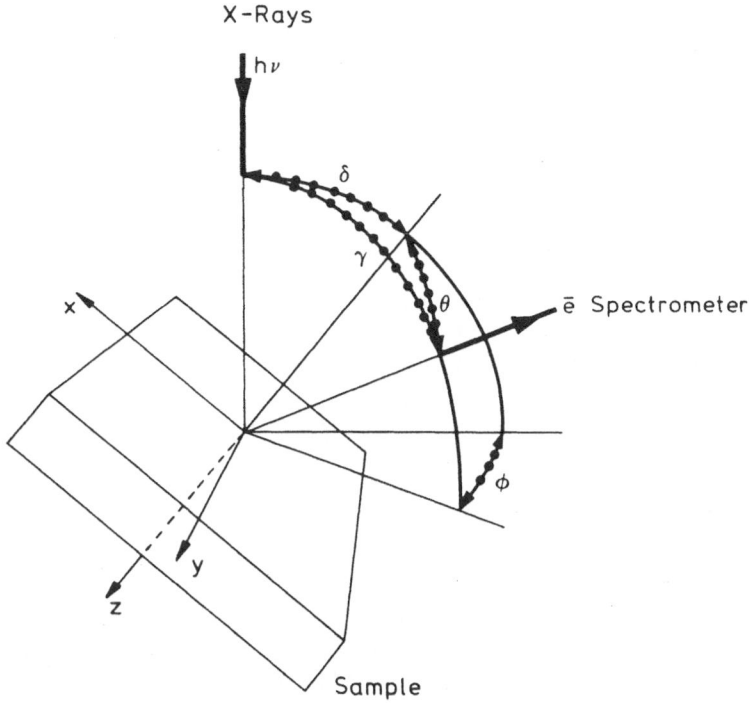

Figure 10. *Geometry of the* XPS *analysis configuration.*

2.5   General equations for XPS

The analogous picture to Figure 5 for XPS shows the characteristic X-rays penetrating the solid to a depth of many microns and ionising atoms in the core level 1 at energy E_1 over this depth. The excited core electrons move through the solid and are finally emitted from the surface with some energy loss, as shown in the spectrum for copper in Figure 3. Only those created in a zone characterised by a depth $\lambda_m(E_1)\cos\theta$ from the surface escape to provide the line intensity in the XPS spectrum. As before, $\lambda_m(E_1)$ is the attenuation length of the characteristic XPS electrons of energy E_1 from level 1 of element A in matrix M, and is the angle of emission of the electron from the surface normal. The energy (E_1) dependence of λ is shown already in Figure 8.

In an analogous manner to Equation 4 we may write:

$$I_i \;=\; \sigma_i(h\nu)D(E_i)\int_0^\pi d\gamma \int_0^{2\pi} d\phi \int_{-\infty}^\infty dx \int_{-\infty}^\infty dy \int_0^\infty dz \; L_i(\gamma)\sec\delta\, J_0(xy)T(x,y,\gamma,\phi,E_i)$$
$$\times N_i(x,y,z)\exp\left\{\frac{-z}{\lambda_m(E_i)\cos\theta}\right\} \tag{19}$$

where the subscript χ for the level χ has been omitted. Here $\sigma_i(h\nu)$ is the cross section for emission of a photoelectron from the relevant inner shell per atom of i, by a photon of energy $h\nu$, $D(E_i)$ is the detection efficiency for each electron transmitted by the electron spectrometer, the co-ordinates $\gamma, \phi, \theta, \delta$, d, y and z are given in Figure 10, $L_i(\gamma)$ is the angular asymmetry of the intensity of the photoemission from each atom, $J_0(x,y)$ is the flux of the X-ray characteristic line per unit area at point (x, y) on the sample, $T(x, y, \gamma, \phi, E_i)$ is the analyser transmission and $N_i(x, y, z)$ is the atom density of the i atoms at (x, y, z). Unlike AES, a large area of the sample is illuminated by the X-rays, there are no backscattering corrections, but there is an angular anisotropy of the emission. Calculations for the cross sections, $\nu_A(h\nu)$, have been made by Scofield (1976) and also by Band $et\ al.$ (1979) and Yeh and Lindau (1985). For the energy levels of concern in XPS these calculations show no significant difference (Seah 1986).

For a homogeneous material the integral over z becomes simply $N_i\lambda_M(E_i)\cos\theta$. If we have reference spectra of the pure elements, I_i^∞, recorded on the same instrument, Equation 19 becomes much simplified by writing:

$$\frac{I_i/I_i^\infty}{I_j/I_j^\infty} = \left[\frac{\lambda_m(E_i)\lambda_i(E_j)}{\lambda_m(E_j)\lambda_i(E_i)}\right] \left[\frac{N_i}{N_i^\infty}\frac{N_j^\infty}{N_j}\right] \tag{20}$$

so that, as before

$$\chi_i = \frac{I_i/I_i^\infty}{\sum_j F_{ji}^x I_j/I_j^\infty} \tag{21}$$

where

$$F_{ji}^X = \frac{\lambda_m(E_i)\lambda_j(E_j)}{\lambda_m(E_j)\lambda_i(E_i)} \frac{a_i^3}{a_j^3} \tag{22}$$

as before we shall assume F_{ji}^X to be independent of χ_i. If, as before the sensitivity factors come from compounds or alloys, I_i^n and not from pure elements, I_i^∞, and if a universal dependence of λ on E is assumed, then $F_{ji}^X = 1$. If not, typically F_{ji}^X will diverge from unity by a factor of 1.5.

2.6   Matrix factors for XPS

As an example of the values of F_{AB}^X we may use the previous systems of carbon in iron and aluminium in nickel. Thus

$$F_{CFe}^X = 1.153 \qquad F_{AlNi}^X = 0.812$$

quite different from the values for AES.

3   Experimental measurements and quantification

3.1   Relating experimental data to basic equations, effects of spectrometers, detectors and electronic efficiencies

In the above it has been assumed that the reference spectra have been recorded on the same instrument under the same conditions as that used for the analysis. Because this

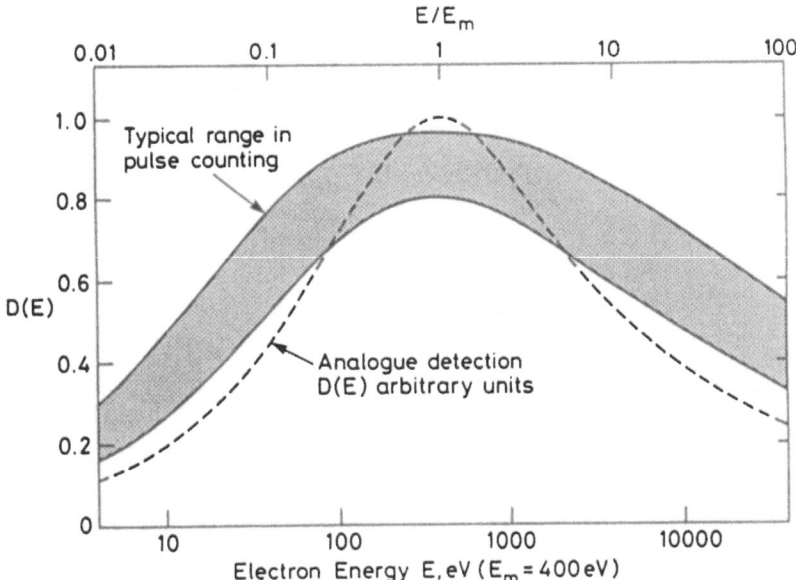

Figure 11. *Channel electron multiplier (CEM) detector efficiencies in the pulse counting and analog modes as a function of the energy of the electron as it hits the horn of the CEM, after Seah (1990a). If the spectrometer has a retardation ratio R this energy may be the electron emission energy E divided by R.*

takes a considerable effort it is customary to use published data banks (*e.g.* Davis *et al.* 1976, McGuire 1979, Shiokawa *et al.* 1979 and Sekine *et al.* 1982). In this case the two terms $T(E_i)$ and $D(E_i)$ need consideration. In electron spectrometers the electron intensity measured, $I(E)$, is related to the spectral intensity, $n(E)$, by (Smith and Seah 1988, Seah *et al.* 1989, Seah 1990a, Seah and Smith, 1991):

$$I(E) = I_0 \, H(E)T(E)D(E)F(E)n(E) = I_0 Q(E)n(D) \qquad (23)$$

where $H(E)$ is a term to allow for stray magnetic fields, design tolerances and aberrations and $F(E)$ represents electronic transfer terms for those components operating between the spectrometer output and the recording of data. Here we shall assume $H(E)$ is unity but note that stray magnetic fields generally cause $H(E)$ to fall in value for low energy electrons, particularly in systems with magnetically focused electron guns. $F(E)$ we shall also treat as unity although, as we shall see in the next section, $F(E)$ must be considered more carefully for differential spectra. $T(E)$, for AES spectrometers operated in the constant $\Delta E/E$ mode (otherwise known as constant or fixed retarding ratio), is simply proportional to E whereas $D(E)$ is a rather more complicated function (Seah 1990a). Figure 11 shows how $D(E)$ is predicted to vary with the electron energy for channel electron multiplier (CEM) detectors used in either the pulse counting or analog modes. Many instruments have been shown to obey this behaviour, particularly those based on single-pass cylindrical mirror analysers (CMAs) (Seah and Smith 1991).

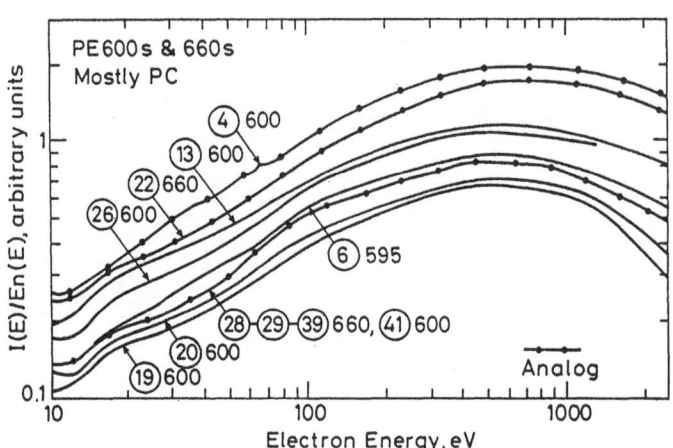

Figure 12. *The average values of Q(E)/E from Equation 23 for a range of Perkin Elmer* CMAs *used for* AES *in the constant* $\Delta E/E$ *mode using pulse counting (*PC*) or analogue amplification after Seah (Seah and Smith 1991).*

Figure 12 shows calibrated $Q(E)$ curves for several Perkin Elmer CMA's operated in the constant $\Delta E/E$ mode. These curves depend on the age and condition of the CEMs and may change significantly over a period of one year (Seah 1992). Unfortunately it is difficult to deduce $Q(E)$ for the instruments used in the AES reference data sets (Davis *et al.* 1976, McGuire 1979, Shiokawa *et al.* 1979 and Sekine *et al.* 1982) and one can only hope to minimise problems by using peaks close in energy or by deriving one's own data sets on the analytical instrument. One should also note that the angle of incidence of the electron beam and its energy also affect I_i^∞ and so these too must be correct.

For those XPS instruments operated in the constant $\Delta E/E$ mode results are similar except that the $D(E)$ term usually becomes $D(E/R)$ where R is the retardation ratio. In the constant ΔE mode $T(E)$ becomes more complex but $D(E)$ reduces to $D(E_p)$, a constant.

In general at high energies $T(E)$ falls as E^{-1} and at low energies $T(E)$ is constant. Thus, in the usual operating range for XPS $T(E)$ will be somewhere between the above limits (Seah 1980). Figure 13 shows calibrated $Q(E)$ curves for several spectrometers but note that each instrument installation of all but the simplest spectrometers will vary from the curves shown even though the instruments may be of an identical specification.

If we wish to use I_i^∞ from a published XPS data bank on instrument 2 we should note, from Equation 18 that for conversion to our instrument 1,

$$\frac{I_{1i}^\infty/I_{2i}^\infty}{I_{1j}^\infty/I_{2j}^\infty} = \frac{L_i(\gamma_1)L_j(\gamma_2)Q_1(E_i)Q_2(E_j)}{L_j(\gamma_1)L_i(\gamma_2)Q_1(E_j)Q_2(E_i)} \tag{24}$$

and that we must know the effect of the angular asymmetry factor $L(\gamma)$ if $\gamma_1 \neq \gamma_2$. This term describes the intensity distribution of the photoelectrons ejected by unpolarised

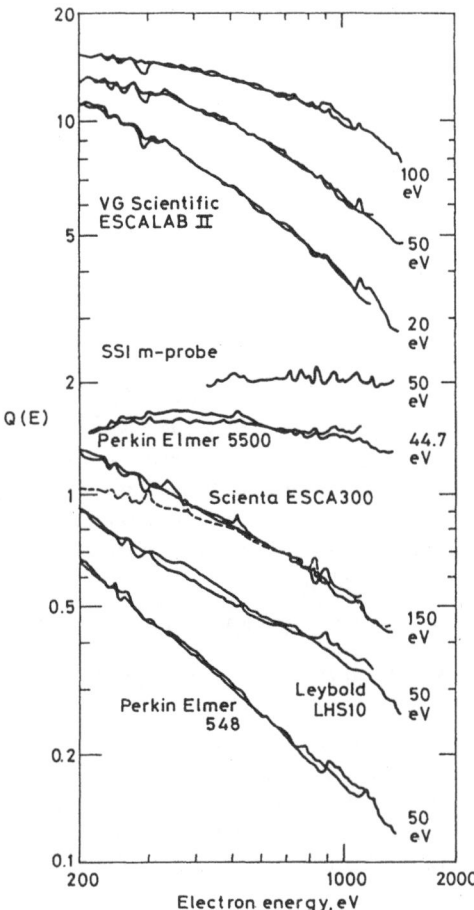

Figure 13. *Typical curves for Q(E) for Equation 23 for a range of* XPS *instruments used in the constant* ΔE *mode, after Seah (1992). Note that these are the results for individual instruments at the pass energies indicated. Another instrument of the same model may give a different function. Note the change with pass energy illustrated by the* VG *Scientific* ESCALAB II. *This type of change occurs for most instruments. Changes also occur with the slits selected and the area of analysis.*

X-rays from atoms or molecules (Reilman *et al.* 1976):

$$L_i(\gamma) \;=\; 1 + \frac{1}{2}\beta_i \left[\frac{3}{2}\sin^2\gamma - 1 \right] \tag{25}$$

where β_i is a constant for a given subshell of a given atom and X-ray photon. This takes no account of the diffraction effects that the electrons may undergo in being transported

through a crystalline lattice. For large crystals the diffracted beams may be less than
10° FWHM with first order intensity changes (Trehan and Fadley 1986, Fadley *et al.*
1979, Sekine *et al.* 1986). However, for finely polycrystalline, sputtered (Sekine *et al.*
1986), or amorphous solids the diffraction effects may be ignored.

Calculated values of β are tabulated by Reilman *et al.* (1976) for the important
transitions of all elements using the commonly used X-ray sources (Al Kα, 1486.6 eV;
Mg Kα, 1253.6 eV; and Zr Mζ, 151.4 eV. Typical values range between 1.0 and 2.0.
Further calculations are given by Band *et al.* (1979) and Yeh and Lindau (1985) but for
the range of parameters relevant to XPS the results show no significant difference (Seah
1986). These calculations give a very good correlation with the data for gases (Krause
1969) but not directly in solids (Vulli 1981 and Baschenko *et al.* 1984).

We should not expect the results to be exactly the same in solids as the electron
trajectories there are somewhat randomised by the effects of elastic scattering. These
effects have been included in a series of detailed calculations by Baschenko and Nevedov
on solids (Baschenko and Nefedov 1979, 1980, 1982 and Baschenko *et al.* 1984), with
and without overlayers, as well as by Jablonski *et al.* (Jablonski and Ebel 1984, Ebel
et al. 1985, Jablonski *et al.* 1986, Jablonski and Ebel 1988). For homogeneous bulk
materials Jablonski (1989) find that, for emission along the surface normal and with
the X-ray source inclined at an angle γ to the surface normal, the elastic scattering
reduces the effect of β to β^* according to

$$\beta^* = (0.781 - 0.00514\,Z + 0.000031\,Z^2)\beta \qquad (26)$$

which reduces from 0.93 at carbon to 0.38 at silver. Equations 19, 23, 24 and 25 should
thus be replaced by their equivalents involving $L_A^*(\gamma)$ and β_A^*.

Figure 14a shows the β values for Al Kα radiation. At the bottom of the figure
are shown some typical commercial arrangements but it is important to note that the
geometry, even for a given manufacturer's model number, may be different from that
shown since the manufacturers may move the X-ray sources to optimise the experimental
configuration. A typical value of β may be 1.4 so that $0.5 \leq \beta^* \leq 1.4$. Figure 14b shows
the angular dependence of $L_A(\gamma)$ for such typical values of β^* with the geometries of
some of the commercial systems noted. The entrance angle about the analyser directions
is typically a few degrees except for the instruments incorporating a cylindrical mirror
analyser (CMA) where the angle extends from 45° to 135° depending on the experimental
set-up. Clearly reference data taken on an instrument with one geometry will show
different relative peak intensities from those recorded on an instrument with different
geometry.

3.2   Direct spectra-peak heights

In early work for XPS peak heights were used for I_i, however it is clear that for a constant
peak area the height will depend on the peak width. If the spectrometer is not operated
at the highest resolution, peaks of different widths will be affected differently so that
apparent intensity changes occur as the spectrometer resolution changes. For a series
of measurements during one experiment in which no chemical state or peak broadening
effects occur the peak height may well prove to have the greatest precision for relative
measurements. This must be assessed by the user.

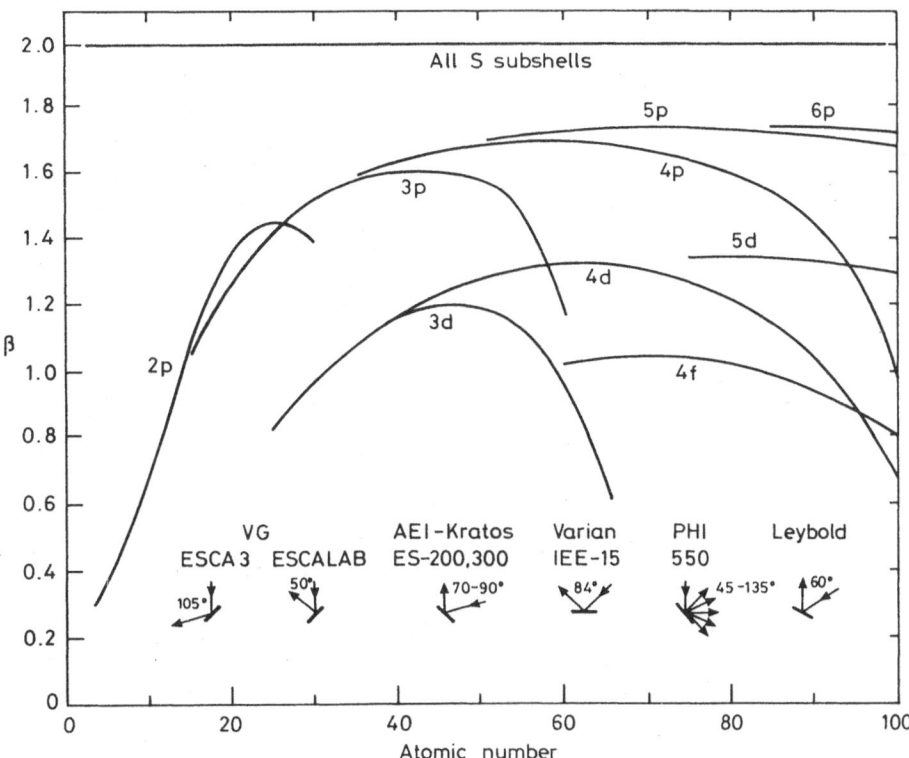

Figure 14. *(a) Calculated values of β, after Reilman et al. (1976). Typical commercial configurations are shown in the lower part of the diagram.*

3.3 Direct spectra-peak areas and background subtractions

The true intensity, I_i, to measure for both AES and XPS is peak area. The problems in removing the background to find the peak area are, in principle, the same in both AES or XPS. Peak area measurements should relate to I_i in Equations 4 and 19 irrespective of the analyser resolution or the chemical state of the species. Below some of the major methods, in use today, are outlined.

Straight line background

In this method a straight line is used to remove the background under the peak so that all of the intensity above this line is summed into the peak area as shown by the dotted line in Figure 15. This leads to a good estimate for the true area in some XPS peaks, particularly for narrow XPS peaks. This method gives good results for polymers and is available on all commercial data processing systems. Sometimes there can be difficulties in defining the precise points on the background from which to draw the line which then leads to gross errors where a small peak of one chemical state sits beside

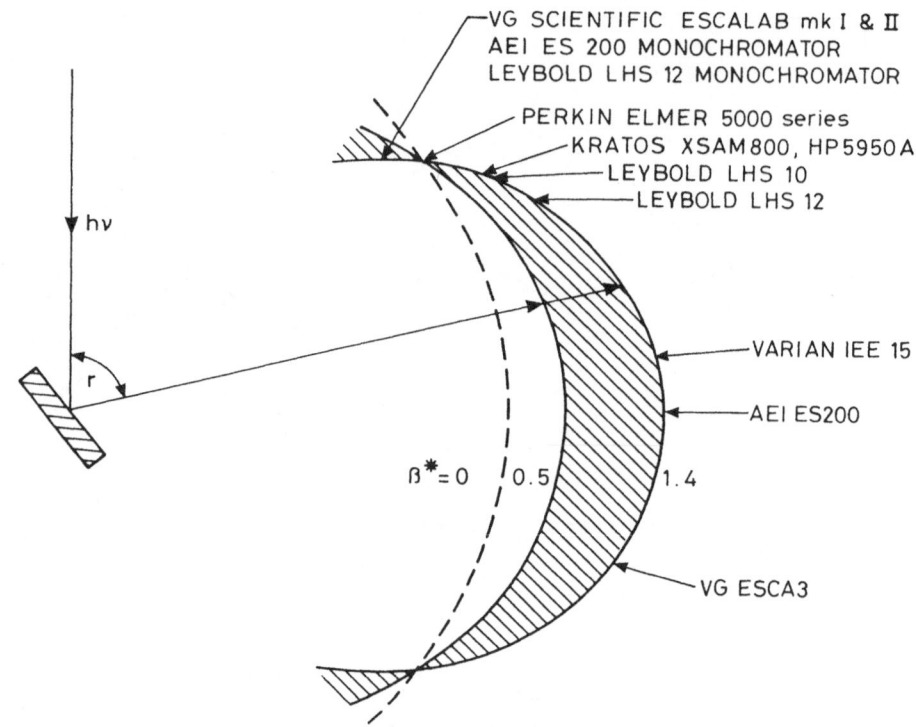

Figure 14. *(b) The angular intensity distributions for β*=0, and 0.5 to 1.4 as a function of the relative orientation of the x-ray source and the spectrometer. General commerical configurations are shown on the right.*

that of another, especially if the peaks are followed by fairly intense losses as in the transition metals. The more complex methods discussed below reduce this problem.

Shirley background

Shirley's method (1972) assumes that each intensity P_q in an energy channel above the background leads to losses of intensity kP_q in all of the lower channels. Thus, first one identifies plateau regions either side of the peak in channels f and g where the background counts are B_f and B_g at higher and lower energy than the peak, respectively. The background B_i in the ith channel between the plateau regions is then given by:

$$B_i - B_f = t \sum_{j=i+1}^{f} (C_j - B_j) \tag{27}$$

where t is a constant found by matching the background at B_g. Figure 15 shows this method applied to the results for Cu and CuO as the solid curve. This method is available on commercial systems and overcomes some of the objections one may raise

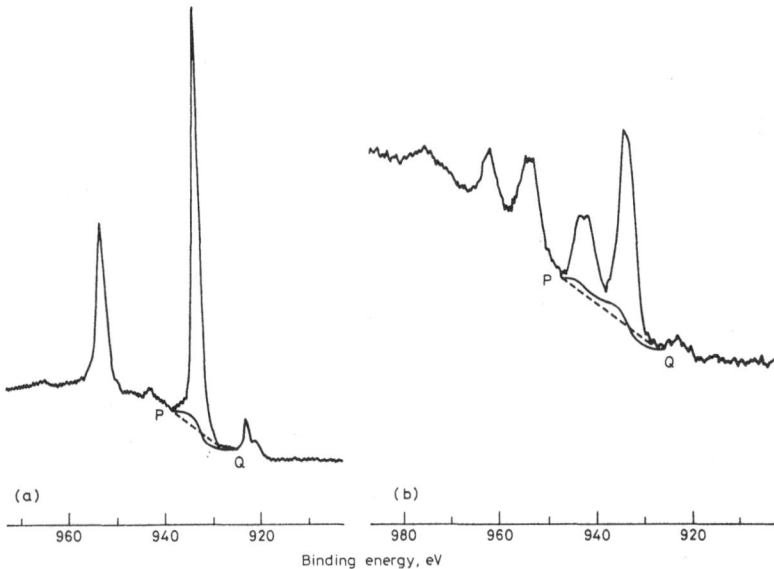

Figure 15. *Copper photoelectron peaks from (a) metallic copper and (b) CuO powder showing background subtraction methods, after Seah (1980).*

in one's mind to the straight line method. However, it is not clear if, in practice, this does give more satisfactory results.

Tougaard's method

It had been recognised for some time that the excess background at lower energies than the peaks was caused by photo or Auger electrons created in the deeper layers which had suffered characteristic losses during their movement to and emission from the surface.

Tougaard and Sigmund (1982) showed that if $j(E)$ is the measured flux of emitted electrons at energy E from a homogeneous solid, the primary excitation spectrum, $F(E)$, is given by

$$F(E) \;=\; j(E) \;-\; \lambda_i \int\limits_{E}^{\infty} K(E' - E)\, j(E') dE' \tag{28}$$

The primary excitation spectrum is our set of unscattered peaks with no background and $K(E' - E)$ is the probability that an electron of energy E shall lose energy $E' - E$ per unit path length travelled in the solid. In the derivation of Equation 28 it is assumed that the cross section for inelastic scattering is identical for all electron energies within the spectrum to be deconvoluted. Multiple inelastic scatterings are also included but the angular scatterings from elastic events are not. If the elastic scattering may be characterised by an exponential attenuation length, λ_a, then λ_i in Equation 28 is replaced by $\lambda_i\lambda_a(\lambda_i + \lambda_a)$ (Tofterup 1986, Tougaard and Chorkendorff 1986). The loss function $K(E' - E)$ may be calculated theoretically from the dielectric theory and excellent results for $F(E)$ have been obtained for many peaks in Al (Tougaard and Chorkendorff 1987, Tougaard 1988), Cu, Ag and Au (Tougaard 1986, 1988). To generalise matters for practical analysis Tougaard proposed (1987) a general function:

$$\lambda_i K(E' - E) = \frac{B(E' - E)}{\{C + (E' - E)^2\}^2} \tag{29}$$

where, averaging the data for Cu, Ag and Au, he finds B=2866 eV^2 and C= 1643 eV^2. Combining the above relations gives

$$F(E) = j(E) - B_1 \int\limits_{E}^{\infty} \frac{E' - E}{\{C + (E' - E)^2\}^2} j(E')dE' \tag{30}$$

Figure 16 (Tougaard 1989) shows a widescan spectrum for Cu, as measured but corrected for the analyser transmission function, $j(E)$, and the primary excitation function, $F(E)$, calculated using Equation 30. The resulting background removal is excellent over a 700 eV range and shows precisely the peak areas to be determined. At the present time we do not know the general applicability of Equation 29 but it is hoped that work in hand will define this.

It is quite clear in Figure 16 that half of the intensity for the Cu 2p$_{3/2}$ peak lies in the part of peak area defined by the Shirley or straight line methods but that a similar amount of intensity is spread over 50 eV to lower kinetic energies due to shake-up or other events. Thus, any attempt to correlate I_i calculated from first principles via Equations 4 and 19 and that measured by experiment must incur errors of a factor of two unless Tougaard's method is used.

Sickafus and Langeron

In the above we have mainly considered XPS peaks. With AES the spectra also include large backgrounds due to (*i*) the cascade of secondary electrons and (*ii*) the backscattered primary electrons as shown in Figure 2a. The cascade electrons are thought to follow a power law dependence on energy such that the background when plotted on the log intensity/log energy axes gives a straight line. If the straight line region from intensities higher than the Auger electron peaks is extrapolated to lower energies the cascade contribution may be removed. When this is done Tougaard's method may subsequently be applied. The concept of studying the Auger peaks on the log/log plots (Seah 1969) was first exploited by Sickafus (1977a,b) who showed that this led to an easy distinction between surface and sub-surface sources.

The energy range over which the above process works depends on the value of the primary beam energy. The inelastically scattered primary electrons exhibit a background shape, B$_P$, given by (Jousset and Langeron 1987)

$$B_P = B_L \exp(E/E_L) \tag{31}$$

in the energy range $0.2 \leq E \leq 0.75E_0$. If this is fitted to the data to deduce E_L and B_L, this may be subtracted first, and then the Sickafus plot may be linearised to higher energy.

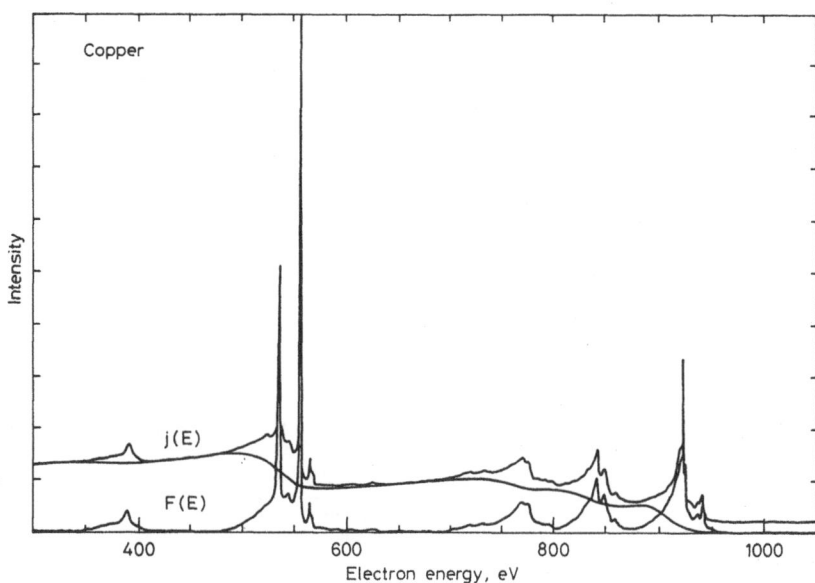

Figure 16. *Experimental Al Kα spectrum of pure Cu corrected for the analyser transmission function, j(E), and the primary excitation spectrum, F(E), deduced using Equation 52 with $\beta_1 = 3010 \ eV^2$, after Tougaard (1989).*

3.4 Differentiated spectra in AES

Traditionally all spectra in AES were differentiated to remove the large background and improve peak visibility. In early work this was done electronically using a sinusoidal modulation on the mirror electrode of a CMA. This method was used for establishing .most data banks (Davis *et al.* 1976, McGuire 1979, Shiokawa *et al.* 1979 and Sekine *et al.* 1982) . However, the intensities generated in this way depend critically on *(i)* the differentiating modulation amplitude and *(ii)* the peak width and hence the analyser resolution. A 25% increase in peak width will lead to a 36% loss in the differential peak-to-peak intensity. Nevertheless, where peaks are unchanged by chemical state or where peaks from different elements have similar shapes so that they are all affected equivalently by changes in the differentiating amplitude or analyser resolution, the errors may be very small. The method therefore has been very popular in metallurgical work and work on semiconductor metallisations *etc.* The measurement of the peak-to-peak intensities shown in Figure 2(b) will be very precise and repeatable whereas a full analysis of peak areas will lead to considerable uncertainties due to the noise in the background and the statistics of background fitting. Therefore, despite its obvious shortcomings, the differential method in complex systems may be rapid and effective.

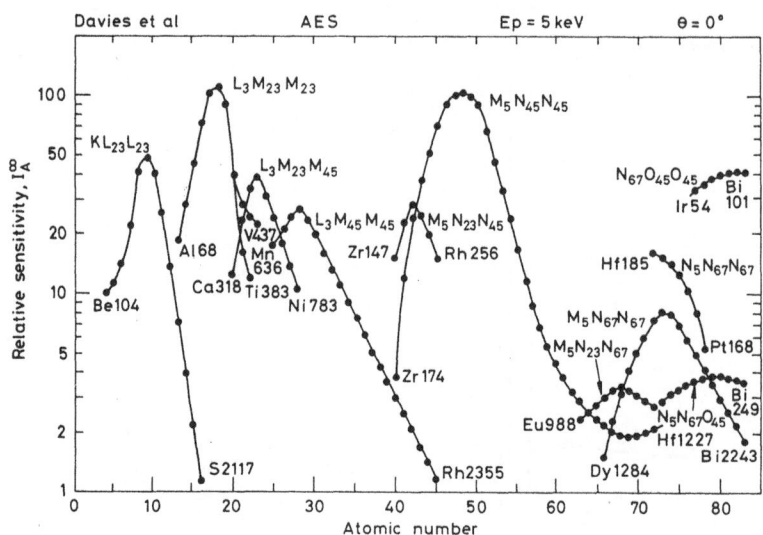

Figure 17. *Relative sensitivity factors for derivative spectra in* AES *compiled from Davis et al. (1976). The numbers at the ends of the curves indicate the energies of the peaks and the elements at these points.*

Modern systems which do not use the traditional modulation method record direct spectra which are then differentiated numerically by the Savitzky and Golay function (Savitzky and Golay 1964, Steiner *et al.* 1972). This is mathematically different from the sinusoidal modulation but, in practice, gives very similar results.

4   Data banks of sensitivity factors

4.1   AES differential mode

Figure 17 shows sensitivity factors for a 5 keV beam as measured in the differential mode with a CMA (Davis *et al.* 1976). The modulation has not been kept constant through this set and low energy peaks are depressed due to gain loss in the CEM detector (Seah 1986). Data are provided at 3 and 10 keV as well and, despite all of the shortcomings, this set of sensitivity factors has proved immensely beneficial to many users. Unfortunately, $Q(E)$ for the instruments used in this work are unknown. Several other data sets exist (McGuire *et al.* 1979, Shiokawa *et al.* 1979 and Sekine *et al.* 1982) which are compared and contrasted elsewhere (Seah 1986). Data from the Anelva Corporation (Shiokawa *et al.* 1979) are less accessible but appear to be slightly better controlled, of slightly higher resolution and rather cleaner through the rare earth metal series.

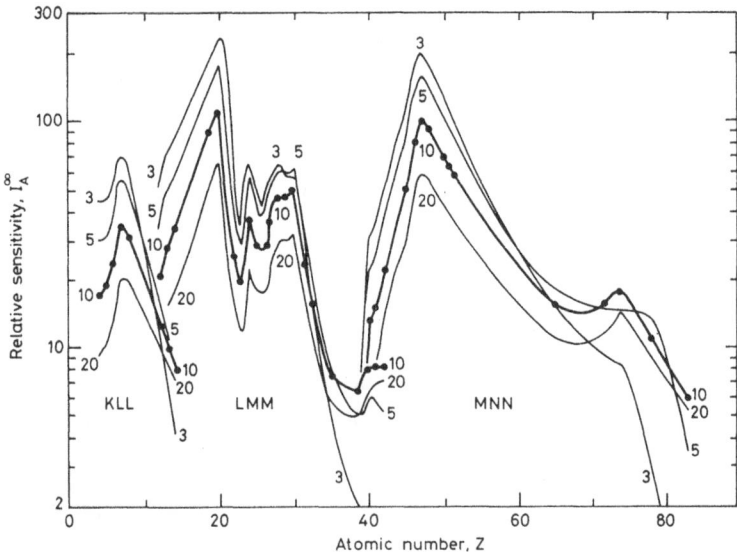

Figure 18. *Relative sensitivity factors for direct spectra in* AES *after Sato et al. (1989). The 10 keV data are for experimental measurements with interpolation by theory whereas the data for other energies are calculated using Shimizu and Ichimura's backscattering equation (Shimizu 1983) and Gryzinski's ionisation cross section (1965). All sensitivities are for normal incidence and a* CMA *spectrometer of 0.5% resolution.*

4.2 AES direct mode

Only one data set exists here with data at 10 keV (Sekine *et al.* 1982). The sensitivity factors are deduced from peak heights for a CMA at 0.5% energy resolution but, again, an unknown $Q(E)$. Results are shown in Figure 18 together with theoretical extrapolations to 3, 5 and 20 keV. Numerical differentiation of this data is also provided but differs somewhat from those in the other handbooks which all differ from each other anyway (Seah 1986).

4.3 XPS direct mode

A few rather inconsistent XPS data sets have been recorded for elemental samples. Instead, the main effort has been to run standards from well defined chemically stoichiometric powders and to measure the relative peak areas. Sensitivity factors are deduced, the XPS counterpart to Equation 10, where matrix factors may largely be ignored and a straight line subtraction method is used. The samples are usually hydrocarbon contaminated so that the low energy peaks are relatively weak. Since many technical samples are similarly contaminated it is assumed that this effect often may be ignored. A comparison of three data sets is shown in Figure 19 and a detailed analysis of the ratio of Wagner's sensitivity factors to theoretical predictions is shown in Fig-

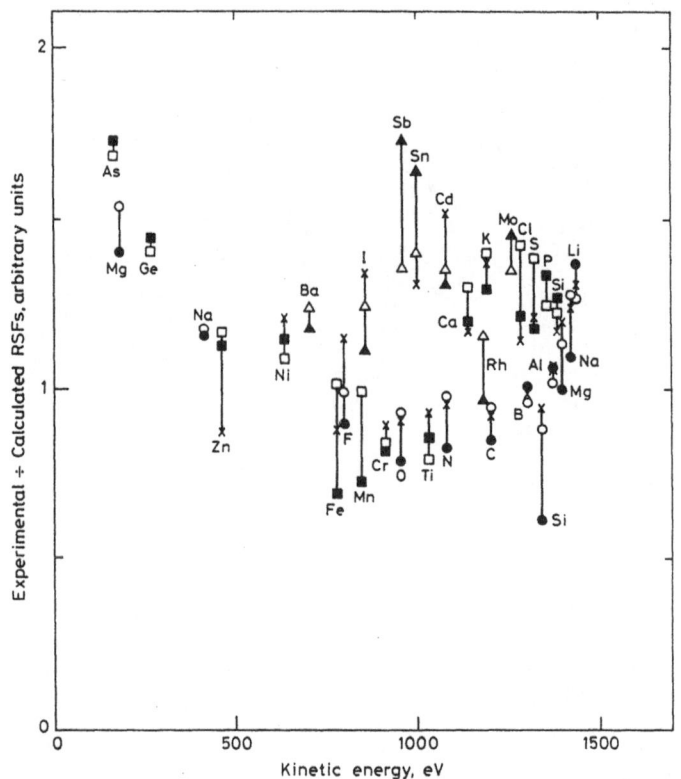

Figure 19. *The ratio of the experimental relative sensitivity factors of Wagner (1972, 1977 and Wagner et al. 1981) (filled symbols), Nevedov et al. (1973, 1975) (open symbols) and Evans et al. (1977, 1978 and Adams et al. 1977) (crosses), corrected for a contamination overlayer, to calculated values.*

ure 20. The consistency of these sets coupled with Wagner's experience indicates that a reproducibility of 10% may be possible using Wagner's data. Wagner's work was performed on instruments with $Q(E) \propto E^{-1}$ so that transfer to more modern instruments with other $Q(E)$ functions is now possible.

5 Quantification for inhomogenous samples

In practice one is only interested in inhomogeneous samples. The above discussion, however, becomes rather confusing if the inhomogeneity is included formally from the start. One may be interested in (*i*) lateral inhomogeneities or (*ii*) depth inhomogeneities or (*iii*) both.

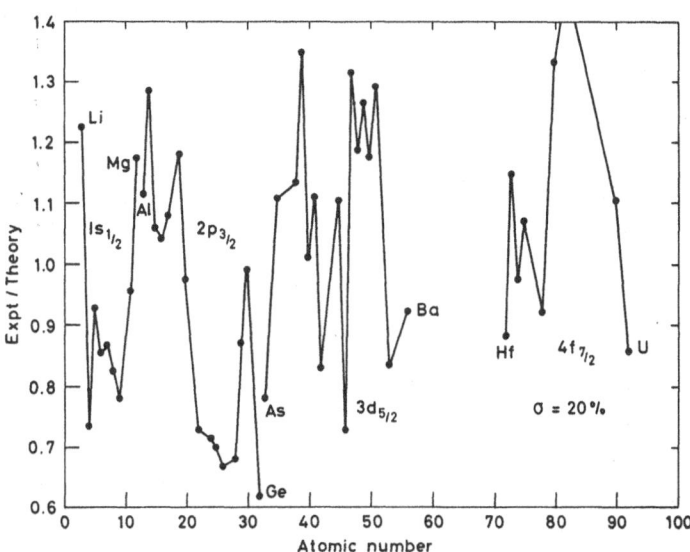

Figure 20. *Differences (%) between Wagner's et al. (1981) empirical relative sensitivity factors and theoretical values, normalised to F 1s, after Seah (1986).*

5.1 Imaging and microscopy in AES and XPS: simple effects, resolution limits and signal-to-noise ratio

Details of imaging in AES are presented by Prutton in this volume. Here we consider some simple basic concepts. As shown in Figure 5 the incident beam in AES excites the fraction 1 of Auger electron from an area equivalent to the beam size. The back reflected primary electrons then excite a fraction $r_m(E_0, \alpha)$ exhibiting a Gaussian profile characterised by a half-width at half maximum which at 20 keV is given by H in Figure 21. For a focused beam entering a flat surface of a homogeneous substrate the radial distribution of the detected Auger electrons is shown in Figure 22. Thus, it is clear that an image using a 5 nm beam will have excellent contrast and 5 nm point-to-point resolution but nevertheless significant spectral intensity contributions may arise from material 1 μm away from the point of focus. If the samples are not flat greater problems arise as discussed by Prutton in this volume.

It is most common to use the imaging to select the area from which to measure full spectra which are then quantified. This is likely to incur error if the image is not used as there is then no record of how close to a region of different material one has been. Considerable efforts are being made to provide fully quantitative images since it is often important to know the relative areas of the different reaction products. This is much more time consuming than the simple approach since a diagnostic image may be satisfactory with a signal-to-noise ratio of 1 in each pixel whereas one would need at least 20 (400 times the time) for quantifying the image.

Imaging in XPS is usually made by using a broad probe of x-rays and focusing the

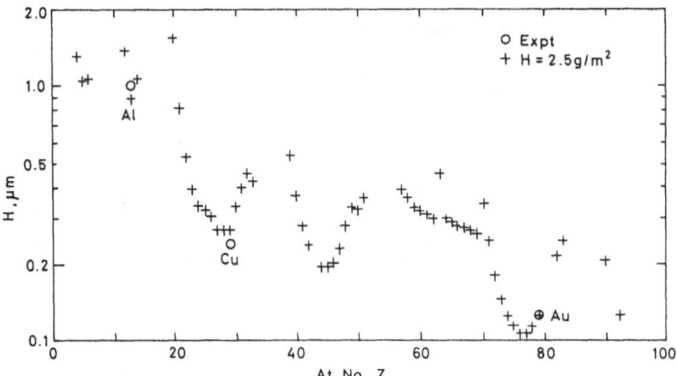

Figure 21. *The half width maximum (H) of the backscattered flux in* AES *using 20 keV electrons, after Seah (1983). The element symbols represent the data of El Gomati and Prutton(1978).*

emitted electrons instead of the reverse method used for AES. There are no backscattered electrons to deal with and so the spatial resolution concepts are much simpler. However, at the present time it does not appear that submicron images will be available for some years except using focused sources from synchrotron beam lines. Sample damage for the materials traditionally studied by XPS will probably limit the usefulness of these approaches. Image interpretation, however should be simpler than in AES.

5.2   Adsorbed and segregated atomic layers

The case of adsorbed or segregated layers on a homogeneous substrate is an ideal example of the first stage of inhomogeneity normal to the sample surface. The two methods below are both popular.

Sputter depth profiling

In many metallurgical studies of segregation it has been necessary to separate elements in segregated layers from those in precipitates. One way of doing this is to follow the Auger or XPS signals as the surface is removed by sputtering. The statistical models of sputtering (Hofmann 1976 and Seah *et al.* 1981) show that the signals should decay exponentially with time as shown in Figure 23. It is difficult to define the removal rate accurately since the sputtering yields in these situations may only be estimated to within a factor of 2. The result in Figure 23 is consistent with a sub-monolayer of phosphorus at the surface of a low alloy steel or of a composition which decays exponentially from the surface. Only the former configuration has been predicted theoretically and these data confirm the monolayer configuration to within the noise level of the measurements.

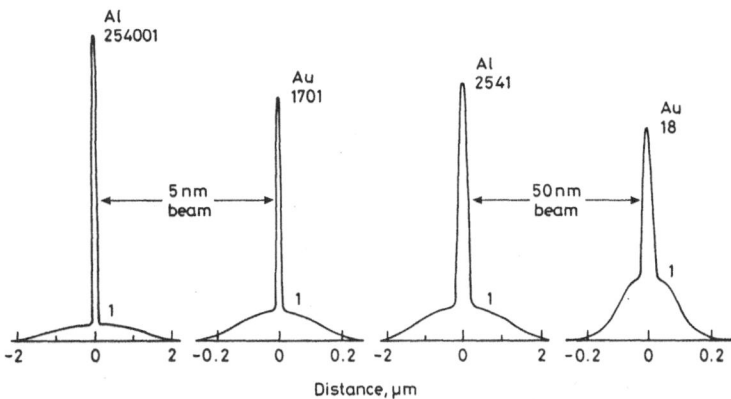

Figure 22. *The point spread functions for electron beams of 5 and 50 nm diameter incident on Al and Au at 20 keV. The numbers show the intensities but the vertical scales are schematic.*

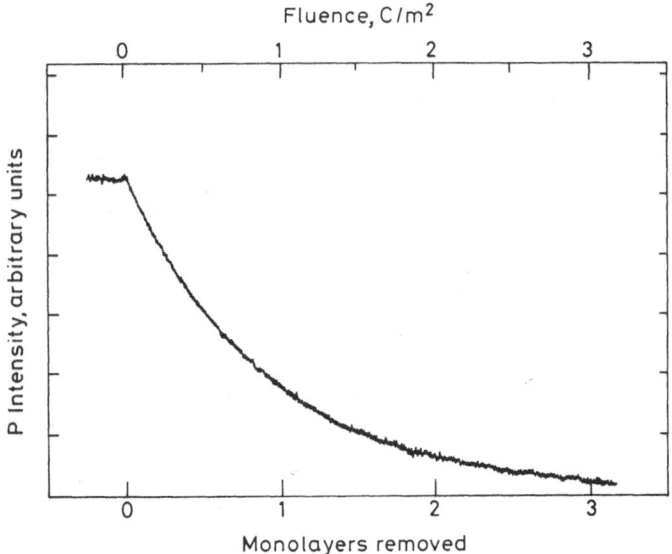

Figure 23. *The P120 differential peak-to-peak height intensity from a fractured grain boundary in a low alloy steel as a function of the sputtering fluence of 3 keV argon ions, after Hunt and Seah (to be published).*

We should now quantify the data recorded before sputtering in terms of fractions of a monolayer. We consider first a high index surface. In the absence of elastic scattering the signal of the substrate B covered by a fractional monolayer, ϕ_A, of A is:

$$I_B = I_B^\infty [1 - \phi_A + \phi_A \exp\{-a_A/\lambda_A(E_B) \cos\theta\}] \tag{32}$$

and if covered by a thickness d_A of overlayer of A:

$$I_B = I_B^\infty \exp\{-d_A/\lambda_A(E_B) \cos\theta\} \tag{33}$$

Here we assume that the correct measure for I_A is being used and that I_A^∞ is measured on the same instrument and settings. If not, the corrections already discussed, must be applied. The signal from the overlayer in each of the cases is, respectively:

$$I_A = \phi_A I_A^\infty \left[\frac{1 + r_B(E_A)}{1 + r_A(E_A)}\right] [1 - \exp\{-a_A/\lambda_A(E_A) \cos\theta\}] \tag{34}$$

$$I_A = I_A^\infty \left[\frac{1 + r_B(E_A)}{1 + r_A(E_A)}\right] [1 - \exp\{-d_A/\lambda_A(E_A) \cos\theta\}] \tag{35}$$

(where the backscattering terms r_A and r_B are, of course, zero for XPS).

Although we notionally take I_A^∞ as a tabulated relative sensitivity factor it must be remembered that, because of sample roughness and all the other changeable settings, it is really only possible to work with intensity ratios such as I_A^∞/I_B^∞ etc. Thus from Equations 32 and 34 the fractional monolayer coverage, ϕ_A is given by:

$$\frac{\phi_B [1 - \exp\{-a_A/\lambda_A(E_A) \cos\theta\}]}{1 - \phi_A [1 - \exp\{-a_A/\lambda_A(E_A) \cos\theta\}]} = \left[\frac{1 + r_A(E_A)}{1 + r_B(E_A)}\right] \frac{I_A/I_A^\infty}{I_B/I_B^\infty} \tag{36}$$

If the Auger or photoelectron peaks are at high energy and if ϕ_A is small, the simple relation below, analogous to Equation 7, is obtained.

$$\phi_A = Q_{AB} \frac{I_A/I_A^\infty}{I_B/I_B^\infty} \tag{37}$$

where the monolayer matrix factors for AES and for XPS are given by:

$$Q_{AB}^A = \left[\frac{\lambda_A(E_A) \cos\theta}{a_A}\right] \left[\frac{1 + r_A(E_A)}{1 + r_B(E_A)}\right] \tag{38}$$

$$\tag{39}$$

$$Q_{AB}^X = \left[\frac{\lambda_A(E_A) \cos\theta}{a_A}\right] \tag{40}$$

respectively and where the attenuation lengths or IMFPs may be calculated from Equations 12 to 17, as discussed earlier, and the backscattering terms from Equation 11.

As an example we can evaluate the Q_{AB}^A factor for the metallic adsorption system of tin on iron from Figure 7, Equation 12 and tabulated data for a_{Sn}:

$$Q_{SnFe}^A = \frac{\lambda_{Sn}(E_{Sn}) \cos\theta}{a_{Sn}} \left[\frac{1 + r_{Sn}(E_{Sn})}{1 + r_{Fe}(E_{Sn})}\right]$$

$$= 0.41 \, (0.29986 \times 430)^{0.5} \, 0.74 \left[\frac{1.87}{1.715}\right]$$

$$= 3.76$$

where a typical value of $\cos \theta = 0.74$ has been assumed (Seah 1972). This value compares very favourably with the experimental value of 3.82 that can be extracted from the data of Seah (1973) using the Handbook of AES (Davis *et al.* 1976). Other experimental values of Q^A have been tabulated by the author (Seah 1979) for grain boundaries where the values are twice as high since only half of the grain boundary segregant remains on the fractured surface being analysed.

Elastic scattering is important, as discussed earlier and, as we have seen, increases in effect with atomic number. Equation 33 describes the reduction in the substrate signal with either the film thickness d of the overlayer or with increasing angle of emission θ. In this equation either λ is the attenuation length or, in the absence of elastic scattering, the inelastic mean free path. Baschenko and Nefedov (1982) show that the effect of elastic scattering is to make λ appear to vary in Equation 33. In one example with an overlayer of $0.2\lambda_i$ the effective value of λ, λ_{eff}, is $0.82\lambda_i$ for $0 \leq \theta < 70°$ but rises to $1.00\lambda_i$ at $80°$. If the overlayer is increased to $2\lambda_i$, λ_{eff} rises from $0.8\lambda_i$ at $0 \leq \theta \leq 30°$ to $1.0\lambda_i$ at $70°$ and $1.67\lambda_i$ at $80°$. For the thinner film this means that the deduced film thickness using Equation 57 is only correct for $0 \leq \theta \leq 70°$ and is 25% too low at $80°$, whereas for the thicker film there is a 16% error already at $70°$ emission angle. As shown in Figure 9, the errors generally grow with increasing film thickness, increasing atomic weight of the overlayer and increasing angle of emission. Similar results have been obtained by Ebel *et al.* (1987).

Angle-resolved measurements

In most commercial instruments angle resolved measurements are popular in XPS but not AES. As the sample is rotated both the angle of emission of the outgoing electrons and the angle of incidence of the exciting radiation change. The latter causes the relative intensities of peaks to change in AES but not, except for incidence angles over $75°$, for XPS. This added complexity leads to greater uncertainty in AES. For XPS, from Equations 33 and 35, if we have, for two chemical states of the same element, one in the overlayer and one in the substrate then

$$\frac{d}{\lambda_A(E_A)\cos\theta} = \log\left\{\frac{I_B I_A^\infty}{I_A I_B^\infty} + 1\right\} \tag{41}$$

so that a plot of the right hand side versus $\sec\theta$ may be used to prove the existence of a uniform overlayer and provide the value of $d/\lambda_A(E_A)$. Calculations by Baschenko and Nevedov (1980), for SiO_2 overlayers on Si, show that elastic scattering destroys this linearity more or less as shown in Figure 9, *i.e.* for very thin films Equation 33 will be valid but for thicker films divergencies occur at higher values of θ. As the atomic numbers fall, the elastic scattering problems weaken.

With layers of even thickness angle resolved measurements give an accurate measure of d providing that $\lambda_A(E_A)$ is known (Olefjord 1990).

5.3   Layered substrates

Equations 32 to 39 ignore the elastic scattering of electrons. More complex situations, relevant to rough samples with particles or stacked layers of catalysts but still ignoring

the elastic scattering, are reviewed by Fulghum and Linton (1988). In that review for
XPS studies all of the equations for the different complex sample geometries derive from
Equations 33 and 35 with, of course, the backscattering terms removed.

Angle-resolved measurements in AES and XPS, as noted above in Equation 40, may
be used to define d, and in principle one measurement is sufficient for this purpose. If two
layers with different d values exist we may make two measurements at different emission
angles to define those d values. For a composition-depth profile one, in principle, needs
as many angles of measurement as the number of points in the depth profile, the latter
being limited to depths of less than $3\lambda_i$.

The intensity measured is simply the integral of the composition as a function of
depth times the exponential decay constant. Thus, the intensity measured is proportional to the Laplace transform of the composition as a function of depth. The composition as a function of depth is, in turn, generated by the inverse Laplace transform.
Unfortunately, many functions have the same Laplace transform so the inversion is not
unique (Bussing and Holloway 1985) and becomes uncertain as the noise in the acquired data increases (Nefedov and Baschenko 1988). By restricting the possible range
of solutions to the problem, unique results may be obtained which may be judged valid
in the light of a $priori$ knowledge of the sample. Nevedov and Baschenko (1988) show
that, if they calculate the intensities using equations such as Equation 33 and add some
noise, they may restore the original depth profile with reasonable accuracy to a depth
of $3\lambda_i$ with some 10 points in the profile. These profiles are defined to be monotonically
changing or to have one maximum and minimum in the profile. Tests are now being
made using real data which include elastic scattering and which no longer conform to
the ideal Laplace transform. Because this method is one of the only ways of defining
the composition-depth profile in the first few atom layers there has been much effort to
try to obtain general regular and stable solutions to the problem (Bussing and Holloway
1985, Nefedov and Baschenko 1988, McCaslin and Young 1987, Yih and Ratner 1987,
Tyler et $al.$ 1989). These methods effectively smooth the result to stop the development of ambiguous oscillations. Figure 24 shows a reconstructed profile from the work
of Tyler et $al.$ (1989) by Smith and Livesey (to be published).

5.4   Sputter depth profiling and thin films

In most work high depth resolution through thin films is sought. With amorphous films
very high resolutions are possible directly and, with crystalline films, the addition of
sample rotation, and inclined ion beams usually ensures excellent results. In two particular regimes the quantification should be particularly considered (i) films comparable
in thickness to λ and (ii) films of the order of 50–500 nm thickness.

With very thin films the overlayer intensity follows Equation 35 if the overlayer is
of a single element. Thus, the thickness in the profile is not given by the time to reach
50% of the plateau intensity, as is often used, but is given by the zero intensity in the fit
to that equation (Hoffmann 1976 and Olefjord 1990). If the overlayer is of more than
one element preferential sputtering occurs which gives the appearance of compositional
changes but which are an artefact of the sputtering. This may occur over the first 2 nm
(Seah and Hunt 1983).

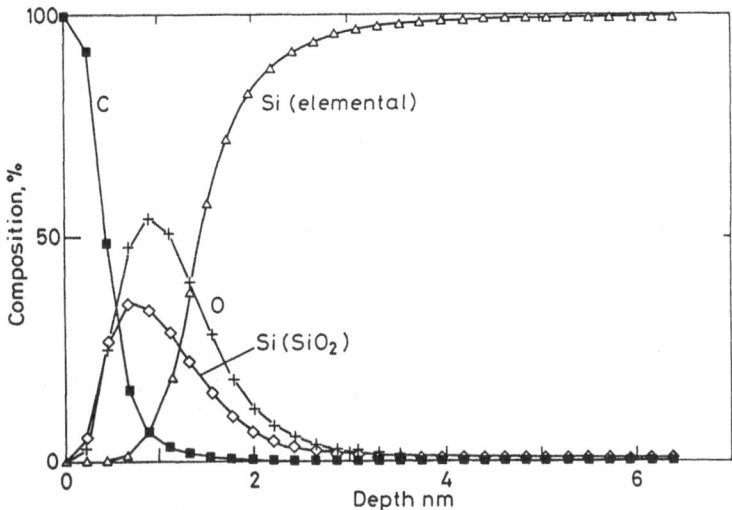

Figure 24. *Reconstruction of the data of Tyler et al. (1989) for a thin oxide layer on silicon, after Smith and Livesey (to be published).*

With the thicker films as sputtering proceeds the electron beam used for AES begins to see deeper into the sample until the backscattered electrons start to become more characteristic of the substrate than the overlayer. Figure 25 shows measurements for 700 nm of polysilicon on W, Mo and Ti substrates using 5 keV electrons and on W using 3, 5 and 10 keV electrons (Sato *et al.* 1989). The ratio of the intensity in the peak just before the interface to that in the plateau region would be

$$(1 + r_W)/(1 + r_{Si}) \tag{42}$$

etc., if the interface were very sharp. Some degradation of the depth resolution has occurred so that the intensity proportional to $(1 + r_W)$ *etc.*, is not quite reached. If the experiment were reversed with deposition of Si, *in situ*, the above would be more clearly measured.

The depth, z_0, over which the above backscattering variation occurs is given at normal incidence by (Seah 1990b)

$$z_0 = \frac{0.46.10^5}{\rho} \{E_0(\text{keV})\}^{1.35} \text{ nm} \tag{43}$$

where ρ is the overlayer film density in kg/m^3. For Si this gives

$$z_0 = 20 \{E_0(\text{keV})\}^{1.35} \text{ nm} \tag{44}$$

as shown by the dots on Figure 25.

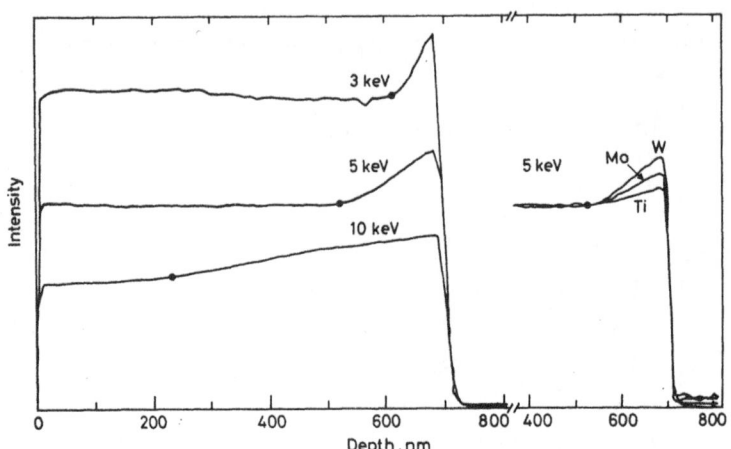

Figure 25. *Sputter-depth profiles of 700 nm of polysilicon on W, Mo and Ti substrates using 5 keV electrons and on the W substrate using electron beam energies of 3, 5 and 10 keV, after Sato et al. (1989).*

6 Systematic errors and uncertainties

6.1 Definition of uncertainties

In all quantitative work we find two major contributions to uncertainties; random and systematic. Random uncertainties have traditionally been those uncertainties for which we can ascribe a statistical value related to the standard deviation or certain confidence limits because we either know the underlying distribution of events from our prior knowledge of the physics of the problem or because we have made repeated measurements of the value. Random uncertainties from different sources may be combined statistically if they are uncorrelated by summing the individual variances.

Systematic uncertainties have to be estimated by an educated guess and the precise values cannot be calculated by any rules. Usually one's knowledge is so limited that one may only sum these and the random errors in quadrature to provide an overall global figure for the final uncertainty in the quantity determined.

If uncertainties are to be given with results it is often useful to separate the random and systematic uncertainties and then to quote them separately with the result. In this way, if, for instance, values are used to evaluate an enthalpy from an Arrhenius plot only the random uncertainties are significant whereas if the entropy is to be determined both uncertainties must be used.

6.2 Major sources of systematic uncertainty

The major sources of systematic uncertainty are reviewed in detail elsewhere (Powell and Seah 1990). In the past one of the major problems has been in the establishment of the relative sensitivity factors I_i^{∞}, or I_i^n, and their transfer from instrument to instrument. This problem is close to solution at the present time. Another problem has been in the calculation of values of inelastic mean free paths and their conversion to attenuation lengths. Further problems have arisen due to compositional variations in the surface to a depth of 3λ. All of these problems are now close to solution with simple analytical equations being developed suitable for the commercial data processing software. Certain problems such as preferential sputtering however are not yet being resolved as the phenomenon includes differences in recoil implantation and ejection probabilities, Gibbsian segregation, radiation enhanced diffusion *etc.*, all of which, individually, cannot be predicted with any accuracy.

6.3 Determination of random uncertainties

From most pulse counting systems the counts exhibit a Poissonian distribution such that if there are N_i counts in the ith relative sensitivity factors channel, the uncertainty at one standard deviation is $N_i^{0.5}$. If we have a spectrometer which views one channel on the peak and one on the background with counts P and B, respectively, the signal S is given by

$$S = P - B \pm (P + B)^{0.5} \tag{45}$$

In measuring a peak height in a spectrum by eye one would do better than this as one would judge a curve through neighbouring points and provide some subjective smoothing. Depending on the width of the peak one could smooth with a Savitzky and Golay (1964) cubic quadratic function with m channels. This would reduce the noise by $0.67\ m^{0.5}$ so that

$$S = P - B \pm 1.5(\frac{P}{M} + \frac{B}{m})^{0.5} \tag{46}$$

In XPS and, more commonly now, in AES we use peak synthesis methods to define the peak area. One is no longer using the m channels around the top of the peak but is fitting the whole ensemble of peaks and the background. In doing this a least squares fitting programme is usually used and at the end a value of χ^2 is generally provided. χ^2 is simply the sum over all channels of the squares of the residual in each channel divided by the uncertainty in each channel. One may now determine the one standard deviation σ_H in any of the fitting parameters H, by increasing H by a small amount, Δ, and, with all other fitting parameters free, note the increase $\Delta\chi^2$ in χ^2. It may be shown if the residuals are dominated by random errors that (Cumpson and Seah 1992)

$$\sigma_H = \frac{\Delta H}{(\Delta\chi^2)^{0.5}} \tag{47}$$

By this type of approach or using the Hessian matrix approach the uncertainties in peak areas may be established.

If we now consider quantification in a binary alloy AB and we have intensities $I_A \pm \varepsilon_A$, and $I_B \pm \varepsilon_B$, we are interested in how these errors propagate into X_A and X_B. Let us assume that we know I_A^∞ and I_B^∞ and either these have been derived from a known composition of the binary alloy so that $F_{AB}^A = 1$ or they are from elemental standards which, by good fortune, give $F_{AB}^A = 1$. There are two calculation routes (Cumpson and Seah 1992):

(a) via relative sensitivity factors

$$\chi_A = \frac{(I_A/I_A^\infty)}{(I_B/I_B^\infty) + (I_A/I_A^\infty)} \pm \sigma_i$$

$$\chi_B = \frac{I_B/I_B^\infty}{(I_B/I_B^\infty) + (I_A/I_A^\infty)} \pm \sigma_i \tag{48}$$

$$\text{where} \quad \sigma_i = \chi_A \chi_B \left\{ \left(\frac{\sigma_{I_A}}{I_A}\right)^2 + \left(\frac{\sigma_{I_B}}{I_B}\right)^2 \right\}^{\frac{1}{2}}$$

(b) via absolute sensitivities

$$\chi_A = \frac{I_A}{I_A^\infty} + \left(1 - \frac{I_A}{I_A^\infty} - \frac{I_B}{I_B^\infty}\right) \frac{\sigma_{I_A}^2 I_B^{\infty 2}}{\sigma_{I_A}^2 I_B^2 + \sigma_{I_B}^2 I_A^2} \pm \sigma_{ii}$$

$$C_B = \frac{I_B}{I_B^\infty} + \left(1 - \frac{I_A}{I_A^\infty} - \frac{I_B}{I_B^\infty}\right) \frac{\sigma_{I_B}^2 I_A^{\infty 2}}{\sigma_{I_A}^2 I_B^{\infty 2} + \sigma_{I_B}^2 I_A^{\infty 2}} \pm \sigma_{ii} \tag{49}$$

$$\text{where} \quad \frac{1}{\sigma_{ii}^2} = \frac{I_A^{\infty 2}}{\sigma_{I_A}^2} + \frac{I_B^{\infty 2}}{\sigma_{I_B}^2}$$

These two routes give very different uncertainties. In the traditional relative sensitivity factor approach the composition χ_A requires a measure of I_B since the composition is deduced by normalising the sum of components to 100%. Thus, the accuracy in I_B is crucial to the uncertainty in χ_A and indeed we see in Equation 49 that the fractional uncertainties in I_A and I_B sum in quadrature. In the absolute sensitivity method we know χ_A merely from the measure of I_a. If we also measure I_B then the accuracy of χ_A improves. The fractional uncertainties now sum in reciprocal quadrature. Very often it is not possible to take advantage of the absolute method since absolute I_i^∞ values are difficult to measure accurately, however there are two situations in which absolute I_i^∞ values may be deduced with very high precision; (i) in depth profiles in which each element appears at 100% composition and (ii) in maps where each element appears at 100% composition. In these situations we may use the ensemble of data to reduce errors as indicated by Equations 50 to 52.

7 Conclusions

Quantification in AES and XPS is complex since very many different types of sample, with elemental and chemical changes possible over very short distances, may be studied. Few other techniques encompass the wide range of possible configurations except perhaps secondary ion mass spectrometry (SIMS). Progress in the last few years has been extremely strong and many of the problems identified over recent years are now being solved. Before the end of the century we expect all of the developments discussed here to be available in commercial data systems and prototype expert systems will be well established.

Acknowledgements

This work forms part of the programme of the National Measurement System Policy Unit of the Department of Trade and Industry.

References

Adams J M, Evans S, Reid P I, Thomas J M and Walters M J, 1977, *Anal Chem* **49** 2001
Band I M, Kharitonov Yu I and Trzhaskovskaya M B, 1979, *Atomic Data and Nuclear Data Tables* **23** 443
Baschenko O A and Nefedov V I, 1979, *J Electron Spectrosc* **17** 405
Baschenko O A and Nefedov V I, 1980, *J Electron Spectrosc* **21** 153
Baschenko O A and Nefedov V I, 1982, *J Electron Spectrosc* **27** 109
Baschenko O A, Machavariani G V and Nevedov V I, 1984, *J Electron Spectrosc* **34** 305
Bethe H, 1930, *Ann Phys* **5** 325 (Leipzig)
Briggs D, 1990, Auger and X-ray Photoelectron Spectroscopy, in *Practical Surface Analysis* **1**, eds Briggs D and Seah M P (Wiley Chichester) p 437
Bussing T D and Holloway P H, 1985, *J Vac Sci Technol* **A3** 1973
Cumpson P J and Seah M P, 1992, *Surf Interface Anal* **18** 361
Davis L E, MacDonald N C, Palmberg P W, Riach G E and Weber R E, 1976, *Handbook of Auger electron spectroscopy* 2nd edn, (Physical Electronics Industries Inc, Minnesota)
Ebel H, Ebel M F and Jablonski A, 1985, *J Electron Spectrosc* **35** 155
Ebel M F, Moser G, Ebel H, Jablonski A and Oppolzer H, 1987, *J Electron Spectrosc* **42** 61
Ebel H, Ebel M F, Baldauf P and Jablonski A, 1988, *Surf Interface Anal* **12** 172
El Gomati M M and Prutton M, 1978, *Surf Sci* **72** 485
Evans S, Pritchard R G and Thomas J M, 1977, *J Phys C: Solid State Phys* **10** 2483
Evans S, Pritchard R G and Thomas J M, 1978, *J Electron Spectrosc* **14** 341
Fadley C S, Kono S, Petersson L G, Goldberg S M, Hall N F T, Lloyd J T and Hussain Z, 1979, *Surf Sci* **89** 52
Fulghum J E and Linton R W, 1988, *Surf Interface Anal* **13** 186
Gries W M and Werner W S M, 1990, *Surf Interface Anal* **16** 149
Gryzinski M, 1965, *Phys Rev A* **138** 336
Hall P M and Morabito J M, 1979, *Surf Sci* **83** 391
Harris L A, 1968, *J Appl Phys* **39** 1419
Harris D W and Nowicki R S, 1990, Auger and X-ray Photoelectron Spectroscopy, in *Practical Surface Analysis* **1**, eds Briggs D and Seah M P (Wiley Chichester) p 257
Hofmann S, 1976, *Appl Phys* **9** 59
Hofmann S, 1990, in *Practical Surface Analysis* **1**, eds Briggs D and Seah M P (Wiley) p 143
Hunt C P and Seah M P, *Mater Sci Tech* - to be published
Ichimura S and Shimizu R, 1981, *Surf Sci* **112** 386

Jablonski A, 1980, *Surf Interface Anal* **2** 39

Jablonski A and Ebel H, 1984, *Surf Interface Anal* **6** 21

Jablonski A, Ebel M F and Ebel H, 1986, *J Electron Spectrosc* **40** 125

Jablonski A, 1987, *Surf Sci* **188** 164

Jablonski A, 1989 *Surface and Interface Analysis* **14** 659

Jablonski A and Ebel H, 1988, *Surf Interface Anal* **11** 627

Jablonski A, Lesiak B, Ebel H and Ebel M F, 1988, *Surf Interface Anal* **12** 87

Janghorbani M , Vulli M and Starke K, 1975, *Anal Chem* **47** 2200

Janssen A P and Venables J A, 1978, *Surf Sci* **77** 351

Jousset D and Langeron J P, 1987, *J Vac Sci Technol* **A5** 989

Krause M O, 1969, *Phys Rev* **177** 151

Lotz W , 1970, *Z Physik* **232** 101

McCaslin P C and Young V, 1987, *SEM* **1** (4) 1545

McGuire G E, 1979 *Auger Electron Spectroscopy Reference Manual* (Plenum Press)

McIntyre N S and Chan T C, 1990 *Practical Surface Analysis 1* eds. Briggs D and Seah M P p.485 (Wiley, Chichester)

Nefedov V I, Sergushin N P, Band I M and Trzhakovskaya M B, 1973, *J Electron Spectrosc* **2** 383

Nefedov V I, Sergushin N P, Salyn U B, Band I M and Trzhakovskaya M B, 1975, *J Electron Spectrosc* **7** 175

Nefedov V I and Baschenko O A, 1988, *J Electron Spectrosc* **47** 1

Olefjord I, Mathieu H J and Marcus P, 1990, *Surf Interface Anal* **15** 681

Palczynski J, Dolinski W and Mroz S, 1991, *Surf Sci* **247** 395

Palmberg P W, Bohm G K and Tracy J C, 1969, *Appl Phys Lett* **15** 245

Powell C J and Seah M P, 1990, *J Vac Sci Technol* **A8** 735

Reilman R F, Msezane A and Manson S T, 1976, *J Electron Spectrosc* **8** 389

Reuter W, 1972, *Proc. 6th Int Conf on X-ray Optics and Microanalysis*, edited by Shinoda G, Kohra K and Ichinokawa T (Univ of Tokyo Press) p 121

Sato T, Nagasawa Y, Sekine T, Sakai Y and Buonaquisti A D, 1989, *Surf Interface Anal* **14** 787

Savitzky A and Golay M J E, 1964, *Anal Chem* **6** 1627

Scofield J H, 1976, *J Electron Spectrosc* **8** 129

Seah M P, 1969, *Surf Sci* **17** 132

Seah M P, 1972, *Surf Sci* **32** 703

Seah M P, 1973, *Surf Sci* **40** 595

Seah M P, 1979, *Surf Interface Anal* **1** 86

Seah M P, 1980, *Surf Interface Anal* **2** 222

Seah M P, 1983, in *Proc 9th Int Vacuum Congr and 5th Int Conf on Solid Surfaces* 26-30 September 1983, Madrid, Spain, *Invited Speakers Volume*, ed J L de Segovia p 63, ASEVA, Madrid (1983)

Seah M P, 1984, *Vacuum* **34** 463

Seah M P, 1986, *Surf Interface Anal* **9** 85

Seah M P, 1989, in *Methods of Surface Analysis*, ed J M Walls (CUP, Cambridge) p 57

Seah M P, 1990a, *J Electron Spectrosc* **50** 137

Seah M P, 1990b, Auger and X-ray Photoelectron Spectroscopy, in *Practical Surface Analysis* **1**, eds Briggs D and Seah M P (Wiley, Chichester) p 201

Seah M P, 1990c, in *Practical Surface Analysis* **1**, eds Briggs D and Seah M P (Wiley) p 311

Seah M P, 1992 *J Elect Spectr and Relat Phenom* **58** 345

Seah M P and Dench W A, 1979, *Surf Interface Anal* **1** 2

Seah M P and Hunt C P, 1983, *Surf Interface Anal* **5** 199

Seah M P and Hunt C P, 1988, *Rev Sci Instrum* **59** 217

Seah M P and Smith G C, 1991, *Surf Interface Anal* **17** 855

Seah M P, Sanz J M and Hofmann S, 1981 *Thin Solid Films* **81** 239

Seah M P, Lim C S and Tong K L, 1989, *J Electron Spectrosc* **48** 209

Sekine T, Nagasawa Y, Kudoh M, Sakai Y, Parkes A S, Geller J D, Mogami A and Hirata K, 1982, *Handbook of Auger Electron Spectroscopy* (JEOL, Tokyo)

Sekine Y, Owari M, Kudo M and Nihei Y, 1986, *Japanese J Appl Phys* **25** 538

Shimizu R, 1983, *Japanese J Appl Phys* **22** 1631

Shimizu R and Ichimura S, 1981, Quantitative analysis by Auger electron spectroscopy, in *Toyota Foundation Research Report* I-006 No.76-0175 (Osaka)

Shiokawa Y, Isida T and Hayashi Y, 1979, *Auger Electron Spectra Catalogue —A Data Collection of Elements* (Anelva Corp, Tokyo)

Shirley D A, 1972, *Phys Rev B* **5** 4709

Sickafus E N, 1977a, *Phys Rev B* **16** 1436

Sickafus E N, 1977b, *Phys Rev B* **16** 1448

Siegbahn K, Nordling C N, Fahlman A, Nordberg R, Hamrin K, Hedman J, Johansson G, Bermark T, Karlsson S E, Lindgren I, Lindberg B, 1967, *ESCA: Atomic, Molecular and Solid State Structure Studied by Means of Electron Spectroscopy*, (Almqvist & and Wiksells, Uppsala)

Smith G C and Seah M P, 1988, *Surf Interface Anal* **12** 105

Smith G C and Livesey A K, 1992 *Surface and Interface Analysis* **19** 175

Steiner J, Termonia Y and Deltour J, 1972, *Anal Chem* **44** 1906

Szajman J, Liesegang J, Jenkin J G and Leckey R C G, 1981, *J Electron Spectrosc* **23** 97

Tanuma S, Powell C J and Penn D R, 1990a, *J Vac Sci Technolog* **A8** 2213

Tanuma S, Powell C J and Penn D R, 1990b, *J Electron Spectrosc* **52** 285

Tanuma S, Powell C J and Penn D R, 1991a, *Surf Interface Anal* **17** 911

Tanuma S, Powell C J and Penn D R, 1991b, *Surf Interface Anal* **17** 927

Tofterup A L, 1986, *Surf Sci* **167** 70

Tougaard S, 1986, *Phys Rev B* **34** 6779

Tougaard S, 1987, *Solid State Commun* **61** 547

Tougaard S, 1988, *Surf Interface Anal* **11** 453

Tougaard S, 1989, *Surf Sci* **216** 343

Tougaard S and Sigmund P, 1982 *Phys Rev B* **25** 4452

Tougaard S and Chorkendorff I, 1986, *Solid State Commun* **57** 77

Tougaard S and Chorkendorff I, 1987, *Phys Rev B* **35** 570

Trehan R and Fadley C S, 1986, *Phys Rev B* **34** 6784

Tyler B J, Castner D G and Ratner B D, 1989, *Surf Interface Anal* **14** 443

Vulli M, 1981, *Surf Interface Anal* **3** 67

Wagner C D, 1972, *Anal Chem* **44** 1050

Wagner C D, 1977, *Anal Chem* **49** 1282

Wagner C D, Davis L E, Zeller M V, Taylor J A, Raymond R H and Gale L H, 1981, *Surf Interface Anal* **3** 211

Wagner C D, Davis L E and Riggs W M, 1986, *Surf Interface Anal* **2** 53

Weber R E and Peria W T, 1967, *J Appl Phys* **38** 4355

Werner W S M, 1991, *Surf Sci* **257** 319

Werner W S M, Gries W H and Stori H, 1991, *Surf Interface Anal* **17** 693

Yeh J J and Lindau I, 1985, *Atomic Data and Nuclear Data Tables* **32** 1

Yih R S and Ratner B D, 1987, *J Electron Spectrosc* **43** 61

Zagorenko A I and Zaporozchenko V I, 1989, *Surf Interface Anal* **14** 438

Surface Analytical Imaging

M Prutton

University of York
York, UK

1 Surface analysis and imaging instruments

1.1 The need for imaging techniques

Much of the microanalysis described in this book is concerned with the measurement and interpretation of spectra of various kinds. The general principle is that particles or radiation emitted from a particular area on a sample are energy or wavelength analysed and the heights or areas of peaks in the observed spectrum are compared with the corresponding quantities from samples with known properties. With the use of the appropriate scattering model the properties of an unknown sample can be deduced. The commonest property required is the chemical composition of a particular area but other information such as chemical coordination or local electronic or vibrational data may also be deduced. If microanalysis in this general sense is carried out in such a way that spatially resolved information from different regions of the sample is obtained then *imaging microanalysis* is being performed.

Spectroscopic imaging of this kind involves the acquisition of a great deal of data. Imagine, for example, the characterisation of the chemical composition of an area of a sample containing 6 chemical elements whose concentrations vary from place to place. At each position on the sample at which an observation is made the heights or areas of at least 6 peaks in a spectrum must be measured. The simplest approach might be to make a crude estimate of the peak heights by subtracting the measured background in the spectrum at an energy above or below the peak from the height of the spectrum at the peak. Thus, for 6 elements, a minimum of 12 measurements must be made. Consider that it is desired to build up a picture with 512 by 512 picture points (pixels). This means that at least 3,145,728 measurements are required to derive the 6 images required. Clearly we are immediately in the realm of experiments controlled by digital computers. Should subsequent numerical processing of these images be required, as is inevitable if the images are to be quantified so as to have contrast representing the concentrations of the 6 elements at each pixel, then it is necessary to be careful about the precision with which these data are stored in the computer. Assume that 32 bit precision is necessary. This means that the data will require about 12.5 million bytes of memory capacity. Some experiments require sets of such images to be collected; for example the sample temperature may be varied to change local concentrations in a study of the progress of a chemical reaction or the fracture of a piece of a metallic alloy. In such cases even relatively cheap random access and hard disk storage facilities on modern computers soon become prohibitively expensive.

The problem of large data sets can be reduced by using efficient schemes of data coding to store the information but many researchers even question the wisdom of acquiring images because of the time required to accumulate all of the information. As will be seen below there are always statistical criteria which determine the precision of each individual measurement and such criteria impose minimum data acquisition times for each pixel. Depending upon the experiment being carried out and the techniques being used, these minima can lead to quite long data acquisition times for each set of images. These times range between minutes and several hours. The question therefore arises as to whether it is more efficient to choose places in or on the sample by some criterion such as a characteristically different contrast in a rapidly measured scanning electron microscope (SEM) image. These places are then carefully observed using spectroscopy and the analysis performed at those places only. A slightly more detailed approach might be to scan a region of a sample along a line using spectroscopic imaging techniques and so reduce the data acquisition time from that required to collect a whole picture (a frame) to that required to collect a single line of that frame whilst ensuring that the line crosses the feature of interest.

A criticism of these faster experimental methods is that it is very easy to miss features of special interest. One example will be shown in a subsequent section for which there is barely any contrast at all in the SEM image and yet the Auger image reveals a distinct layer structure in the sample. In another example *unexpected* regions or phases occur in the sample and they just happen to have identical contrast in whatever fast scanning technique is used to try to classify the number of different kinds of regions in the sample. Often it is the occurrence of unexpected effects that are both interesting and valuable in science! Finally, it should be stressed that the human eye-brain combination is very good at spotting spatial correlations. For example it is often easy to spot a pattern of faint lines in a diffuse optical image. By analysing at points or along lines the spatially correlated information about the sample may be entirely or largely thrown away— 'a picture is worth a thousand words'!

This chapter will address some of the issues that arise in imaging microanalysis by placing particular emphasis upon scanning beam techniques. These are very widespread in use and are conveniently adaptable to computer controlled equipment. Further, Auger imaging of the surfaces of solids will be of central concern because this has been developed to an advanced level and because it illustrates the ways in which images can be combined to obtain more information than was readily apparent in the raw data. The methods which are described here can be extended to other microanalytical techniques and this extension has begun in some fields. It will be assumed that the reader is reasonably familiar with the material in the article by Seah in this book and with some of the textbooks in the field. An excellent review of the principles and methods of surface spectroscopy can be found in the book by Briggs and Seah (1990) and the references therein. Finally, the methods of surface analysis require that experiments are carried out under ultra-high vacuum (UHV) conditions and so this is taken for granted throughout the material described here.

1.2 Scanning and direct microscopies

There are two modes of acquisition of analytical images — scanning and direct. The scanning modes utilise a focussed probe of particles or radiation which is scanned in an analogue or digital fashion across the sample under study. Some of the scattered particles or radiation intended for use in microanalysis are collected by a spectrometer/detector arrangement and the signal measured is either stored in the memory of a control computer or is used to modulate the brightness of the beam in a cathode ray tube display which is being scanned synchronously with the motion of the probe on the sample. This is the principle used in the well known SEM. The general arrangement is outlined in Figure 1a. The subject has been extensively reviewed in a number of textbooks and research monographs. The SEM and the physics of the scattering of an incident electron beam into a variety of signals have been described by Reimer (1985). Examples of analytical techniques which use a scanned probe and detect scattered particles or radiation to form an image are:

- *secondary-ion mass spectroscopy* (SIMS) in which the probe is a beam of focussed, monochromatic ions and the signal is formed by selecting a particular mass of secondary scattered ion using a mass spectrometer.

- *scanning transmission electron microscopy* (STEM) in which the probe is a focussed beam of high energy electrons (perhaps 100 keV) and the signals detected are those arising at the bottom surface of a thin sample through which the electrons have been transmitted. These signals might be the true secondary electrons, those electrons which have undergone many energy loss processes in the sample but still emerge with considerable kinetic energy (say greater than about 50 eV), characteristic x-rays excited by ionisation of the atoms in the sample, Auger electrons also excited by ionisations, incident electrons which have lost energy to a single ionisation of the atoms of the sample (this is the basis of electron energy loss spectroscopy or EELS), incident electrons which have lost energy to the excitation of plasmons or phonons in the sample. In this kind of instrument there are many processes causing electron scattering and many kinds of signals which can be measured. Some STEMs are equipped with multiple detectors so that several different signals can be measured from the same place on the sample simultaneously. This allows subsequent use of image correlation techniques to learn more about the sample than can be gleaned from individual measurements.

- *imaging energy or wavelength dispersive x-ray analysis* (EDX or WDX). Here an electron probe is scanned across the sample and x-ray emission is detected. The wavelengths or energies of the x-rays are characteristic of the types of atoms in the sample.

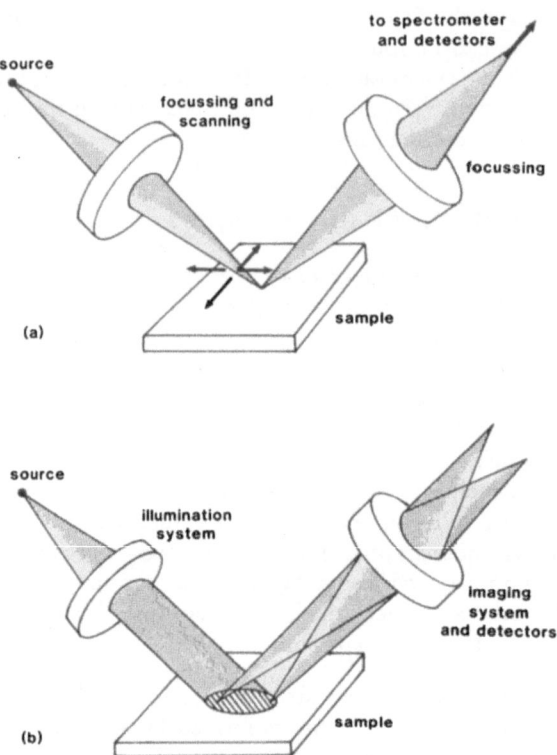

Figure 1. *(a) Scanning imaging: the incident beam may be focussed onto the sample and scanned across its surface as shown with spectrometer collecting electrons from the place which is illuminated. Alternatively, the surface may be flooded with electrons and a spectrometer has a small field of view scanned across the surface. (b) Direct imaging: The surface is flooded with electrons and the spectrometer is designed to image the surface at the detector using electrons of a selected energy.*

- *scanning Auger microscopy* (SAM). This is analogous to EDX except that the Auger electrons ejected from a sample after ionisation by energetic electrons are used to form the signal. This technique is very sensitive to the distributions of atoms in the top few atomic monolayers of the sample.

- *scanning low energy electron loss microscopy* (SLEELM). This is a scanned electron beam technique in which the spectrometer is set up to pass electrons which have lost rather small energies in their interactions with the sample. Examples of such excitations are phonon, plasmon and exciton processes and interband transitions in insulators and semiconductors.

Direct imaging methods of microanalysis (Figure 1b) are designed so that the sample is illuminated over a large area instead of the incident radiation being focussed into a fine probe. The spectrometer is of a type which images the emitting region of the sample onto a detector array. This kind of instrument has the substantial advantage that all regions in the area of interest are imaged simultaneously and so it is usually much faster than the scanned probe techniques. However, there is always a price to pay for improved performance and in these cases it is usually that the spectrometer is more complex (and thus expensive and more sophisticated in use) than those used in scanning instruments. Examples of such direct methods are:

- *parallel electron energy loss spectroscopy* (PEELS). Here a high energy resolution electron spectrometer collects electrons from an illuminated area on a sample and images those electrons which have lost a selected amount of energy through ionisations or other excitations onto an array of solid state electron detectors.

- *low energy electron microscopy* (LEEM). Ingenious lens designs are used in this technique to illuminate a sample with low energy electrons (say 100 eV kinetic energy) and then to collect an image of the elastically scattered electrons backscattered from the surface of the sample. This is a powerful method for detecting crystallographic variations from place to place.

- *photoelectron microscopy or imaging* XPS (PEEM). In this imaging technique x-rays or ultraviolet photons illuminate an area of the sample and the spectrometer images a selected photoelectron energy onto a detector or detector array.

This list of techniques, although incomplete, is given to indicate the range of incident probes and detected signals that are in use in both scanning and direct methodologies. Acronyms abound! (see Appendix 1) Although attention will be focussed hereafter upon the scanning methods and upon Auger imaging in particular many of the data manipulation and interpretation techniques which will be described are applicable to the other kinds of images.

1.3 Signal-to-noise ratios in imaging

Consider a beam of electrons with energy E_p and current i_b incident upon a solid sample. The atoms of the solid are ionised in a process with cross-section ϕ and emit Auger electrons into a spectrometer accepting a solid angle Ω. The current of Auger electrons passing through the spectrometer will be i_A, where:

$$i_A = i_b \phi N r \lambda \Omega / 4\pi \tag{1}$$

In this equation N is the number of atoms per unit volume generating Auger electrons of the correct energy to pass around the spectrometer and be detected, and r is the Auger backscattering factor (see Seah in this volume). The quantity λ is the *inelastic mean free path* (imfp) of the Auger electrons and is a measure of the depth from which Auger electrons have escaped from the solid into the vacuum. This general form of the yield of Auger electrons was first proposed by Bishop and Riviere (1969).

Equation 1 can be used to estimate the current of Auger electrons that may be detected in a particular situation. Consider for example a beam current of 10 nA incident upon a monolayer of oxygen atoms adsorbed upon a silicon surface. Auger electrons with 505 eV kinetic energy are emitted from the oxygen atoms which are present with a density of about 10^{15} atoms per square cm. The cross-section is about 10^{-21} cm^2. If, say, 1% of the electrons emitted enter the spectrometer and are detected and if the values of r and λ are taken as 1 and 10^9 m respectively then the current collected is about 10^{15} A. This is about 6,000 electrons per second. If a measurement of this current is made by counting electrons for τ secs then 6000τ electrons are counted. Should repeated measurements of this count be taken then the set of measurements will have a standard deviation of $(6000\tau)^{0.5}$ because of the Poisson statistics associated with the random arrival rate of electrons at the detector. Thus the signal-to-noise ratio will be about $6000\tau/(6000\tau)^{0.5}$ *i.e.* $(6000\tau)^{0.5}$. Thus, for example, if a measurement is made for 17 ms then about 100 electrons will be detected and the signal-to-noise ratio will be 10:1.

Such arguments always provide the basic limit to the sensitivity of an experiment in which particles are counted— provided that it is the statistics of particle detection which sets the noise and not the performance of any associated detection electronics. For experiments which detect particles or photons other than Auger electrons all that is required is to replace Equation 1 with the appropriate relationship between the exciting particles or radiation and the yield of detected particles or radiation. In scanning imaging experiments it is these signal-to-noise considerations that lead to the inconveniently long data acquisition times for a single image. Using the numbers given above a signal-to-noise ratio of 10:1 will require 17 ms per pixel per energy. If 6 energies are being acquired and a 512 by 512 pixel image is required then the total acquisition time must be at least 45 minutes. Less time can be taken only by sacrificing signal-to-noise ratio or by reducing the pixel density. Even when making such sacrifices the improvement in acquisition time scales only with the square root of the quantity changed because the standard deviation of the data scales with the square root of the number of particles counted for each pixel.

Alternatively, one might be free to increase the beam current i_b. The difficulty here can be that a small beam size is required to obtain a good spatial resolution in the image and an increase in the beam current may cause an increase of the size of the spot into which the beam is focussed and so a loss of spatial resolution. Further, an increase in beam current will result in an increase in beam current density for a fixed beam size and this may lead to an increase in the rate at which the sample is damaged by the electron beam. (Practical instruments use beam energies in the range 5 keV to 100 keV, beam currents in the range 1 to 10 nA and beam diameters in the range 1 nm to 500 nm. Beam current densities of the order of 1000 A cm^{-2} are common). Therefore, one is usually faced with the need to accept long times for image acquisition. A consequence of this fact is that it is often worth having a strategy for the conduct of an experiment. This will be discussed further below.

1.4 Spatial resolution

Monte Carlo modelling of the electron-solid interaction

Theoretical evaluation of the spatial resolution of a scanning electron probe technique can be carried out using Monte Carlo modelling of the interaction of the electrons with the sample. This method involves tracing the trajectory of each incident electron as it is scattered by the atoms in the sample, recording the generation of the particle or photon which is going to be detected and then tracking the path of that entity out of the solid and into the detector. This is called a Monte Carlo method because a random number generator is used to determine the probabilities of various events actually occurring — the ionisation of an atom, the choice of a particular direction of emission and so on. A typical sequence in such a model calculation is:

1. The energy and direction are chosen for an electron incident upon the surface of the solid.

2. The range of the primary electron beam is computed and this range is divided into steps depending upon the detail required in the calculation of the electron trajectories. 100 steps is a typical value.

3. Look up tables for the energy losses and the ionisation cross-sections are set up. The energy loss mechanism often used is that due to Bethe (1930) (see also Reimer 1985) in which electrons are assumed to lose energy continuously along their paths through the solid. The ionisation cross-sections used may be those due to Gryzinski (1965).

4. Each electron is allowed to move in a random direction after the ith scattering event in the solid having had a randomly chosen probability of ionising an atom with the cross-section appropriate to the current kinetic energy.

5. Record is kept of the position, energy and direction of motion of the each electron as well as the corresponding quantities for any electron generated in a scattering process.

6. Depending upon the quantities being modelled, the Auger electrons emerging from a given point on the surface, the x-rays emerging or the secondary electrons generated are logged and the scattering is continued until the original electron emerges from the solid or has penetrated into the solid by its range—at which point it is regarded as having lost so much energy that it can be neglected as a generator of events.

7. The calculation is repeated at step 4 until a sufficient number of scattered particles have been recorded for statistics of adequate accuracy to have accumulated. It is quite typical to count up the effects due to 10^5 or 10^6 incident electrons. Such calculations can require considerable computer power.

An example for the trajectories of 20 keV electrons striking the surface of a copper sample at 3 angles of incidence is shown in Figure 2. It can be seen that most of the electrons execute a random walk into the sample and finally reach their range where

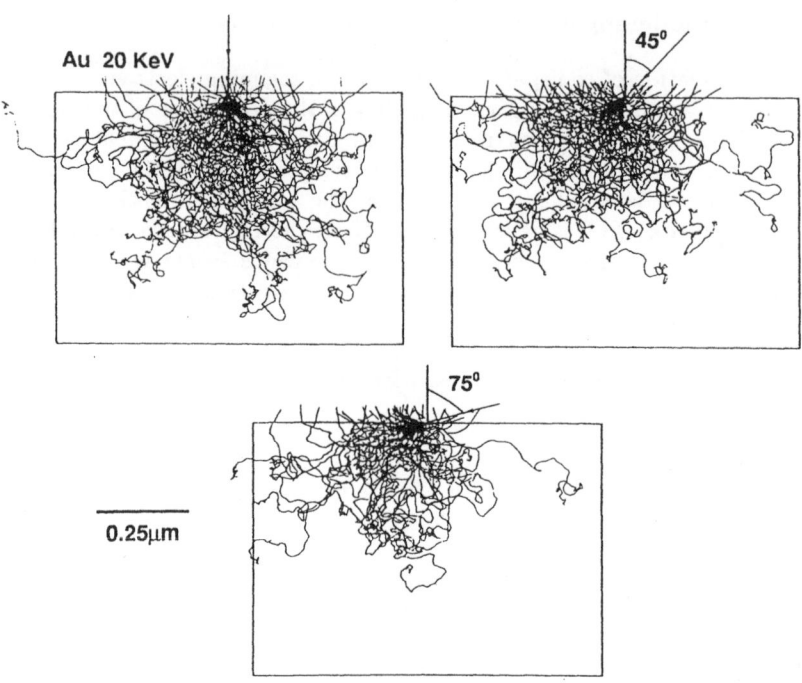

Figure 2. *Pictorial view of the random walk of a 20 keV electron beam incident on gold at angles of 0°, 45° and 75°. (Courtesy M M El Gomati)*

they contribute to the current flowing from the sample to ground. Some, however are backscattered towards the surface where they can ionise atoms to produce Auger electrons which are sufficiently near to the surface that they can escape into the vacuum and may be detected. Some electrons inside the sample will ionise atoms and characteristic x-rays will be produced. These too can escape but from greater depths than the Auger electrons because the scattering cross-sections for x-rays are smaller than those for electrons. It can also be seen in Figure 2 that the scattering of the incident electrons is forward peaked and that the number of electrons reaching the surface rises as the angle of incidence increases. The method has been reviewed recently by Shimizu and Ding Ze-Jun (1992).

Beam energy, working distance and spot size

The Monte-Carlo simulation outlined above was carried out with a stream of electrons striking the sample in an infinitesimally narrow beam. In practice the incident beam has a finite width and it is usually taken as having an approximately Gaussian distribution. The beam has this finite width because of the Coulomb repulsions between

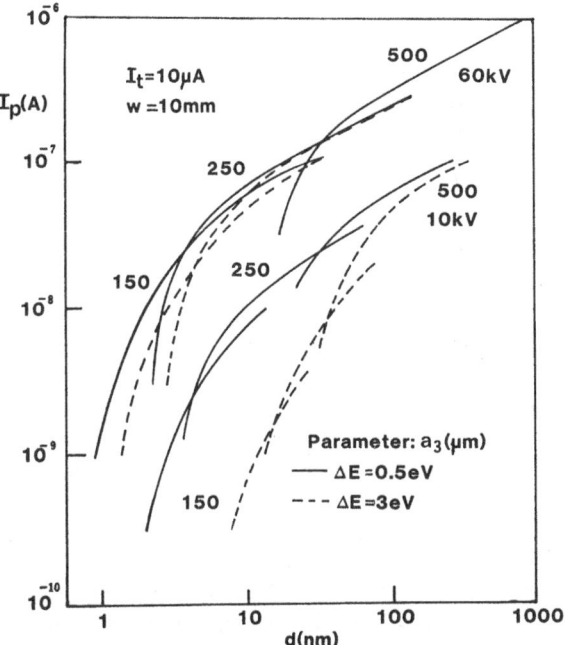

Figure 3. *Beam current versus spot size for a particular electron column. (Courtesy Venables and Archer) The calculations are made for an emission current of 10 mA with an energy spread of 0.5 eV (solid lines) or 3 eV (dashed lines). The calculations include a beam angle limiting aperture whose diameter is marked in mm beside each curve. Both 60 keV and 10 keV calculations are shown.*

its electrons, because of the spherical, chromatic and astigmatism aberration effects in the components of the beam forming system and because of diffraction effects at the apertures in these electron optical components. These effects are described in detail in many textbooks on electron optics — see for example Grivet (1972) and Wells *et al.* (1974). The net effect of these aberrations is that the current which can be focussed into a beam of electrons at a chosen energy always decreases as the apertures are modified to produce smaller beam sizes. Since a small beam size is what is required for high spatial resolution (see below) the higher the resolution required the smaller the beam current will be and the longer will be the acquisition time for data of given statistical precision. An example of the variations of beam current with beam size for a particular electron column modelled and measured by Venables and Archer (1980) is shown in Figure 3.

In order to try to use as high a beam current as possible whilst maintaining a small beam diameter it is important to use the brightest source of electrons that can be found. There are two kinds of sources which have brightnesses greater than that of the simple tungsten thermionic emitter. These are lanthanum hexaboride (LaB_6) and field emission sources . LaB_6 sources can have brightnesses of the order of $5x10^6$ A $sr^{-1}cm^{-2}$ and field electron emitters based upon very sharp tips of W(100) coated with Zr or tips of clean W(310) can have brightnesses of the order of 10^{10} A $sr^{-1}cm^{-2}$. The field emitters are the favoured sources for analytical electron microscopes because they have the highest brightness obtainable and they can be fabricated as Schottky (Danielson and Swanson 1979) emitters in a form which is very reliable and can result in an electron beam which has a stable current over very long periods of time (many 100's of hours). Using such sources it is possible to generate, for example, a beam of electrons with a diameter of 50 nm, a current of 10 nA and an energy of 5 keV using an all electrostatic column (Todd, Poppa and Veneklasen 1979) or a beam with energy 100 keV, diameter approximately 4 nm and current 1.5 nA using a column with magnetic lenses, (*e.g.* Bleeker and Kruit 1990) which have lower spherical aberration effects than electrostatic lenses.

Monte-Carlo modelling near to an abrupt chemical edge in a flat surface can be carried out with an infinitesimally narrow electron beam striking the surface at a variety of distances away from the edge. This calculation leads to an *edge resolution function* for that combination of materials, angle of incidence and beam energy. The spatial resolution for a sharp edge with a practical beam of finite width can then be found by convoluting the beam intensity profile with the edge resolution function. This calculation was reported by El Gomati *et al.* (1979) who showed that the spread of a sharp edge, Δ_{50}, from 25% to 75% of the total change in Auger signal occurring across the edge is given by:

$$\Delta_{50} = 2rb \tag{2}$$

In this equation r is the Auger backscattering factor and b is the full width at half maximum of the Gaussian electron beam. Since r is usually in the range 1 to 2.5 the edge resolution is somewhere between 2 and 5 times the beam width depending upon the atomic number of the sample. This expression has been tested experimentally by El Gomati *et al.* (1979) and appears to provide a good working guide to the spatial resolution to be expected.

In order to obtain the best spatial resolution there have been two extreme approaches. (a) Reduce the spreading effects of electron backscattering by reducing the primary beam energy. This was the tactic adopted by Todd, Poppa and Veneklasen (1979) who demonstrated a miniature electron column based upon a field emission source which focussed a beam of electrons at 5 keV into a spot about 50 nm wide. (b) At the other extreme, the primary energy can be raised to reduce the spot size by reducing the aberrations in the column. This has been demonstrated by Cazaux (1989) who used a STEM working at 100 keV and by Venables and Hembree (1991) using a very special electron spectrometer also in a STEM (see later). At the time of writing the instrument described by Venables and Hembree has demonstrated the best spatial resolution obtained to date, approximately 4 nm at 100 keV using small Silver particles on a Si(100) substrate (Figure 4).

The spatial resolution observed with a practical surface which may be very rough and may even contain sharp or re-entrant steps is worse than that estimated from Equation 2 or observed in idealised flat surface experiments. There are a variety of complicating

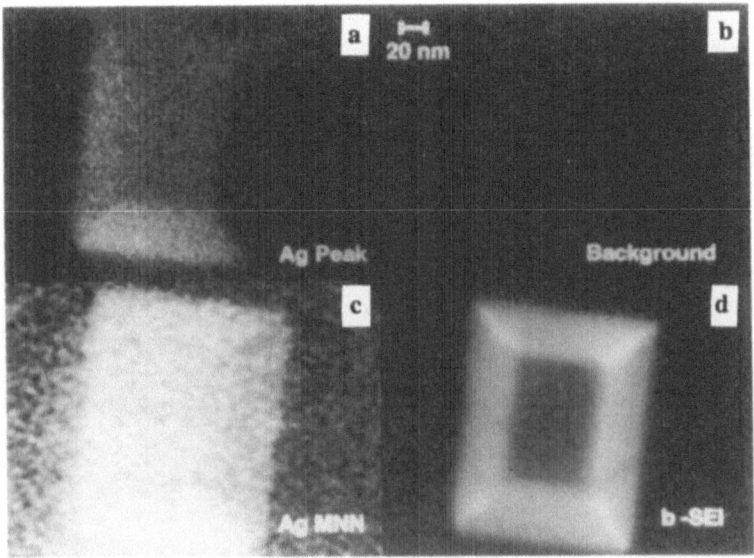

Figure 4. *A silver crystal on a silicon (100) substrate at 500°C imaged in the* MIDAS *instrument (see later). (a) Energy analysed image on silver peak. (b) Energy analysed image on the background just above the silver peak. (c) The silver MNN Auger image produced from (a) and (b). (d) The secondary electron image of the same crystal. (Courtesy J A Venables)*

scattering processes occurring at such steps which can degrade the resolution. These will be discussed later.

Detailed considerations as to the physical limits applying to the spatial resolution have been presented by Cazaux (1984) who predicts that chemical identification of a single atom will be possible using AES for those samples that are able to tolerate a very large electron dose.

1.5 Spectrometers in imaging instruments: CMA & CHA

There are two kinds of electron spectrometer in widespread use in scanning Auger microscopes. They are the cylindrical mirror analyser (CMA) and the concentric hemispherical analyser (CHA) which usually incorporates a transfer lens (Figure 5).

The CMA is attractive for its mechanical and electrical simplicity — it consists of a pair of coaxial cylinders with appropriate entrance and exit slits. Further, it can be constructed to collect a reasonably large fraction (upto 10%) of the electrons emitted from the sample into 2π steradians. Its disadvantages are that it operates at constant energy resolving power as the spectrum is swept which means that electron counts are lost at low energies and it can have a rather small field of view — the maximum area of the surface under study which can be scanned and used to form an image without unacceptable loss of transmission around the spectrometer.

The CHA has the advantages that it can be operated in different modes in order to optimise the measurement of the quantity being observed. The entrance aperture may be reduced to increase the energy resolution or increased to improve the solid angle of collection of electrons leaving the sample. The potentials used may be arranged to sweep through a spectrum with constant energy resolving power or with constant energy window. With appropriate design of the transfer lens between the sample and the hemispheres the field of view can be arranged to be quite large compared to that of a CMA — perhaps a few hundred microns across. The disadvantages are that a CHA is substantially more complex than a CMA because there are many electron optical elements whose potentials have to be varied as a spectrum is swept. The solid angle of collection is usually smaller than that of a CMA; it may be in the range 0.05 – 2% of 2π sr. The flexibility of the CHA and its adaptability for multi-channel operation (see later) make it the favoured choice at the moment.

Transmission functions T(E)

The transmission function $T(E)$ of an electron spectrometer is discussed earlier in this book by Seah. It can be defined as the ratio the number of electrons actually counted over the number of electrons entering the entrance slit of the spectrometer with the correct energy to reach the detector. Clearly, an ideal spectrometer would have a transmission of unity for all energies but in practice this is impossible. One reason is that the Helmholtz-Lagrange law is at work in the transfer lens between a sample and the hemispheres of a CHA. This means that as the retardation of the transfer lens is increased so as to slow the electrons to the correct energy to pass around the gap between the hemispheres so the angular divergence of electrons leaving the transfer lens must rise. This divergence can become so large that even electrons with the correct energy to be detected can strike the metal parts of the system and are lost to the detector. Thus $T(E)$ is in general a decreasing function of E. A second reason is that the aberrations in the transfer lens also spread the beam and allow any apertures or sections of restricted diameter to cut off some the electrons destined to reach the detector. This too results in $T(E)$ being a decreasing function of E.

Why should this decreasing $T(E)$ be of consequence in an imaging context? In the first place, the arguments above about signal to noise ratio have shown how the analyst is always working to keep the signal as high as possible. At higher kinetic energies the low value of $T(E)$ means that electrons are being lost in their path through the spectrometer and this represents lost sensitivity and increased frame scan times. Secondly, $T(E)$ is a decreasing function of E because some electron trajectories intercept metal parts in the spectrometer and the extent to which this occurs is changed as the source of electrons is moved about on the surface as the primary beam is scanned. This means that $T(E)$ can be different at different sample positions and is the origin of the finite field of view of a spectrometer. If the spectrometer is being used to collect images which are going to be quantitatively interpreted then this variation will add to the complexity of analysis. Finally, if the spectra obtained using the spectrometer in the microscope are to be

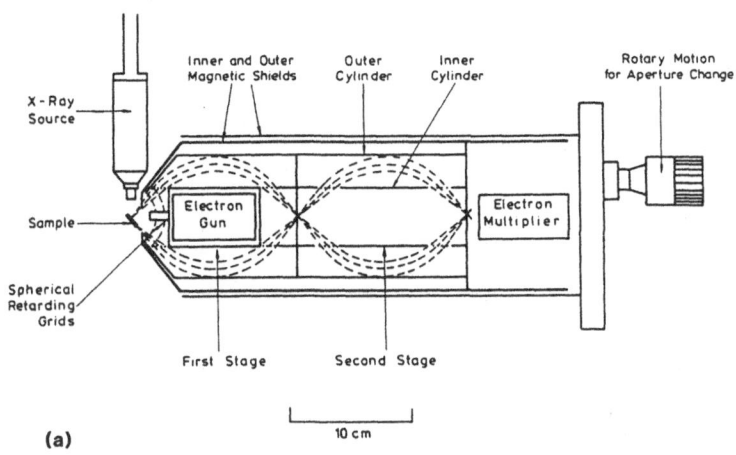

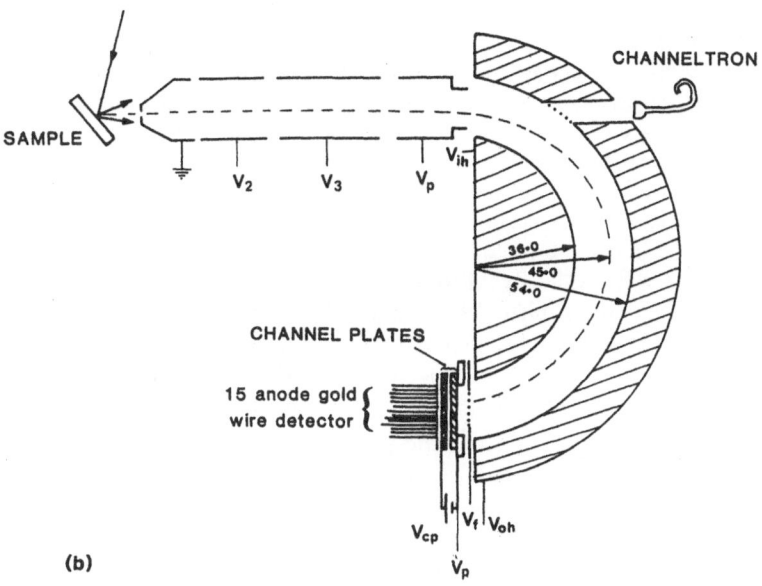

Figure 5. *(a) A double-pass cylindrical mirror analyser (CMA) which consists of two CMA's in series. (Courtesy of P W Palmberg and Elsevier Scientific Publishing Company) (b) A concentric hemispherical analyser (CHA). This is the CHA in the MULSAM instrument referred to later in the text. The accelerating voltage across the channel plates is referred to the kinetic energy of the electrons leaving the sample. The radii are in mm.*

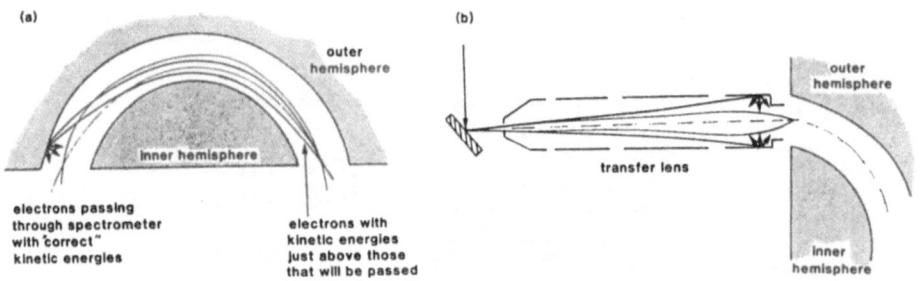

Figure 6. *Internal scattering originating (a) at the outer hemisphere and (b) in the transfer lens of a* CHA.

compared with spectra obtained in other instruments then it is essential to correct the measured spectra for the distortion caused by $T(E)$ for each instrument.

A further degradation of the observed spectrum which affects the extent to which images have acceptable signal-to-noise ratios and can be quantitatively interpreted is due to scattered electrons inside the spectrometer structure reaching the detector and adding to the background under the peaks being measured. Electrons travelling around the centre of the gap between the hemispheres of a CHA are focussed on the detector and counted. However, electrons with kinetic energies just above this value can strike the outer hemisphere near the exit plane of the CHA (Figure 6a). These scattered electrons will clearly contribute an amount $I(E)$ to the background under any peak in the spectrum and reduce the signal-to-noise ratio. This mechanism has been identified and studied by Seah and Smith (1991). Secondly, electrons with energies above those focussed by the transfer lens of the spectrometer will strike metal parts on this lens and there generate secondaries (Figure 6b). Some of these secondaries may also reach the hemispheres, pass around and be detected, also contributing to the background.

A third complication arises from the fact that the channel electron multipliers or microchannel plates, usually employed to detect individual electrons successfully navigating the spectrometer, have gains which can vary with kinetic energy of the electrons arriving at their front faces, their state of cleanliness and the total dose of electrons to which they have been exposed. This whole subject has been studied carefully by Seah and Tosa (1992). The kinetic energy variation can be overcome by accelerating the electrons from the exit plane of the hemispheres to the front face of the channeltron or channel plates such that they arrive with a kinetic energy independent of their initial kinetic energy. The other two effects are harder to quantify. One approach is to mount a standard elemental sample on the specimen manipulator alongside the sample under study. If this is always the same material (say gold or silver) then the area under the spectrum of the standard can be normalised to the same quantity at some defined time. This ratio can then be used to scale the spectra or pixel values of the unknown to the sensitivity at the same defined time. This is not a time consuming process and provides a useful guide as to the extent of ageing effects in the detectors.

Some sensitivity considerations

The solid angle Ω subtended by the spectrometer entrance aperture at the sample surface determines the fraction of all emitted electrons which are collected: it is therefore a crucial parameter determining the sensitivity. As discussed above it has a maximum value which is usually determined by the energy resolution required.

Another way of raising the sensitivity is to increase the current in the incident beam (see Equation 1). However, this too has disadvantages. Firstly, increasing the current in the beam increases the beam diameter (because of Coulomb repulsion within the beam, *i.e.* space charge effects) and as a result the spatial resolution is degraded. Secondly, an increased current density in the beam can result in increased damage of the sample surface—see Pantano and Madey (1981). Insulating materials like oxides and many organic materials undergo dissociation or desorption in an electron beam which, of course, causes a time dependent variation in the surface chemical composition. In many cases the damage mechanism appears to be dependent upon the total dose of electrons striking a point on the surface. Pantano and Madey suggests that noticeable damage to adsorbed monolayers on metals will occur with a dose of 5 keV electrons of 1.6×10^{-4} to 1.6×10^{-3} C/cm^2. For damage to bulk materials they suggest that noticeable damage occurs at about 10 times this dose. These figures are applicable to those materials that damage rather easily in the electron beam such as Al_2O_3, SiO_2, alkali halides and most organic materials. Metals and semiconductors can be exposed to many orders of magnitude higher doses and not show any sign of damage.

As the rate of arrival of electrons at a channeltron or microchannel plate rises, because of increases in the solid angle Ω or the beam current, a new source of distortion of the spectrum can occur. This is due to the non-linearity of these detectors which exhibit a dead time effect causing the measured count rate to saturate as the incident rate is increased. This can be corrected for provided the deviation from non-linearity is not too large. The correction used in the York MULSAM instrument (see below) is of the form:

$$n = \frac{m}{S - \pi m} \tag{3}$$

In this expression n is the corrected count rate, m is the observed count rate, S is the sensitivity of the detector and π is the dead time for the detector. The quantities S and π are determined in a calibration experiment.

Field of View $S(E, x, y)$

Every spectrometer has a field of view $S(E, x, y)$ within which electrons at energy E can be detected with a sensitivity proportional to $T(E, x, y)$ for a particular range of positions (x, y) of the electron beam on the surface. This is an important property when imaging because $S(E, x, y)$ determines the area of the sample that can be viewed and analysed for any particular setting of the spectrometer potentials. If the area being scanned is large (low magnifications) then the field of view can often be seen in the image contrast. An example is shown in Figure 7a where the bright region in the centre is the area of the surface that can be imaged. The field of view of the double pass

CMA has been investigated by Erickson and Powell (1986) and that of a CHA system by Peacock *et al.* (1984). Their results are shown in Figure 7b and c. Because of the Helmholtz-Lagrange law, the field of view of a CHA accessed by a transfer lens becomes smaller as the electron optical retardation of the transfer lens is increased. Thus, if a spectrum is scanned with constant pass energy of the electrons around the hemispheres then the field of view becomes smaller as the spectrum is swept up to higher kinetic energies. This can be a very important consideration when imaging because incorrect positioning of the sample can result in the scanned area falling outside the field of view of the spectrometer . It is important to be able to centre the field of view when using a large flat sample in the field of view before moving the sample manipulator to bring the sample under study into the same position before acquiring data which is to be quantitatively interpreted.

Examples of microscope configurations

Three distinctive examples of UHV, energy analysing, scanning electron microscopes designed for scanning Auger work are shown in Figure 8.

The principle of Auger microscopy was first demonstrated by Macdonald and Waldrop (1971). An early (1979) practical instrument was a compact assembly of a miniature field emission electron gun and column mounted coaxially in a CMA was described by Todd, Poppa and Veneklasen (1979) of Stanford University and is sketched in Figure 8a. The gun delivered a beam of about 5 nA into a spot about 50 nm diameter at 5 keV. The specimen manipulator was mounted directly onto the gun/CMA assembly in order to minimise the effects of mechanical vibrations. The whole assembly is supported on a single 10 inch diameter UHV flange. The instrument was subsequently placed under computer control and modified to detect and image spin polarised electrons by VanZandt *et al.* (1989).

A different philosophy was adopted at York where emphasis was placed upon the technique required to quantify images in order to map the variations of the chemical composition in the surface. Spatial resolution was a second priority. The first instrument here was reported in 1977 (Browning *et al.* 1977) and this developed into the instrument sketched in Figure 8b. The distinctive quality of this design is that there is an array of detectors around the sample which can be used to acquire sets of simultaneous images of the same region on the sample. This instrument is referred to as MULSAM or *Multi-spectral Scanning Auger Microscope*. The beam energy is 20 keV, the beam current is about 7 nA and the beam focuses into a 200 nm diameter spot. The method of combining images from the different detectors using models of the physics of the scattering processes for each detector signal in order to derive information about the area being imaged will be described in the next sections. This instrument has been described by Prutton *et al.* (1991).

The microscope which has demonstrated the highest spatial resolution has been built and used by workers at the Universities of Sussex and of Arizona State and is based upon a heavily modified STEM. The instrument is sketched in Figure 8c and is known as MIDAS (Microscope for Imaging, Diffraction and Analysis of Surfaces). The arguments above have outlined how it becomes crucially important to collect as many of the scattered electrons as possible when the spatial resolution is improved. This was done

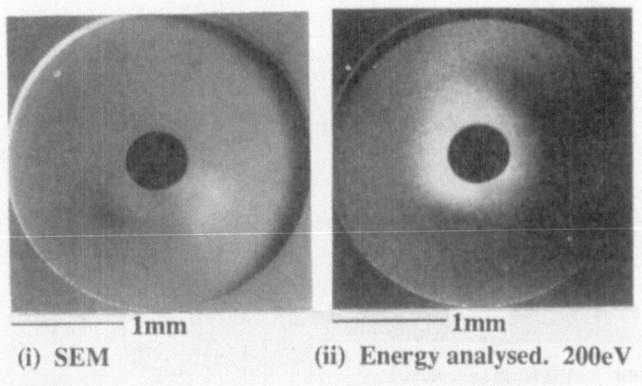

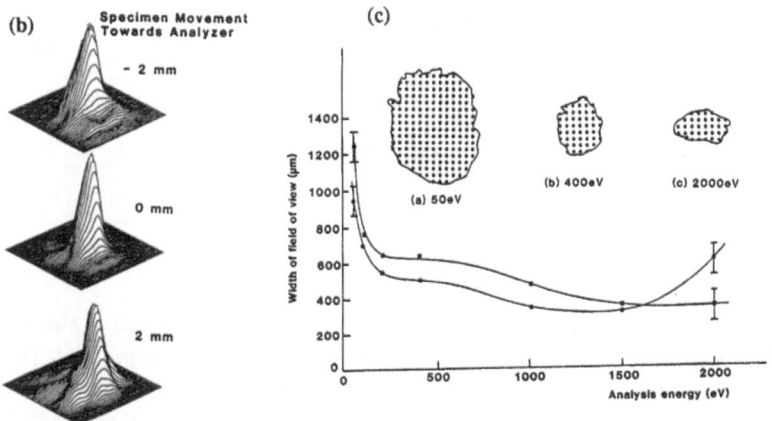

Figure 7. *(a) Image of a field of view obtained using the* MULSAM *instrument. The sample is the stainless steel face of a Faraday cup. The dark circle in the* SEM *image (i) is about 300 mm diameter. Spectrometer pass energy 100 eV, electrons analysed with 200 eV kinetic energy. The bright region surrounding the Faraday cup aperture in (ii) is due to the electrons passing around the* CHA *and being detected. Most of the electrons are entering the Faraday cup. (b) Images of the fields of view of a double-pass* CMA *recorded with an analysis energy of 50 eV, an incident beam energy of 500 eV and the analyser operated in an* XPS *mode with a pass energy of 50 eV. The sample is mounted with its surface in a vertical plane and the 500 eV electrons are scanned through 13 mm horizontally (bottom of the images) and 15 mm vertically. The signal passing the analyser is represented in three dimensions and shows the effects on the field of view of moving the sample away from the optimum position (0 mm) by +2 mm (away from) and −2 mm (towards) the analyser. (Courtesy of Erickson and Powell). (c)* CHA: *The width and height of the field of view as a function of the kinetic energy of the electrons leaving the sample and with 50 eV pass energy around the hemispheres. York* SAM.

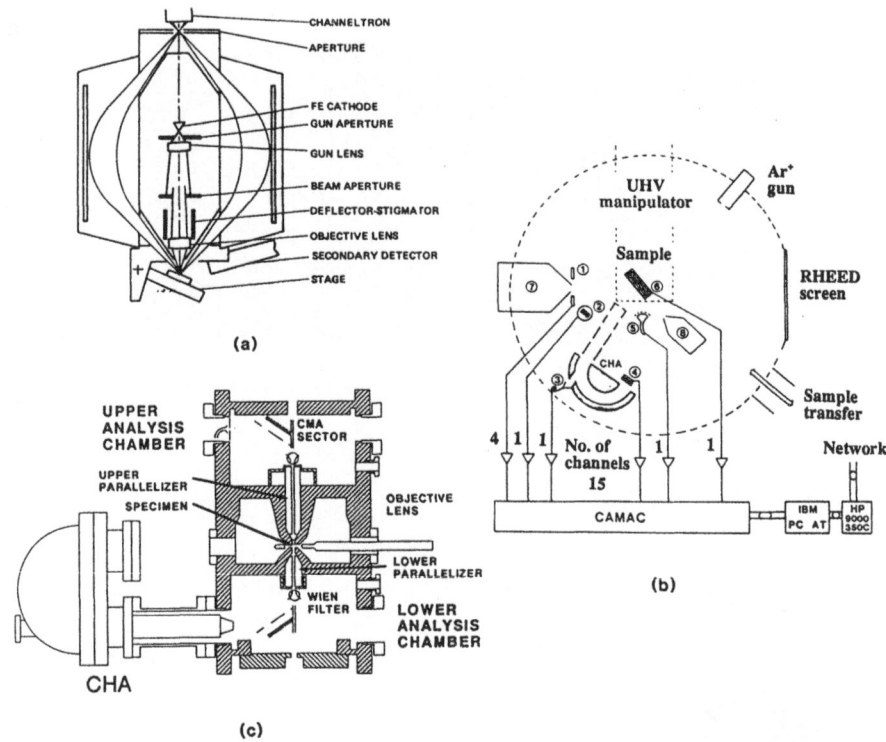

Figure 8. *Three arrangements of components in* SAM. *(a) Poppa: a single pass* CHA *with coaxial field emission electron gun and integral specimen manipulator. (Courtesy H Poppa); (b) Prutton (MULSAM): a* CHA *with seperate field emission gun and and array of detectors working in synchronism. (c) Venables (MIDAS): A modified STEM with a paralleliser, CAM section as a deflector and an external CHA. (Courtesy J A Venables)*

by Hembree, Luo and Venables (1990) by using a novel electron spectrometer structure which incorporates a beam paralleliser before a CHA. The column in this instrument (MIDAS) delivers about 1.5 nA into a 4.3 nm spot at 100 keV. In order to focus the beam to this small spot size and yet to be able to collect a large fraction of the scattered electrons Venables et al have designed a magnetic immersion lens which focuses electrons onto the sample and extracts emitted electrons to pass to the spectrometer. Part of this immersion lens structure is a magnetic field paralleliser (Kruit and Venables 1988). This device provides a field parallel to the surface normal of the sample which falls off as the reciprocal of the square of the distance from the sample. It causes emitted electrons to move in spirals about the field lines and converges them into a cone with about 6° semiangle having collected almost all of those emitted. This microscope is near the beginning of its development and more interesting observations are likely.

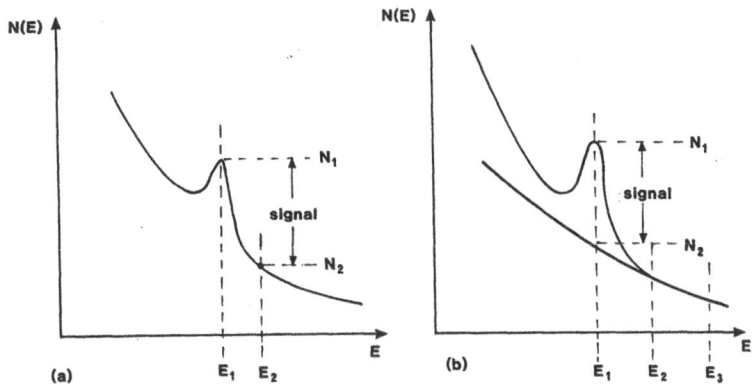

Figure 9. *The estimation of spectral peak height: a) Peak minus background; b) Peak minus extrapolated background.*

2 Auger images, data manipulation and image correlation

2.1 Auger images

The peaks — heights, areas and shapes

Several features of the peaks in the electron spectrum can be used to modulate the brightness of a digital image. The simplest to measure is the peak height which can be derived from the difference between the height of the peak, N_1, and the height, N_2, of the spectral background at some energy just above the peak (Figure 9a). This measurement has to be made at each point on the surface which is going to be mapped into an image pixel. The procedure amounts to collecting two energy analysed images for each peak to be measured. Alternatively each peak height can be estimated by measuring the height of several points on the spectral background above each peak and then using some analytical expression to extrapolate the background N_2 at the energy of the peak. $(N_1 - N_2)$ is then a signal given by the peak minus the extrapolated background nnheight (Figure 9b). The latter is clearly a more accurate estimate of the size of the peak but it requires the acquisition of more data — at least three measurements per pixel for linear background extrapolation. The advantages of such an extrapolation are discussed later.

In order to construct images in which the contrast is proportional to the concentrations of the various chemical elements in the surface it would be advantageous to acquire a signal proportional to the area under each peak in the electron spectrum. In general this is rather difficult and has not been reported. In the first place, the estimation of peak area requires that a rather large set of energy analysed images is collected so that the whole of each peak is spanned by many images in the set. Secondly, there is always a secondary electron cascade underneath each peak in the spectrum and this would need to be modelled and then subtracted from the raw data. This is necessary so that the area of the peak above the background can be calculated at each pixel. Therefore a great deal of data has to be acquired. Perhaps high sensitivity parallel spectrometers which collect all energies in the spectrum simultaneously will be invented in the future and this approach will become feasible. At present the acquisition times needed for this estimation of area are prohibitively large.

The shape of the peak in the spectrum contains information about the chemical state of the emitting atoms. The simplest effect is that the Auger peaks associated with a given kind of atom are shifted in energy depending upon the chemical environment in which the atom is situated. These *chemical shifts* can be quite large in Auger electron spectroscopy. Thus for instance, there is a shift of 18 eV in the energy of the Mg LVV line between metallic Mg and the insulator MgO. The Auger image may be formed with contrast to reflect this chemical shift by subtracting an image formed at the metallic Mg energy (45 eV) from one formed at the oxide energy (27 eV). The detailed shape of a peak also changes (Madden 1983) as the chemical environment changes. Another oxidation example of this kind of change is the shape of the LMM peak of metallic titanium at 416 eV which has a sharp peaked feature at its high energy extreme. This feature vanishes rapidly as the Ti surface is exposed to small amounts of contamination, particularly oxygen. Again, this change can be imaged by acquiring and post-processing a pair of energy analysed images.

The spectral background: Sickafus and AE^{-m}

The shape of the spectral background in an electron excited Auger spectrum has been alluded to above. The general shape of the spectrum is sketched in Figure 10 where the curved background $B(E)$ is shown underneath the peaks. This background is important to the imaging analyst because it contributes to the statistical noise in a measurement of the peak height and because it has to be subtracted from the spectrum before a peak height or area can be estimated. In an important series of papers Sickafus (Sickafus 1977a,b, and Sickafus and Kukla 1977) studied $B(E)$ at low primary beam energies and found that it could be described as function of kinetic energy E using the power law:

$$B(E) = AE^{-m} \qquad (4)$$

In this equation A and m are attributes of the material being studied but are functions of the primary beam energy and the angles of incidence and the take-off to the spectrometer entrance aperture. This expression is applicable for kinetic energies below about half the primary beam energy. Above this threshold the scattering of energetic electrons (often called *backscattered electrons* or *rediffused primary electrons*) can make a significant contribution to the size of $B(E)$. The contribution of backscattered electrons yields a background which is a rising function of kinetic energy. The background described by Equation 4 will be referred to as the *secondary-electron cascade*. One advantage of working with high primary energies is that the electron spectrometer is usually designed to detect electrons with kinetic energies up to a few keV and so the background observed is dominated by the secondary-electron cascade and Equation 4 is a good physical description of its shape (Matthew *et al.* 1988). Of course, if this

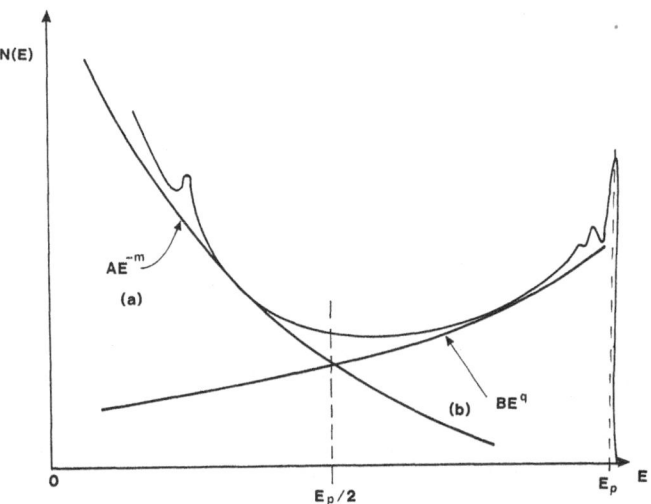

Figure 10. *The background under a peak: (a) secondary-electron cascade; (b) rediffused primary electrons.*

is to be exploited in any background subtraction scheme it is essential to know the spectrometer functions $T(E)$ and $I(E)$ which distort the true spectrum and to correct the observations to remove this distortion prior to background removal operations.

If the kinetic energies to be observed are above about half the primary beam energy then another power law representation can be used and has been described by Peacock and Duraud (1986). This has the approximate form:

$$B(E) \;=\; CE^q \tag{5}$$

For low primary energies (< 15 keV) the general form of $B(E)$ below 2 keV is the sum of Equations 4 and 5. For high primary energies Equation 4 is a good approximation to the background shape for the lower kinetic energies (< 2 keV) of electron emission.

The signal—removing B(E) and topography

Approximate corrections are required for the following complicating factors.

- The background $B(E)$ beneath each peak in the spectrum,

- The variation of the Auger electron yield from place to place because any real surface will be rough. This means that the angles of incidence and take-off will vary with incident beam position and the Auger yield will vary even if the concentration of the element being imaged does not vary.

- Given the long frame scan times discussed above there may be fluctuations from time to time in the current in the incident beam. This will result in contrast in the image unrelated to concentration variations because the Auger yield (and all other yields of emitted particles and radiation) scales linearly with incident beam current.

The simplest scheme for deriving a signal containing some correction for these effects is to calculate the peak to background ratio P/B from the quantities N_1 and N_2 using $(N_1 - N_2)/N_2$. This was proposed by Todd and Poppa (1978), evaluated in a quantitative analytical context by Langeron *et al.* (1984) and is in widespread use in commercial instruments. Although this is a useful measure of the signal in many applications El Gomati *et al.* (1985) have shown that it can sometimes result in contrast artefacts due to possibility that sub-surface concentration variations can lead to variations in the signal due to changes in N_2 when there have been no changes in the surface composition. An effect of this kind is demonstrated in Figure 11c where it can be seen that the brightness of the silver Auger image is greater in the unsupported regions than over the gold bars of the grid underneath. We might expect that the Auger image should be brighter over the bars because of the backscattered electrons from the underlying gold.

The next simplest scheme which was devised in order to compensate for topographical effects and which also corrects for beam current fluctuations, is to derive the signal from two energy analysed images N_1 and N_2 but to form the ratio $(N_1 - N_2)/(N_1 + N_2)$. The proposal for the use of this ratio is due to Janssen *et al.* (1977) who derived a signal from the ratio of the first derivative of $N(E)$ with respect to E evaluated at the energy of the maximum positive excursion of this derivative to the value of $N(E)$ at the same energy. This is essentially a logarithmic derivative of the spectrum $N(E)$ and it is proposed as a compensator for local topographical effects because it seems reasonable to assume that the angular dependence of the background count is approximately the same as that of the peak height because they are measured at only slightly separated energies. A simple digital approximation to the logarithmic derivative is the ratio $(N_1 - N_2)/(N_1 + N_2)$. Prutton *et al.* (1983) have shown (see Figure 12) that this ratio is an effective corrector for surface topography. However, El Gomati *et al.* (1985) have shown that this ratio has to be used with care because it may also overcompensate for contrast effects arising from sub-surface composition variations, as can be seen in Figure 11d.

In order to correct for the variations in subsurface composition, and the way in which they cause the slope of the background just above the peak to change from place to place, it is advantageous to extrapolate the background under the peak and then subtract the background at the peak energy from the peak height. This too was first suggested by Harland and Venables (1985). The simplest extrapolation scheme is to acquire three energy analysed energies at equally spaced energy intervals with one energy on the peak and at two energies on the background above the peak. In each image pixel the two background images N_2 and N_3 can be used to extrapolate linearly under the peak N_1 and to form a signal, S, given by:

$$S = N_1 - 2N_2 + N_3 \qquad (6)$$

Such an extrapolated background signal is shown in the image of Figure 11b. The contrast now has the expected sign. Of course, more accurate extrapolation schemes

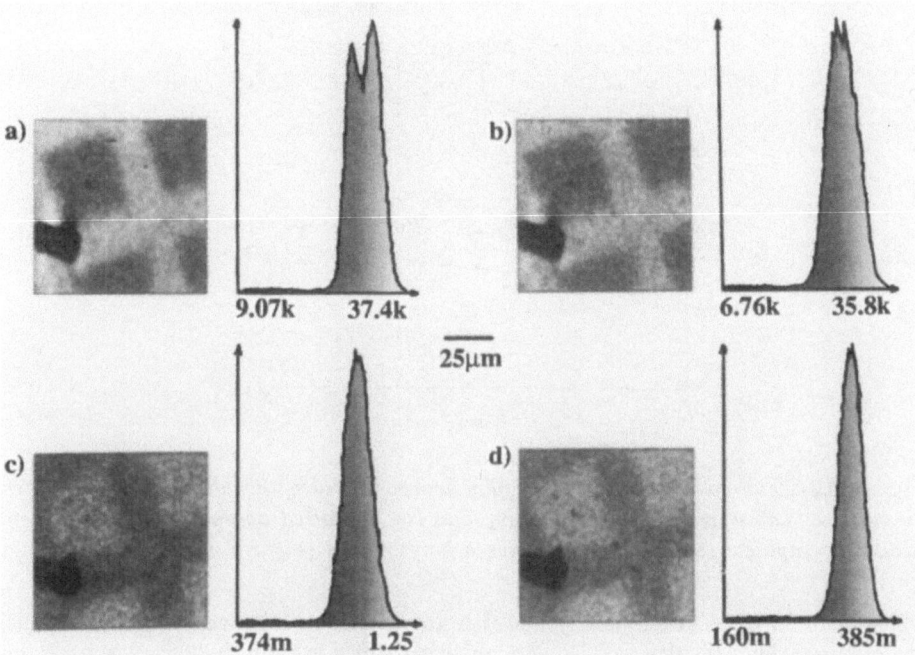

Figure 11. *Imaging of a 200 nm thick film of silver supported on a gold electron microscope grid using various combinations of measured signals: (a) (N_1-N_2), a simple approximation to the peak height; (b) $(N_1-2N_2+N_3)$, background extrapolated under the peak; (c) $(N_1-N_2)/N_2$, the peak to background ratio; (d) $(N_1-N_2)/(N_1+N_2)$, an approximate logarithmic derivative. Note that the ratio images (c) and (d) are brighter where the silver film is unsupported. The simple difference (a) and the peak minus extrapolated background image (b) have the correct sign of contrast.*

are possible in which, for instance, the background is measured at three energies and a quadratic function extrapolated under the peak but these schemes require more energy analysed images and so take more time to acquire. Equation 6 is not normalised in any way as are the two signals discussed previously. It could be normalised to the background under the peak by dividing the right hand side of Equation 6 by $(2N_2-N_3)$ so as to reduce the effects of beam current fluctuations.

Sequential image acquisition

The long frame scan times associated with the serial acquisition of Auger images are common in commercial instruments which are presently available. This has tended to discourage researchers from obtaining and interpreting images in spite of the advantages

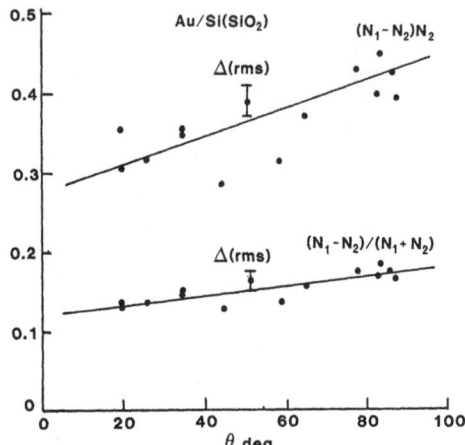

Figure 12. *The correction of topography demonstrated with an anisotropically etched Si sample. The appropriate ratio of N_1 and N_2 is plotted against angle of incidence. Ideal topographical correction would give a horizontal straight line in this plot.*

which were outlined in Section 1. Apart from matters of inconvenience and expensive use of time, the difficulty with long acquisition times is that there may be drift both in the mechanical arrangements for holding the sample in a fixed position with respect to the electron column and in the voltages applied to the scanning electrodes which determine the position of the beam on the sample. Thus, a set of Auger images intended to be of the same region in a surface may not be spatially registered with each other. If corresponding pixels in each image of a set do not correspond to the same place on the surface then quantitative analysis of the composition at that place becomes impossible. Such images are then reduced in value because although they may indicate the spatial distribution of the chemical elements in the surface and thus be a useful guide to means for solving some particular problem they do not allow exploitation of the huge amount of information that they contained before the confusion introduced by the drift. A solution to this problem is to try to collect information from the sample in parallel rather than sequentially.

Parallel image acquisition

A desirable solution to the problem of parallel acquisition would be to use a spectrometer which collected electrons with all energies in the spectral range required and to disperse and detect them simultaneously. No such spectrometer exists and the best compromise that can be made at present is to collect all the electron energies required by switching the spectrometer energy rapidly whilst the beam is stationary on the surface. Thus the energy analysed images are still acquired sequentially but the time interval between the various energies at a given pixel is kept as small as possible. If drift occurs it is likely to be slow compared to the acquisition time per pixel and so

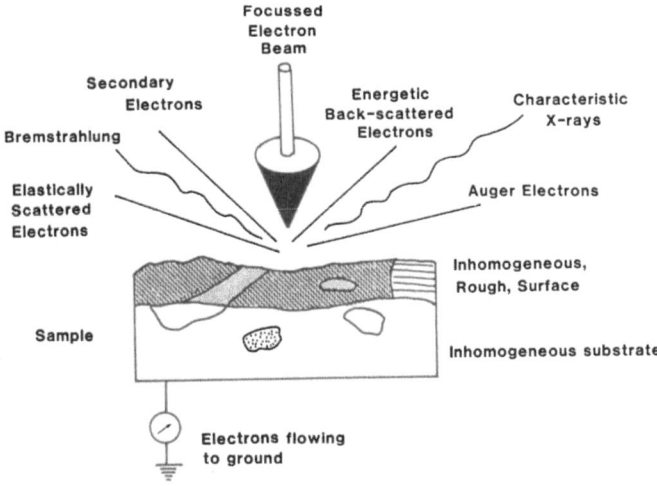

EMITTED PARTICLES AND RADIATION

Figure 13. *The various emissions from an inhomogeneous solid sample bombarded by energetic electrons.*

the information in corresponding pixels of the images in an image set comes from very nearly the same place on the surface. Drift will cause the whole image to be distorted compared to the real surface but the spatial registration is preserved. However, other detectors of SEM, backscattered, x-ray and other signals (Figure 13) can be arranged around the sample and connected to electronics which does acquire the data coming from them simultaneously. Since the physics of the electron-solid interaction has been very thoroughly studied, the correlation of the energy analysed images and the images obtained from other detectors can be exploited using this physics.

The LANDSAT satellite system (Moik 1980) operated by NASA is an analogous system to that used in multi-spectral Auger microscopy. The general disposition of components for LANDSAT is indicated in Figure 14 and the images that have been obtained from it are published in many textbooks on digital image processing (Moik 1980, Gonzalez and Wintz 1977, Pratt 1978). The idea of viewing the surface of a planet through many spatially registered telescopes each admitting a different spectral band of radiation is very similar to the idea of having several different detectors admitting different electron energies or different radiation coming from an electron bombarded surface.

The MULSAM system of analytical electron microscopy was proposed by Browning *et al.* (1984) and has been described in detail in a series of papers by Browning (1984, 1985) and by Prutton *et al.* (1990, 1991). Similar principles have also been used in SIMS by Bright and Newbury (1991) (who refer to their technique as being *composition histogramming*), in energy dispersive x-ray imaging by King *et al.* (1990), in STEM by Jeanguillaume (1985) and in EELS by Bonnet *et al.* (1988). The arrangement of the

LANDSAT Satellite – NASA

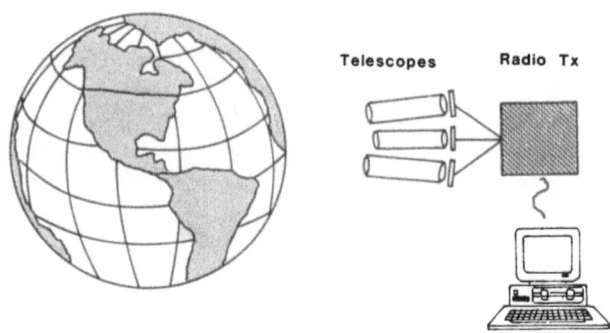

Telescopes   Radio Tx

Figure 14. *Sketch of the principles underlying the* LANDSAT *image acquisition system for viewing and analysing the surfaces of planets.*

York MULSAM instrument is shown in Figure 8b. Some examples are shown below of the use of multivariate statistical techniques and of the exploitation of image correlations that become possible once spatially registered image sets are collected.

2.2 Data manipulation

In order to remove the spectral background beneath the peaks used to form each Auger image in a set and then go on to try to cause the brightness of each image to be quantitatively related to the surface composition it is necessary to carry out the arithmetic operations outlined above and then perform iterative matrix corrections as described earlier by Seah. Further, in order to enhance the images by increasing the signal-to-noise ratio of what is always noisy data it is useful to be able to be carry out two-dimensional smoothing — perhaps with a polynomial least squares fit to each small segment of each image (*e.g.* an extension to two dimensions of the fits described by Savitsky and Golay 1964) or a two-dimensional Fourier transform and noise filtering operation as described by Kosarev and Pantos (1983). Thirdly, it is often necessary to use peak to background ratios to form the image contrast in order to reduce topographical effects as outlined above. This means that provision needs to be made to add, subtract, divide or multiply corresponding pixels in an image pair. Therefore, Auger microscopes are provided with a set of mathematical tools to process the raw image data. Sometimes these tools are provided by commercial mathematical software packages and sometimes they are purpose written. Some of the useful tools have been described by Prutton and Peacock (1982).

Scatter diagrams

If several spatially registered images are available from the same area on a surface then it is possible to exploit multi-dimensional histogram techniques to assist with the classification of the number of different types of region within that area. These histograms are really correlation plots but it is a convenient shorthand to refer to them as scatter diagrams. Suppose one has an image set consisting of two images obtained using two different detectors to collect simultaneous data. If the ith pixels in each image have values N_1^i and N_2^i then a scatter diagram is a plot in which points are marked with coordinates (N_1^i, N_2^i). Thus, a pair of identical and uniform images (*i.e.* without contrast) will have all pixel pairs coincident and a scatter diagram of all the pixel pairs which will contain a single point. Two noisy, but otherwise identical and uniform images will have a scatter diagram consisting of a single cluster of points whose centre corresponds to the mean values of N_1 and N_2. Two perfectly correlated images will have a scatter diagram with all points on a line making an angle of $+45°$ to the horizontal (N_1) axis. Two perfectly anti-correlated images will show a diagram with all points on a straight line at an angle of $-45°$. The dimensionality of the scatter diagram is equal to the number of images in the set. Finally, the number of clusters in the scatter diagram is equal to the number of different combinations of the signals there are in the image set. Thus, the number of clusters is just the number of different kinds of regions (or phases) that there are in that area of the surface. This diagram is therefore a powerful tool for making an objective decision about *how many* regions are needed to characterise a surface and *where* spectra must be acquired to estimate the composition of the region corresponding to each cluster. Examples of the use of scatter diagrams have been given by El Gomati *et al.* (1986) and Figure 15 shows an example of a simple situation with two anti-correlated images.

Once scatter diagrams can be constructed and displayed then *false colour imaging* becomes a clear possibility. Software can be written to enclose a selected cluster in the scatter diagram such that a new image is constructed which has a particular pixel colour attributed to all points which lie in that cluster. This means that the new false colour image has a contrast which identifies the different regions in the surface detectable in the original image set. Thus the new image is a *phase map* of the surface properties contained in the raw image set. An example of such a phase map for a simple case is shown in Figure 16 which is made up of two images formed from backscattered electron detector signals, their scatter diagram and their false colour image. In this case the contrast is dominated by the topography of the sample and the false colour image contains 'phases' which are in fact the differently inclined facets on a single crystal polyhedron sticking out of a surface. This example demonstrates how the definition of just what is meant by a 'phase' is determined by the physics of the particular signals collected by the detectors that happen to be used. This is a powerful technique often used in the presentation of images from space vehicles and some examples of its application in surface imaging will be shown in Section 3.

Principal component transforms

There are some useful statistical techniques associated with the interpretation of sets of spectra or of images which are variously known as the *Hotelling Transform*, the

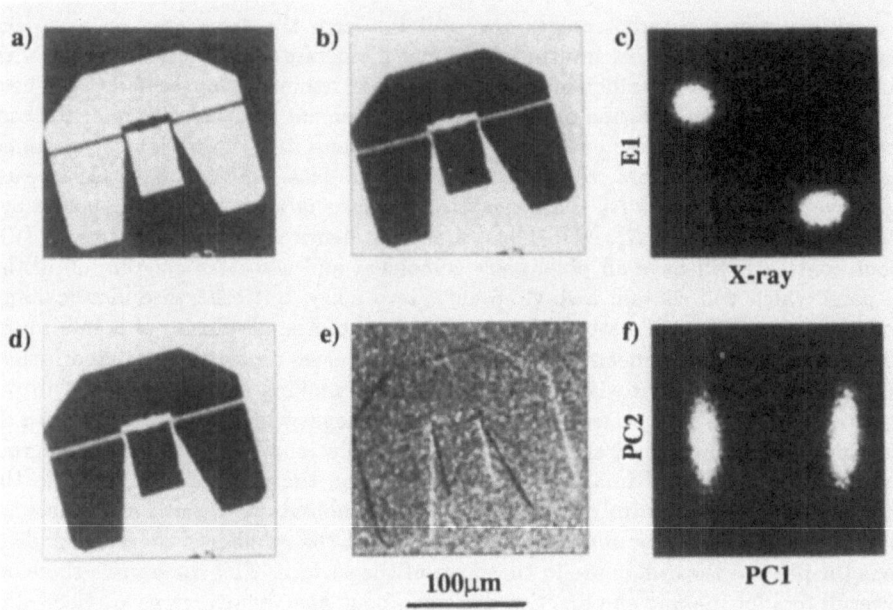

Figure 15. *Two images of a tungsten film overlay structure on a silicon substrate which are clearly anti-correlated (by inspection) and their two-dimensional scatter diagram. All the points corresponding to the pixel pairs in the original image set (within the spread determined by the signal-to-noise ratio) lie on an anti-correlation line making an angle of -45° to the x-axis. (a) A silicon* EDX *image; (b) An image formed with energetic scattered electrons (> 2 keV). (c) The scatter diagram formed from (a) and (b). The principal component images and their scatter diagram produced by application of the Hotelling Transform to the raw images. (d,e) The first and second principal component images of (a) and (b). (f) The scatter diagram formed from (d) and (e). Note that almost all the contrast has been condensed into the first principal component and the scatter diagram has been rotated with the line through the centres of the two clusters twisting through -45°.*

Eigenvector Transform, the Karhuenen-Loeve Transform, Factor Analysis or *Principal Components Analysis* (PCA). These transforms can be applied to any data that can be represented as a column vector of numbers and so they are applicable to spectra, which are clearly just tables of numbers versus kinetic energy. Images are also just tables of numbers but now they are tabulated versus the position of the beam on the sample. The methods are described in many books about image processing and in books on multivariate statistics (*e.g.* Malinowski 1991 and Krzanowski 1988). Gaarenstroom (1981) was the first to report the use of factor analysis for electron spectroscopy and this has subsequently been extended it (Gaarenstroom 1986) to a method of quantita-

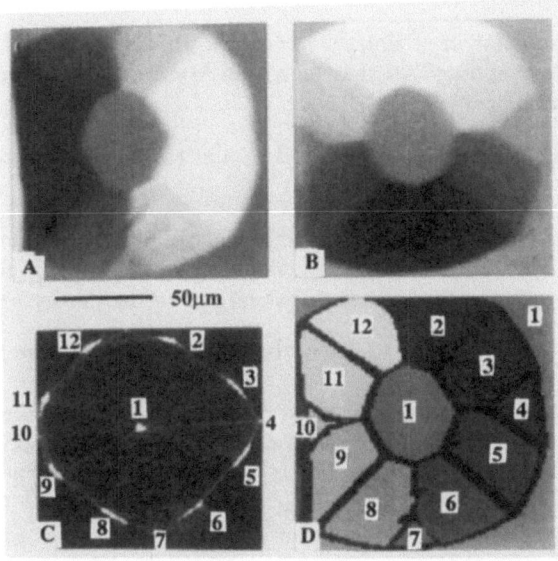

Figure 16. *Two difference images derived from four backscattered electron detectors symmetrically disposed about the incident electron beam, their scatter diagram and their false colour image (grey scale representation) . In this case the false colour image delineates the regions with different inclinations to the incident electron beam. A and B are the difference images, C is the scatter diagram and D the grey scale representation of the false colour image.*

tive analysis called target factor analysis. Applications to surface imaging have been described by Prutton *et al.* (1990).

If all or some of the individual images in an image set are correlated with each other then there is redundant information in the set which can be exploited to improve the signal-to-noise ratio, increase the contrast or to reduce the total amount of information being stored. The Hotelling transform does this. The output of the transform is a set of images, the same in number as raw images but the information is compressed into a smaller set and the noise tends to be distributed in the higher members of the set. The pixel values in the transformed image set — *the principal components* — are linear combinations of the corresponding pixel values in the raw images and the scatter diagram is rotated about its centre of gravity. Such a rotation applied to the data in Figure 15 would place the clusters along one axis of a new scatter diagram and only noise along the second axis This shown in Figure 15d–f. Thus, an image set containing two correlated (or anti-correlated) images would be transformed into one new principal component image containing information about the correlated components in the raw data and a second principal component image containing only noise.

The procedure for carrying out this transform is as follows:

1. Each raw data image is 'centred' by subtracting the mean pixel value of that image from each of its pixels and then by normalising it by that mean.

2. The covariance matrix of this set of centred images is calculated. If there are M raw images in the image set then this is an M×M matrix of cross-correlation and auto-correlation coefficients.

3. Standard numerical routines are used to find the eigenvalues and eigenvectors of this matrix.

4. These eigenvectors are sorted by size of eigenvalue into rows in a new MxM matrix, the *Hotelling Matrix*, in which the eigenvector corresponding to the largest eigenvalue forms the first row the next largest forms the second row and so on.

5. Operate on the raw image set by multiplication by the above matrix.

The transform defined by these steps rotates the scatter diagram because it is a linear operation. Further, it orthogonalises the image set as though it is separating out a set of independent simultaneous linear equations. The rotation of the scatter diagram places clusters with the largest separations along the first axis, those with the next largest separations along the second axis and so on. In other words it has sorted the raw image set into a new set of orthonormal components in order of decreasing variance. The transform is useful to the analyst because:

- It reduces the amount of information that must be stored.

- The number of statistically significant principal components in the transformed set tells the analyst how many different kinds of region there are present in the solid and where they are. Spectroscopy of 'typical' places can be carried out subsequently to identify the differences between these regions. The selection of what is typical is then completely objective.

- Although the overall signal-to-noise ratio is unchanged as a result of applying the transform, the contrast in the first few principal component image is increased because the rotation of the scatter diagram has increased cluster separation along the principal axes.

- The principal component image set can reveal unexpected features of the sample surface for which specific spectral information was not sought. These regions can be analysed retrospectively to find out what is special about them.

- The transform can be applied to data which has been pre-processed using some physical model of electron-solid scattering intended to try to separate the confusing effects of composition variations in the surface and in the bulk as well as the topography. The transformed images and their eigenvectors reveal the extent to which the pre-processing has been successful in separating the various effects.

Examples of some of these advantages are demonstrated below. This discussion is not confined to Auger imaging — PCA is a useful method for the examination of all kinds of spectral and image data in all the techniques outlined at the start of Section 1.

A recent paper describing the method of estimating the signal-to-noise ratio in an image set and applying the Hotelling Transform to EDX images has been published by Browning (1992).

2.3 Image correlations

Some examples of the use of scatter diagrams and Hotelling Transforms are given here using data obtained using the MULSAM instrument mentioned above. There are large numbers of correlations between the signals which can be measured from the array of detectors in that system but attention will be confined here to those which help the analyst with the removal of image contrast artefacts which can prevent the accurate quantification of the surface chemical composition.

SEM and absorption current

The long acquisition times associated with particle counting in the electron microscope lead to a need to try a derive an image containing contrast dominated by any fluctuations in the current in the incident electron beam. Since all other signals are linear functions of the beam current the effects of such fluctuations can be corrected out of the raw data using simple division by a beam current image. Unfortunately, it is not technically easy and fast to measure the beam current (usually a Faraday cup has to be manoeuvred into the beam) at the same time as all other signals are being detected. Therefore it is useful to be able to exploit some of the images collected to derive a beam current image which can then be used to correct, for instance, the chemically specific images which can be collected using the electron spectrometer or backscattered electron detectors. Such an image has been derived using the anti-correlations between simultaneous SEM and absorption current images (Barkshire *et al.* 1991a).

The SEM signal is anti-correlated to the sample current flowing to ground because the incident beam current is normally constant. Thus, if the secondary-electron yield rises then conservation of the total number of electrons means that the current to ground must fall. An example of the anti-correlation of these two images is shown in Figure 17. The scatter diagram shows points on lines at −45° to the x-axis as expected. If the beam current varies slowly compared to the time between pixel measurements then a set of such anti-correlation lines may be observed in the scatter diagram of these two images. By normalising all pixel pairs to lie on a single line at 45° the normalising factor map must be proportional to the instantaneous beam current. The beam current image can then be divided into any other image measured at the same time as the SEM and sample current images and the effects of beam current fluctuations are thus removed. Such normalisation against beam current variations is shown in the Auger images of Figure 17e, f and g.

This is not a perfect correction for these fluctuations because the sample current corresponds to the beam current minus all the electrons which are emitted from that point on the sample but the SEM detector collects only a fraction of the solid angle of emission and only electrons with low energies. This means that the anti-correlation cannot be perfect. However, it is often adequate because the beam current fluctuations are not large (generally a few percent over many hours with field emission guns and

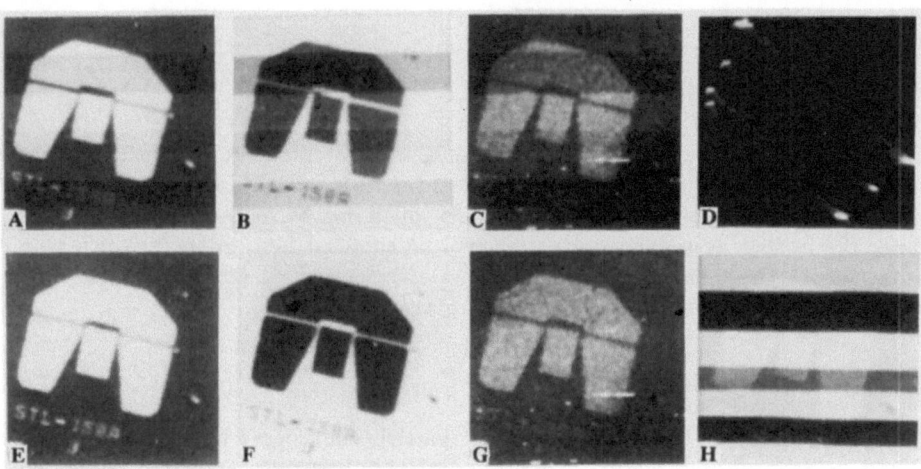

Figure 17. *Images from the same sample of Figure 15 measured while the beam current was deliberately changed in sudden steps. A, B and C are the* SEM, *absorption current and an energy analysed signal collected simultaneously. The horizontal bars show the effects of changing the beam current. D shows the scatter diagram between A and B. H is the beam current fluctuation image derived from D. E, F and G are images A, B and C after division by H.*

less with thermionic guns) and because the variations in the angular distribution of secondary electrons with local atomic number may not be large.

Backscattered electron signals

The use of backscattered electron (BSE) detectors to obtain atomic number contrast (Z contrast) and topographical contrast in conventional SEM is well established (Barkshire *et al.* 1992). A set of four quadrants of a circle of backscattered electron detectors positioned around the surface normal of a sample can be used to obtain Z contrast by adding the signals from all four, and topographic contrast by subtracting signals from opposing members of the four and normalising to the sum. Because the angular distribution of the backscattered electron yield is rather well understood (*e.g.* Niedrig 1982, 1983) and repeatable it is possible to calibrate a set of detectors once and for all so that the local angle of incidence of the electron beam can be calculated from the positions of points or clusters of points in the scatter diagram formed from two difference images derived from the four BSE detectors. This is demonstrated in Figure 18 which is derived for a deliberately rough sample of anisotropically etched Si(100). The sample

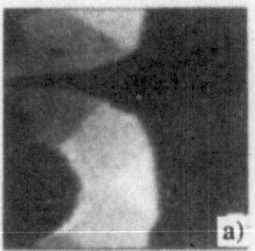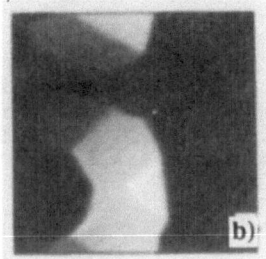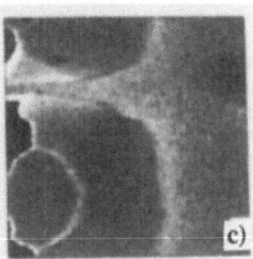

Figure 18. *Correction of topography from the carbon covered etched Si(100) sample of Figure 16. (a) The carbon KVV Auger image (peak minus extrapolated background); (b) The correction image derived from the sum of the four signals from the* BSE *detectors; (c) The ratio of (a) to (b). Edge effects remain.*

has been coated with a thick layer of carbon. The local angle of incidence can be estimated from the scatter diagram of the BSE difference images using the calibration of the detectors. This estimate will be used in the example described below to correct an Auger image for the effects of topography. It should be noted that this is an experiment using correlations between four images and some physical model in order to derive the local angle of incidence map.

On the other hand, the sum of the signals from the four detectors is dominated by the Z contrast in the sample. The information depth for this contrast is that corresponding to the range of the incident beam into the sample (Section 1) which can be from just below a micron to several microns. Therefore, the BSE sum signal can be used to derive an image in which the contrast is dominated by the variations in atomic number of the substrate with negligible contribution from a surface layer or two with different composition. An example of such Z contrast is shown in Figure 19. Note that the horizontal scales of the linescans across the bevel in this backscattered electron image are marked in atomic number because the sum of the BSE detector signals has been calibrated to relate the size of the sum signal to the subsurface atomic number. This will be used below to calculate an Auger correction image to remove subsurface backscattering effects from the surface Auger contrast.

ASE/BSE correlations

Once the BSE detectors have been calibrated it is then possible to use them in combination with energy analysed images to produce Auger images which have been corrected for topographical artefacts or for variations in the substrate composition.

The sample shown in Figure 16 is coated with such a thick layer of carbon that, in the absence of any topographic effects, the Auger images should show no contrast and only a magnitude of Auger signal corresponding to carbon on top of silicon. However, as expected, the Auger yield does vary with angle of incidence and the raw Auger

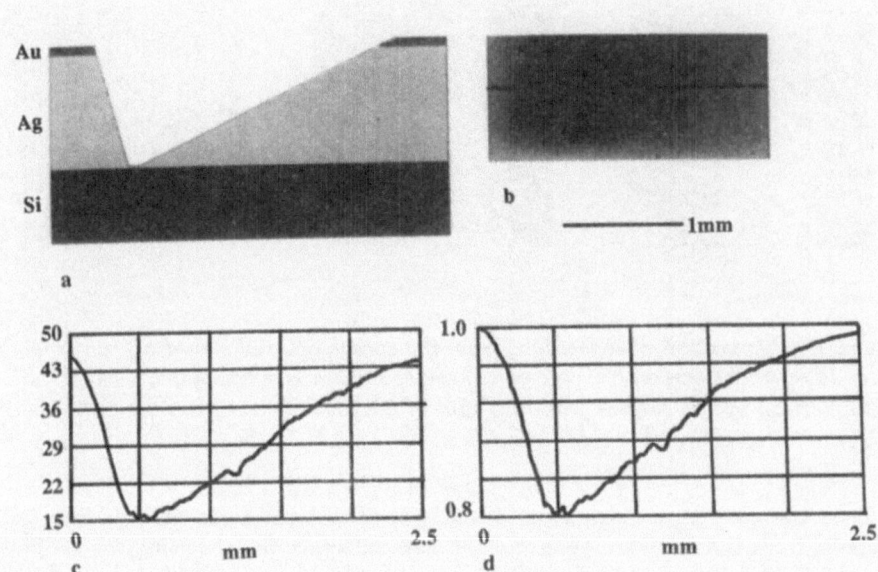

Figure 19. *The summed images from a bevelled sample of silver upon silicon after quantification using the sum of the 4 BSE signals. (a) A sketch of the bevel cutting through a thin Au marker film on top of the Ag film which is on a silicon substrate. (b) A quantitative Ag MNN Auger image of a plan view of the bevel. The dark line represents the position of the retrospective linescans shown in (c) and (d). (c) A retrospective linescan of the effective atomic number versus horizontal position across the bevel. (d) A retrospective linescan across the same part of the image corresponding to (c) but using the Auger factor image computed from the sum of the 4 BSE detector signals.*

image shows contrast corresponding to the different facets on the sides of the polyhedra of anisotropically etched Si (Figure 18a). By using the difference signals from the BSE detectors an Auger correction image can be computed from a model of the angle dependence of the Auger yield. The raw Auger image is then divided (pixel by pixel) by this correction image leading to the topography corrected Auger image in Figure 18c. It can be seen that the contrast due to the inclination of the facets has disappeared and the only topographic features remaining are at the corners where facets join each other or where they join the flat silicon surface.

These sharp corner effects are quite complicated. In both experimental and Monte-Carlo modelling studies of Auger linescans across the sharp sidewalls of etched metallic overlayers on top of silicon, El Gomati *et al.* (1988) were able to demonstrate enhancement effects which lead to the kind of contrast seen at the corners of the polyhedron in Figure 18(c). The most exaggerated form of this artefact has been observed by Umbach *et al.* (1989) who examined linescans across narrow (0.4–1.2 μm) gold lines on top of silicon. The results of this study are shown in Figure 20. The enhancement of the *substrate* Auger signal which arises when the beam strikes the overlay near to a sidewall appears

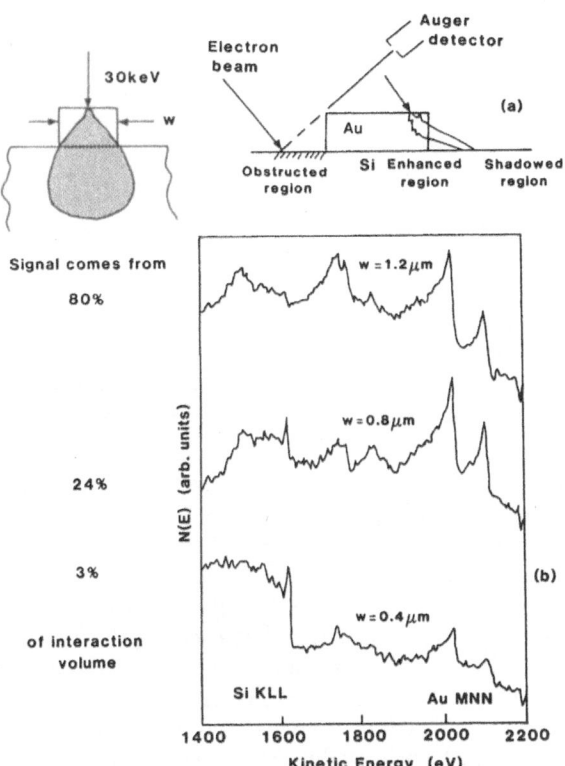

Figure 20. *The edge effects that can occur at the sharp edges or corners of an overlay structure or very rough surface. (a) Sketch depicting the origin of shadowing, obstruction and enhancement effects. (b) The extreme edge enhancement observed by Umbach et al. (1989) for a narrow Au strip on a Si substrate. Here, a 30 keV electron beam strikes a 0.6 mm thick Au layer at normal incidence. Overlayer widths of 0.4–1.2 mm were used as indicated. Emitted electrons entered the spectrometer with a take-off angle of 60°. The inset shows an overlay strip and indicates the profile of the electron beam as it enters the sample. Not only does the Si signal dominate the spectrum when the beam is striking only the 0.4 mm Au strip but also Umbach et al. find that the spatial resolution is degraded by a factor 10. (Courtesy of Umbach, Hoyer and Brunger)*

to happen because electrons are inelastically scattered in the overlay and emerge from the sidewall. They then strike the substrate both with a lower energy than that of the beam and with a more grazing angle to the surface. These two changes result in a rise in the Auger yield from the substrate above that due to the direct impingement of the beam because both the ionisation cross-section and the escape probability rise. To date there are no model calculations which simulate this effect for a general combination of materials and a general geometry. Thus correction of these edge artefacts out of Auger images is not yet possible.

The sum of the signals from the four BSE detectors can be used to derive an image in which the contrast is dominated by the fluctuations in the subsurface atomic number. Given a model of the effect of this variation upon the Auger backscattering factor it is possible to use this Z contrast image to compute a correction image to be applied to the raw Auger image. This has been demonstrated by Barkshire *et al.* (1991c) for a silicon/germanium alloy film on top of a silicon substrate in some places and on top of a thick gold layer on silicon in others. The results are shown in Figure 21. Silicon KLL and Ge LMM peak minus extrapolated background Auger images were corrected by two Auger backscattering factor images derived from the BSE detector signals and the model of Ichimura and Shimizu (1981) for the Z dependence of the Auger backscattering factor. The resulting corrected Auger images are shown in Figure 21e, f and g where the contrast is now quantitatively related to the concentrations of silicon and of germanium in the surface. It should be noted that this is an example using correlations (with modelling) within a 10 image set — three energy analysed images for each of the peak minus extrapolated background Auger images and four BSE images all collected simultaneously.

The summed BSE signals can also be used to quantify Auger depth profiles. In this case advantage is taken of the assumption that the Auger backscattering factor in a layer structure will vary in proportion to the backscattering coefficient. Since the latter is proportional to the summed BSE signal then these detectors can be used to calculate the modification to the effective Auger backscattering factor as the depth profile proceeds and this can be divided into the Auger signal at each depth to remove the broadening of the depth resolution due to substrate backscattering. This has been demonstrated by Barkshire *et al.* (1991b). Moir *et al.* (1989) have also reported techniques for the removal of backscattering effects from conventional depth profiles. Their method uses spectroscopic data combined with models of the scattering process rather than the image correlation procedure outlined here.

Quantitative imaging — pixel by pixel iterative corrections

The matrix corrections described earlier by Seah can be applied to a set of Auger images in a pixel by pixel fashion with a view to producing a new set of Auger images in which the pixel values correspond to the concentration of that element. Since the computer codes which calculate the matrix corrections usually converge very rapidly (Walker *et al.* 1988) this operation does not carry an excessive penalty in computing time. Of course, it does require that Auger images for each element in the surface are acquired and are spatially registered. Further, it requires that the corresponding Auger signals for bulk samples of each of the elements are available because entry into a matrix correction procedure demands that the raw Auger images be converted into ratios of the signal from the unknown to the corresponding signal from the elemental standard for each pixel in each image. An application of this approach to quantification of surface concentration maps of NiCrAl alloys is given by Walker *et al.* (1988).

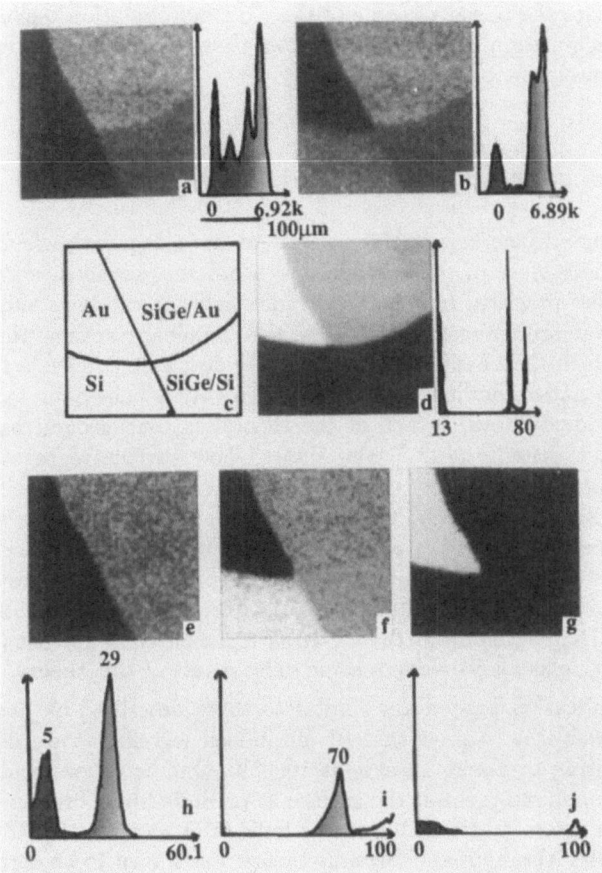

Figure 21. *The correction of Auger images for substrate backscattering effects. (a,b) The Ge LMM and Si KLL Auger images of a sample partially covered with an SiGe alloy film which is on top of Au in some places and on top of Si in others. The intensity histograms are shown alongside each Auger image. (c) A sketch of the arrangement of the materials. (d) An effective atomic number image calculated from the sum of the 4 BSE detector signals acquired at the same time as the Auger images. The histogram alongside this image has an abscissa calibrated in effective atomic number as are the intensities of the pixels in the image. (e,f,g) Quantitative Auger images of Ge, Si and Au obtained by combining (a), (b) and (d) with a model for the dependence of the Auger backscattering factor on Z. (h,i,j) The intensity histograms of images (e), (f) and (g) respectively. Again, the horizontal scale is in units of effective atomic number.*

3   Some applications of quantitative surface imaging

This section contains several examples of the use of surface analytical microscopies, chosen to illustrate the power of some of the data manipulation techniques and simultaneous image acquisition described in the two previous sections. It is not intended to be an exhaustive survey.

3.1   Surface segregation in alloys

In a binary alloy surface segregation is said to have occurred when, in the absence of impurities, the surface of the material has a different composition from that of the bulk. It is of great practical importance in such fields as catalysis and corrosion where it can bring about pronounced effects upon the chemical reactions that can occur and the speeds at which they happen. Therefore it is desirable to have a theoretical model which both aids understanding and enables segregation behaviour to be predicted for previously unstudied alloys. Much of the theoretical background has been reviewed by Abraham and Brundle (1981) who showed how surface segregation is driven by both the relief of bulk lattice strain and the lowering of the surface free energy. In a pair of papers Peacock (1986a,b) described the application of Auger spectroscopy and imaging to two alloys (Ni-5%Pt and Cu-10%Pd) which had not been studied at the time of Abraham and Brundle's paper. They were particularly interesting binary alloys because the Abraham and Brundle theory predicted that the Ni-5%Pt alloy should show solute (Pt) segregation but the Cu-10%Pd should show solvent (Cu) segregation. Thus, these particular alloys were in some sense a test of this theory.

Using quantification procedures similar to those described by Seah earlier in this book, Peacock was able to show that Pt did indeed segregate towards the surface, the concentration rising to about 15 at% at 1050°K. On the other hand, the Cu-10%Pd alloy showed Cu enhancement at the surface as predicted by Abraham and Brundle. In this case the Cu concentration fell from its bulk value at about 930°K to 920±0.5% at 1300°K. Thus, the Abraham and Brundle theory was shown to be correct for these two test materials.

Multiple simultaneous images collected from the area that had been used for the spectroscopic analysis were particularly interesting for the Cu-10%Pd alloy. At each temperature of conducting the segregation experiment the surface composition was found to vary from grain to grain but was relatively constant across individual grain surfaces. This variation is shown in Figure 22. It is an example of anisotropic surface segregation. In this case the chemical compositions of individual grain surfaces could be computed by averaging the pixel values over windowed image regions containing between 20 and 70 pixels. The temperature dependence of the equilibrium surface segragation could then be measured for each individual grain surface. This varied quite extremely from one grain orientation to another. One particular grain (number 6 in Figure 22b) showed almost no temperature dependence of segregation and a 1.5 at% enhancement of surface copper. Grain number 3, at the opposite extreme, showed a pronounced temperature dependence with surface copper changing from about 88.5% at 950°K to about 7.5 at% at 1300°K.

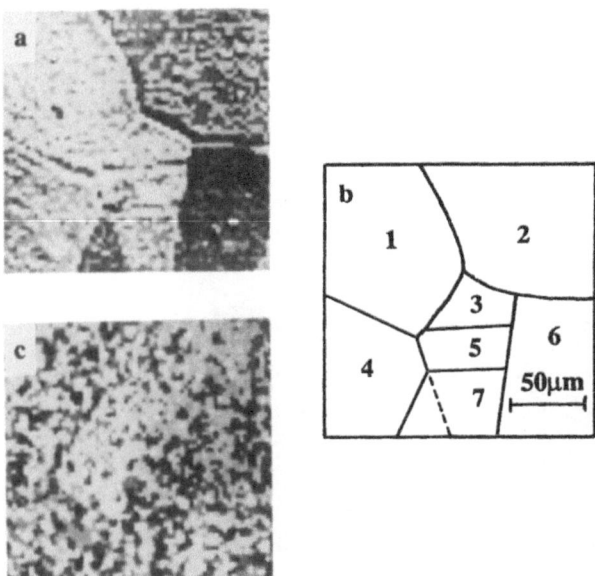

Figure 22. *An area of a Cu-10%Pd specimen used for a study of the anisotropy of surface segregation. (a) A digital* SEM *image showing electron channeling contrast between grains and across a twin boundary within one grain. (b) A sketch of the grains showing the labels attached to them. (c) A grey scale representation of the original false colour quantitative Auger map collected after heating the specimen to 1220°K. The grey levels represent the Pd atomic percentages varying from 7.7% (grain 4, bottom left) to approximately 9% (grain 2, top right).*

This example is interesting in the current context because it required a multi-imaging approach and the production of quantitative Auger maps in order to obtain all the data which is required to examine the anisotropy of surface segregation from one grain to another. The high sample temperatures involved lead to considerable thermal drift in the sample position during the measurement time and simultaneous multi-imaging becomes essential in order that the pixels in the images from each chemical component of the alloy are spatially registered. The test of the theory would not have been feasible without a multi-imaging approach.

3.2 Superlattice bevels and depth profiling

Ion beam erosion is commonly used in combination with AES or SIMS in order to estimate the concentration distribution of elements in a solid as a function of depth. Such measurements are usually referred to as depth profiles. One way of using a scanned ion beam to obtain a depth profile is to cut a shallow angle bevel into the solid (Skinner 1989). The bevel can be imaged subsequently using SEM and MULSAM techniques,

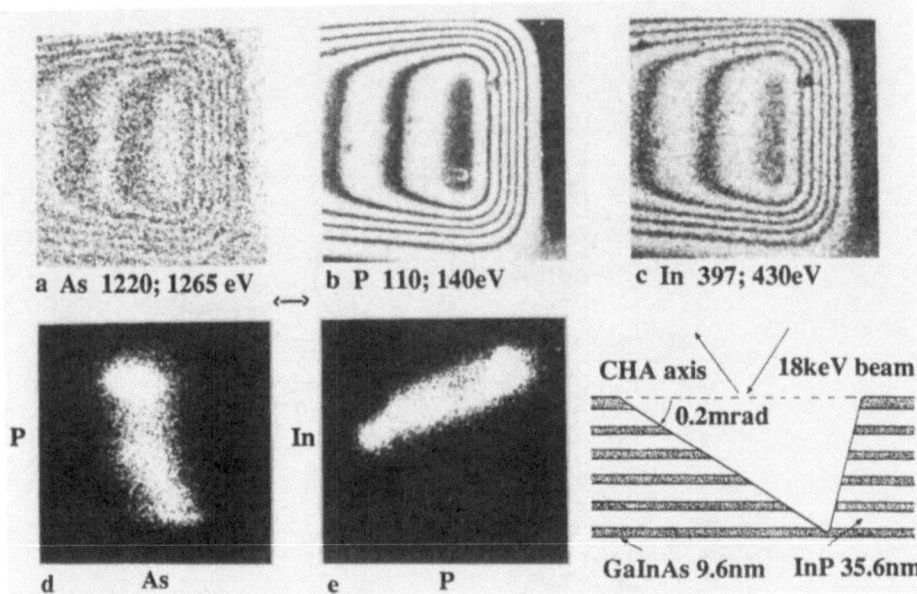

a As 1220; 1265 eV   b P 110; 140eV   c In 397; 430eV

P   In   CHA axis / 18keV beam

0.2mrad

d   As   e   P   GaInAs 9.6nm   InP 35.6nm

Figure 23. *N1–N2 images for As, P and In in an MBE-grown structure (kindly provided by D K Skinner, GEC-Marconi Materials Development Ltd, Caswell). Beam energy is 18 keV, beam current is 6 nA and constant pass energy is 100 eV. Dwell times are 200 ms per pixel for (a) and 100 ms per pixel for (b) and (c). The magnification marker is 150 mm long. Two scatter diagrams and a sketch of the cross-section of the layers are shown below the Auger images.*

features of interest can be selected and spectroscopy or linescans carried out using measurement times appropriate to the data required for analysis. This is a particularly technique for analysis of materials containing thin layers— as in many quantum well and superlattice structures. If the angle between the bevel and the free surface is very small (0.1–1 mrad is possible in practice) then the depth information is spread out laterally on the bevel surface and extremely high depth resolution can be obtained with moderately large electron beam diameters. The technique has been demonstrated for such layer structures by Tatlock *et al.* (1987) and by Prutton *et al.* (1991).

An example of an Auger image set from a superlattice containing InP and InGaAs layers which are respectively 35.6 nm and 9.6 nm thick is shown in Figure 23. In this case the bevel angle was 0.2 mrad and the depth resolution obtained was about 2 nm. The intrinsic resolution determined by the beam size and the bevel angle was 0.4 nm and so the observed resolution probably originates from ion beam mixing effects occurring when the bevel was being cut with 2 keV Xe^+ ions. The scatter diagrams in Figure 23d and e show the anticorrelation expected (from the known bulk compositions) between the phosphorus and arsenic Auger signals as well as the correlation between the indium and phosphorus signals.

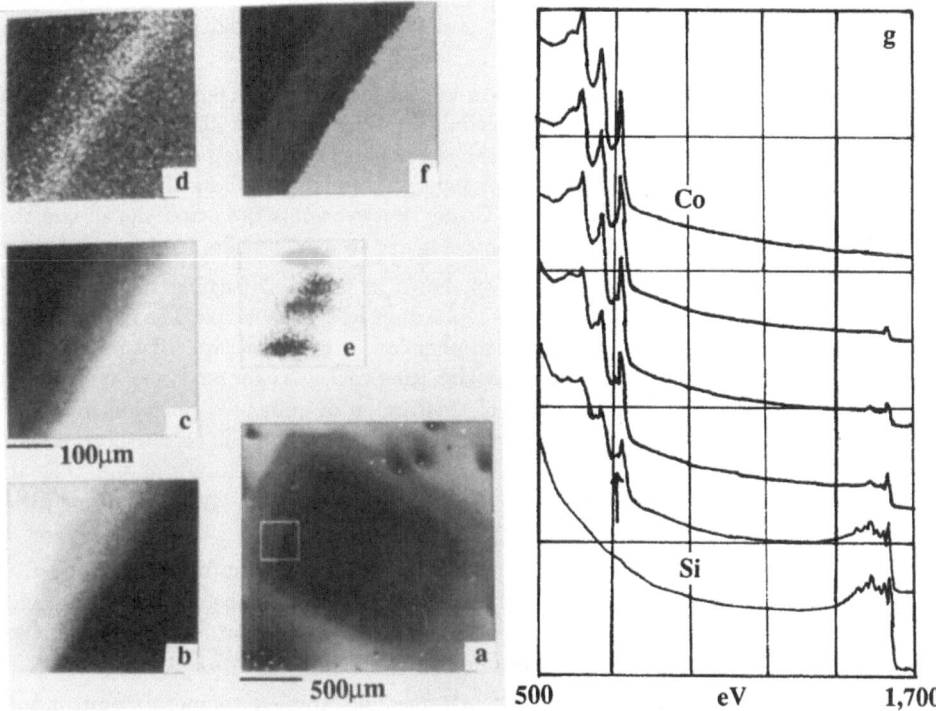

Figure 24. *Ion beam bevel section of a Co/Si structure after rapid thermal annealing. (a) An* SEM *image of the whole bevel. The energy analysed images were obtained from the area marked with a white square; (b) Co MVV Auger image. (c) Si KLL Auger image. (d) A image of the same area obtained using the loss peak marked with an arrow in (g). (e) The scatter diagram of the loss image (horizontal) versus the Si image (vertical). (f) A grey scale representation of a false colour image obtained using the four cluster in (e). (g) Spectra from different regions along a line perpendicular to the lines of contrast in (f).*

This example is chosen not only because the depth resolution is good but also because this a case of the kind mentioned in Section 1 in which the contrast between the layers is barely perceptible in the SEM at 20 keV. Thus, the layer structure could easily have been missed if the prelimenary study of the sample was only by SEM. The Auger images could be acquired in about ten minutes and show clear contrast corresponding to the layer structure.

3.3   Metal-semiconductor contacts

The transition metal silicides, in particular those of titanium and cobalt, have been of some technological interest for their use as conductors in the interconnect patterns of VLSI circuits. These compounds are formed as thin layers on the surface of a silicon device structure by rapid thermal annealing (RTA). The objective is to grow thin layers of the disilicides $CoSi_2$ and $TiSi_2$ which are known to have high (metallic) electrical conductivities.

An example of an ion bevelled section cut through a cobalt/silicon layer structure is shown in Figure 24. This data is described by Greenwood *et al.* (1992). Analysis of this bevel (Figure 24) reveals that between the cobalt layer on top of this structure and the silicon substrate there are two other layers. Immediately beneath the surface Co there is a layer of uniform composition Co_2Si. Between this layer and the silicon there is a layer of varying composition. $CoSi_2$ is not apparent anywhere in the structure.

An example of false colour imaging is shown in Figure 24f using this sample. The layer of varying composition contained a plasmon loss peak below the Co LMM peak which was at the silicon loss energy but enhanced in height compared to pure silicon. This peak height has been used together with the Si KLL Auger peak height to produce the false colour image. The thickness of this region of enhanced plasmon intensity is less than the layer thicknesses and is localised near to the silicon surface. Perhaps there is some reduction in the localisation of the electron dynamics of the Si near to the Si/CoSi interface when there are Co atoms in the environment. This needs further investigation.

The whole subject of metal (or metal-like)/semiconductor interfaces is particularly appropriate for SAM methods because the analyst is examining the majority components (>1 at%) of the materials so that the sensitivity of AES is adequate whilst quantification is practical and yet the electron beam damage effects are not too extreme.

A number of examples of the MULSAM method applied to metal/semiconductor systems are given in Prutton *et al.* (1992) and El Gomati *et al.* (1992).

3.4   EDX and geological samples

Geological samples very often contain many minerals with multiple element compositions and so form a particularly important class of inhomogeneous materials. The demands upon a complete quantitative analysis of such materials are quite considerable because the analysis should yield the number of phases present, the volume occupied by each phase in the bulk sample, the average and the dispersion of compositions in each phase and the bulk composition of the sample. Paque *et al.* (1990) and Browning (1992) have shown how energy dispersive x-ray microanalysis combined with PCA can address this problem. In this case the analysis is not of the surface but it is an excellent example of the use of multivariate statistical techniques to provide quantitative analytical information about a complex multi-image data set.

The example given by Paque *et al.* is from EDX imaging of the mineral Efromovka. This is a relatively simple mineral containing components in the oxide system CaO-MgO-Al_2O_3-SiO_2-TiO_2. A scatter diagram of the pixel intensities in the Al and Si images reveals only four clusters of points although there are five major mineral phases and one minor phase known to be in the sample. This is because some clusters are swamped by the number of points and spread of data in the others. This is exactly the situation which can be improved by PCA (Section 2) because it is procedure which partitions the images so as to separate the orthonormal components. If this is done with

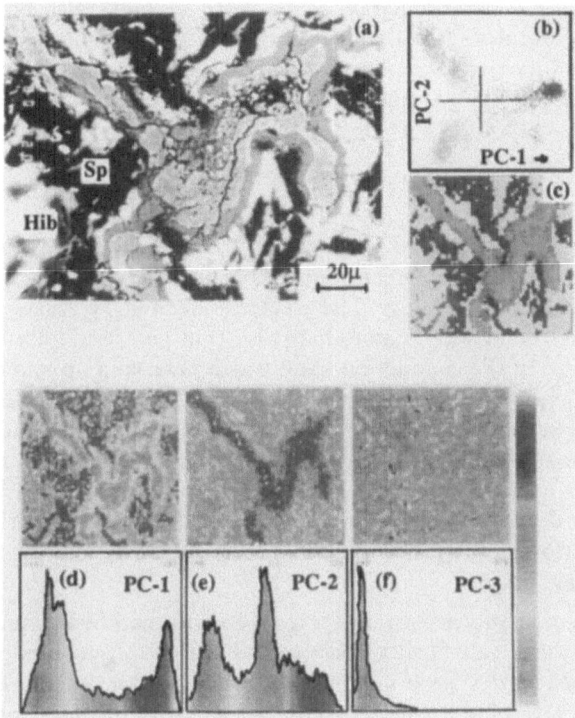

Figure 25. *An* EDX *study of the mineral Efremovka. (a) A* BSE *image of the area sampled using* EDX *imaging. (b) A scatter diagram of the pixel intensities of the first two principal component images formed by* PCA *from the original 7 elemental channels collected. (c) A false colour image of the five clusters in (b). (d,e,f) The first three principal component images and their intensity histograms. (Courtesy R Browning, Stanford)*

the EDX images on the elements listed in Table 1 then the scatter diagram between Al and Si pixel intensities becomes as shown in Figure 25b. There are now five clear clusters which can be false coloured as shown in Figure 25c. The spatial distribution of these five phases is immediately revealed and is more easily interpreted than the BSE image of Figure 25a. Further, from a count of the number of pixels within each cluster it is possible to calculate the exposed area fraction of each phase and the average intensity associated with each element within that phase. Using matrix correction techniques (*e.g.* ZAF analysis) these intensities can be converted to the average composition of each mineral.

This particular example shows how multi-imaging combined with statistical methods can give much more information about a sample than point spectroscopy or inspection of a single EDX image. Very similar conclusions could be drawn from Auger or SIMS image sets for surface analysis or PEELS image sets for bulk analysis.

The application of scatter diagram methods to EDX has also been developed and demonstrated by Newbury (1992) and by Bright and Marinenko (1992) who have presented some beautiful three dimensional scatter diagrams (which they refer to as composition histograms) of SiMgAl and CoVMg alloys.

3.5   Magnetic contrast

If energetic electrons are scattered from a heavy metal sample like gold the angular distibution of the scattering is dependent upon the spin polarization in the electron beam. This is known as Mott Scattering. The electrons excited by Auger or photoemission processes from a ferromagnetic material are also spin polarised because of the unequal numbers of electrons in the conduction band which have spin up and down. The polarisation of electrons leaving a ferromagnetic material thus depends upon the direction of the magnetisation vector in the material. Browning *et al.* (1990) and VanZandt *et al.* (1990) have described the modification of a CMA in combination with a Mott detector to observe the vector magnetisation in a magnetic domain wall in Fe-4%Si. By using the techniques of PCA described in Section 2 they were able to improve the contrast in the images from the Mott detectors so that magnetic contrast was significantly improved over that in the raw images.

The CMA system shown in Section 1 Figure 8a was modified to turn the instrument into what the authors call SEMPA, *Scanning Electron Microscopy with Polarisation Analysis.* The CMA optics were modified to accelerate low energy electrons into the analyser. About 45% of the electrons emitted from the sample with 2.5 eV kinetic energy were collected by the CMA with an energy resolution of about 1.25 eV. A lens accelerated the electrons passing through the CMA to 50 keV and transferred them to a gold Mott scatterer with four planar silicon diode detectors. The spin polarisation is determined by measuring the asymmetry in these backscattered electron currents. In a way similar to the BSE detectors in the MULSAM instrument described above, the difference signals from pairs of detectors can be formed after image acquisition. These two differences then reveal the polarisations in two orthogonal directions corresponding to components of the magnetisation of the sample parallel to its surface plane. The two difference signals can be used to plot a scatter diagram in which clusters reveal the 'phases' present in the sample. In this case a phase is a region with a constant direction of magnetisation in the plane of the surface. Thus, each cluster originates in a magnetic domain. If PCA is applied to this data then the scatter diagram is rotated as described in Section 2 and the contrast between the regions in the principal component images is enhanced over that in the raw data.

The scatter diagram of the raw pair of images is shown for a region of the FeSi sample in Figure 26a. There are four clusters corresponding to the four magnetisation directions in this region of the sample. The one-dimensional histograms of the two polarisation images shown in Figure 26b and c show two peaks each which are rather poorly resolved. Applying PCA to this data produces a scatter diagram between the two principal component images shown in Figure 26d with the associated one dimensional intensity histograms in Figures 26e,f. The rotation has produced the expected enhancement of contrast as seen in the clear separation of the pairs of peaks in Figures 26e,f. False colour imaging can now be applied to the data in the scatter diagram

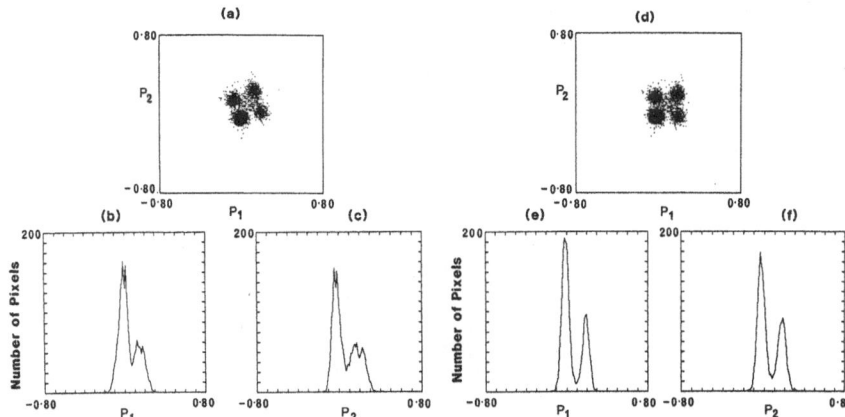

Figure 26. *Statistical data from four spin polarised images of a region in an Fe-4%Si magnet. (a) The scatter diagram of the two difference images from the spin polarised electron detectors. (b) The intensity histogram of one polarised electron image P1. (c) The intensity histogram from the other image P2. (d) The scatter diagram of the principal component images (a 19° rotation has been generated). (e,f) The histograms of the two principal component images. (Courtesy of R Browning, Stanford)*

of Figure 26d and a magnetic image of the sample produced with intensity levels which are quantitatively related to the magnetisation directions in the domains (Figure 27).

By increasing the magnification of the SEMPA instrument so as to obtain pixel points within the width of a domain wall (a few hundreds of nm across) VanZandt *et al.* (1990) were able to produce images of the magnetisation distribution within a wall: see Figure 28 where each pixel point is plotted as an arrow in the direction of the magnetisation and with length proportional to its magnitude. Figure 28a shows a Bloch wall containing Neel sectors in regions along its length. A Neel wall is one in which the magnetisation vector is confined to a plane parallel to the surface whilst a Bloch wall contains magnetisation vectors which spiral about the normal to the wall.

This is quite a different kind of microanalysis in that it is providing quantitative information about the magnetisation vector in a material. Nevertheless, it uses the same data manipulation and interpretation methods needed for Auger microscopy.

3.6 SIMS imaging

Secondary-ion mass spectroscopy can be carried out using equipment which allows the parallel acquisition of several images of the same area of a sample with each image being formed by a different mass of secondary ion. Therefore it is a good candidate for application of the techniques described in Section 2. This has been reported by Newbury and Bright (1990).

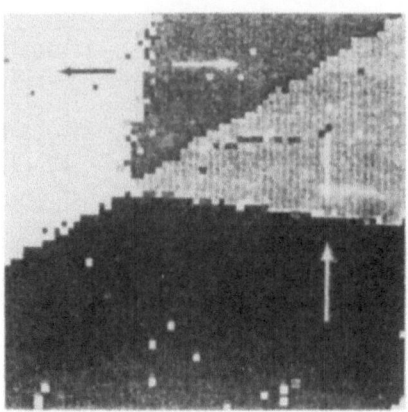

Figure 27. *A monochrome representation of the false colour image generated from the scatter diagram of Figure 26d. The image is of an area 550mm square. (Courtesy of R Browning, Stanford)*

a)   b)

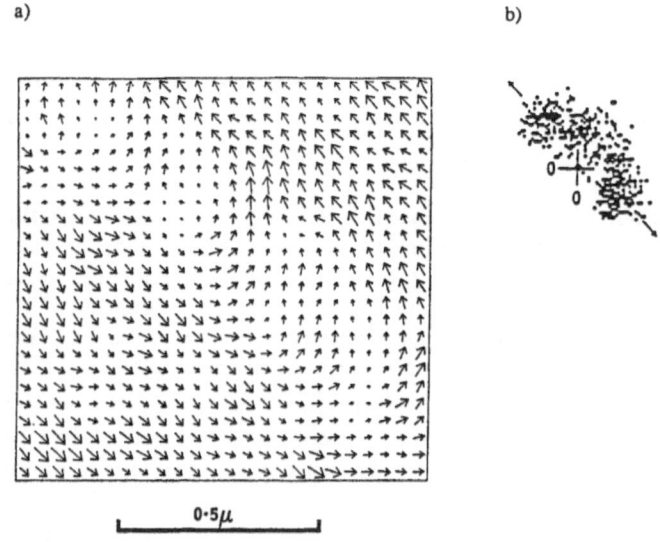

Figure 28. *A high resolution image of a 180° domain wall in an FeSi alloy. (a) Vector plot of the spatial arrangement of the magnetisation vector. (b) Scatter diagram of the polarisation data. The curvature in pattern corresponds to the positive helicity of the Neel wall segments. (Courtesy of R Browning, Stanford)*

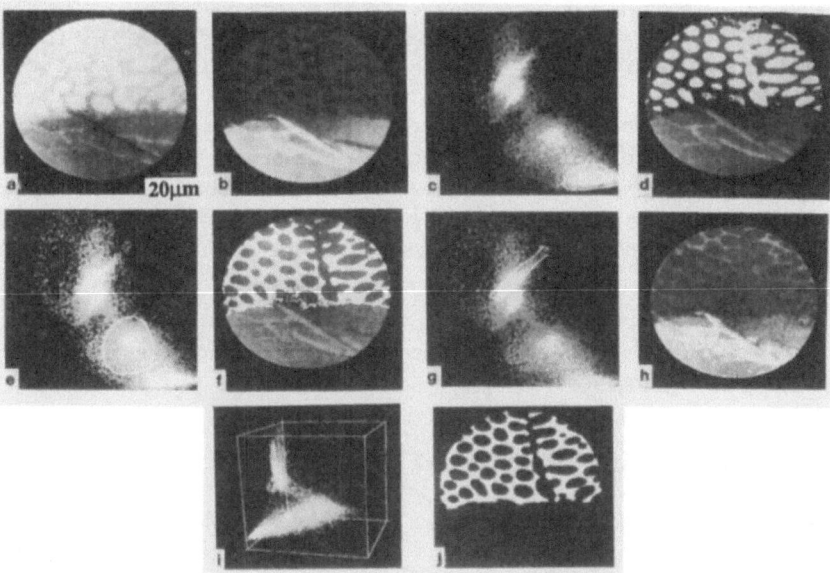

Figure 29. *Some* SIMS *compositional mapping of a CuAlLi alloy. (a) Al processed compositional map. (b) Li processed compositional map. (c) The scatter diagram of (a) (vertical scale) versus (b) (horizontal scale). (d) The false colour image of the area encircled in (c). (e) Another cluster in the bivariate scatter diagram selected for imaging. (f) The false colour image of the pixels of the image pair situated in the cluster encircled in (e). (g) Another cluster encircled. (h) The false colour image of the cluster encircled in (g). (i) A three dimensional scatter diagram of the compositional maps (a), (b) and The Cu image in (j). The Li axis is vertical, the Cu axis is horizontal and the Al axis is out of plane. (Courtesy of D.E.Newbury, NIST)*

The scatter diagram method described above in the MULSAM technique has been applied by Newbury and Bright quite independently. First they have corrected the raw digital images, one for each mass of scattered ion, for the instrumental and matrix effects which apply to their SIMS instrument. This results in a set of images which contain pixel values corresponding to the concentration of each ion species in the surface. The correlations are then performed on these modified images by using the scatter diagram method. Newbury and Bright refer to the scatter diagram of this processed set of images as a concentration histogram.

An example of their method is shown in Figure 29 for a CuAlLi alloy sample. Figures 29a,b show the processed ion images for Al and Li and the scatter diagram of this pair is shown in Figure 29c. The contrast in Figures 29a,b is not particularly good because there is Li in both regions at the top and the bottom of the circular area of the images. Figure 29c reveals that the Al and Li signals are anticorrelated and at least four different clusters can be clearly identified. By windowing each cluster and tracing back to the images to find the pixels in each window an image can be constructed of the regions in the surface that correspond to that cluster. Such a high Li, low Al region is enclosed by a line in Figure 29c and the false colour image traced back from it is shown in Figure 29d. The elliptically shaped regions in the upper half of the field of view are clearly places which are rich in Li. Figures 29e to h show the windowed regions of the composition histogram and their corresponding images. The contrast in these images is quantitative in the sense that the raw data was corrected for the instrumental and matrix effects before the composition histogram was constructed. Therefore the values of the pixels correspond to the atomic concentrations present in the false colour images. Finally, the Cu image for this area was also incorporated and the three-dimensional composition histogram of Cu, Al and Li constructed as shown in Figure 29i. The Cu compositional map added to the maps 29a and b in order to make this diagram is shown in Figure 29j.

This is a clear illustration of the way in which the compositional relationships can be numerically identified and mapped in order to extract as much information as possible from a complex heterogeneous sample.

3.7   Multi-imaging strategies

The material presented in Sections 1 and 2 and the examples of multi-imaging given above illustrate that quite large amounts of data have to be collected and processed if an attempt at quantification of the images is an objective. The quantity of data to be handled and the algorithms for carrying out the processing are not a problem given the power of modern digital computer techniques. However, the time needed for data acquisition often has to be long in order to achieve sufficient statistical precision in the quantification. Therefore it is advantageous to have some experimental strategy to help to reduce the probability of having to discard data and repeat experiments.

With instruments which acquire data sequentially the procedure after preparation of the sample often contains the following steps:

1. A general view of the surface is taken using a fast acquisition method like SEM.

2. From this general view a set of places is selected which are thought to be of special interest or to be typical in some sense. This is a subjective step depending for its effectiveness upon the experience and expertise of the analyst.

3. Spectra are acquired at points or from small areas within the selected regions. These spectra are used to identify the chemical elements for which analysis must be performed.

4. Chemically specific images, one for each element present, are acquired for each of the selected regions. These are collected sequentially and so are likely not to be in spatial registration with each other.

5. The contrast in these images is assessed visually to try to make judgements as to the places with high or low concentrations of each element. If the images are judged to be in sufficiently precise registration then an attempt may be made to quantify the data to concentrations using a matrix correction calculation on a pixel by pixel basis.

This is a difficult and time consuming procedure. There is no objective criterion that all places of interest have been found. For example, the SEM image may not show any contrast for some of the regions that may be important in determining other properties of the sample. The sample may have drifted with respect to the electron beam so that spatial registration is lost and quantification becomes impossible. The contrast visible for a small area of some composition — an important precipitate or inclusion critical to the determination of the properties of the sample — may be very poor because it is swamped by the noise fluctuations in the signals from the surrounding and larger area. Elements may be missed altogether because they happened not to be present in the small areas selected for the preliminary spectroscopic survey.

Instruments which collect several signals simultaneously overcome several of these problems and make possible entirely different analytical procedures. The drift problem is solved because the information in each pixel comes from the same place on the surface under study as the corresponding pixel in other images in the set. A procedure which reduces the importance of the other problems is:

1. A multi-image set is acquired first. This can have a rather sparse pixel density, say 64×64 pixels in each image of the set.

2. A scatter diagram of this set is formed and the presence of clusters is noted. It is important to observe that each point in a cluster diagram may be statistically significant. After all many measurents have contributed to the position of that point in the diagram. A PCA operation may be carried out on this set to increase cluster separation and ensure that all statistically independent phases have been identified.

3. The regions in the sample corresponding to each cluster can be back-tracked to the images by false colouring the image set as described above.

4. Spectroscopy in each of the regions can then be carried out with some confidence that every type of region characteristic of the area imaged has been identified.

5. If the area so evaluated remains interesting then an image set can now be acquired using energy analysed images together with any other data channels available and the pixel density can be chosen to be as high as is required for the desired quality of data. The time taken for this lengthy step will not be wasted because the images are spatially registered and so can be quantified and steps have been taken to ensure that every element present has been included.

6. The possibility of missing small regions swamped by the noise statistics of larger regions can be reduced by using PCA procedures to partition the raw data set and produce principal component images of the statistically independent features of the original images.

This is a powerful and objective approach to image analysis which can save time because the room for missing important features of the data is reduced and which yields more information about the sample than be derived from sequential approaches.

Acknowledgements

The author has been very fortunate to work with many talented people who have contibuted to the material described here in all sorts of important ways. Special thanks go to Ray Browning and Dave Peacock who, together with the author, started down the multi-imaging path. To Mohammed El Gomati for his long standing positive collaboration and his help and ideas with many aspects of what is reported here. To Chris Walker and Peter Kenny for their ingenuity, persistence and skills with the specification, invention, and hard graft of creating the thousands of lines of computer code needed to do this kind of work. To John Greenwood and Ian Barkshire for their efforts in making many of the detectors in MULSAM, devising ways to make them really work well and for obtaining many of the fascinating first results concerning correlations between images. To Ron Roberts who, in several long visits to York from Australia, contributed to the careful and thorough characterisation of the whole MULSAM instrument and imbued everyone with a clear idea of how to be a first class experimentalist (even if they cannot quite get there). To Jack Dee who has provided so much technical support in the design of the instrument and keeping it all running. To the technical staff of the Physics Department at York who have suffered my persistent pressures and worrying and responded with beautifully made electronic and mechanical components.

The work at York has been supported with the financial assistance of first the Alvey Project and then the Silicon Towards 2000 initiative of the SERC and the DTI whose help is gratefully acknowledged.

References

Abraham F F and Brundle C R, 1981, *J Vac Sci Tech* **18** 506

Barkshire I R, Greenwood J C, Kenny P G, Prutton M, Roberts R H and El Gomati M M, 1991a, *Surf Interface Anal* **17** 209

Barkshire I R, Prutton M and Skinner D K, 1991b, *Surf Interface Anal* **17** 213

Barkshire I R, Roberts R H, Greenwood J C, Kenny P G, Prutton M and El Gomati M M, 1991c, *Inst Phys Conf Ser* **119** 185

Barkshire I R, Greenwood J C and Prutton M, 1992, *Appl Surf Science* **55** 245

Bethe H, 1930, *Ann Phys* **5** 325

Bishop H E and Riviere J C, 1969, *J Appl Phys* **40** 1740

Bleeker A J and Kruit P, 1990, *Proceedings of the XIIth International Conf for Electron Microscopy* **2** 380 (San Fransisco Press)

Bonnet N, Colliex C, Mory C and Tence M, 1988, *Scanning Electron Microscopy Supplement* **2** 351 (Scanning Microscopy International Chicago AMF O'Hare)

Briggs D and Seah M P, 1990, *Practical Surface Analysis* **1** (Wiley, Chichester)

Bright D S and Newbury D E, 1991, *Anal Chem* **63** 243A

Bright D E and Marinenko R B, 1992, *Microscopy: The Key Research Tool* **22** 21 (Electron Microscopy Society of America, Milwaukee)

Browning R, 1984, *J Vac Sci Tech* **A2** 453

Browning R, 1985, *J Vac Sci Tech* **A3** 1959

Browning R, 1992, *Surf Interface Anal* **18** (in press)

Browning R, Bassett P J, El Gomati M M and Prutton M, 1977, *Proceedings of the Royal Society London* **A357** 213

Browning R, Peacock D C, Prutton M and Walker C G H, 1984, *Inst Phys Conf Ser No 68* **EMAG83** 127 (Institute of Physics, London)

Browning R, VanZandt T, Helms C R, Poppa H and Landolt M, 1990, *J Electron Spectros Rel Phen* **51** 315

Cazaux J, 1984, *Surf Sci* **140** 85

Cazaux J, 1989, *Surf Interface Anal* 14 354

Danielson L R and Swanson L W, 1979, *Surf Sci* **88** 14

El Gomati M M, Janssen A P, Prutton M and Venables J A, 1979, *Surf Sci* **85** 309

El Gomati M M, Matthew J A D and Prutton M, 1985, *Applied Surf Sci* **24** 147

El Gomati M M, Peacock D C, Prutton M and Walker C G H, 1986, *J Microscopy* **147** 149

El Gomati M M, Prutton M, Lamb B and Tuppen C G, 1988, *Surf Interface Anal* **11** 251

El Gomati M M, Barkshire I R, Greenwood J C, Kenny P G, Roberts R H and Prutton M, 1992, *Microscopy: The Key Research Tool* **22** 29 (Electron Microscopy Society of America, Milwaukee)

Erickson N E and Powell C J, 1986, *Surf Interface Anal* **9** 111

Gaarenstroom S W, 1981, *Appl Surf Sci* **7** 7

Gaarenstroom S W, 1986, *Appl Surf Sci* **26** 561

Gonzales R C and Wintz P, 1977, *Digital Image Processing* (Addison-Wesley, London)

Greenwood J C, Lamb B and Prutton M, 1992, *Surf Interface Anal* (to be submitted)

Grivet P, 1972, *Electron Optics* (Pergamon Press, Oxford)

Gryzinski M, 1965, *Phys Rev* **A138** 336

Harland C J and Venables J A, 1985, *Ultramicroscopy* **17** 9

Hembree G, Luo F C H and Venables J A, 1990, *Proceedings of the XIIth Inter Conf for Electron Microscopy* **2** 382 (San Fransisco Press)

Ichimura S and Shimizu R, 1981, *Surf Sci* **112** 386

Janssen A P, Harland C J and Venables J A, 1977, *Surf Sci* **62** 277

Janssen A P and Venables J A, 1978, *Surf Sci* **77** 351

Jeanguillaume C, 1985, *J Microscop Spectros Electronique* **10** 409

King P L, Browning R, Paque J M and Pianetta P, 1990, *Proc XIIth Int Cong Electron Microscopy* **1** 464

Kosarev E L and Pantos E, 1983, *J Phys E: Sci Instrum* **16** 537

Kruit P and Venables J A, 1988, *Ultramicroscopy* **25** 183

Krzanowski W J, 1988, *Principles of Multivariate Analysis* (Oxford University Press)

Langeron J P, Minel L, Vignes J L, Bouquet S, Pellerin F Lorang, G Ailloud P and Le Hericy J, 1984, *Surf Sci* **138** 610

MacDonald N C and Waldrop J R 1971 *Appl Phys Letts* **19** 315

Madden H H 1983 *Surf Sci* **126** 80

Malinowski R, 1991, *Factor Analysis in Chemistry* (Wiley, Chichester)

Matthew J A D, Prutton M, El Gomati M M and Peacock D C, 1988, *Surf Interface Anal* **11** 173

Moik J G, 1980, *Digital Processing of Remotely Sensed Images* (NASA Special Publications, Washington DC)

Moir P A, Fitzgerald A G and Storey B E, 1989, *Surf Interface Anal* **14** 295

Newbury D E, 1992, *Microscopy: The Key Research Tool* **22** No1 11 (Electron Microscopy Society of America, Milwaukee)

Newbury D and Bright D E, 1990, *Secondary Ion Mass Spectroscopy: SIMS VII* 929 (Wiley, Chichester)

Niedrig H, 1982, *J Appl Phys* **53** R15

Niedrig H, 1983, *Electron Beam Interactions with Solids* 51 (SEM Inc AMF O'Hare, Chicago)

Pantano C G and Madey T E, 1981, *Appl Surf Sci* **7** 115

Paque J M, Browning R, King P L and Pianetta P, 1990,
  Proc 12th Inter Cong for Electron Microscopy **2** 244
Peacock D C, 1986a, *App Surf Sci* **26** 306
Peacock D C, 1986b, *App Surf Sci* **27** 58
Peacock D C, Prutton M and Roberts R H, 1984, *Vacuum TAIP* **34** 497
Peacock D C and Duraud J P, 1986, *Surf Interface Anal* **8** 1
Pratt W K, 1978, *Digital Image Processing* (Wiley-Interscience, New York)
Prutton M, Walker C G H, Greenwood J C, Kenny P G, Dee J C, Barkshire I R,
  Roberts R H and El Gomati M M, 1991, *Surf Interface Anal* **17** 71
Prutton M and Peacock D C, 1982, *J Microscopy* **127** 105
Prutton M, Larson L A and Poppa H, 1983, *J Appl Phys* **54** 374
Prutton M, El Gomati M M and Kenny P G, 1990, *J Electron Spectros Rel Phen* **52** 197
Prutton M, Walker C G H, Greenwood J C, Kenny P G, Dee J C, Barkshire I R,
  Roberts R H and El Gomati M M, 1991, *Surf Interface Anal* **17** 71
Prutton M, Barkshire I R, El Gomati M M, Greenwood J C, Kenny P G and Roberts R H,
  1992, *Surf Interface Anal* **18** 295
Reimer L, 1985, *Scanning Electron Microscopy: Phys of Image Formation and Microanal*
  (Springer-Verlag, Berlin)
Savitsky A and Golay M J E, 1964, *Analyt Chem* **36** 1627
Seah M P and Smith G C, 1991, *Surf Interface Anal* **17** 855
Seah M P and Tosa M, 1992, *Surf Interface Anal* **18** 240
Shimizu R and Ding Ze-Jun, 1992, *Rep Progress in Physics* (in press)
Sickafus E N, 1977a, *Phys Rev B* **16** 1436
Sickafus E N, 1977b, *Phys Rev B* **16** 1448
Sickafus E N and Kukla B, 1977, *Phys Rev B* **19** 4056
Skinner D K, 1989, *Surf Interface Anal* **14** 567
Tatlock G J, Beahan P G, Dare D, Hetherington C J D, Eaglesham D J and Kvam E P,
  1987, *Inst Phys Conf Ser 90* 19
Todd G and Poppa H, 1978, *J Vac Sci Tech* **15** 672
Todd G, Poppa H and Veneklasen L, 1979, *Thin Solid Films* **57** 213
Umbach A, Hoyer A and Brunger R, 1989, *Surf Interface Anal* **14** 401
VanZandt T, Browning R, Helms C R, Poppa H and Landolt M, 1989, *Rev Sci Insts.* 3430
VanZandt T, Browning R, Helms C R, Poppa H and Landolt M, 1990,
  J Electron Spectros Rel Phen **51** 321
Venables J A and Archer G D, 1980, *Electron Microscopy 80* **1** 54 (The Hague)
Venables J A and Hembree G G, 1991, *Inst Phys Conf Ser No 119: Section 1* 33
Walker C G H, Peacock D C, Prutton M and El Gomati M M, 1988, *Surf Interface Anal*
  11 266
Wells O C, Boyde A, Lifshin E and Rezanowich A, 1974, *Scanning Electron Microscopy*
  (McGraw-Hill, New York)

Electronic Structure and Electron Spectroscopy

G van der Laan

SERC Daresbury Laboratory,
United Kingdom

1   Introduction

Study of the electronic structure of solids with high-energy spectroscopies, such as x-ray photoemission, Auger electron, and x-ray absorption spectroscopy has become an important part of materials science. The fundamental principle behind all these methods is the photoelectric effect observed by Hertz (1887) and explained by Einstein (1905). However, it was only after ultra-high-vacuum techniques were well enough developed to maintain a clean surface (about 1970) that electron emission measurements became useful to materials scientists. Since then, electron spectroscopy has developed rapidly, stimulated by the advent of electron analysers with high-energy resolution, single electron counting methods, and synchrotron radiation sources and their insertion devices.

In the first instance, one would expect that the emission of an electron from a material measured as a function of kinetic energy results in the one-electron density of states for that material. However, it has become clear over the years that electron correlation effects play an important role, especially in transition metal and rare earth compounds, and they are responsible for the break-down of the one-electron model. In this chapter electron emission will be shown to be a sensitive probe for these electron correlations. The experimental results obtained from photoemission and Auger spectroscopy lead inevitably to the requirement of a localised description for the electronic structure. The presence of electron correlation effects becomes clearer in Auger photoelectron coincidence spectroscopy, and really overwhelming in the resonant photoemission spectroscopy of core levels. The results of the latter technique are so convincing that it is nowadays difficult to imagine that so much discussion of the importance of correlation effects was needed during several decades.

In this chapter electron spectroscopy and electronic structure will be discussed on an equal footing, demonstrating their close relationship. The emphasis will be on understanding the results of the experiments. Special treatment will be given to the metals copper and nickel and their compounds because of their importance in materials science (*e.g.*, high T_c superconductors and magnetic materials), but also because they serve as simple illustrations. The underlying principles and phenomena also pertain to analogous materials, such as other transition metal compounds, rare earths and actinides. Finally, it should be mentioned that this chapter is no more than an introduction to the field of electron spectroscopy for materials research. Comprehensive reviews on photoemission from transition metal compounds in relation to the electronic structure have been given by *e.g.* Wendin (1981), Almbladh and Hedin (1983), Davis (1986) and Sawatzky (1988a, 1988b, 1990, 1991).

2   Electron spectroscopy

2.1   Auger and photoelectron emission

The Auger process is usually given in a two-step model, in which the assumption is made that Auger decay starts from a fully relaxed initial core-hole state, without including the history of the hole. Thus the Auger process is disconnected from the primary photoionisation. In general, however, the total core hole spectrum is more complicated and it can display a lot of interesting structure related to the electronic structure of the system. This structure is directly related to the $N - 1$ electron (core hole) eigenstates of the system, and Auger decay can originate from any of these intermediate states. Therefore, it is useful first to review some basic aspects of x-ray photoemission spectroscopy (XPS). (Electron excited Auger spectroscopy which shows important differences from x-ray excited Auger spectroscopy due to different selection rules and the Pauli principle will not be discussed here.)

XPS involves the measurement of the electron kinetic energy distribution upon x-ray photoionisation. Although, the characteristic energy for the exciting radiation in XPS is usually of the order of 1 keV, the photoionisation process does not differ fundamentally from that in optical spectroscopy. The final states in the photoexcitation process form a continuum of states. The energy of the continuum final states can be separated into the energy $E_j(N - 1)$ of the $N - 1$ electron system left behind (j labels the possible different final states) and the kinetic energy e_p of the outgoing photoelectron. As a consequence of energy conservation between initial and final states, the energy of the emitted electron carries the information of the $N - 1$ electron states

$$h\nu + E_i(N) = e_p + E_j(N - 1) \tag{1}$$

where $E_i(N)$ is the initial state energy and $h\nu$ the energy of the photon. The binding energy of the electron ejected is defined as

$$E_B(j) = E_j(N - 1) - E_i(N) \tag{2}$$

The excited hole can decay by an Auger transition under emission of a second electron with kinetic energy e_a

$$E_j(N-1) = e_a + E_{q,r}(N-2) \tag{3}$$

where $E_{q,r}(N-2)$ is the energy of a $N-2$ electron final state with holes in the levels q and r. If there was no interaction between the holes, the energy of the two-hole state would simply be the sum of the one-hole states.

2.2   Sudden approximation

A central problem in XPS in general (and of all high-energy spectroscopies) is the nature of the $N-1$ electron states and the relation to the N electron state. The simplest relation was proposed by Koopmans (1933), who stated that in a one-electron model, the energy for removing an electron to form the jth ion state, $E_B(j)$, is given by the negative value of the calculated one-electron orbital energy ϵ_j of the N electron system (Koopmans theorem) The implication is made that all the other passive orbitals are frozen, and that no rearrangement of the remaining electrons occurs during the photoionisation process. Consequently, the photoemission spectrum of an electron in an orbital i consists of a single line at a binding energy $E_B(i)$ equal to $-\epsilon_i$.

In reality, the removal of the electron alters the effective potential seen by the remaining electrons, which readjust to the new potential. This effect is called relaxation and not only includes the processes in the ionised atom itself but also has a contribution from the surrounding atoms in the solid state, called extra-atomic relaxation. The relaxation of the $N-1$ orbitals about the hole will lower the lowest energy $E_0(N-1)$ of the $N-1$ electron system, and thus its binding energy estimated from Koopmans theorem is always too large. The concept of a relaxation energy arises from the relaxation between the calculated one-electron orbital energy in the initial state and the energy of the lowest final state with a hole in that orbital.

In the sudden approximation, the assumption is made that the emitted electron has no interaction with the core hole left behind. In principle many lines are seen in the photoelectron spectrum corresponding to various possible eigenstates of the $N-1$ electron system. The relative intensities of these components is given, in the sudden approximation, by the square of the overlap integral of the $N-1$ electron eigenfunction with that of the frozen wave function of the N electron system with the core electron removed. This is similar to the Franck-Condon factors in optical spectroscopy of molecules but where the vibrational degrees of freedom are replaced by electronic ones. The sudden approximation accounts for the fact that various excitations are made simultaneously with the core ionisation (final state effect), *i.e.* the system does not always end up in the lowest possible final state after the photoelectron has been emitted. In such cases, the kinetic energies of the photoelectrons are lower, and the corresponding lines in the photoelectron spectrum are called satellites or shake-up lines.

Within the sudden approximation, there exists a sum rule essentially based on energy conservation (similar to the Franck-Condon principle), stating that the average energy weighted with the intensities is equal to the Koopmans theorem value:

$$-\epsilon_i = \frac{\sum_j E_j(N-1,i)I_j}{\sum_j I_j} \tag{4}$$

In this equation a connection is made between a ground state calculation in the one-electron approximation and the photoelectron spectrum. In connection with this a total relaxation energy can be defined as $R = E_0(N-1) - \epsilon_i$, which can be as large as 20 eV in some materials.

3 Electronic structure

3.1 Band structure

It is well established that the electronic structure of many transition metal compounds cannot be satisfactorily described by standard band structure theory, in which the influence of exchange and correlation are replaced by an effective one-particle potential and the many-electron wave functions are assumed to be single Slater determinants of one-electron Bloch states. This includes the density functional (DF) (Hohenberg *et al.* 1964, Kohn *et al.* 1964) and local density approximation (LDA) (Williams *et al.* 1983). The problem is that the wave function used in DF calculations is a single Slater determinant of one-electron Bloch states, and therefore does not explicitly include any correlation and is not an eigenfunction of the Hamiltonian including the electron-electron interaction but is a mathematical tool to describe the electron density and total energy of the ground state. Although DF theory describes the electron density and total energy of the ground state exactly, the wave function has no physical significance formally and therefore need not give a good description of the density of states below and above the Fermi energy (E_F), which is not a ground state property.

3.2 Localised approach

There is considerable justification for starting with a localised ionic ansatz rather than a one-electron Bloch-wave approach in describing the electronic structure.

Mott (1949) and Hubbard (1964) proposed that materials containing atoms with open shell orbitals with a radial extent small compared to the interatomic spacing can be described with localised electrons rather than putting the electrons in Bloch states. In this way the atomic limit can be approached. The solid is considered to consist of a lattice of atoms each with exactly n, d or f electrons neglecting for the moment the presence of the more free electron-like outer shell s and p electrons. This is quite different from a band theory approach for which the probability that an atom has n electrons is given by the binomial distribution (Hubbard 1964).

$$P(n) = \frac{z!}{(z-n)!\,n!} c^n (1-c)^{z-n} \tag{5}$$

where z is the degeneracy i.e. 10 for d or 14 for f and zc is the electron concentration per atom. This approach therefore assumes that polarity fluctuations are completely suppressed. To determine the stability of this ansatz we consider the energy required to move an electron from atom i to atom j. This will turn out to be our definition of

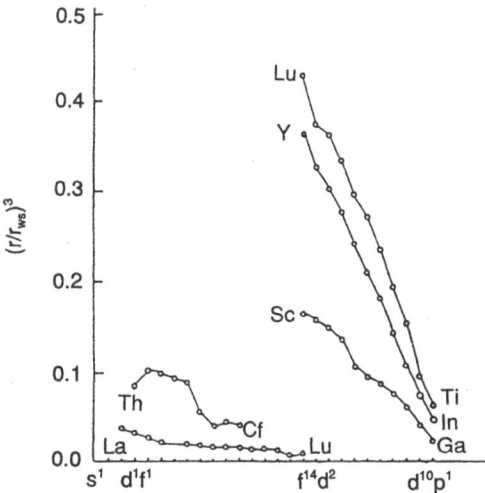

Figure 1. *Ratio of the l shell volume to Wigner-Seitz volume of the 3d, 4d and 5d elements (l = 2) and 4f and 5f elements (l = 3) as a function of their position in the period system. (van der Marel and Sawatsky 1988).*

the electron-electron interaction U.

$$U = E(d^{n+1}) - E(d^n) + E(d^{n-1}) - E(d^n) \tag{6}$$

We see that U involves the energy difference between the d^{n+1} and d^{n-1} states, which differ by two electrons. Since all the one particle potential energy terms cancel, any theory, including DF, which replaces exchange and correlation by a one-particle potential in effect assumes that $U = 0$.

The state with one extra electron on site j and a hole on site i will in fact form a band of width $2W$ because both the hole and the electron can now freely propagate. If $U > W$ we will have a Mott-Hubbard insulator with close to n electrons on each atom. As expected the explicit inclusion of correlation in the wave function reduces polarity fluctuations. More generally each atom has n, and d or f-electrons which will interact just as in the free atom yielding a lowest energy electronic configuration with a maximum spin and a maximum orbital angular momentum. The materials with $U \gg W$ will have local magnetic moments.

It is clear that if the radius of an atomic orbital is small compared to the interatomic spacing we can to good approximation use the atomic orbitals to describe these states. In Figure 1 the radial extent of the charge distribution for the 3d, 4d, 5d and rare earth and actinide f states are shown by comparison to the Wigner-Seitz radius of the elemental solid (van der Marel and Sawatsky (1988). For the rare earth the radial extent is so small compared to the interatomic distance that an atomic behaviour is expected. Also the late 3d lie well above the rare earths as do the 4d, 5d and early actinides. Obviously, the smaller the radial extent of the wave function, the smaller the band width and the larger is U (Harrison 1980).

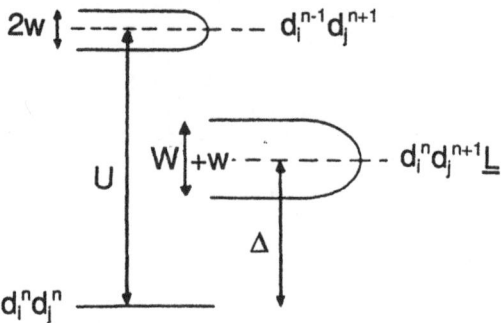

Figure 2. *Total energy level diagram corresponding to an ionic ground state with excitations $d^n \rightarrow d^{n+1}\underline{L}$ and $d^n d^n \rightarrow d^{n-1}d^{n+1}$ (from Sawatzky 1990).*

3.3 Phase diagram

In transition metal (TM) compounds two kinds of charge fluctuations can be distinguished. The first one is given by U in Equation 6, which is the energy required to move a d electron from one TM ion to another. The second one is given by the charge-transfer energy

$$\Delta = E(d^{n+1}\underline{L}) - E(d^n) \tag{7}$$

which is the energy required to move an anion p (often referred to as a ligand) electron resulting in a ligand hole $\underline{L}$ to a TM state d state (van der Laan *et al.* 1982). A total energy diagram based on this ionic ansatz can be drawn as shown in Figure 2 for $U \gg w, U > \Delta$, and $\Delta > W$. The various types of band gaps that might occur can be seen. For $U > \Delta$, the gap is of a charge transfer type. So even for $U \rightarrow \infty$, we can get a metallic ground state if $\Delta < (W + w)/2$. Because generally $w \ll W$, these materials are p-type metals as, for example CuS (Folmer *et al.* 1980). For $U < \Delta$, we are in the Mott-Hubbard regime with *a d-d gap* for $U > w$, and a d-band metal for $U < w$. It is generally accepted that the early $3d$ transition metal oxides belong to this regime. All of this information can be put into a simple phase diagram, as shown in Figure 3, which is a simplified version of the original Zaanen-Sawatzky-Allen (ZSA) diagram (Zaanen *et al.* 1985).

4 Atomic structure and spectra

4.1 Multiplet structure

An important aspect of electron spectroscopy as well as electronic structure is related to the multiplet structure. As is well known from atomic physics a particular electronic configuration like d^n splits up into a multitude of energy levels corresponding to various possible occupations of the orbital and spin quantum numbers (m_l and m_s).

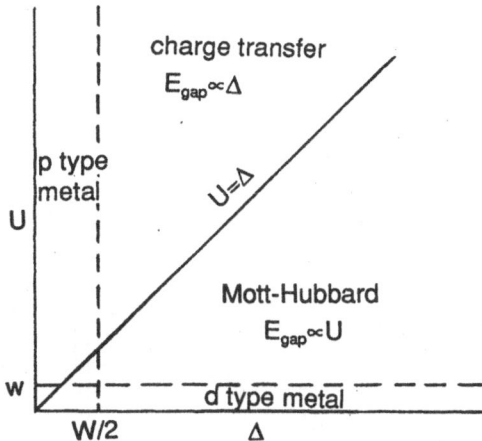

Figure 3. *Simplified Zaanen-Sawatzky-Allen phase diagram showing the various types of insulating and metallic states in transition metal compounds (from Sawatzky 1990).*

The multiplet splitting is determined by the Slater (Coulomb and exchange) integrals F^0, F^2, F^4, G^1, G^3, where F^0 is the monopole and the others are the multipole contributions. The range of this multiplet splitting can be as large as 10 eV and is therefore of great importance. The lowest energy state, or ground state, of an ion in a d^n configuration is given by Hund's rule, which states that we must first maximise the spin, then the orbital angular momentum, and subsequently couple this to the lowest (highest) possible total angular momentum for a less (more) than half-filled shell.

In the presence of a crystal field, the d levels are split, which often quenches the orbital angular momentum since the crystal field splittings are large compared to the spin-orbit coupling, which is of the order of 50 meV. The effect of spin-orbit coupling reduces then to a magnetic anisotropy. These effects are important in understanding the magnetic properties, but can only be observed by using circularly polarised x-rays (Thole *et al.* 1992) or from spin resolved photoemission experiments (Thole and van der Laan 1991).

To get an impression of the magnitude of the multiplet splittings we show in Figure 4 the splitting for $d^n (2 \leq n \leq 8)$. The configurations d^1 and d^9 have no d–d interaction, and give a single term. From the diagram we see *e.g.* that the ^3F state for a d^8 configuration is 2 eV lower in energy than the first singlet state ^1D. This has important consequences for the electronic and magnetic properties of the ground state. In nickel, which is a mixture of d^8, d^9 and d^{10}, the small d^8 weight in the ground state determines the ferromagnetic properties of this metal. The lowest state of this configuration, the ^3F with parallel spins, imposes a spin alignment on adjacent Ni atoms which fluctuate between $d^9 + d^9$ and $d^8 + d^{10}$.

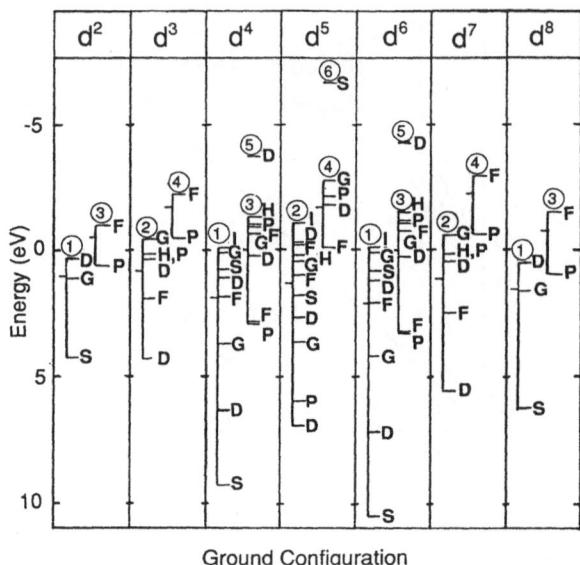

Ground Configuration

Figure 4. *Energy distribution of the terms in the initial state configurations* $3d^n$. *The terms are collected in spin manifolds, where the labels give the values of* $2S + 1$ *and L. The lowest energy state, or ground state, is at the top of the diagram (from Thole and van der Laan 1988).*

4.2 Relaxation effects

The monopole part (F^0) of the Coulomb interaction is strongly screened in the solid since it is the direct interaction between two charges. Say, it costs us $\epsilon - E_{pol}$ to produce a local charge where E_{pol} is the reduction in the ionisation potential due to polarisation and screening. To create two such charges spatially well separated would cost $2\epsilon - 2E_{pol}$. But to create two charges on the same atom would cost $2\epsilon - 4E_{pol}$ because the polarisation or screening energy goes as Z^2 resulting in a screening of F^0 equal to $2E_{pol}$. However, the system does not care what the relative spin direction of the two holes (or electrons) is or what their orbital (m_l) quantum number is if the radial extent is small compared to the interatomic distance. Thus for the states in materials which have radial extents much smaller than the interatomic spacing we have atomic wave functions with atomic multiplet structure but a strongly screened F^0 Coulomb integral. Thus we expect F^0 to be strongly screened but the F^2, F^4, *etc.* are expected to be atomic-like. This is a quite general and for magnetism it is an extremely important result. If the screening of F^2 and F^4 was as large as that for F^0 there would be few, if any, magnetic $3d$ transition metal systems.

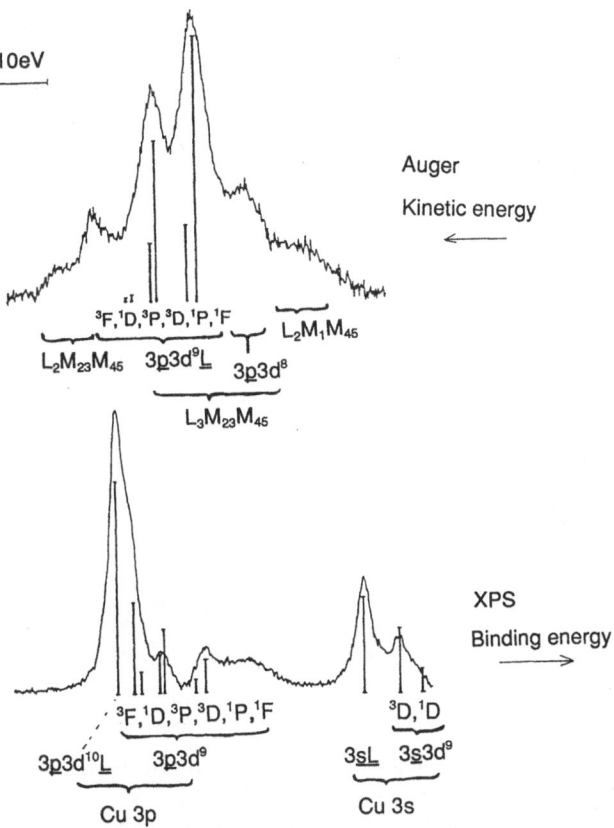

Figure 5. *Comparison of the Cu $L_{23}M_{23}M_{45}$ Auger (top) and the Cu $3p$ spectrum (bottom) of $CuCl_2$. The intensities and positions of the various terms are given by the height and separation of the vertical lines, respectively. The Auger spectrum is due to $\underline{3p}3d^9\underline{L}$ and the XPS is due to $\underline{3p}3d^9$ final states (from van der Laan et al. 1982).*

4.3 Fingerprints

In this section we will give a detailed example (van der Laan *et al.* 1982) which shows how the multiplet structure can be used to identify specific XPS and AES structures. The lower curve in Figure 5 shows the Cu $3p$ XPS spectrum of $CuCl_2$. The ground state of this divalent Cu compound is a mixture of d^9 and $d^{10}\underline{L}$, where $\underline{L}$ denotes a ligand hole. The creation of a $3p$ core hole in the photoemission process results in the final states $\underline{3p}3d^9$ and $\underline{3p}3d^{10}\underline{L}$, where $\underline{3p}$ denotes the core hole. The $\underline{3p}3d^{10}\underline{L}$ has a lower energy than the $\underline{3p}3d^9$ configuration, because the $3p$ core hole is screened by the extra d electron. The energy difference between these configurations is equal to $U(3p, 3d) - \Delta$, where Δ is the energy required to transfer an electron from the ligand to the metal.

For the multiplet structure we have to consider the open (incompletely filled) shells.

Completely filled shells, such as the $3d^{10}$, will not contribute to the splitting. We have to couple spin and angular momenta of the open shells together; and when the spin-orbit interaction is small we can treat spin and orbital compounds separately. The $3p3d^{10}\underline{L}$ gives a single line, because we can neglect the interatomic Coulomb interactions between the $3p$ and the ligand orbitals. In the $3p3d^9$ configuration, where the $3p$-$3d$ Coulomb interactions are large, the orbital momentum $l_p = 1$ of the $3p$ core hole couples with the orbital momentum $l_d = 2$ of the $3d$ valence hole. This results in a total orbital momentum of $L = | l_d - l_p |, ..., l_d + l_p = 1, 2, 3$, indicated as P, D, and F states, respectively. Furthermore, the spin momenta $s_p = 1/2$ and $s_d = 1/2$ couple to a total spin of $S = 0$ (singlet) and $S = 1$ (triplet). Thus, the $3p3d^9$ configuration consists of the terms ${}^1P, {}^1D, {}^1F, {}^3P, {}^3D$, and ${}^3F$. (When the coupling conditions correspond closely to pure LS coupling, then the quantum states of the atom can be accurately descibed in terms of LS coupling quantum numbers. Giving values of L and S specifies a *term*; giving values of L, S, and J specifies a *level*; giving values of S, J and M specifies a *state*.)

Next we have to know the energy positions of these terms as given by the Slater integrals. From the Slater theory of atomic structure (Slater 1960) it can be shown that the relative energy positions are

$$
\begin{aligned}
E({}^1F) &= F^0 + (2/35)F^2 + (6/15)G^1 + (3/245)G^3 &= F^0 + 7.39eV, \\
E({}^1P) &= F^0 + (7/35)F^2 + (1/15)G^1 + (63/245)G^3 &= F^0 + 5.36e0, \\
E({}^3D) &= F^0 - (7/35)F^2 + (3/15)G^1 - (21/245)G^3 &= F^0 - 0.22eV, \\
E({}^3P) &= F^0 + (7/35)F^2 - (1/15)G^1 - (63/245)G^3 &= F^0 - 1.02eV, \\
E({}^1D) &= F^0 - (7/35)F^2 - (3/15)G^1 + (21/245)G^3 &= F^0 - 5.11eV, \\
E({}^3F) &= F^0 + (2/35)F^2 - (6/15)G^1 - (3/245)G^3 &= F^0 - 6.24eV.
\end{aligned}
\tag{8}
$$

The right-hand side was obtained by taking the atomic values of the $3p$-$3d$ Slater integrals from Mann (1967): $F^2 = 13.34$; $G^1 = 16.53$; and $G^3 = 9.94eV$. The calculated final state terms are given in Figure 5 as vertical lines. The upper curve in this figure shows the $L_3M_{23}M_{45}$ Auger spectrum, which is due to the decay of a $2p(L_3)$ core hole to a final state with holes in the $3p(M_{23})$ and the $3d(M_{45})$ core levels. From the $2p3d^9$ and $2p3d^{10}\underline{L}$ excited states in CuCl$_2$ we obtain the $3p3d^8$ and $3p3d^9\underline{L}$ final states. The $3p3d^9\underline{L}$ structure in Auger has the same multiplet splitting as the $3p3d^9$ state in XPS, since the interaction with the ligand hole on the neighbouring site is small. In Figure 5 is seen that the relative positions of the vertical bars in the $3p3d^9\underline{L}$ structure of the $L_3M_{23}M_{45}$ Auger are the same as in the $3p3d^9$ structure of the $3p$ XPS. However, the intensities are completely different. In XPS the intensities are given by the probability to create a $3p$ hole in the $3p^63d^9$ configuration with $3d$-$3p$ interaction. These probabilities are equal to the square of the coefficients of fractional parentage, which turns out to be just the term multiplicities $(2L + 1)(2S + 1)$ in the two particle case. E.g., the ${}^3F$ term with a multiplicity of 21 gives the strongest line; the ${}^1D$ term has a multiplicity of 5; *etc.* The total multiplicity is equal to $(2s_p + 1)(2l_p + 1)(2s_d + 1)(2l_d + 1) = 60$. The Auger intensities are determined by the Coulomb matrix elements R($2p, 3p; 3d, e_a$), which have been tabulated by McGuire (1971) or can be calculated using Cowan's code (Cowan 1968, 1981). It is seen from Figure 5 that the ${}^3F + {}^1D$ peak, which is the strongest peak in the $3p$ XPS, is the weakest peak in the $L_3M_{23}M_{45}$ Auger.

The given example demonstrates the importance of the multiplet structure for the assignment of the peak structures. Generally, the analysis is complicated by the pres-

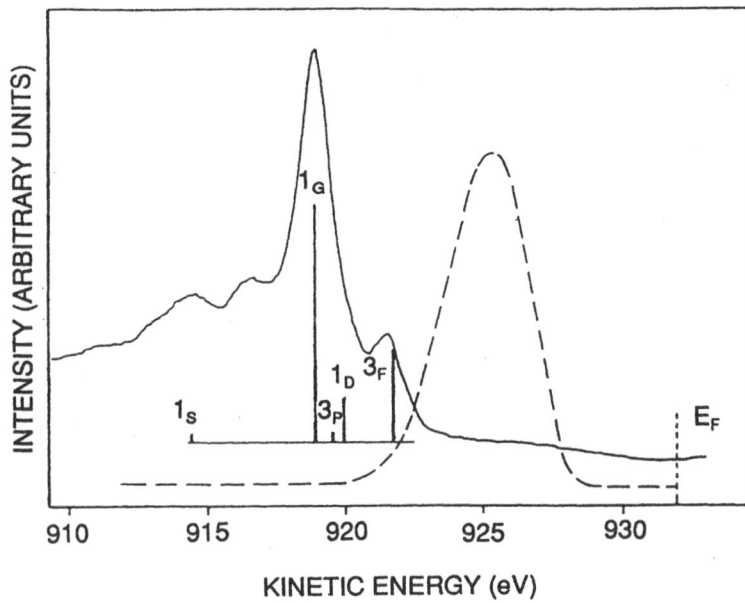

Figure 6. *The $L_3 M_{45} M_{45}$ Auger spectrum of Cu (solid line). The dashed line represents the self-convolution of the 3d band, E_F is the Fermi energy (from Sawatzky 1988b).*

ence of different final state configurations in the XPS and Auger spectra of localised materials. From the energy positions of these final state configurations one can, in principle, obtain U and Δ, which are important parameters for the ground state properties of the material.

5   Coulomb correlation effects

5.1   Cini-Sawatzky model

The value of $U(d, d)$ can be obtained by measuring the energy shift between the localised two-hole state obtained from Auger and the convoluted one-hole state obtained from XPS. This is known as the Cini-Sawatzky model (Sawatzky 1977, Cini 1975 and 1979), which is formally only valid for a d^{10} initial state, such as in Cu metal.

Figure 6 shows the $L_3 M_{45} M_{45}$ Auger spectrum of copper metal, which is due to the decay from the $2p(L_3)$ core hole to a final state with two holes in the $d(M_{45})$ band. The multiplet terms are obtained by the coupling of the two d holes which have $l_d = 2$ and $s_d = 1/2$. This yields the orbital states S, P, D, F, G with $L = 0, 1, 2, 3, 4$ respectively, and the singlet and triplet spin states with $S = 0$ and 1, respectively. Not all terms will be allowed, because the two electrons share the same shell. According to the Pauli exclusion principle, the total wave function must be antisymmetric, which means that

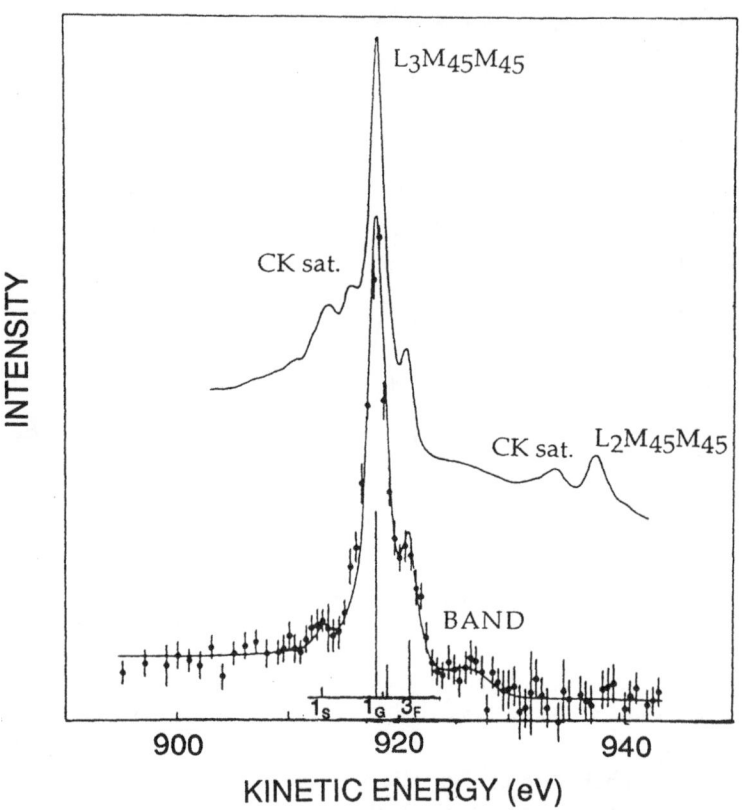

Figure 7. *The normal $L_{23}M_{45}M_{45}$ Auger spectrum (top) and the Auger spectrum in coincidence with the $L_3(2p_{3/2})$ XPS shown as dots (bottom). The solid line fit to the experimental APECS spectrum shows both the atomic-like multiplet split part and the band-like part separated in energy by $U(dd)$ (from Sawatzky 1988b).*

L has to be odd (even) if S is integer (half-integer). Thus the allowed terms of d^8 are $^1S, ^3P, ^1D, ^3F, ^1G$, which are shown in Figure 6. The intensity around 915 eV is not solely due to the ^1S term, but is also due to a satellite structure which will be discussed in Section 6.

The agreement between the experiment and the multiplet calculations seems to justify an atomic-like approach. However, we might ask ourselves what we would have obtained from a one-electron bandstructure approach. In a one-particle theory the local two-hole density of states is just the self-convolution of the one-hole density of states which in turn is closely approximated by the self-convolution of the XPS spectrum. The resulting broad (5 eV) structure together with the Auger spectrum is shown in Figure 6. This Auger spectrum is also shown in Figure 7 together with the Auger photoelectron coincidence spectrum (APECS) (Haak *et al.* 1978). As will be sees in Section 6, in APECS the satellite structure can be eliminated resulting in a more direct picture of the local two hole density of states. Two regions, one marked 'band' and the other consisting of the high intensity sharp structures can be seen. The low intensity band region corresponds closely to the self-convolution of the d density of states shown by the dashed line in Figure 6. The atomic part which contains most of the intensity is sharply structured and shows the multiplet structure of an atomic d^8 state as indicated by the vertical bars. This behaviour is explained by the Cini–Sawatzky theory in which account is taken of the intra-atomic Coulomb interaction $U(dd)$. The separation of the d^8 part and the band like part where the two holes are in different sites is a measure of $U(dd)$ although there are small shifts due to hybridisation (Sawatzky and Lenselink 1980). Obviously, the one-particle band theory breaks down in describing the Auger spectrum of Cu. This is a result of the strong repulsion of the two final state holes if they are on the same atom. In a single particle approach these two sets of states cannot be distinguished since the particles are assumed not to be correlated. This problem does not appear in the photoelectron spectrum of Cu where the d band is initially full. The removal of a d electron results in only one d hole and therefore is a one-particle problem, if we can neglect the correlation with other $4s, 4p$, *etc.*, holes.

5.2   6 eV satellite of Ni

Since Cu metal behaves according to a one-electron model as far as XPS is concerned, it is more interesting to look at the XPS of Ni metal, where the d band is not full in the ground state. In analogy to the Auger spectrum of Cu metal we then expect a breakdown of the one-electron model. Upon removal of one d-electron we can reach states with two hole (d^8) or one hole (d^9) on the same atom if we consider the ground state to have on the average around 9.4 d electrons per atom corresponding to a magnetic moment of $0.6\mu_B$ (Bohr magnetons). This results in the famous '6 eV satellite' in the XPS spectrum of nickel (Davis 1986) which is shown in Figure 8. Also shown is the Ni $L_3M_{45}M_{45}$ Auger spectrum which corresponds to primarily d^8-like final states since it originates from mainly the $2p^5 3d^{10}$ (fully screened) core. The correspondence of the Auger spectrum and the XPS satellites suggest that the interpretation of the origin of the satellite is correct. In Section 7 we will show that even more conclusive evidence can be obtained from resonant photoemission which gives a strong resonance enhancement of the satellite at the $2p$ absorption threshold photon energy.

6   Auger coincidence spectroscopy

So far we have assumed that Auger emission can be described by a two-step model in which the interaction of a photon with a system creates a photoelectron which leaves the ionised system in an excited state with a core hole. In the second step, this core hole state decays with the emission of a second electron, the Auger electron, and the system left behind is in a doubly ionised state. (We ignore the fact that this process can continue by emitting a third and fourth electron, *etc.*)

The correlated nature of the kinetic distributions of the Auger electron and the photoelectron can be seen by considering the total process of the interaction of the photon in a single step model. Assuming that the two-hole state does not further

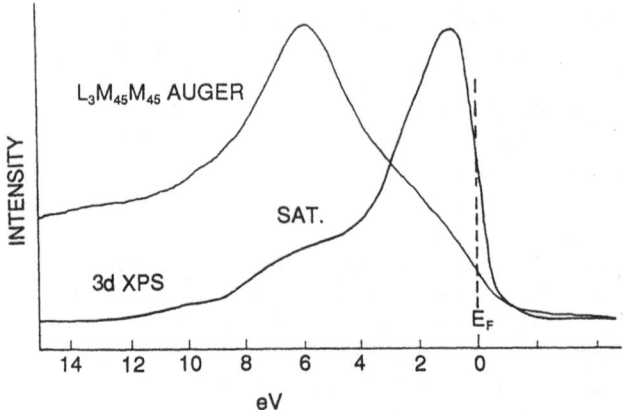

Figure 8. *The $L_3 M_{45} M_{45}$ Auger spectrum and the 3d valence band* XPS *spectrum of Ni, the Auger spectrum is shifted by the $2p_{3/2}$ core level binding energy so that E_F coincides with that of the* XPS *spectrum (from Haak et al. 1978).*

decay, it follows from the law of energy conservation that the sum of the initial state energy and the photon of energy $h\nu$ is equal to the final state energy of the system consisting of a doubly ionised material plus two electrons with energies e_p and e_a.

$$E(N) + h\nu = E(N-2) + e_p + e_a \tag{9}$$

The intermediate (one-hole state) enters only via e_p. Equation 9 shows that the sum of the Auger electron plus the photoelectron kinetic energy has a width which is only determined by the lifetime broadening due to decay of the two-hole state. In other words, in a coincidence measurement, the lifetime broadening due to decay of the intermediate one-hole state can be removed. In a coincidence experiment, only those events are measured which result from one photon, and therefore the events must come from the same atom and in fact involving the same intermediate state. A consequence of the latter is that if we have a multicomponent photoemission spectrum and a corresponding multicomponent Auger spectrum, we can in coincidence separate the various contributions to the total photoemission spectrum and vice versa. If only a narrow range of $E(N-2)$ exists (*i.e.* narrow band width and long lifetime) then the sum $e_p + e_a$ can be specified with an uncertainty smaller than the lifetime energy width of the core hole. Thus, within the manifold of each single feature, high (low) energy photoelectrons give rise to low (high) energy Auger electrons.

The Cu $L_{23} M_{45} M_{45}$ APECS spectrum of Figure 7 was already mentioned in the discussion of the Cini-Sawatzky model. It shows a sharp structure which is due to a final state d^8. However, there is also a structured shoulder at low kinetic energies $(910 - 918eV)$ which apparently does not belong to this d^8-like final state and which masks out the small peak expected from the ^1S final state. This structure is due to a Coster-Kronig $L_2 L_3 M_{45}$ preceded transition which subsequently decays via an

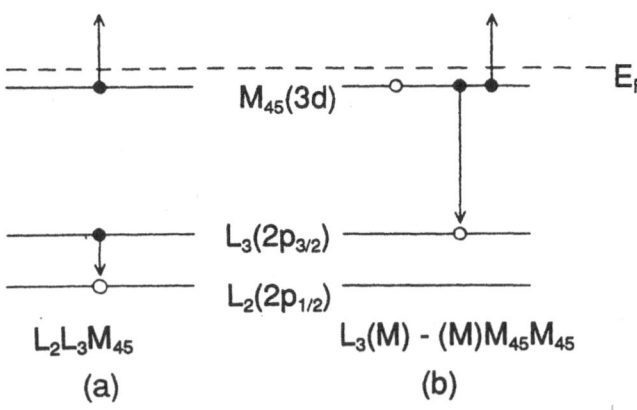

Figure 9. *(a) Schematic representation of the $L_2L_3M_{45}$ Coster-Kronig process; and (b) the $L_3(M) - (M)M_{45}M_{45}$ Auger process (from Sawatzky 1988b).*

$L_3M_{45}M_{45}$ channel, but where the final state now has three holes (d^7). This mechanism is shown schematically in Figure 9.

The $L_{23}M_{45}M_{45}$ coincidence spectra for the XPS energies corresponding to the L_3 and L_2 lines are shown in Figures 7 and 10, respectively. As can be seen in Figure 7, the $L_2M_{45}M_{45}$ part of the Auger spectrum is completely missing for the $L_3M_{45}M_{45}$ APECS spectrum. This part of the spectrum appears in the $L_2M_{45}M_{45}$ APECS spectrum (Figure 10), since its origin is an L_2 hole.

Compared to the Auger spectrum the coincident spectrum shows a reduced background. The effective escape depth λ in APECS is given by $1/\lambda = 1/\lambda_1 + 1/\lambda_2$, where λ_1 and λ_2 are the escape depths for the single events. This gives an increased surface sensitivity and reduces the inelastically scattered electrons originating from somewhat deeper in the solid. This is the reason why the 'band'-like portion of the $L_3M_{45}M_{45}$ Auger spectrum, which should always accompany the atomic part as predicted by the Cini-Sawatzky model, is better visible in the APECS spectrum than in the normal Auger spectrum.

It has been recognised that APECS can be used to: *i)* reduce the core level lifetime broadening in photoemission; *ii)* isolate the individual sites in a solid and probe their local electronic structure; *iii)* distinguish between intrinsic and extrinsic secondary electron emission; *iv)* explore correlated photoexcitation/Auger relaxation events; *v)* separate overlapping spectral features; *vi)* probe electronic structure with improved depth resolution; and *vii)* eliminate uncorrelated secondary background (Haak *et al.* 1978), Bartynski *et al.* 1992a,b).

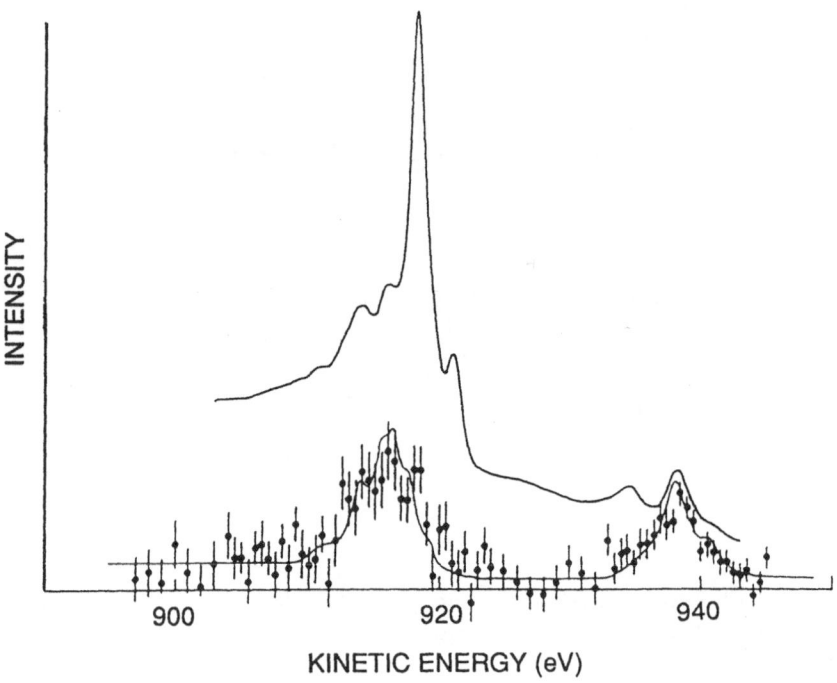

Figure 10. *The normal $L_{23}M_{45}M_{45}$ Auger spectrum (top) and the Auger spectrum in coincidence with the $L_2(2p_{1/2})$ xps shown as dots (bottom) (from Haak et al. 1978).*

7   Resonant photoemission

7.1   3p resonance decay

Electron spectroscopy is applied routinely in the study of covalent mixing in $3d$ transition metal, rare earth and actinide compounds. We have seen that due to the electrostatic core-valence interactions the different final state configurations are separated in energy and display characteristic multiplet structures. The number of parameters needed for a detailed analysis requires the combined information of several core-level spectroscopies, such as x-ray absorption (XAS), x-ray photoemission (XPS) and Auger. However, the different numbers of holes in the final states produced by these methods introduce new uncertainties, hampering a direct comparison. Therefore, important problems remain unsolved, such as whether peak splittings are due to configuration interaction or exchange interaction. Resonant core level spectroscopy has the advantage that the photoemission and x-ray absorption are combined in a single analysis. Despite energy shifts and multiplet splitting the common origin of related peaks in the two spectra can be deduced directly by qualitative reasoning without complete calculations. The technique is element selective in the photon energy as well as in the binding energy. It requires a variable photon energy *e.g.* from a synchrotron radiation source.

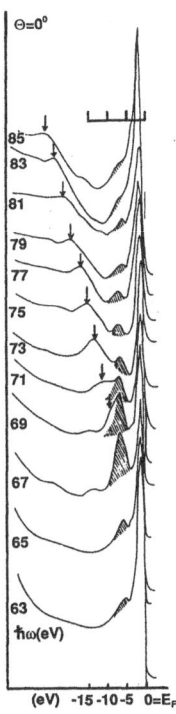

Figure 11. *Photoemission spectra of Ni(100) near the 3p → 3d resonance. For each photon energy the distribution of emitted electrons as a function of initial energy is shown. The peak at E_F corresponds to the d bands; the dashed area is the 6 eV satellite associated with d^8 final states. Arrows indicate the MVV Auger peak which is at fixed kinetic energy of the emitted electrons (from Guillot et al. 1977).*

Resonant photoemission (RESPEC) was first observed in nickel metal (Guillot *et al.* 1977). The results of these measurements are reproduced in Figure 11. The valence band photoemission spectrum shows a resonance in the 6 eV satellite as a function of photon energy in the region of the 3d absorption edge. The 3p resonant photoemission of an initial state $3d^n$ is given by the transition

$$3d^n \rightarrow 3p^5 3d^{n+1} \rightarrow 3d^{n-1} + e_k \tag{10}$$

in resonance with

$$3d^n \rightarrow 3d^{n-1} + e_k \tag{11}$$

The first step of the resonant transition (Equation 10) is the 3p → 3d photoabsorption, and the second step is a super-Coster-Kronig decay. The result is a d^{n-1} final state and a photoelectron e_k. This process involves the same initial and final states as the direct photoemission process (Equation 11), resulting in interference effects. Therefore, resonant photoemission enhances the $3d^7$ and $3d^8$ final states in nickel. The $3d^9$ final

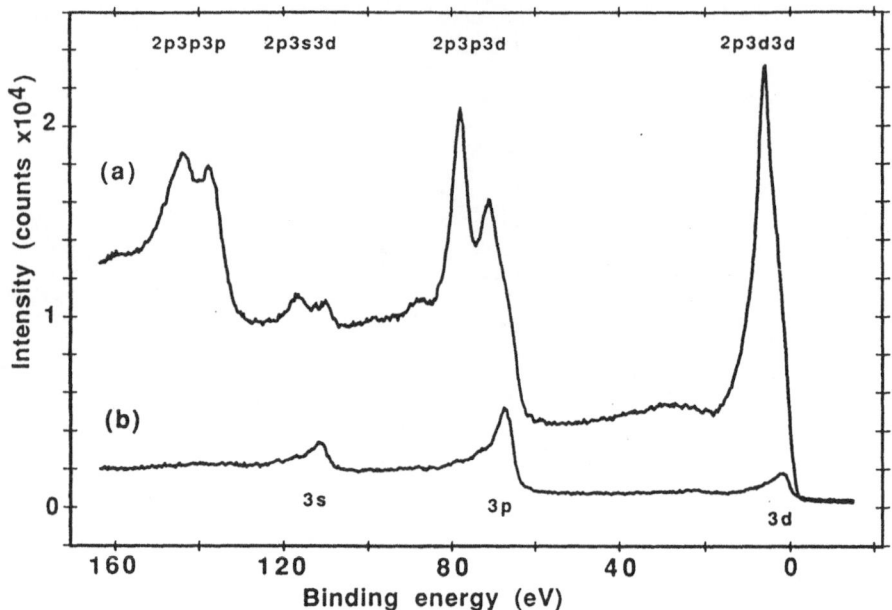

Figure 12. *The 3d, 3p and 3s photoemission of nickel metal (a) at hν = 846 eV; (b) in resonance with the Ni 2p$_{3/2}$ absorption edge at hν = 852 eV. (from van der Laan et al. 1992b).*

state is not enhanced because the transition $3d^{10} \rightarrow 3d^9 + e_k$ cannot interfere with the $3p \rightarrow 3d$ absorption process (the 3d shell is full). The 6 eV satellite is mainly due to the localised d^8 state, which is split from the d^9 band by the $3d$–$3d$ Coulomb interaction.

7.2   2p resonance decay

The analysis of the resonant spectrum is hampered by the large interference term between the 3p de-excitation and the 3d photoemission, which are of comparable magnitude. However, recently Tjeng *et al.* (1991) observed a giant enhancement of the valence band photoemission in CuO at resonance with the Cu 2p absorption edge. At such a deep edge the ratio of the cross-section for the resonant to direct photoemission is considerably larger than at the 3p absorption edge. The interference term is then negligible, and a straightforward interpretation of the resonant spectrum is possible.

Furthermore, we can use the 2p core level resonance decay to study the 3s and 3p photoemission. Although the nature of the satellite structure of the valence band photoemission is well established, the peak splitting in the 3s photoemission of 3d transition metal compounds and alloys has not been assigned unambiguously. This splitting has been ascribed to both d-mixing and exchange interaction. For example van Acker *et al.* (1988) have studied a large number of iron compounds, but found no simple relation between the magnetic moment of the iron and the $3s$ peak splitting and intensity ratio. The resonance enhancement allows us to give clear assignments to the peak structures in the photoemission spectra.

Figure 12b shows the direct photoemission of the $3d, 3p$ and $3s$ levels at a photon energy of 6 eV below the Ni $2p_{3/2}$ absorption edge (van der Laan 1992a,b). Figure 12a shows the resonant photoemission at $852.6eV$, which is the maximum of the Ni $2p_{3/2}$ absorption peak. The strong increase of the $3d, 3p$ and $3s$ photoemission is due to the $2p3d3d$, $2p3p3d$, and $2p3s3d$ decay, respectively. The structure around 140 eV is due to the $2p3p3p$ decay. The direct and resonant photoemission spectrum are also shown in Figures 14 and 15, respectively, where they are compared with calculated results from an Anderson impurity model.

Ni $2p \rightarrow 3d$ absorption

First we will discuss the $2p$ x-ray absorption spectrum (XAS) of nickel because this absorption precedes the decay process. The calculated Ni $2p_{3/2}(L_3)$ absorption, which is comparable to the measured spectrum, is shown at the top of Figure 13. We will not discuss the decay from the $2p_{1/2}$ core hole, which decays dominantly by a super-Coster-Kronig process $L_3L_2M_{45}$. The $2p_{3/2}$ spectrum consists of a main peak ($852.6eV$) and a small satellite structure (857 to 859 eV). The main peak has predominantly $2p^53d^{10}\underline{L}$ character, whereas the satellite structure has mainly $2p^53d^9$ character.

Ni $2p3d3d$ decay

The direct photoemission (Figure 14) of the valence band shows a main peak with predominantly $3d^9$ character. At 6 eV below the Fermi level is the two-hole (d^8) satellite, which is visible as a small shoulder. In the resonance spectrum (Figure 15) the final states with two or more d holes are enhanced. The largest increase is in the d^8 satellite at 6 eV. The intensity of the $3d^7$ final state at $\sim 15eV$ is small due to the low amount of $2p^53d^9$ character in the excited state (Figure 13) and the interference with the two-hole final state.

The enhancement of the two-hole state, which is found in RESPES confirms the Cini-Sawatzky model (*c.f.* Figure 8). However, the RESPES measurement is more direct than the comparison of the $L_{23}M_{45}M_{45}$ Auger spectrum with the $3d$ photoemission.

Ni $2p3p3d$ decay

We can also use the $2p$ decay to study the photoemission from shallow core levels, such as the $3p$ level. The $3p$ direct photoemission spectrum (Figure 14) shows a $3p^5d^{10}$ main peak, and a high energy tail which corresponds to the $3p^5d^9$ state. The latter state is split by the $3p$–$3d$ electrostatic interaction into three more or less distinct structures, *viz.* $^1D + ^3F$, which is hidden under the main peak, $^3P + ^3D$ and $^1F + ^1P$ (*c.f.* Figure 5). The $2p3p3d$ resonant decay is given by the transition

$$3d^n \rightarrow 2p^53d^{n+1} \rightarrow 3p^53d^n + e_k \qquad (12)$$

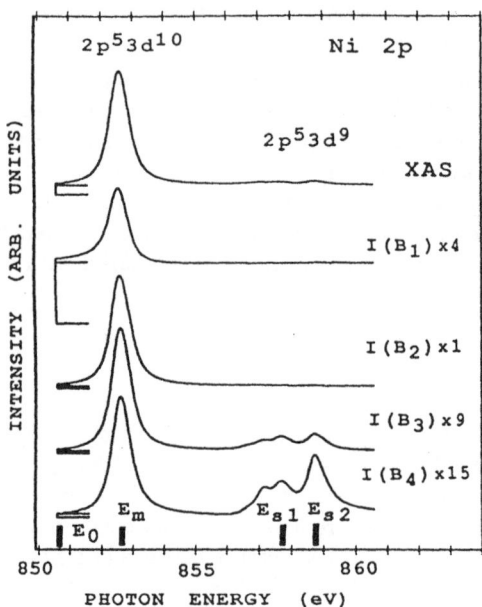

Figure 13. *Calculated Ni $2p_{3/2}$ absorption spectrum (XAS) and the 2p3p3d Constant Initial State (CIS) spectra corresponding to the binding energies B_1, B_2, B_3 and B_4 given in Figure 16. The direct photoemission contribution is shown on the left side. E_0 (off-resonance), E_m (main peak), E_{s1} and E_{s2} (satellites) indicate the photon energies of the photoemission spectra in Figure 16 (from van der Laan et al. 1992c).*

which is in resonance with the transition

$$3d^n \rightarrow 3p^5 3d^n + e_k \tag{13}$$

The $3d^{10}$ state does not resonate. In the resonance spectrum (Figure 15) the $^3P + ^3D$ and $^1F + ^1P$ structures are strongly enhanced. The intensity increase of the main line is due to a small amount of $3p^5 d^9$ character. The weak structure around 90 eV is due to the $3p^5 d^8$ final state. This is clear if we look at the calculated $3p$ photoemission spectra in Figure 16, where $I(E_0)$ gives the direct photoemission and the other curves give the photoemission intensity at the photon energies E_m and $E_{s1,2}$ corresponding to the main and satellite absorption peaks, respectively, of the XAS in Figure 13. Knowing that these peaks correspond to $2p^5 d^{10}$ and $2p^5 d^9$, respectively, it can be seen immediately that the peaks in the middle of the photoemission spectrum must be $3p^5 d^9$ because they are apparently formed by decay of the $2p^5 d^{10}$ states produced at E_m. The higher binding energy peaks, but also hidden structure in the middle of the photoemission spectrum are enhanced at E_{s1} and E_{s2}, and are therefore $3p^5 d^8$. The lowest binding energy peak hardly resonates, confirming its $3p^5 d^{10}$ character.

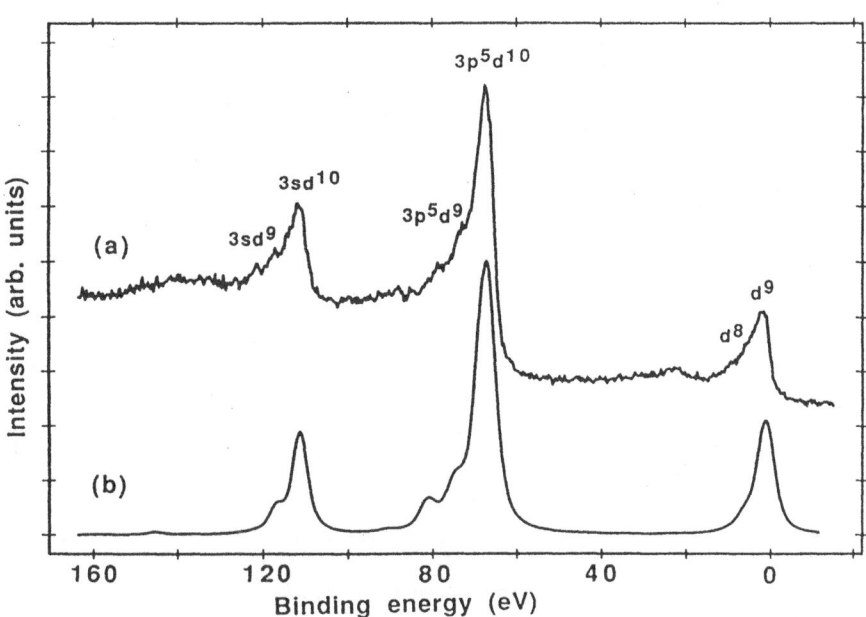

Figure 14. *Photoemission of nickel metal at $h\nu = 846$ eV (a) experimental results; (b) calculational results from a cluster model (from van der Laan et al. 1992b).*

Constant initial state spectroscopy

Conversely, the photoemission can be used to analyse the x-ray absorption spectrum by collecting the emitted electrons at a constant binding energy (photon energy minus kinetic energy). This is called constant initial state spectroscopy (CIS). The CIS spectrum integrated over the whole binding energy range gives the XAS spectrum in total electron yield (Figure 13). Thus we can interpret CIS as partial electron yield at constant binding energy. Figure 13 shows the CIS spectra corresponding to the binding energies B_1, B_2, B_3 and B_4 from Figure 16. The direct photoemission contribution is shown on the left side. The $3p^5d^{10}$ CIS (B_1) gives a small resonance due to admixture of $3p^5d^9$. The $3p^5d^9$ CIS (B_2) only gives the main absorption line, confirming its $2p^5d^{10}$ character and the $3p^5d^8$ CIS ($B_{3,4}$) shows a very enhanced $2p^5d^9$ satellite structure. The latter confirms the d^8 origin of the $2p$ absorption satellite and shows the increased singlet character at higher energy.

Ni 2p3s3d decay

The $2p3s3d$ resonant decay is given by the transition

$$3d^n \rightarrow 2p^5 3d^{n+1} \rightarrow 3s3d^n + e_k \tag{14}$$

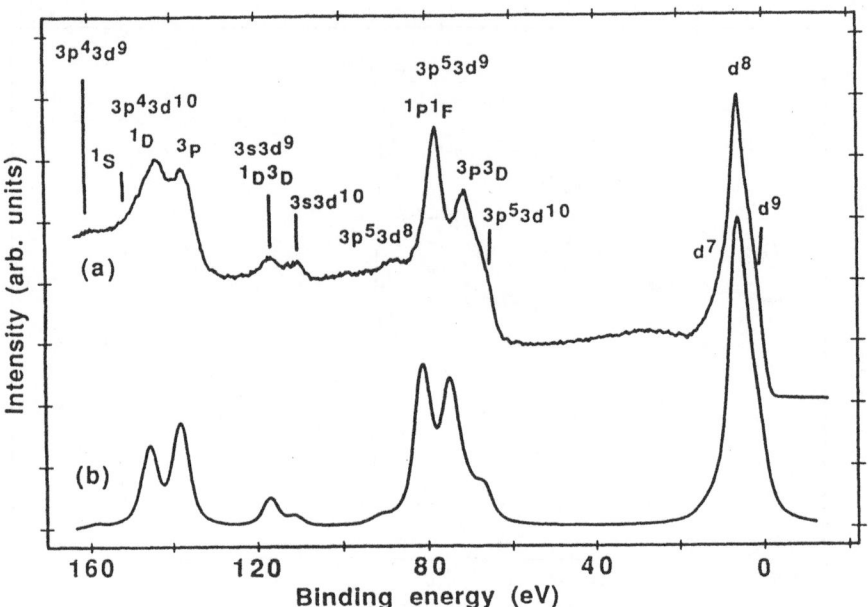

Figure 15. *Photoemission of nickel metal in resonance with the Ni $2p_{3/2}$ absorption edge at $h\nu = 852$ eV (a) experimental results; (b) calculational results from a cluster model (from van der Laan et al. 1992b).*

in resonance with

$$3d^n \rightarrow 3s3d^n + e_k \tag{15}$$

However, the $3s3d^9$ final state has a configuration interaction with the $3p^43d^{10}$ state, which is produced in the $2p3p3p$ decay

$$3d^9 \rightarrow 2p^53d^{10} \rightarrow 3p^43d^{10} + e_k \tag{16}$$

The direct $3s$ photoemission spectrum (Figure 15) shows a main peak, which has $3sd^{10}$ character, and a small satellite with $3sd^9$ character, which is split into a ^3D and ^1D state. In the resonant spectrum (Figure 15) only the satellite structure at 118 eV is enhanced, which demonstrates that the $3s$ peak splitting is due to d-mixing and not to exchange interaction. The $3p^4d^{10}$ configuration is split by the $3p - 3p$ electrostatic interaction into a $^3P, {}^1D$ and ^1S state. The $3p^4d^{10}$ state can be reached by decay from the $2p^53d^{10}$ intermediate state, as well as by intra-atomic configuration interaction with the $3sd^9$ final state of the $3s$ photoemission. This configuration interaction pushes the ^1D peak of the $3sd^9$ structure towards the ^3D peak and the ^1D peak of the $3p^4d^{10}$ structure towards the ^1S peak, and destroys the usefulness of the ^1D–^3D splitting to determine the $3s$–$3d$ exchange.

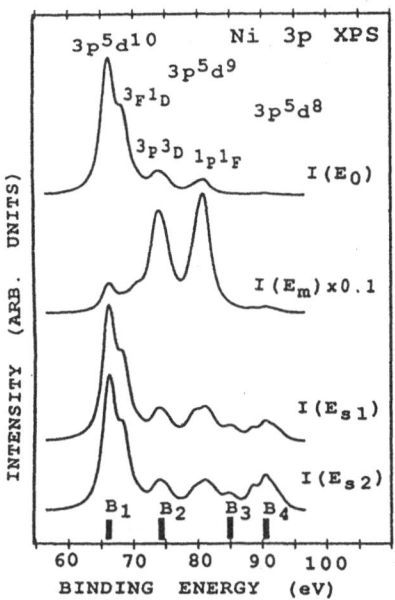

Figure 16. *Calculated Ni 3p photoemission intensity at the photon energies E_0, E_m, E_{s1} and E_{s2} given in Figure 13 (from van der Laan et al. 1992c).*

8 Conclusions

Electron spectroscopy has significantly contributed to the understanding of the electronic structure of transition metals. The phenomenology of the 6 eV satellite in Ni is now well understood. Coincident spectroscopy and resonant photoemission have clearly demonstrated that the Ni satellites are due to the d^8 character in the ground state. There is strong evidence that the hole-hole correlations have a significant influence on the electronic structure, and that one-electron bandstructure models, such as density functional methods, are less suitable in explaining the observed photoemission spectra of correlated materials. Although, the Mott-Hubbard-Anderson model has been applied successfully to explain the observed phenomena, it does not account for the full details of the band structure if translational symmetry is not considered. The unification of band structure and localised models is still an enormous challenge to theorists. For the moment the localised model seems to be the better option for the interpretation of photoemission spectra from transition metal and rare earth compounds. In these materials the correlated electronic structure manifests itself in the form of atomic-like multiplet structure and satellite structure due to the charge transfer and configuration interaction.

Acknowledgements

This chapter has strongly benefitted from the reviews by L.C. Davis and G.A. Sawatzky. I would also like to thank B.T. Thole, A. Kotani, H. Ogasawara, and Y. Seino who were my collaborators in the resonant photoemission work.

References

Almbladh C-O and Hedin L, 1983, *Handbook on Synchrotron Radiation*, ed by E.E. Koch (North- Holland, New York) **1b** 607

Bartynski R A, Jensen E and Hulbert S L, 1992a, *Physica Scripta* **T41** 168

Bartynski R A, Yang S, Hulbert S L, Kao C-C, Weinert M, Zehner D M, 1992b, *Phys Rev Lett* **68** 2247

Cini M, 1975, *Solid State Comm* **53** 716

Cini M, 1979, *Surface Science* **87** 483

Cowan R D, 1968, *J Opt Soc Am* **58** 808

Cowan R D, 1981, *The Theory of Atomic Structure and Spectra* (University of California Press, Berkeley)

Davis L C, 1986, *J Appl Phys* **59**, R25

Einstein A, 1905, *Ann Phys, Leipzig* **17** 132

Folmer J C W and Jellinek F, 1980, *J Less Common Met* **76** 153

Guillot C, Ballu Y, Paigné J, Lecante J, Jain K P, Thiry P, Pinchaux R, Petroff Y and Falicov L M, 1977, *Phys Rev Lett* **39** 1632

Haak H, Sawatzky G A and Thomas T D, 1978, *Phys Rev Lett* **41** 1825

Harrison W, 1980, *Electronic Structure and the Properties of Solids* (Freeman, San Francisco

Hertz H, 1887, *Ann Phys, Leipzig* **31** 983

Hohenberg P and Kohn W, 1964, *Phys Rev B* **136** 864

Hubbard J, 1964, *Proc Phys Soc Ser. A* **277** 237

Kohn W and Sham L J, 1964, *Phys Rev A* **140** 1133

Koopmans T, 1933, *Physica* **1** 104

Mann J B, 1967, *Los Alamos Scientific Laboratory Report* No. LASL-3690

McGuire E J, 1971, *Sandia Laboratory Report* No. SCRR-710075

Mott N F,1949, *Proc Phys Soc Ser. A* **62** 416

Sawatzky G A, 1977, *Phys Rev Lett* **39**, 504

Sawatzky G A, 1988a, in *Core-Level spectroscopy in Condensed Systems*, ed by Kanamori J and Kotani A, (Springer, Berlin) p 99

Sawatzky G A, 1988b, *Treatise on Materials Science and Technology* **30, 167** (Academic Press, New York)

Sawatzky G A, 1990, in *Earlier and Recent Aspects of Superconductivity*, ed. by Bednorz J G and Müller K A (Springer, Berlin) p 345

Sawatzky G A, 1991 in*High Temperature Superconductivity: Proceedings of the 39th Scottish University Summer School in Physics*, edited by Tunstall D P and Barford W (Institute of Physics Press, Bristol UK)

Sawatzky G A and Lenselink A, 1980, *Phys Rev B* **21** 1790

Slater J C, 1960, *Quantum Theory of Atomic Structure* Parts I, II (McGraw-Hill, New York)

Thole B T and van der Laan G, 1988, *Phys Rev B* **38** 3158

Thole B T and van der Laan G, 1991a, *Phys Rev B* **44** 12424

Thole B T and van der Laan G, 1991b, *Phys Rev Lett* **67** 3306

Thole B T, Carra P, Sette F and van der Laan G, 1992, *Phys Rev Lett* **68** 1943

Tjeng L H, Chen C T, Ghijsen J, Rudolf P and Sette F, 1991, *Phys Rev Lett* **67** 501

van Acker J F, Stadnik Z M, Fuggle J C, Hoekstra H J W M, Buschow K H J and Stroink G, 1988, *Phys Rev B* **37** 6827

van der Laan G and Thole B T, 1992a, *J Phys: Condensed Matter* **4** 4181

van der Laan G, Westra C, Haas C, and Sawatzky G A, 1982, *Phys Rev B* **23** 4369

van der Laan G, Surman M, Hoyland M A, Flipse C F J, Thole B T, Seino Y, Ogasawara H and Kotani A, 1992b, *Phys Rev B*(in press)

van der Laan G, Thole T, Ogasawara H, Seino Y and Kotani A, 1992c, *Phys Rev B* **46** 7221

van der Marel D and Sawatzky G A, 1988, *Phys Rev* **37** 10674

Wendin G, 1981, in *Structure and Bonding*(Springer, New York) **45, 1**

Williams A R and von Barth U, 1983, in *Theory of the Inhomogeneneous Electron Gas*, eds. Lundqvist S and March N H (Plenum, New York) p 189

Zaanen J, Sawatzky G A and Allen J W, 1985, *Phys Rev Lett* **55** 418

Auger Electron Spectroscopy in the STEM

P Kruit

Delft University of Technology
The Netherlands

1   Introduction

Microanalysis starts with the incidence of a well focused stream of electrons, ions, photons or positrons on a target. The question of how that is obtained might easily be dismissed by the user of a fancy, expensive instrument with the argument that getting such a microbeam is what he paid for: let instrument designers worry about the problem. However, when we look at the most successful scientists, we find that they have a more than average knowledge of how the instrument works. With that knowledge they are able to push the performance of the instrument to its limits, sometimes even beyond the specifications of the manufacturer. A second reason for wishing to understand microbeam optics is to be able to judge the manufacturer's specifications. Are they realistic? How can they be compared to the specifications of competing manufacturers who quote totally different parameters? A third reason may be that we are not satisfied with what is for sale and we may wish to make our own improvements or modifications.

2   Basic electron optics for microbeam analysis

In the following I shall concentrate on electron optics, but much of the material is equally applicable to ions or positrons, only the section on sources is specific to electrons.

2.1 Electron sources

Characterisation of sources

Most parameters relating to an electron beam can be manipulated by the optics of the instrument: beam size can be demagnified, current can be cut off by an aperture, *etc*. So it is only of secondary interest to know these parameters at the source. The only parameter that is not influenced by any action of the instrument is the reduced brightness.

Brightness
The definition of brightness is $B = I/dOd\Omega$, where I is the current through an area dO into a solid angle $d\Omega$. Reduced brightness B_r is obtained on dividing by the acceleration voltage V. The convention is to express B in A/cm²srV. Although the brightness concept is applicable anywhere in the beam, it is most easily evaluated in cross-overs or source images. A lens can magnify an area dO from one cross-over to the next, but at the same time $d\Omega$ will be demagnified. A similar thing happens when aperturing or accelerating a beam: the reduced brightness is a constant. Only individual electron-electron interactions can decrease B_r. Lens aberrations can decrease the effective brightness of a beam, that is the brightness obtained when taking the envelope of the area and the angle instead of a local, infinitesimal area dO and solid angle $d\Omega$.

Virtual source size
The size of the source as viewed from the lens system is of importance for the choice of optical system. Usually this is not the size of the electron emitting area, but is influenced by the extraction optics.

Total source current
The total current from the emitting area is only of importance if it is too small to deliver the required beam current.

Energy spread
This is the FWHM value of the energy distribution of the electron beam produced by the source. The energy spread is a result of a number of processes:
  1: thermal energy distribution of the electrons emitted by the cathode

  2: statistical interactions of the electrons in the beam

  3: variations of the accelerator voltage.

Life time
The average time an electron source can operate without replacement of the cathode.

Ambient requirements
Vacuum conditions, robustness, *etc.* are important for the ease of operation of a source.

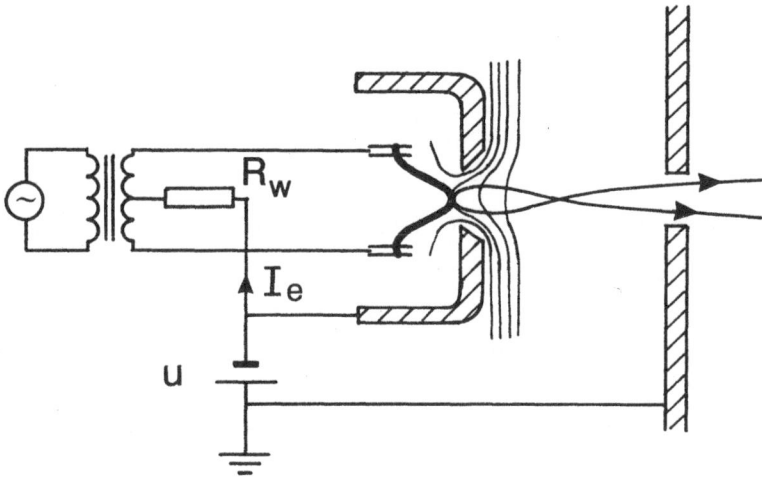

Figure 1. *Thermionic source. The equipotentials around the Wehnelt electrode which limit the emission to a small area are indicated.*

Thermionic electron source

A typical electron source consists of a tungsten wire of 125 μm diameter bent in the shape of a V, heated to about 2800 K by a 3 A a.c. current. To restrict the emission to a small area a control electrode, the 'Wehnelt cylinder' is used. Figure 1 shows how the accelerating field from the anode can only reach through the hole in the Wehnelt cylinder to an area of about 50 to 100 μm. Electrons emitted from the rest of the tungsten wire return to the wire. Usually the voltage between the cathode and the Wehnelt cylinder is obtained by forcing the total source current to go through a resistor, indicated in Figure 1 by R_w. The effect of this is that at a low source current, the voltage difference is small and emission occurs from a large area. Increasing the cathode temperature, there is a fairly sudden point where the emitting area contracts and the source current stays almost constant with increasing temperature. This sudden change indicates the work point. The total source current at the work point can be varied by varying the Wehnelt resistor. A larger resistor gives a smaller source of slightly higher brightness. Typically the reduced brightness is about 1–10 A/cm^{2}srV at a total current of 100 μA and a life time of 10 to 50 hours. Note that the accelerating voltage of the beam is, in a set-up as in Figure 1, not equal to the voltage of the power supply, but influenced by the total current and the resistor value.

A Lanthanum hexaboride (LaB$_6$) crystal, or a material with a lower work function than tungsten, can be used as a cathode. This yields a higher brightness at a lower temperature and results in an electron source with a longer life time. The only disadvantage is the more severe requirement on vacuum conditions.

	Thermionic		Schottky	Field emission
	tungsten	LaB6	zirc-tung	tungsten
Reduced brightness A/cm².sr.V	1 – 10	10 – 100	$10^3 - 10^4$	$10^3 - 10^4$
Energy spread eV	0.6 – 1.5	0.4 – 1.5	0.5 – 1.5	0.25 (cold) 0.5 (1200 K)
Short term current stability %	1	0.4	1	3 (cold) 10 (warm)
Source size μm	20	10	0.02 – 0.05	0.002 – 0.01

Table 1. *Comparison table for some electron sources*

Field emission source

By applying a very high electric field at the surface of a metal (about 10^{10} V/m) field emission or cold emission occurs. Electrons tunnel through the potential barrier. In practice the high field strength needed for field emission is realised by locating a sharply pointed emitter opposite to an extraction electrode. The emitter is generally fabricated from a tungsten single crystal. The tip is etched to produce a point with a radius of approximately 100 nm.

Although tungsten has a high work function, it is the preferred material because it is rigid, has a high melting point and good etching characteristics. The high melting point is important for the cleaning procedure which consists of heating the tip up to a high temperature (2500 K) for a short period to evaporate all contaminants. Since the FEG tip is spot welded to a standard microscope filament, this temperature is simply achieved by applying a heating current through the filament. Since the presence of contaminants can critically affect the work function of the tip, the FEG must operate in a very clean high vacuum, (10^{-10} Torr). A very small effective source diameter of about 2 nm is obtained due to the lens effect of the curved emitting area. The small source size makes the FEG quite sensitive to stray fields and vibration.

Schottky emitter

The Schottky effect is the effective decrease of the potential barrier when an external field is applied at a metal surface. The barrier does not not reduce to the extent that electrons tunnel out as in field emission, but the operating temperature of a cathode using the Schottky effect can be lower than in thermionic emitters. The most successful emitter of this type uses a zirconiated tungsten tip in a set-up similar to a field emission gun, although often an additional electrode at negative potential is added to suppress emission from the shaft of the tip.

Other emitters

The electron emitter most frequently found is the 'oxide cathode' type used in most television tubes: a small block of low work function material is indirectly heated to emit a large current, but with not quite the characteristics which are required to form a microbeam.

Many experimental electron sources exist. Examples are photoemission sources, avalanche electron emitting diodes close to the semiconductor surface, and negative electron affinity photoemitters which are used as sources of polarized electrons. Some typical parameters are shown in in Table 1.

2.2 Electron lenses

Characterization of lenses

Focal distances and principal planes
In describing a lens the first characteristic that comes to mind is its strength. However, electron lenses can be weak or strong depending on the excitation, so maximum strength or minimum focal length would be a better parameters. Still we must be more precise: an accelerating or decelerating electrostatic lens has a different focal length at each side, related through $f_1/f_2 = [\theta_1/\theta_2]^{\frac{1}{2}}$: obviously, at the side where the electrons have the largest energy θ, the focal length is the largest. A further complication is that lenses cannot always be considered to be thin: the effective position and action of the lens then has to be described by the position of two principal planes, which will not be discussed here. It is important to realise that an electron lens does not always gets stronger when the excitation of the magnet or the voltages on the electrodes are increased. Imagine a parallel beam entering a lens. Increasing the excitation pulls the cross-over closer until it enters the field of the lens. A further increase of the excitation makes the lens effectively weaker: the field behind the cross-over makes the beam parallel again, until at some point a second cross-over is formed far from the lens. Only if the target is positioned in the lens at the first cross-over, can the lens be considered to get stronger and stronger with increasing excitation.

Spherical aberration
In a perfect lens, electrons experience a deflection proportional to the distance from the axis. In every electron lens, however, there is spherical aberration, causing an additional deflection proportional to the distance to the third power. This causes a parallel beam

not to be focused to a point but to a disc of radius $r = C_s \alpha^3$, where α is the limiting semi-angle at the focus and C_s is the coefficient of spherical aberration. However, at a defocus given by $\Delta z = 0.75 C_s \alpha^3$, all electrons are within a radius of $0.25\ C_s \alpha^3$. (If the electrons do not enter the lens as a parallel beam the effective spherical aberration is larger: in order to find the size of the aberration disc one must use a modified coefficient $C_s(M) = C_s(1 + M)^4$ where M is the magnification.) A rule of thumb for the value of the aberration coefficient $C_s(\infty)$ is, that it is about equal to the focal distance f, if the lens is excited to maximum strength. At other strengths $C_s \approx f^3 / f_{\min}^2$. Values obtained with this rule are usually correct to within a factor of 2.

Chromatic aberration

When the electrons have a different energy the effect of the electron lenses may change, *e.g.* the focal plane may depend on the energy. The coefficient of chromatic aberration is defined as follows: if D is the distance between the focal plane at electron energy V and energy $V + \Delta V$ then

$$D = C_c \frac{\Delta V}{V} \tag{1}$$

So at an opening angle α, a parallel incoming beam with energy $V + \Delta V$ is focused at the focal plane for electrons energy V into a disc of radius $r_c = C_c\,\alpha\,\Delta V / V$. The chromatic aberration constant is usually of the order of the focal distance.

Mechanical astigmatism

Elliptical electrode bores or pole piece bores, magnetic inhomogeneities, misalignment of electrodes can all cause the lens to be stronger in one plane than in the perpendicular plane. The effect can be compensated for by a quadrupole element which has positive strength in one direction and negative in the other. Because the rotation of the correction must be adapted to the effect of the lens, it is always necessary to have two quadrupole elements.

Image aberrations

For an electron beam that does not go through the center of the lens, other aberration terms have to be considered.

Further lens characteristics

Obviously numerous non-optical characteristics are important: size, necessary power supplies, free space around the target, electrostatic or magnetic field in which the target is immersed, *etc.*

Electrostatic lenses

Most electrostatic lenses consist of three electrodes: cylinders or apertures or more complicated forms. Figure 2 shows an example. To explain the lens effect we follow the electron through the electrostatic field, visualised with the equipotential planes. The forces on the electron are always perpendicular to these planes, so the following happens. At electrode 1 the electron gets an impulse in the outward direction. Between 1 and 2 the electron moves further outwards and is decelerated. At electrode 2 it gets an inwards directed impulse which is large compared with the impulse at electrode 1 because of the following reasons:

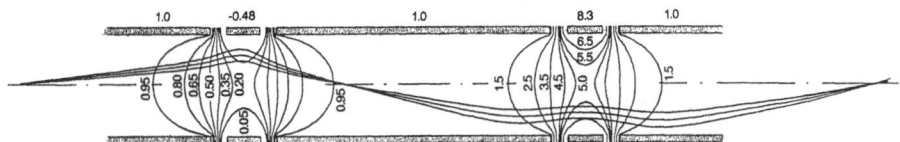

Figure 2. *Decelerating and accelerating electrostatic lens with calculated equipotentials and electron paths through the lenses (distance to the axis has been enlarged for clarity).*

- The electron is closer to the electrode and hence the lines of force are more nearly perpendicular to the axis.

- The size of the area where the force is acting is twice as long compared with the area of electrode 1, because at electrode 2 the field is at both sides of the electrode.

- The electron is slower, so the force acts longer.

It is evident that the impulse at electrode 2 is more than twice as high as at electrode 1. The result is that the electron will be closer to the axis at electrode 3 than at electrode 1. Although at electrode 3 the electron gets an impulse directed away from the axis, the remaining total effect remains towards the axis, and the total lens effect is positive. The same reasoning can be applied to a three-electrode lens with an accelerating middle electrode. At the outer electrodes the lens effect is positive and at the inner electrode the effect is negative. Although the electron trajectory where a negative lens effect exists is longer, this effect is weaker: the electrons are closer to the axis, and move faster. It can be shown that any electrostatic lens with the object and the image in a field free region has a positive lens effect.

The book *Electrostatic Lenses* by Harting and Read (1976) contains a useful collection of tables of the optical properties of two and three electrode lenses. Nowadays there are many computer programs available to calculate the properties of a lens of one's own design.

Magnetic lenses

A magnetic lens consists of a coil around the optic axis, giving a field parallel to the axis. Usually this field is concentrated in a small volume by a magnetic circuit. Figure 3 shows a simplified cross section of two lenses. The magnetic field is visualized with the aid of magnetic flux lines.

In order to understand the action of a magnetic lens we divide the magnetic field into a homogenous part where the field is parallel to the axis of the lens and the end sections, where the field is radial. This is a good approximation if the lens bore D is much smaller than the gap S. An incoming electron at distance R_0 from the axis receives, because the field at the entrance is radial, such a tangential impulse that

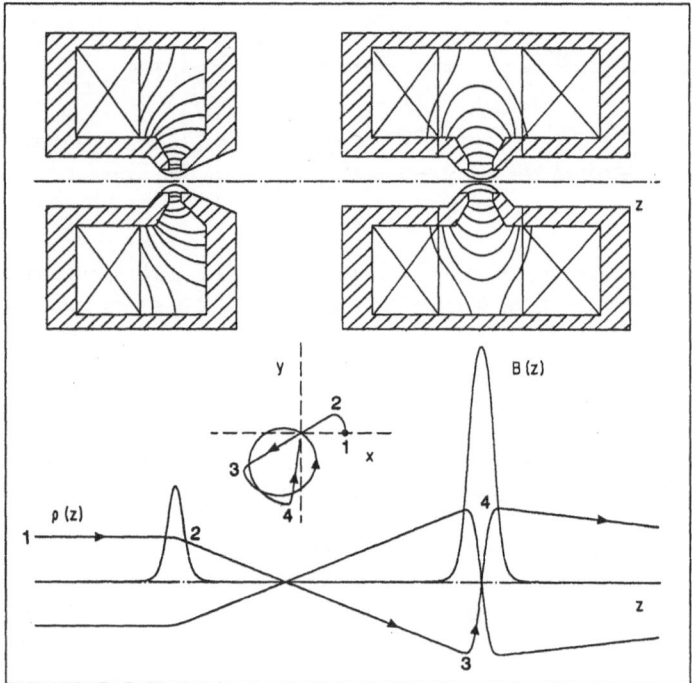

Figure 3. *Weak and strong magnetic lens as in the condenser and objective lenses of transmission electron microscopes. Fluxlines indicate the form of the magnetic field. The calculated electron trajectory is also shown in x–y projection to demonstrate the lens action.*

the radius of curvature in the constant field is $R_0/2$. At the field exit the tangential impulse is compensated, while the radial impulse remains, which is then the net lens effect. This also explains why a magnetic lens rotates the beam. The focal length of a weak magnetic lens, that is a lens where the focus is outside of the field, is given by

$$\frac{1}{f} = \frac{e}{8m\theta} \int\limits_{-\infty}^{\infty} B_z^2(z)\, dz \tag{2}$$

where e and m are the charge and mass of the particle, θ the relativistically corrected acceleration voltage and $B_z(z)$ the magnetic field strength on axis. This equation shows why magnetic lenses are not good for ions: they are usually too weak. Targets which are not magnetic can be placed inside the magnetic field. The lens can then be quite strong, even for high accelerating voltages, and the aberration coefficients proportionally low. With pole piece distances and gaps of a few millimeters, the focal length, C_s and C_c are also a few millimeters—or even less than 1 mm. If the target is outside the field, the focal length is not at minimum value and C_s is usually more than 20 mm.

2.3   Probe forming optics

Contributions to the probe size

The diameter of a focused electron beam has contributions from the geometric image of the electron source, from the spherical and chromatic aberration discs, from the diffraction error and from stray fields and mechanical vibration displacements. The latter displacements shall be neglected although in practice they can be a give problems.

The problem in evaluating the total diameter is that there is no exact convention on what should be taken as a measure for these contributions, so care has to be taken when comparing different instruments. Addition of the contributions is usually done in quadrature:

$$d_{tot}^2 = d_{geo}^2 + d_{C_s}^2 + d_{C_c}^2 + d_{dif}^2 \tag{3}$$

which would be allowed if the current distributions of the individual contributions is Gaussian, which is not the case here. Also different contributions such as spherical aberration and diffraction are correlated. The correct method, however, is too complicated for general use. We have recently compared the quadratic addition method with wave optical theory and found that the best agreement is obtained when using FW_{50} values, that is the diameter within which 50% of the electrons are contained:

$$d_{dif} = 0.54 \frac{\lambda}{\alpha} \tag{4}$$

where α is the limiting half angle, and λ the electron wavelength.

$$d_{C_s} = 0.18 C_s \alpha^3 \tag{5}$$

at a defocus of $\Delta z = 3/8 C_s \alpha^2$.

$$d_{C_c} = 0.34 C_c \alpha \left(\frac{\Delta V}{V} + 2 \frac{\Delta I}{I} \right) \tag{6}$$

where ΔI is the current fluctuation of the probe forming lens.

$$d_{geo} = M d_{gun} = \left(\frac{4 I_p}{\pi^2 B_r V} \right)^{\frac{1}{2}} \frac{1}{\alpha} \tag{7}$$

where I_p is the probe current.

C_s and C_c are theoretically the combined aberration coefficients of the total optical system. However, if the aperture angle at the probe is much larger than anywhere else in the system, only the aberration of the probe forming lens is of consequence. This condition is certainly valid in instruments with thermionic emitters since the source is demagnified by a large factor. In systems with a field emission gun, the aberration of the gun lens, formed by the extraction and acceleration electrodes, has to be taken into account. This can be done by attributing an effective brightness to the gun, determined by the gun lens aberration. At a given aperture angle at the probe, the probe current I_p can only be increased by accepting a larger aperture α_g at the gun, which is effectively done by changing the magnification of the system. With an angular current density J_g from the tip:

$$\alpha_g = \left[\frac{I_p}{\pi J_g} \right]^{\frac{1}{2}} \tag{8}$$

When for convenience, we now make the approximation that the virtual source size is pointlike, the effective brightness is limited by the spherical aberration C_{sg}:

$$B_{\text{reff}} = \frac{I_p}{\frac{\pi^2}{4} \left[\frac{1}{2} C_{sg} \alpha_g^3 \right]^2 \alpha_g^2 V_g} \tag{9}$$

and with the equation for d_{geo} given above:

$$d_{\text{geo}} = \frac{1}{2} C_{sg} \left(\frac{V_g}{V} \right)^{\frac{1}{2}} \frac{I_p^2}{\pi^2 J_g^2} \frac{1}{\alpha} \tag{10}$$

This approach using an effective brightness is not recommended for a precise calculation of probe size. The fact that Equation 10 uses a factor 0.5 in the size of the aberration disc, while Equation 5 uses a factor 0.18 is an indication of the lack of precision. Note that C_{sg} and J_g are properties of the gun lens and of the electron beam when the acceleration is not yet complete: the beam voltage at that point is V_g, usually about 5 kV.

Figure 4 gives the total probe diameter for some typical instruments: a scanning transmission microscope at 100 kV acceleration with the specimen inside the magnetic objective lens and a scanning microscope or scanning Auger instrument at 5 kV with the specimen at a working distance of about 15 mm behind the final lens. The curves for field emission guns are only valid for very small tips. For Schottky emitters, which have larger radius tips, one must expect that the curves will be pushed upwards in the right hand bottom corner of Figures 4c and 4d.

Conclusions on probe formation

Examining the contributions to the probe diameter a number of important practical conclusions can be drawn:

- The diffraction spot and the geometrical image of the gun (at a required probe current I_p) decrease with α and the aberration discs increase with α, so there is a best choice of aperture angle. An instrument which does not give control of this angle, either optically or by mechanical change of diaphragm, will only give optimum performance at one probe size.

- The probe current obtainable in a given probe diameter increases with the acceleration voltage.

- At low accelerating voltage the chromatic aberration dominates over the spherical aberration.

- Although a field emission gun is for small probe sizes clearly superior over a thermionic source, it cannot deliver large currents without serious increase of the probe size.

- Having the specimen behind the final lens at a large working distance may be convenient for placing detectors, manipulating the specimen, etc. ; this luxury is paid for by a decrease of current at a given probe size or an increase of probe size at a given current.

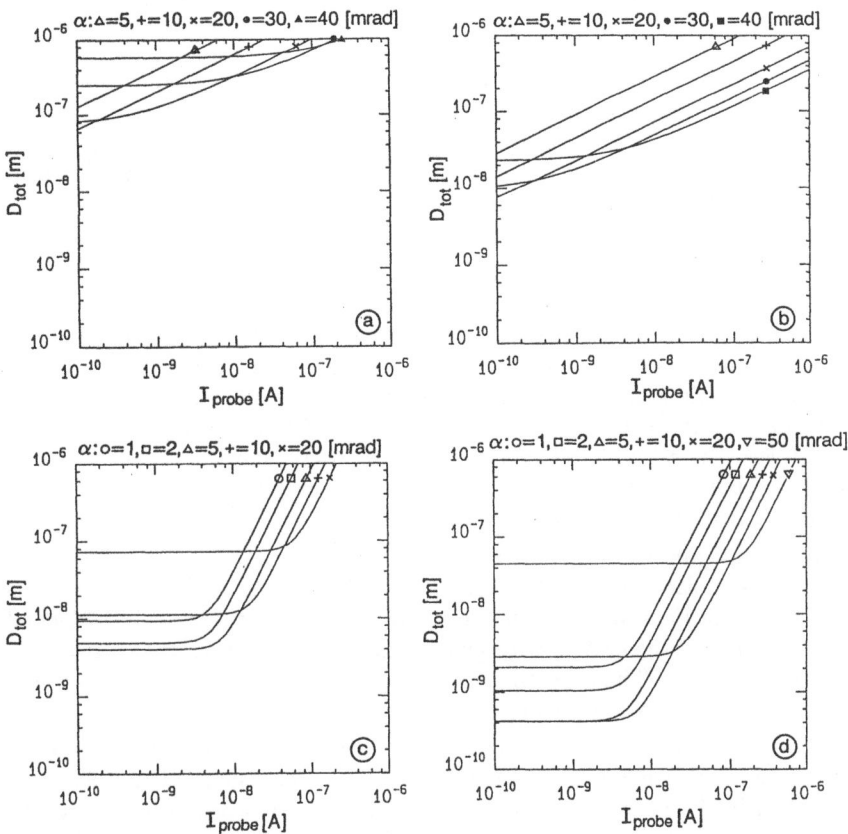

Figure 4. *(a) Total probe diameters for a typical Auger system with thermionic source:* $V = 5$ kV, $B_r = 2$ A./cm²srV, $C_s = 50$ mm, $C_c = 20$ mm, $\Delta V = 1$ V; *(b) a typical* TEM/STEM *system with thermionic source:* $V = 100$ kV, $B_r = 2$ A/cm²srV, $C_s = 2$ mm, $C_c = 2$ mm, $\Delta V = 1V$; *(c) a typical* SEM *system with field emission gun:* $V = 5$ kV, $C_s = 50$ mm, $C_c = 20$ mm, $\Delta V = 0.5$ V, $C_{cg} = 20$ mm, $J = 50$ μA/sr, $V_g = 5$ kV; *(d) a typical* STEM *system with field emission gun:* $V = 100$ kV, $C_s = 2$ mm, $C_c = 2$ mm, $\Delta V = 0.5$ V, $C_{cg} = 20$ mm, $J = 50$ μ A/s_r, $V_g = 5$ kV.

3 Optics for efficient collection of Auger electrons

The first demand on a high spatial resolution scanning Auger instrument is, of course, a small electron probe. However, there is no point in having a 1 nm probe if the Auger signal is down to a few counts per second. So collection efficiency is closely related to resolution.

When signal rate is a problem, the first question which should be asked is what is the most suitable spectrometer: a cylindrical mirror analyser or a hemispherical deflection analyser, or possibly a retarding field analyser? The last is usually ruled out because of the large statistical noise in the spectrum which results from detecting the total current of all electron energies above the selected energy. How do we compare? For the case of an extended source the comparison of etendue, as done by Heddle (1971), is appropriate. Etendue is the product of the accepted solid angle and the accepted source area, so a spectrometer with a large etendue at a given energy resolution is a good spectrometer. The fact that etendue is limited, is a result of the aberrations of the spectrometer: the spot at the detector has a contribution from the geometrical size of the source area and from the size of the spectrometer aberration disc. In an optimized situation these contributions are about equal and determine the energy resolution of the measurement.

In order to find the transmission of the system the etendue must be compared with the properties of the electron source. For extended sources as in photoelectron spectroscopy this is not a problem because the system selects an angle and an area (with the product equal to ε) from the source and the transmission follows from the percentages that are selected. For a very small source only an acceptance angle would be selected but not a source area, so the obvious thing to do is demagnify the emission angle such that the system accepts a larger angle.

This demagnification is however limited by the aberrations of the first lens, because the aberration disc can become so large that the system selects an area which is smaller than the aberration disc. Because the size of the aberration disc totally depends on the acceptance angle, the system with given etendue now only selects an angle and not an area at the source. Thus the transmission of the instrument depends strongly on the aberrations of the pre-spectrometer optics.

Figure 5 shows schematically an electron optical system consisting of a spectrometer with characteristic dimension (usually a radius) R and dispersion dx/dE, an aberration free set of lenses for retardation and magnification and the crucial first lens for the demagnification of the accepted angle at the source. Because this lens demagnifies the angle, only the aberrations of this lens determine the effective source size d_i at the entrance of the spectrometer.

3.1   Characterisation of analysers

The 180° hemispherical analyser

The 180° concentric hemispherical analyser (CHA) is an often used instrument because of its double focusing properties from the entrance to the exit. Similar analysers with just less than 180°, or only 90°, or cylindrical analysers with 127° all have roughly similar performance, say to within a factor 2. The energy resolution of CHA is:

$$\Delta E = E \frac{d_i}{2R} \tag{11}$$

The second order aberration of the spectrometer then sets a maximum to the aperture angle α_i:

$$\alpha_i \leq \left(\frac{d_i}{2R}\right)^{\frac{1}{2}} \tag{12}$$

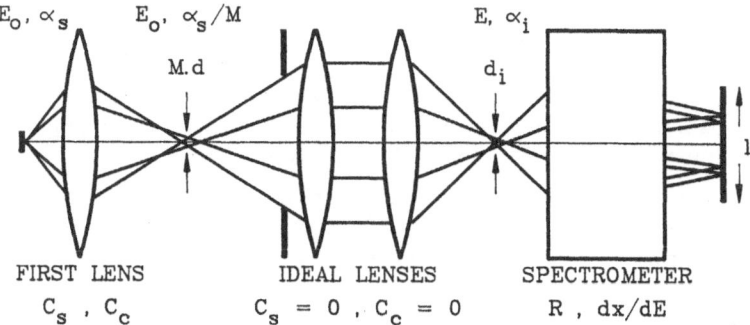

Figure 5. *Schematic division of a spectrometer, an aberration free set of lenses for retardation and magnification and a lens for the demagnification of the accepted angle, which introduces aberrations.*

For the case of a circular entrance spot and identical aperture angles in the dispersion plane and perpendicular to the dispersion plane, the etendue of a spectrometer placed directly after a source of electrons with energy E_0 is:

$$\varepsilon = \pi^2 R^2 \left(\frac{\Delta E}{E_0}\right)^3 \tag{13}$$

If a set of aberration free lenses would decelerate the electron from energy E_0 at the source to energy E at the spectrometer, the etendue would be:

$$\varepsilon = \pi^2 R^2 \left(\frac{\Delta E}{E}\right)^3 \frac{E}{E_0} \tag{14}$$

Without demagnification of the entrance angle, the transmission would simply be determined by the aperture angle α_i. Assuming, for simplicity, a homogeneously emitting source, then

$$T = 2\left(\sin \frac{\alpha_3}{2}\right)^2 \approx \frac{1}{2}\alpha_s^2 = \frac{d_i}{4R} = \frac{\Delta E}{4E_0} \tag{15}$$

which illustrates that the transmission of a system without pre-spectrometer lenses is rather low.

The cylindrical mirror analyser

Aksela *et al.* (1970) and Heddle (1971) have calculated the transmission T of a cylindrical mirror analyser with both the source and the detector on axis. For electrons leaving the source at an angle $42.3° \pm \alpha_s$ and a spectrometer with cylinder of inner radius R_1 and distance between source and detector $L = 6.13R_1$, the acceptance solid angle is $4\pi\alpha_s \sin 42.3°$ so that $T = 1.35\alpha_s$ when α_s is small. Considerable improvement in energy resolution can be obtained by positioning the detector slightly off axis to intercept the aberrated focus at its minimum trace width (Sar-el 1970), such that

$$\Delta E/E_0 \approx 2.1\alpha_s^3 \tag{16}$$

for α_s up to approximately 14°. This gives the transmission

$$T = 1.04 \, (\Delta E/E_0)^{\frac{1}{3}} \tag{17}$$

neglecting the effect of grids, which reduce the constant by approximately 20%. Axis-to-ring designs have several advantages (Gerlach 1980), and a shorter length of approximately $5R$, but the energy resolution and transmission are similar.

Pre-retardation can be done with spherical grids (Gerlach 1973). Apart from the reduction in transmission implied by the grids, Equation 17 is valid with E_0 replaced by E. Clearly the transmission is far superior to that of a CHA without pre-spectrometer lenses.

3.2   The effect of pre-spectrometer optics

To improve the transmission of a CHA, it is possible to demagnify the emission angle α_s to a value α_i in front of the analyser. In doing so, the aberration disc must not increase to a value where it limits the energy resolution. The chromatic aberration effectively only causes a defocus spread proportional to ΔE and can be neglected. When using a parallel detection system, the defocus has to be taken into account, but here this will be done. The spherical aberration causes an effective, virtual source size

$$d_s = \frac{1}{2} C_s \alpha_s^2 \tag{18}$$

The luminosity Λ of an electron source is defined as the source area multiplied by the emission solid angle, so effectively the source now has a luminosity

$$\Lambda = \frac{\pi^2}{16} C_s^2 \alpha_s^8 \tag{19}$$

For an optimized system, α_s must be chosen in such a way that the luminosity equals the etendue of the analysis system. From that equality, a value for α_s and thus for T follows:

$$T = \left[\frac{R^2 \, (\Delta E)^3}{C_s^2 \, E^2 E_0} \right]^{\frac{1}{4}} \tag{20}$$

We shall now compare three different pre-spectrometer systems whose geometries are schematically shown in Figure 6.

Conventional system of electrostatic lenses

The coefficient of spherical aberration C_s of a lens depends strongly on the distance between this lens and the specimen, the minimum obtainable value usually being about equal to this working distance. A normal value would be 50 mm. To get an idea of the transmission, some other characteristic values can be substituted in Equation 20. For example when $R = 100$ mm, $E_0 = 1000$ eV, $E = 100$ eV, $\Delta E = 3$ eV the value of T is 0.057. From Equation 11 the slit width under these conditions d_i must be 6 mm.

It is conceivable to use decelerating grids before the first input lens in which case the retardation effect on the etendue is different from Equation 13. It is also possible,

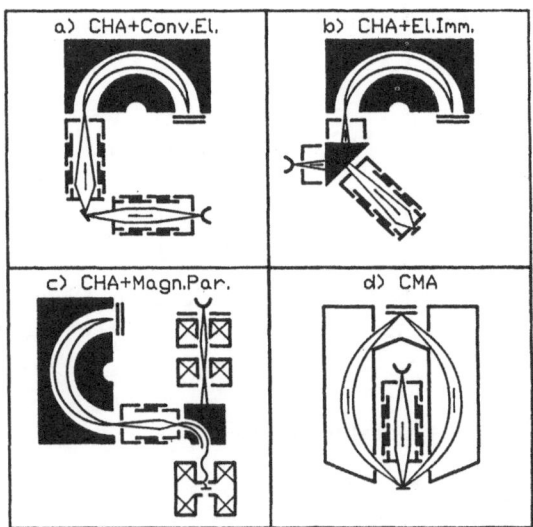

Figure 6. *The four configurations for Auger analysis discussed in this paper: a) 180° hemispherical analyser with conventional electrostatic pre-spectrometer lenses; b)* CHA *with electrostatic immersion lens; c)* CHA *with magnetic field paralleliser; d) cylindrical mirror analyser with co-axial electron gun inside the analyser.*

and applied in several commercially available analyser systems, to use lenses with input meshes to reduce effective C_s values. However, grids introduce other aberrations, so that the transmission of such lens-analyser combinations is typically of the above order or less.

Electrostatic immersion lens

An electrostatic immersion lens, as used in low energy reflection and emission microscopy (Tielieps and Bauer 1985) can also be used as a coupling lens to a spectrometer. In such a lens there is a strong accelerating field at the surface of the specimen. Such a field has a spherical aberration

$$C_s = z \frac{E_0}{U_0} \left[1 - \frac{1}{\sqrt{1 + U_0/E_0}} \right]^2 \tag{21}$$

in which z is the distance between the specimen and the plate opposite the specimen, which is at potential U_0. In order to make the accelerating field a real lens, there must be a hole in the plate and some additional focusing electrodes, which will add to the effective aberration. Equation 17 shows that the aberration of the hole is relatively small and the aberration of the focusing lens is effectively reduced by a factor which depends on the value U_0/E_0. If this factor is large and the design of the focusing lens careful, we can use the value of Equation 21. Numerical calculations of Chmelik *et al.* (1989) show that such design is quite difficult. For large values of U_0/E_0, Equation 21 simplifies to

$$C_s = z \frac{E_0}{U_0} \tag{22}$$

This leads to a transmission, T, of the immersion lens/CHA combination given by

$$T = \left[\frac{R^2}{z^2} \frac{U_0^2}{E_0^2} \frac{(\Delta E)^3}{E^2 E_0} \right]^{\frac{1}{4}} \tag{23}$$

The field strength U_0/z must be limited in practice to approximately 25 kV/cm; this value, together with $R = 100$ mm, $E_0 = 1000$ eV, $E = 100$ eV and $\Delta E = 3$ eV a $C_s = 0.4$mm give a transmission $T = 0.64$.

Magnetic immersion lens/paralleliser field

A third approach to pre-spectrometer optics is to create a magnetic field around the specimen which diverges in the direction of the spectrometer (Kruit and Read 1983, Kruit and Venables 1988, and Kruit 1991). The first advantage is that this is also the best situation for probe formation: all high resolution scanning and transmission electron microscopes have magnetic immersion objective lenses (Bleeker and Kruit 1990), a condenser objective lens with asymmetric pole pieces to facilitate the extraction of secondary and Auger electrons. Bleeker and Kruit (1991) have shown that a diverging magnetic field can be considered as a repetitive focusing lens for the Auger electrons. If the field resembles a magnetic monopole field, the multi-focusing does not increase the aberrations, and practical designs with a spherical aberration of only a few mm's are possible, yielding a high transmission as Equation 20 is applied. However, there is an additional effect: large emittance angles which would be excluded when using a simple lens, are included by the paralleliser field. They can be thought of as being focused into the spectrometer with one, two or three intermediate foci more than the small emittance angles. For the case of a large number of foci, a description in terms of adiabatic theory is possible: the electron trajectories are trapped in magnetic flux tubes. While the radii of these tubes and thus the radii of the electron trajectories increase, the relative angles between the electrons decrease. When the electrons exit from the weak magnetic field, the angles can be small enough that aberrations of subsequent pre-spectrometer lenses are small.

The effective source size is then equal to twice the cyclotron diameter of the electrons in the field B_0 around the specimen. For an acceptance angle α_s:

$$d_s = 4 \sqrt{\frac{2m}{e}} \frac{E_0^{\frac{1}{2}}}{B_0} \sin \alpha_s \tag{24}$$

With further approximations the following approximation for the transmission of the paralleliser/CHA combination can be derived:

$$T = \sqrt{\frac{e}{8m}} B_0 R \frac{(\Delta E)^{\frac{3}{2}}}{E E_0} \tag{25}$$

For a spectrometer of $R = 100$ mm, magnetic field $B_0 = 1$ T, $\Delta E = 3$ eV, $E = 100$ eV, $E_0 = 1000$ eV, the transmission T is found to be larger than 1, so all emitted electrons are accepted.

3.3 Conclusions on Auger electron collection optics

In Figures 7 and 8 the transmission is given as a function of electron energy E_0 for the four different analyser systems discussed in this chapter. Figure 7 is for high resolution spectroscopy conditions: fixed energy resolution $\Delta E = 0.5$ eV and pass energy 50 eV (this would call for a slit width behind a CHA with $R = 100$ mm of about 2 mm). Figure 8 is for high collection efficiency conditions with $\Delta E/E_0 = 0.03$ and $E = E_0$. In both figures we use the following: $C_s = 50$ mm for the conventional electrostatic lens, $U_0/z = 25$ kV/cm for the electrostatic immersion lens, $B_0 = 1$ T for the paralleliser field and $C_s = 2$ mm of the paralleliser field when used as a multifocusing lens, which is appropriate at large values of E_0.

If a parallel detector (a linear diode array or a TV camera or a position sensitive counter) can be employed, a number of channels can be detected at the same time, thus obviously reducing the time necessary to obtain an Auger spectrum. The use of such a detector will also influence the choice of operation parameters of the lens spectrometer combination. With just one detector behind an energy defining slit, retardation of the electrons always increases the transmission at a given energy resolution, because the increase in spot size is simply accompanied by an opening of the slits. With a parallel detector there is a second effect when retarding: the accepted range of energies decreases. At large values of E_0, one may not expect a dramatic improvement with a parallel detector, since the slits are usually already open wide. At low values of E_0, especially when the transmission is already 100%, parallel detectors may be very useful for fast spectrum acquisition. Reasoning along the lines set out in this paper one can deduce quantitative estimates of the improvement in acquisition time.

4 Auger spectroscopy in the STEM

The previous sections on the electron optics of probe formation and the comparison of electron spectrometer systems were distinctly biassed to lead the reader to one conclusion: if one sets out to perform Auger spectroscopy at the highest possible spatial resolution, the best choice is to work with a scanning transmission electron microscope with the specimen immersed in a strong magnetic field. The magnetic immersion lens and the high acceleration voltage (*e.g.* 100 kV) gives the smallest probe with a reasonable current, and the paralleliser concept yields a high transmission of the spectrometer system. In the following I shall give some more information on instruments of this type which have been built. Differences between these systems and typical scanning Auger systems working at lower voltage will be discussed, and results obtained in recent years will be reviewed.

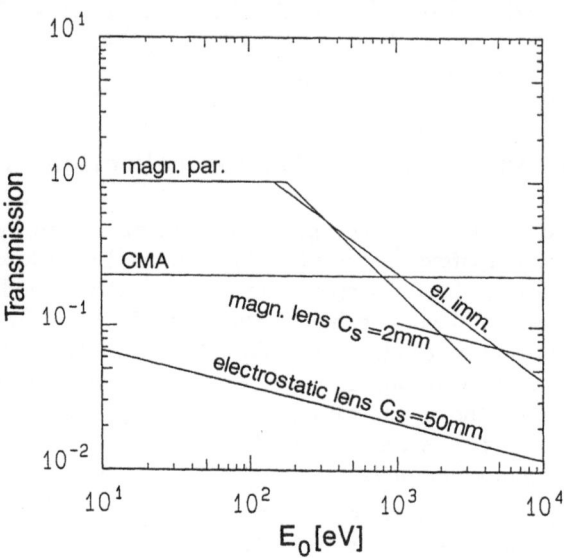

Figure 7. *Transmission as a function of electron energy E_0 for $\Delta E = 0.5$ eV at a pass-energy of 50 eV.*

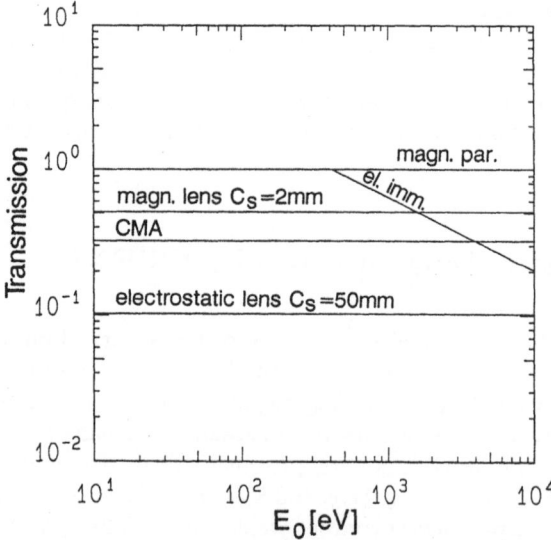

Figure 8. *Transmission as a function of electron energy E_0 for $\Delta E / E_0 = 0.03$ at a pass energy $E = E_0$.*

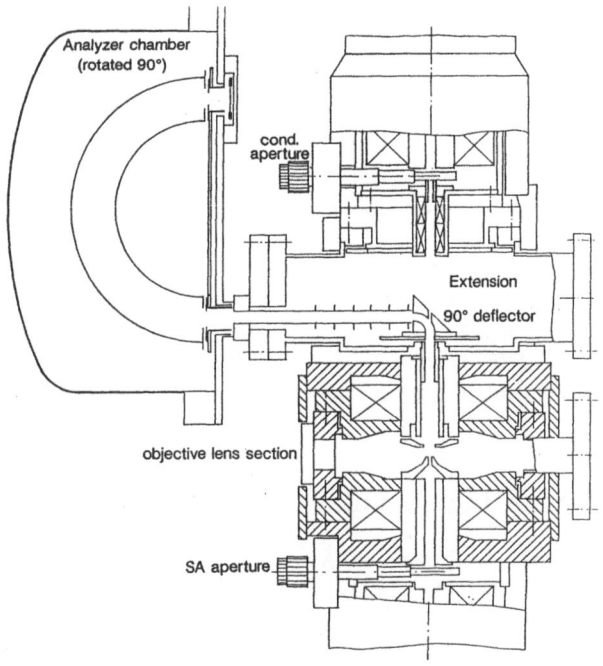

Figure 9. *Schematic view of the Auger microscope based on a Philips EM430* STEM.

4.1 Instrumentation

Instruments employing the paralleliser concept have been built at Delft University and at Arizona State University, the latter in conjunction with VG Microscopes Ltd. The first instrument in Delft (Kruit 1980, Kruit and Venables 1988) was designed to test the optical principles and perform some preliminary experiments where ultra high vacuum around the specimen is not essential. The smoothly decreasing paralleliser field is obtained by using a pole piece with a wide bore and adding an extra current carrying coil. More recent insights in the adiabatic, or rather non-adiabatic movement of electrons in magnetic fields (Bleeker and Kruit 1990, Kruit 1991) lead to a different design of the paralleliser in the second instrument in Delft, now under construction.

Figure 9 shows a schematic cross section of this microscope based on a Philips EM 430. The pole pieces of the objective lens are formed in such a way that the field of a magnetic monopole is created (Bleeker and Kruit 1991). The straight flux lines guide the Auger electrons to the entrance of a 90° deflector, at which point the paralleliser field suddenly stops. From here onwards the Auger electrons are guided by fairly conventional optical elements. The 90° deflector is not meant as a dispersive element, in fact it would be preferable if all Auger electrons were deflected through

90°, whatever their energy is: in that case the deflection field, also experienced by the primary electron beam would not change during the acquisition of a spectrum. Bleeker and Kruit (1991) designed a deflector with superimposed electrostatic and magnetic fields which is non-dispersive in first order. Electrons with an energy range between 300 and 800 eV can pass without a change of field strength. An electrostatic transfer lens system decelerates and focuses the electrons with variable magnification towards a 140 mm radius 180° concentric hemispherical analyser.

The instrument in Arizona is based on a Vacuum Generators HB 501S STEM, with paralleliser coils added both under and above the pole pieces of the objective lens, (Kruit and Venables 1988, Venables *et al.* 1987). The parallel beam (half angle approximately 6°) of Auger electrons is separated from the primary beam by a Wien (E×B) filter, where the magnetic field is now used to compensate the electrostatic deflection of the primary beam. The Wien filter, deflecting over 15° is followed by a gridless cylindrical mirror analyser which deflects the electrons through a further 75° (Hambree *et al.* 1988). A 100 mm radius concentric hemispherical analyser with electrostatic transfer optics completes the instrument.

4.2   The effect of high primary beam voltage

The beam voltage commonly used in Auger spectroscopy is 5–20 kV. The use of a 100 kV beam has several effects: different excitation cross sections for both signal and background, a different range of backscattered electrons and different parameters of the electron probe. Some effects are advantageous, others are not. At energies well above threshold, the cross-section, σ, for Auger production varies according to the Bethe loss law. For example, the parametrization of Powell (1976) gives:

$$\sigma_n \sim (E_n E_0)^{-1} \ln (c_n E_0 / E_n) \tag{26}$$

where E_n is the core excitation energy and $c_n \approx 0.6$. Thus the cross-section decreases with increasing E_0, but this is compensated by an increase in the brightness of the electron source, and by a reduction in background.

A good parameter to consider is the time t it takes to obtain a certain signal-to-noise ratio in a spectrum. The statistical noise is proportional to the square root of the background cross-section σ_b; thus:

$$t \sim \frac{\sigma_b}{\sigma_n^2 I_p}. \tag{27}$$

In thin films, we expect that the background is primarily secondary electrons with a cross-section σ_s which follows Equation 26 with $E_s < E_n$.

Thus at a given probe current, t would increase slowly with increasing E_0. However, the probe current I_p increases with E_0 at a given spot size, the exact relation depending on the instrument and the regime (see the section on probe formation). At a given spatial resolution, we expect hardly any influence of the primary beam voltage on the acquisition time of a spectrum with a certain signal-to-noise ratio.

Experimental data to support these conclusions are relatively scarce. Using a non-optimized STEM configuration, Chazelas *et al.* (1988) have shown that the signal-to-noise ratio decreases by 30% for Ag (MNN–350 eV) and 15% for Au (MNN–2100 eV) on going from 30 to 100 kV. Batchelor *et al.* (1989) has shown that the background especially for Ag is dominated by secondary electrons at high E_0 (10–30 kV), and that in this regime peak-to-background ratio is independent of E_0 and of incidence angle θ_0, even for bulk samples. The peak-to-square-root of the background ratio at constant charge collected, *i.e.* constant $(I_p t)$, decreases slowly as E_0^{-m} with $m \leq 1/2$. Although more data are needed for the STEM geometry proposed, these findings are in qualitative agreement with our simple predictions.

An advantageous effect of high primary beam voltages is that the radius of the circle within which the background electrons appear increases with E_0, and at 100 kV is measured in microns rather than nanometers; thus their influence on a scanned Auger image is limited to a slowly varying background. Monte Carlo calculations on this effect have been performed by Glezos and Nassiopoulos (1991).

4.3 Transmission samples versus bulk samples

Not every practical sample to be analysed needs to transmit the primary beam: typical sample thicknesses in transmission microscopy are 10–100 nm. However, using transmission samples has several advantages. For the Auger spectrum it gives a substantial reduction of background from backscattered electrons. Also, it enables other analysis modes: energy loss spectroscopy, high resolution x-ray fluorescence spectroscopy, diffraction and imaging microscopy. Although these modes all yield information on the full thickness of the sample, and not just the surface, the information does contribute to the interpretation of the Auger results.

A very different advantage of transmission samples is the possibility of coincidence measurements between the primary electrons responsible for the emission of Auger electrons and the Auger electrons themselves. The most direct application of that technique is expected to be a reduction of background: an Auger spectrum with only counts that resulted from primary electrons which suffered an energy loss E is equivalent with an x-ray generated Auger spectrum from x-rays of energy E. Other prospects of the coincidence method (Kruit and Pijper 1990) include the study of Auger decay from selected excitations.

4.4 Review of results

Experimental evidence that Auger spectroscopy in a STEM at 100 keV primary beam energy has spatial resolution limited by the probe size, not by the backscatter range, was given by Cazaux *et al.* (1988). In a VG HB 501A equipped with a 150° spherical analyser which can collect electrons from a sample placed outside the objective lens, they were able to demonstrate a line resolution of 8 nm from a GaAs/Si edge.

The feasibility of extracting Auger electrons from the high magnetic field of the STEM immersion objective lens was demonstrated by Kruit *et al.* (1988). A gas cell was mounted in the microscope and Auger spectra of N_2 and Ar contained peaks narrower than 1 eV at energies of a few hundred eV. The high spatial resolution obtainable with the sample in the immersion objective lens has now been demonstrated in several studies. Hembree *et al.* (1990) used Ag deposits on Si as a test system. Deposits three tenths of monolayer thick were clearly detected. Images obtained by selecting the Ag MNN lines demonstrate that 5 nm edge resolution could be achieved. A study of the nucleation of Ge deposited on Si (Hembree *et al.* 1991) shows similar spatial resolution. In an application directed at the understanding of catalytic processes (Liu *et al.* 1991), silver was deposited on α-Al$_2$O$_3$. The Auger peak images show Ag particles as small as 2.5 nm. The edge resolution is estimated at 3 nm.

Not only the Auger electrons contain information on the sample surface: spectroscopy of the low energy secondary electrons can also be useful. The secondary electron spectrum changes even at submonolayer deposits (Krishnamurthy *et al.* 1990, Bauer and Seiler 1986), resulting in high contrast images. In secondary electron images with the sample biassed, Krishnamurthy *et al.* (1990) observed images of atomic steps on bulk Si (100) samples.

Auger spectra in coincidence with a specific energy loss of the primary electrons have only been obtained in very preliminary experiments. However, a number of studies have been performed on the coincidence between secondary electrons and energy loss events (Pijper and Kruit 1991, Müllejans *et al.* 1991). So far these studies have not been used for the microanalysis of materials but rather to increase our understanding of the production mechanism of secondary electrons.

5   Conclusions

Auger spectroscopy in the STEM can attain a similar spatial resolution to EELS or x-ray analysis in STEM: it is now down to a few nm's. The surface sensitivity of Auger spectroscopy results in the analysis of a volume of only several nm^3, which is only a few hundred atoms. With a detectable mass fraction of the order of several percent, this makes Auger spectroscopy in the STEM the most effective instrument for the detection of very small amounts of material: the minimum detectable mass approaches one atom.

Aknowledgments

Much of the material presented in Section 2 is adapted from a course on electron optics in Delft written by J E Barth, E Koets and K D van der Mast. Material in Section 3 is based on a paper by P Kruit and J A Venables in *Scanning Microscopy*, Supplement 1 (1987).

References

Aksela S, Karras M and Suoninen E, 1970, *Rev Sci Instruments* **41** 351
Batchelor D R, Bishop H E and Venables J A, 1989, *Surface and Interface Analysis* **14** 709
Bauer H E and Seiler H, 1986, *Proc XIth Int Cong on Electron Microscopy* (Kyoto, Japan)
Bleeker A J and Kruit P, 1990, *Nuclear Instruments and Methods in Physics Research* **A298** 269
Bleeker A J and Kruit P, 1991, *Rev Sci Instruments* **62** No 2 350
Cazaux J, Chazelas J, Charasse M N and Hirtz J P, 1988, *Ultramicroscopy* **25** 31

Chazelas J, Friederich A and Cazaux J, 1988, *Surface and Interface Analysis* **11** 36

Chmelik J, Veneklasen L and Marxl G, 1989, *Optik* **83** 155

Gerlach R L, 1973, *J Vac Sci Tech* **10** 122

Gerlach R L, 1980, *VIIth Eur Con on El Micr* **3** 210

Glezos N M and Nassiopoulos A G, 1991, *Surface Science* **254** 309

Harting E and Read F H, 1976, *Electrostatic Lenses* Elsevier

Heddle D W O, 1971, *J Phys E: Sci Instrum* 4 589

Hembree G G, Luo F C H, Bennett P A and Venables J A, 1988, *Proc 46 of the Annual Meeting of EMSA* (San Francisco Press)

Hembree G G, Luo F C H and Venables J A, 1990, *Microbeam Analysis* (San Francisco Press) 249

Hembree G G, Drucker J S, Luo F C H, Krishnamurthy M and Venebles J A, 1991, *Appl Phys Lett* **58** (17) 1890

Krishnamurthy M, Drucker J S and Venables J A, 1990, *Proc of the XIIth Int Congr for Electron Microscopy, Seattle* 308

Kruit P, 1986, *Proc XIth Int Cong on Electron Microscopy* (Kyoto,Japan) 593

Kruit P, 1991, *Advances in Optical and Electron Microscopy* (Academic Press Limited) **12** 93

Kruit P and Read F H, 1983, *J Phys E: Sci Instrum* **16** 313

Kruit P, Bleeker A J and Pijper F J, 1988, *Inst Phys Conf Ser No 93* (Proc EUREM 88, York) **1** 249

Kruit P and Venables J A, 1988, *Ultramicroscopy* **25** 183

Kruit P and Pijper F J, 1990, *Inst Phys Conf Ser No 98* (Proc EMAG-MICRO 89, London) 271

Lenc M, 1992, *Immersion objective lenses in electron optics*, PhD thesis, Delft University of Technology

Liu J, Hembree G G, Spinnler G E and Venables J A, 1991, *Proc of the 49th Annual Meeting of the Electron Microscopy Society of America* (San Francisco Press) 690

Müllejans H, Bleloch A L, Howie A and McMullan D, 1991, *Inst Phys Conf Ser* (Proc EMAG 91) **119** 117

Pijper F J and Kruit P, 1991, *Phys Rev B* **44** no 17 9192

Powell C J, 1976, *Rev Mod Phys* **48** 33

Sar-el H Z, 1970 *Rev Sci Instruments* **41** 561

Telieps W and Bauer E, 1985, *Ultramicroscopy* **17** 57

Venables J A, Cowley J M and von Harrach H S, 1988, *1987 Inst Phys Conf Ser* **90** 85

Electron Energy-Loss Spectroscopy (EELS)

R F Egerton

University of Alberta
Canada

1 Why tangle with EELS?

Given the plethora of microanalysis techniques currently available, the student of quantitative microanalysis might well ask: why bother with electron energy-loss spectroscopy? While human choices are always subjective, one can nevertheless offer the following justifications for getting involved with EELS.

1. The energy-loss spectrum combines the information content of spectroscopy in the x-ray, ultraviolet, visible and (sometimes) the infra-red regions of the electromagnetic spectrum, and is obtainable from a single instrument.

2. Due to the refinement of electron lenses, transmission EELS has the highest spatial resolution of any microanalytical technique, excepting perhaps the field-emission atom probe. Given the right specimen and instrumentation, it is capable of identifying single atoms (of certain elements) and of analysing concentrations down to several parts per million.

3. Because of the relatively straightforward physics involved, the technique is capable of giving quantitative answers without the need for elaborate corrections or calibration specimens.

In this chapter, we outline some basics of the EELS technique and illustrate procedures for extracting quantitative information. A fuller account of fundamentals is given in Egerton (1986) and Colliex (1984); recent developments and applications are covered in Disko *et al.* (1992), Hofer (1991), Cockayne *et al.* (1991), Krivanek *et al.* (1991) and McComb *et al.* (1991).

2   Species of EELS

Electron energy-loss spectroscopy is performed by analysing the energy distribution of initially monoenergetic electrons, after they have interacted with a specimen. This interaction may take place at the surface of the specimen, for example, when an primary beam is 'reflected' (scattered) from the surface of a sample. Surface sensitivity is optimum for electrons with energy of the order of 100eV, so the apparatus does not require high voltages and is relatively small, but involves ultrahigh vacuum otherwise the information comes mainly from carbonaceous or oxide layers on the specimen's surface. At these energies, it is quite easy to use a monochromator to reduce the energy spread of the primary beam to a few meV (Ibach, 1990); provided the energy analyser has a similar resolution, energy exchange due to the excitation of vibrational modes of the surface atoms can be studied, besides the excitation of valence electrons of these atoms (Ibach, 1982). This type of energy-loss spectroscopy is usually designated high-resolution EELS (HREELS). While it is an important technique in surface science, it falls outside the range of our present discussion.

Near-surface sensitivity is also obtained at higher electron energies, provided the electrons arrive at a glancing angle to the surface, such that they penetrate only a short distance below the surface before being scattered out. Thus, it is possible to do reflection EELS (REELS) at energies as high as 100keV (Wang and Cowley, 1988).

As an alternative, interaction can take place throughout the thickness of the specimen (transmission geometry), provided the latter is sufficiently thin and the electron energy sufficient for complete penetration. For 100keV incident energy, the specimen must be less than one micrometer thick, and preferably less than 100nm. Since such specimens are self-supporting only over limited areas, we need to focus the electrons onto a small area, but in doing so we gain the advantage of analysing small volumes. The electron-optical requirements are well fulfilled by using a transmission electron microscope (TEM) for the spectroscopy, with the additional advantage that we can form a high-resolution image or a diffraction pattern of the specimen. In what follows, we will concentrate on transmission EELS carried out with the aid of a conventional (CTEM) or scanning transmission (STEM) electron microscope, where the incident energy is typically in the range 80–400keV.

3   Elastic and inelastic scattering

When fast electrons travel through a solid, they are scattered elastically, from the Coulomb field of each atomic nucleus, and inelastically, by Coulomb interaction with the atomic electrons. In elastic scattering, very little energy is exchanged in most cases, but in inelastic events there is a gain of energy by the scattering atoms and an equal loss of energy by the fast electrons. Since the energy gains are dependent on the atomic or solid-state properties of the specimen, we can obtain information about the specimen by directing the transmitted beam into an electron spectrometer. We are analysing the primary process of electron scattering, rather than secondary events as in x-ray emission and Auger spectroscopy, for example.

A typical energy-loss spectrum (Figure 1) contains a zero-loss peak, which represents

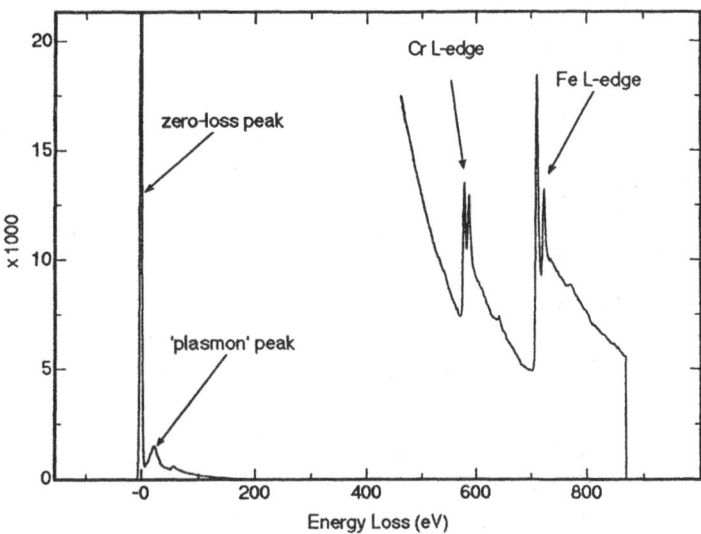

Figure 1. *Energy-loss spectrum of a Fe/Cr alloy, showing the zero- loss peak, valence-loss (plasmon) peak and L-ionisation edges for chromium and iron. At 470eV loss, the sensitivity of the detector has been increased by a factor of 3000.*

elastically scattered electrons, as well as those which can be considered as unscattered. Since the energy exchange in elastic scattering is small, the width of this peak is a measure of the energy resolution of the spectrometer system. In electron-microscope systems, this resolution is usually limited by the energy width of the electron gun: 1 to 2 eV for a thermionic (tungsten or lanthanum hexaboride) source and 0.5 to 1 eV for field-emission. Resolution of 2meV has been achieved (with 25keV incident electrons) by use of a monochromator (Geiger *et al.* 1970), allowing the study or vibrational modes, but is not yet available for the TEM.

Inelastic scattering from valence electrons (the conduction electrons, in the case of a metal) is visible as one or more peaks below 50eV. In some crystalline specimens, observed fine structure can be related to a joint (conduction-valence band) density of states. In other cases, particularly ionic compounds, sharp peaks are observed which are characteristic of excitons. In addition, all solid specimens show peaks which arise from the excitation of collective oscillations of the valence electrons. These 'plasmon' peaks are sharp (width < 4eV) for 'free-electron' metals such as aluminum and certain semiconductors (Si, GaAs etc) but rather broad (5eV–20eV) for most materials. Except in extremely thin specimens, there is a reasonable probability that transmitted electrons undergo more than one inelastic event, so the plasmon peaks are repeated at multiples of the plasmon energy, which ranges from a few eV in alkali metals to 34eV in diamond. The relative areas under the peaks reflect the probabilities P_n of n-fold inelastic scattering, which are given by Poisson statistics:

$$P_n = \frac{1}{n!} \left(\frac{t}{L}\right)^n \exp(-t/L) \tag{1}$$

Here, t represents specimen thickness and L is a mean free path for inelastic scattering, which depends on the elemental composition of the specimen but is of the order of 100nm for 100keV electrons. At higher energy loss, the spectral intensity falls rapidly, typically as a third or fourth power of energy loss. But superposed on this smooth background are ionisation edges which correspond to the excitation of inner-shell (core) electrons. These edges take the form of an abrupt rise in intensity followed by a more gradual return to the background. They are directly analogous to the edges observed in x-ray absorption spectroscopy. The edge threshold corresponds to the binding energy of a particular atomic shell, which depends on the type of shell (K,L etc.) and the atomic number of the excited atom. The edges can therefore be used for elemental identification, like the emission peaks in EDX spectroscopy.

The inner-shell ionisation edges have characteristic shapes which can be calculated on the basis of atomic physics. K-edges are rapidly rising (sawtooth-shaped) whereas L- and M-edges are generally rounded (delayed maximum) but, in the case of transition metals and rare earths, they have sharp peaks at the ionisation threshold (so-called white lines, first seen in x-ray absorption spectra); see Figure 1. Superposed on these basic shapes is a fine structure, consisting of pronounced oscillations of intensity in the near-edge region (ELNES) with more minor oscillations extending well beyond the edge (EXELFS).

Besides being characterised in terms of an energy dependence, the inelastic scattering has an angular distribution which reflects the dynamics of the scattering process. Provided the scattering angle θ is not too large, the inelastic intensity per unit solid angle has a Lorentzian form:

$$\frac{dI}{d\Omega} \propto \frac{1}{\theta^2 + \theta_E^2} \tag{2}$$

where θ_E is a characteristic angle, equal to the half-width at half maximum of the angular distribution, and given approximately by $\theta_E = E/E_0$ where E is the energy loss and E_0 the incident energy. For valence excitation (e.g. plasmon losses) by 100keV incident electrons, θ_E is of the order of 0.1mrad, so the scattering is sharply peaked about the unscattered beam. For the ionisation edges, where E is typically in the range 100–2000eV, the angular width may be a few mrad. Since the angles of elastic scattering in a crystalline specimen (twice the Bragg angle) are typically at least 15mrad, one can imagine each Bragg beam surrounded by a halo of inelastic scattering, as shown in Figure 2. This difference in angular width allows us to analyse the elastic and inelastic scattering separately, to a first approximation. In the CTEM, we can control the range of scattering angles entering the imaging lenses by means of an objective aperture. Depending on the aperture diameter, the maximum scattering angle beta may be as small as 2mrad or (with no aperture inserted) as high as 150mrad (limited by the bore of the objective lens).

The electron spectrometer is usually mounted below the CTEM camera chamber, in which case a spectrometer entrance aperture (SEA) can be used to further control which electrons enter the spectrometer. If the microscope forms a diffraction pattern on

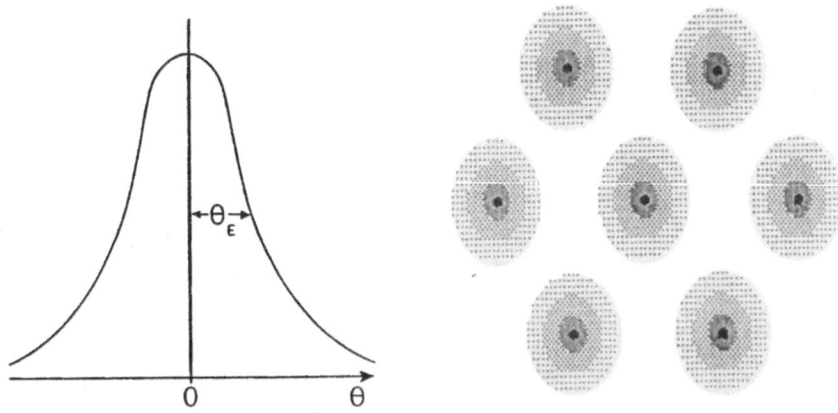

Figure 2. *Angular distribution of inelastic scattering, shown as a plot of intensity per unit solid angle and as intensity on the diffraction plane.*

the viewing screen, the angular range is determined by either the SEA or the objective aperture, whichever is smaller in angle.

If the microscope forms an image (magnification $= M$) of the specimen on the viewing screen, the SEA can in theory be used to select a region of the image for analysis, functioning similar to the selected-area aperture in traditional diffraction work. But the same concerns that apply to selected-area diffraction apply here also. Spherical aberration of the objective lens (coefficient C_s) causes the image to be blurred by a factor $MC_s\beta^3$, effectively limiting the spatial resolution to $r_s = C_s\beta^3$ in the specimen plane. By using a 10mrad objective aperture to limit b, we ensure that the resolution remains better than 2nm.

Of more concern is *chromatic* aberration of the objective (coefficient C_c), which degrades the spatial resolution by an amount $r_c = C_c\beta E/E_0$ which can exceed 10nm for energy losses of several hundred eV (Yang and Egerton, 1992). Moreover, if the image blurring $R_c = Mr_c$ exceeds the SEA radius R, the inelastic signal entering the spectrometer will be reduced. Since R_c is a function of energy loss E, this chromatic aberration will affect the elemental ratios deduced from measurement on different ionisation edges. But in a modern CTEM, the incident energy E_0 can be changed in small (*e.g.* 50eV) steps; if E_0 is raised by an amount approximately equal to energy loss being analysed *without* changing the excitation of imaging lenses, the electrons involved will suffer little chromatic aberration, so the collection efficiency of the spectrometer and the spatial resolution will not be degraded. Alternatively, the microscope can be operated in diffraction mode and the region of analysis defined by the incident beam. Provided the diameter of the latter is below $1\mu m$, the effect of chromatic aberration is negligible

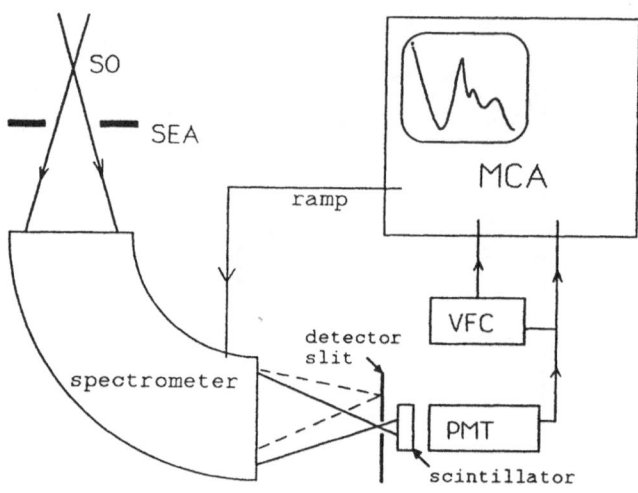

Figure 3. *Serial-recording spectrometer system. Dashed lines show the exit trajectories for electrons of higher energy loss.*

(Yang and Egerton 1992, Titchmarsh and Malis 1989).

4   Electron energy-loss spectrometers

The illumination system of an electron microscope provides the capability for irradiating a variable area of the specimen, from many micrometers to (in some cases) below 1nm diameter. The imaging lenses are not essential for EELS but are useful in providing a small-diameter spot which forms the object point (*SO* in Figure 3) for the spectrometer. For a spectrometer mounted beneath a CTEM, this point is the projector-lens crossover. For a spectrometer attached to a STEM, the virtual image of the (stationary or mobile) electron probe on the specimen forms the object point.

Electron spectrometers take many different forms but the only kind which has fractional resolution (1 in 100,000) sufficient for EELS at high incident energy is the magnetic-prism design. We will concentrate on the single-sector magnet, as manufactured by the Gatan company. A uniform magnetic field bends the electron trajectories through 90 degrees, as shown in Figure 3. Since the magnetic force (Bev) is equal to the centripetal force mr^2/R, the bending radius R is smaller for electrons of lower kinetic energy and the exit beam is dispersed, with the less energetic electrons deflected through a larger angle. The field also has a focusing action: electrons of the same energy are returned to a single point after the spectrometer, so the different energy losses are all brought to a focus at the spectrometer image plane. As with all magnetic lenses, the focusing is imperfect. Most of the second-order aberration can be corrected by curving the entrance and exit boundaries of the magnet. Quadrupole coils at the entrance of the spectrometer are used to make fine adjustments to the focusing, to obtain the best energy resolution (sharpest zero-loss peak). Two further controls adjust the magnitude and phase of a mains-frequency signal fed into dipole deflector coils, to compensate for stray a.c. fields (also known as wavering-current fields).

The spectrometer is provided with an entrance aperture (SEA) of variable diameter (1mm–5mm). Although located several centimetres below the CTEM screen, this aperture selects intensity from a well-defined area of the image (or diffraction pattern) because of the very large depth of field of the final image. Increasing the aperture size increases the number of electrons entering the spectrometer, and therefore the output signal, but may degrade the energy resolution because of spectrometer aberrations.

5 EELS serial recording

To be useful, the spectrometer should give a digital output which can be fed into a multichannel analyser (MCA) or computer of some sort, for display and subsequent processing. The simplest scheme for doing this is serial recording. A narrow slit is placed at the spectrometer image and the magnet current is varied slightly to sweep the spectrum (vertically) across the slit. The ramp signal is supplied by the MCA, after pressing the 'acquire' button. The speed of the ramp (or dwell time per MCA channel) can be varied; a few ms per channel is sufficient for recording the intense peaks at low energy loss, whereas 50ms/channel or more gives lower noise when recording higher energy losses. A sensitive electron detector, usually a scintillator followed by a photomultiplier tube (PMT), gives a pulse for each electron which passes through the slit or, at high intensities, a current output which is digitised via a voltage-to-frequency converter (VFC); see Figure 3.

A mechanical control varies the width of the detector slit. A wide slit gives a larger signal which is relatively free of shot noise, but the energy resolution is poor. To increase resolution, the slit width is reduced while observing the zero-loss peak; the optimum setting is reached when the *height* of the zero-loss peak starts to fall. Focusing and aberration correction (to minimise the width of the zero-loss peak) is made more convenient by scanning the electron beam rapidly across the slit (by applying a ramp signal to the entrance dipole coils) and observing the output of the PMT directly on an oscilloscope screen. The large inductance of the main spectrometer scan coils limit the usual mode of scanning to no more than 1/sec.

One significant advantage of serial-recording detectors is that, since only one detection channel is involved, the sensitivity (gain) of the system is the same (neglecting transient effects in the PMT) across the entire spectrum. The gain can be varied by changing the PMT supply voltage or switching to pulse counting during the acquisition, producing a 'gain change' which is usually introduced between the low-loss and core-loss regions of the spectrum. If necessary, the sensitivity can also be changed by varying the dwell time per channel, and the sensitivity factor will be simply the ratio of the two dwell times.

The linearity of response is good in the analogue mode of recording, provided the anode current drawn from the PMT is small compared to the dynode current (Craven and Buggy 1984). Overloading causes a transient increase in PMT dark current, giving a 'tail' on intense peaks such as the zero-loss peak. This tail is less troublesome if the spectrum is scanned downwards in energy loss (Disko 1986, Joy and Mayer 1980) or if the spectrum is recorded (at low incident intensity) in pulse-counting mode (Craven and Buggy 1984).

In pulse-counting mode, linearity is within 15% for observed count rates f_{ob} up to 1MHz. At higher count rate, the true count rate f_t is given by: $f_t = f_{ob}/(1 + f_{ob}T)$ where T is the deadtime of the detector, typically in the range 150-200ns (Cheng et al. 1987, Craven and Buggy 1984).

In the pulse-counting mode, the scintillator/PMT combination provides a very-low-noise detector which, despite the low signal intensity, is capable of recording ionisation edges at least up to 2000eV loss. At higher energies, and particularly for very thin specimens, the signal tends to get lost in a background which comes mainly from zero-loss electrons which arrive at the detector after being backscattered from the slits (Joy and Maher 1980, Egerton 1986). Procedures have been devised to correct for this background (Craven and Buggy 1984, Egerton 1986).

6 EELS parallel recording

is that it makes inefficient use of the electrons passing through the spectrometer, since a large fraction (typically 99% or more) of them are absorbed by the energy-selecting slit. If this slit is removed and replaced by a position-sensitive detector, all energy losses are recorded simultaneously; the signal is stronger and more free of shot noise.

Convenient forms of position-sensitive detector include the photodiode and charge-coupled-diode (CCD) arrays . Although these arrays respond directly to electrons, they are designed to respond to photons; there is less risk of damage by interposing a thin scintillator and transferring the light-optical image by a lens or fibre optics onto the array. In the Gatan PEELS spectrometer, the thin yttrium aluminum garnet (YAG) scintillator is directly mounted on a fibre-optic plate; see Figure 4. A set of quadrupole lenses (Q1- Q4) precedes the detector, in order to magnify the dispersion in the vertical direction and achieve an energy resolution of 1 eV. Each photodiode behaves like a leaky capacitor, and before recording a spectrum, all the diodes are charged to the same potential. During spectrum acquisition, electron-hole pairs created by light emission in the scintillator cause the diodes to discharge by an amount proportional to the local electron intensity. At the end of the acquisition period, the charge remaining on each diode is sampled in turn and the spectrum is read into the MCA as a series of pulses, as in the case of serial recording.

The diodes also discharge by thermal generation of electron-hole pairs, giving rise to a background component in the energy-loss spectrum. This background is not constant, since the leakage current is slightly different for each diode in the array, giving rise to an additional source of 'noise' in the data. These effects can be corrected for by taking a readout with no electrons incident on the scintillator, then subtracting this 'bias spectrum' from the EELS data inside the computer.

In addition, the individual diodes in the array (and different parts of the scintillator)

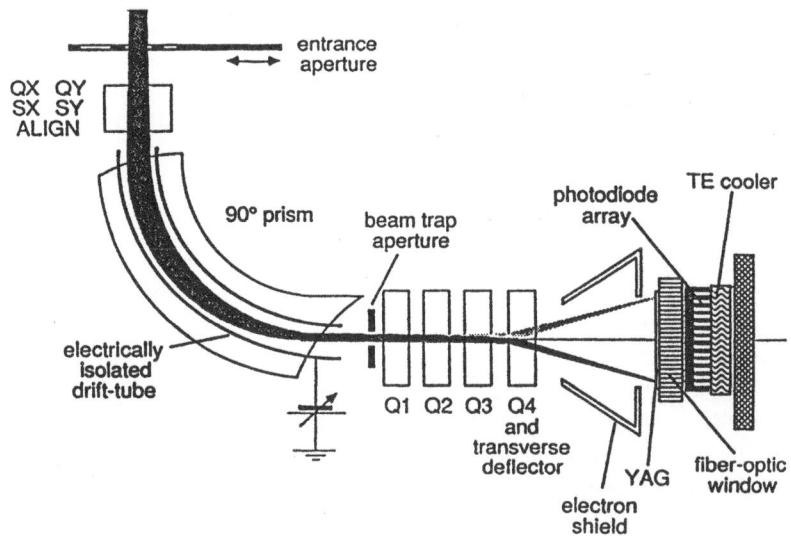

Figure 4. *Gatan model 666 parallel-recording spectrometer*

have slightly different sensitivities to irradiation. To remove this modulating effect of the detector, a 'gain spectrum' is recorded by uniformly illuminating the array; the computer then divides each background-subtracted energy-loss spectrum by the gain spectrum.

Currently-available parallel-recording detectors suffer from memory effects. Incomplete readout in the photodiode array causes a faint replica of previous spectra to appear in subsequent ones, but this can be eliminated by interposing a sufficient number of readouts with the electron beam excluded, for example by lowering the viewing screen in the CTEM. Being an electrical insulator, the YAG scintillator accumulates internal charge in strongly irradiated regions, which has the effect of locally increasing its conversion efficiency. This artifact can be reduced by 'annealing' the scintillator with a broad (undispersed) beam for several minutes.

The intensity response of the Gatan PEELS detector is found to be somewhat nonlinear, the sensitivity being increased at higher electron intensity. Although this increase is only around 10% per decade of intensity, the effect is cumulative and may be substantial if several segments of a spectrum are 'spliced' together to overcome the limited dynamic range of the detector.

Stray scattering of the zero-loss beam is a severe problem in parallel-recording detectors, due to the absence of a defining slit which discriminates against electrons travelling far from the optic axis. The effect has been reduced in the Gatan spectrometer by means of a beam-trap aperture and an automatic deflection system which deflects the zero-loss beam when it falls outside the array.

Despite their limitations, parallel-recording detectors are preferable in many situations involving the recording of ionisation edges, particularly those at high energy loss where shot noise must be minimised to achieve an acceptable signal. Their advantage in terms of collection efficiency is greatest in the case of analysis of very small volumes of material and low concentrations. The reduced recording time needed to get adequate statistics results in less drift of, and beam damage to, the specimen.

7   Elemental analysis by EELS

As discussed earlier, analysis for specific elements is most easily done from the ionisation edges at higher energy loss. Provided the energy-loss scale has been calibrated to within a few percent (by the manufacturer or by reference to a standard specimen), elements present in the specimen are identified by comparing the energies of observed edges with tabulated values (Egerton 1986, Ahn and Krivanek 1983).

The height of an edge gives only a rough indication of the amount of the element, on account of the near-edge fine structure which depends on the spectrometer resolution and on the environment (chemical and crystallographic) of the atoms being detected. But by integrating the core-loss intensity over an energy region (width Δ) which exceeds that of the fine structure (50eV is sufficient), we obtain a quantity $I_c(\beta, \Delta)$ which relates to elemental concentration.

I_c is written as a function of collection angle β, as well as integration width Δ, because we detect only core-loss electrons whose scattering angles lie within the angular range defined by the SEA or objective aperture. For quantitative analysis, the angle-limiting apertures need to be calibrated in terms of scattering angle. In the case of CTEM objective apertures, this calibration is done by comparing the 'shadow' of the aperture with the diameters of diffraction rings from a crystalline test specimen. To ensure that $I_c(\beta, \Delta)$ represents only the inner-shell component of scattering, we must make allowance for the background underlying an ionisation edge (dashed line in Figure 5). This background consists of several components: (1) the tails of any preceding ionisation edges; (2) the tail of the valence-electron intensity (low-loss region); (3) plural inelastic scattering involving a large energy loss (the previous two components) and a limited number of low-loss (valence-electron excitation) events; (4) instrumental background arising from stray scattering in the spectrometer and any detector-circuit bias (if not already subtracted).

From atomic and solid-state physics, components (1) and (2) are expected to have approximately a power-law dependence on energy loss, and this relation is observed experimentally if the other two components are negligible. So we model the background intensity as:

$$J_b = AE^{-r} \tag{3}$$

The adjustable parameters A and r cannot be found directly within the core-loss integration region but can be deduced from least-squares fitting of the intensity over a region of width Γ directly preceding the edge, on the assumption that these parameters remain approximately constant. Because of this assumption, and noise present in the data, the range Δ of extrapolation should not be too large. Values of the order of 50eV are suitable for lower-energy edges but bigger values (one or two hundred eV) may give

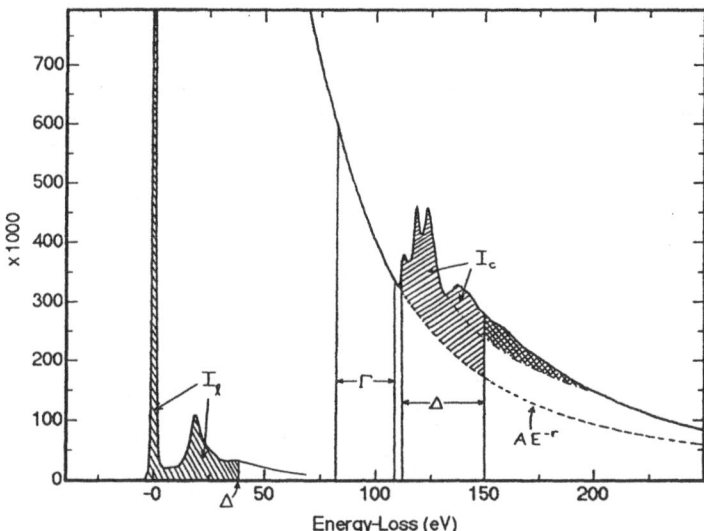

Figure 5. *Elemental analysis by integration of core-loss intensity.*

better accuracy for higher-edges (where the data is comparatively noisy) provided other ionisation edges do not occur within this region.

As for any transmission measurement, the core-loss signal will depend on the number N of atoms per unit area of the specimen. The concept of a cross-section $\sigma_c(\beta, \Delta)$ for scattering within our defined range or angle and energy loss enables us to write:

$$I_c(\beta, \Delta) = NI\sigma_c(\beta, \Delta) \qquad (4)$$

where I is the incident-beam current, in the same units as I_c.

For routine microanalysis, Equation 4 is inconvenient because it requires us to measure I (via a Faraday cage) and the sensitivity of the system (MCA counts per incident electron, which will vary with beam energy, for example). More importantly, Equation 4 is correct only for vanishingly thin specimens; for a typical TEM specimen, elastic scattering outside the angle-limiting aperture will cause I_c to be considerably reduced, while plural inelastic scattering will cause some of the core-loss intensity to appear outside the integration window (cross-hatched area in Figure 5). The situation is improved if we can make the following assumptions: (1) that the probability of elastic scattering (outside the collection aperture) is the same for core-loss electrons and those in the low-loss region; (2) that the probability of inelastic scattering out of the core-loss integration window is the same as the probability of low-loss scattering beyond an energy loss Δ. We can then replace I in Equation 4 by the intensity $I_l(\beta, \Delta)$ within the low-loss region, giving:

$$I_c(\beta, \Delta) = NI_l(\beta, \Delta)\sigma_c(\beta, \Delta) \qquad (5)$$

This approximation breaks down if the specimen is too thick (when elastic and plural inelastic scattering are predominant) but it remains accurate up to a much larger specimen thickness ($t \approx L$) than Equation 4.

Assumption (2) can be avoided by removing plural scattering from the spectrum by a process referred to as deconvolution, but the results are similar provided $\Delta > 50\text{eV}$ (Egerton 1986). In the case of a very small probe of large convergence angle, Equation 5 requires correction by a factor whose value depends on the edge energy as well as on the convergence and collection angles. This factor is easily found by running a short FORTRAN program (Egerton 1986).

As an example of the use of Equation 5, Leapman (1992) has measured the number of Fe atoms in a single ferritin molecule using a 12nm-diameter probe of 100keV electrons. Absolute quantification is also useful for measuring thicknesses of oxide or contamination layers.

In most situations, however, we require only the ratios of elements (x and y), in which case Equation 5 can be written:

$$\frac{N^x}{N^y} = \frac{I_c^x(\beta, \Delta)}{I_c^y(\beta, \Delta)} \frac{\sigma_c^y(\beta, \Delta)}{\sigma_c^x(\beta, \Delta)} \tag{6}$$

In this case, the low-loss region need not be measured, unless it is required for deconvolution. Equation 6 is directly comparable to the measurement of elemental ratios in thin specimens by EDX analysis, except that in EELS we deal with atomic rather than weight ratios. The cross-section ratios are equivalent to k-factors (Titchmarsh 1992), and it is quite feasible to determine these factors experimentally by EELS of binary standards of known composition (Malis and Titchmarsh 1985, Hofer 1991).

However, we can also calculate the individual cross-sections directly, on the basis of atomic physics. From a particular model of the atomic wavefunctions, one can evaluate the generalised oscillator strength (GOS) for a particular shell and element, as a function of momentum and energy transfer to the atom. The GOS is then integrated over the required range of energy loss and scattering angle, for a particular incident-electron energy. Commercial EELS software packages use hydrogenic theory, since the GOS is known in analytic form and can be evaluated rapidly on-line by the data-analysis computer. More accurate results, particularly for L- and M-shells are obtainable from Hartree-Slater calculations, but the latter require care and are time-consuming.

A third option is to tabulate the oscillator strengths for different elements, on the basis of experimental data and/or calculations. A great simplification is possible if we can equate the GOS to the dipole oscillator strength, as used in optical theory. Dipole conditions apply provided we measure energy losses not too far from an ionisation edge and over a restricted angular range, such that the angular distribution of core-loss scattering is Lorentzian, as in Equation 2. These conditions apply well to EELS analysis, provided we use an angle-restricting collection aperture. The energy-loss cross-section can then be evaluated from:

$$\sigma(\beta, \Delta) = A \left(\frac{\gamma}{\gamma + 1/\gamma} \right) \left(\frac{1}{E_0 \langle E \rangle} \right) \left(\ln(1 + \beta^2/\theta_E^2) + G \right) f(\Delta) \tag{7}$$

where $A = 6.50 \times 10^{-14} \text{cm}^2 \text{ eV}^{-2}$, $\langle E \rangle = [E_c(E_c + \Delta)]^{1/2}$, $\theta_e = (\langle E \rangle / E_0)(\gamma/1 + \gamma)$, $\gamma = 1 + E_0/511 \text{ keV}$ and G is a further relativistic correction ($G \approx 0$ for $E_0 < 200 \text{ keV}$);

$f(\Delta)$ is the integrated optical oscillator strength, which is tabulated for useful values of the integration window Δ and can be obtained for intermediate values by interpolation (Egerton 1992).

8 Fitting to spectral standards

The above method of elemental quantification is difficult to apply to noisy data, or if the edges are very weak or occur in close proximity to each other. If statistical noise in the spectrum is dominant, the uncertainty dI_c in the measurement of the core-loss integral I_c can be written (Egerton 1986) as:

$$dI_c = \sqrt{\mathrm{var}(I_c)} = \sqrt{(I_c + hI_b)} \qquad (8)$$

Here intensities are measured in terms of the number of energy-loss electrons; h is a number which depends on the width of the fitting and integration regions, and on the energy-dependencies of the signal and background. If the background were negligible, dI_c would be simply the Poisson shot noise in the signal I_c, but for small elemental concentrations the second term within parentheses in Equation 8 is dominant. For comparable fitting and integration widths, h is typically in the range 5 to 10, reflecting the fact that the pre-edge background noise becomes 'amplified' in the process of extrapolation.

The situation can be improved if the fitting procedure is extended across the edge; in other words, we use multiple least squares (MLS) techniques to fit the observed intensity $J(E)$ to an expression of the form:

$$F(E) = AE^{-r} + B_1 S_1(E) + \quad \dots \qquad (9)$$

where $S_1(E)$ is a standard edge profile stored in the MCA. This procedure also has the advantage that further standards, $S_2(E)$ *etc.* (representing other ionisation edges) can be added to the equation and values found for each fitting coefficient B. It does not matter if the edges are closely spaced, since we do not attempt to model the background for higher-energy edges as a power law.

If the specimen is very thin, such that plural scattering can be ignored, the standard spectra can be calculated differential cross-sections (Steele *et al.* 1985) and the coefficients B are simply the product (NI_0) of areal density of the appropriate element and the zero-loss intensity. Alternatively, each $S(E)$ can be measured from a test specimen containing the appropriate element (Leapman and Swyt 1988), in which case the atomic ratio of two elements can be obtained from the fitting coefficients by:

$$\frac{N_1}{N_2} = \frac{B_1}{B_2} \frac{I_1(\beta, \Delta)}{I_2(\beta, \Delta)} \frac{\sigma_2(\beta, \Delta)}{\sigma_1(\beta, \Delta)} \qquad (10)$$

where I_1 and I_2 are integrals of the standard spectra (over some convenient energy window Δ); σ_1 and σ_2 are the corresponding core-loss cross-sections.

Unfortunately, the region just above the edge (where the core-loss intensity is highest) contains prominent ELNES fine structure which depends on chemical environment.

Either the reference edges should be obtained from a chemically-similar standard or this region should be excluded from the MLS fitting. Also, if the reference spectra are obtained from standards whose thickness is different from that of the specimen being analysed, it may be necessary to add terms to Equation 9 to allow for plural scattering. Despite these complications, the procedure has been successfully applied to determine small concentrations of elements in biological matrices (Leapman and Swyt 1988, Shuman and Somlyo 1987).

To remove the need for a background term in Equation 9, it is possible to fit first or second differentials of the data, as in EDX analysis. In the case of a parallel-recording spectrometer, these differentials can be obtained directly from the detector, by recording the spectrum two or three times with a small (few eV) energy shift applied between acquisitions, then subtracting the spectra to form a first or second difference. This procedure has the advantage (over numerical differentiation in the computer) that, for a small edge superposed on a large background, gain variations between individual channels of the detector array largely cancel. Using this technique, concentrations below 25ppm of calcium and phosphorus have been measured (Shuman and Somlyo 1987).

9   Sensitivity of EELS elemental analysis

The detection limits for microanalysis can be specified in two ways: as a minimum number of detectable atoms MDN (or minimum detectable mass MDM) and as the minimum detectable atomic fraction f (or minimum mass fraction MMF). The best indication of these quantities is obtained experimentally, by pushing EELS to its limits with the best available technique and instrumentation. But it is also useful to calculate expected detection limits, to provide a guide to which experimental parameters are important and what limits arise from fundamental physics rather than instrumentation. A number of such calculations have been done (Isaacson and Johnson 1975, Colliex 1984, Leapman 1992, Egerton 1986). From Equation 5, the core-loss signal is:

$$I_c = I_l N_x \sigma_c \left(\beta, \Delta\right) \tag{11}$$

where N_x is the number of atoms (per unit area of specimen) to be detected. By analogy, the background I_b (due mainly to other 'matrix' atoms, N_t per unit area) is:

$$I_b = I_l N_t \sigma_b \left(\beta, \Delta\right) \tag{12}$$

where σ_b is a cross-section for inelastic scattering into the background. As a result of statistical noise in the signal and background, the standard deviation of our measurement of I_c is, from Equation 8:

$$dI_c = \left(\frac{hI_b}{\eta}\right)^{1/2} \tag{13}$$

where we have assumed $I_b \gg I_c$, for a small detectable mass and η is the detective quantum efficiency (DQE) of the spectrometer system. We require that the signal I_c should be three times the standard deviation (Rose visibility criterion) to ensure that the signal I_c is genuine. Using a mean free path L_e to allow for elastic scattering (outside the collection aperture) by matrix atoms:

$$I_l = (IT) \exp(-t/L_e) \tag{14}$$

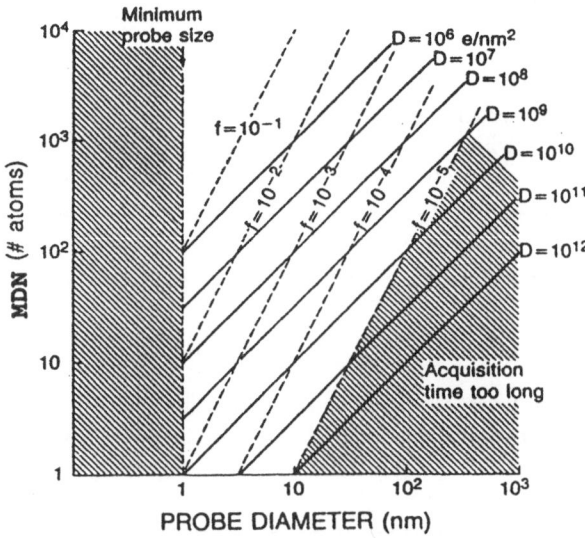

Figure 6. *Number of iron atoms detectable in a 10nm carbon matrix by parallel-recording* EELS, *as calculated by Leapman (1992). Dashed lines give the corresponding atom fractions of iron; solid lines show the electron dose to the specimen, in electrons/nm². Shaded regions represent conditions which have been excluded on the basis of electron optics or excessive recording time.*

where IT is the number of electrons (D per unit area) received by the specimen during the acquisition period T. Combining these equations, we arrive at expressions for the minimum detectable atomic fraction f and the corresponding number of atoms MDN in a probe of diameter d:

$$f = \frac{3}{\sigma_c} \left(\frac{h\sigma_b}{\eta IT N_t} \right)^{1/2} \exp(t/2L_e) \tag{15}$$

$$MDN = \frac{1}{4}\pi d^2 f N_t \tag{16}$$

These quantities are given in Figure 6 for the case of detection of Fe atoms on (or in) a 10nm-thick carbon film, assuming $h = 8$, $\eta = 1$ (parallel detection), E_0=100keV and hydrogenic cross-sections (Leapman 1992). It is seen that detection of single atoms should be possible, but the radiation dose is high ($D > 10^{10}$e/nm² $= 1.65 \times 10^5$C/cm²) and would severely damage some types of specimen.

10   Comparison with EDX spectroscopy

For elemental analysis, the technique most directly comparable to EELS is energy-dispersive x-ray spectroscopy carried out using a windowless, ultra-thin-window (UTW) or low-Z atmospheric-window detector. Both techniques take advantage of our ability to focus electrons into a very small probe, allowing analysis with a spatial resolution of the order of 10nm in many cases.

The two techniques can even be performed simultaneously on the same TEM specimen. In CTEM, this aperture should be removed to avoid errors in EDX quantification arising from electrons which are backscattered from this aperture and which generate characteristic x-rays from regions of the sample far outside the probe. Since EELS is best performed with an angle-selecting aperture, the CTEM should be operated in diffraction mode, with β limited by the spectrometer entrance aperture.

The EDX signal (number of x-ray photons) is given by:

$$I_x = N \left(\frac{IT}{e}\right) \sigma(\pi, E_0)\, \omega \eta_x$$

where $\sigma(\pi, E_0)$ is the total ionisation cross-section for a particular atomic shell and ω is the corresponding x-ray fluorescence yield; η_x is the collection efficiency of the x-ray detector. The ratio of the two signals can therefore be written as:

$$\frac{I_c}{I_x} = \frac{\sigma(\beta, \Delta)}{\omega \sigma(\pi, E_0)} \frac{\eta}{\eta_x} \exp\left(-t/L_e\right) \tag{17}$$

The exponential term, which is typically 0.3, represents loss of the EELS signal as a result of elastic scattering outside the collection aperture. Also, the electron spectrometer collects only part of the angular range of core-loss scattering and we analyse only a range Δ of the ionisation edge, so the cross-section ratio in Equation 17 is appreciably less than unity; a typical value is 0.1. However, the x-ray fluorescence yield ω, while close to unity for K-lines of heavy elements, is below 0.05 for photon energies below 2keV and falls to 0.002 for carbon K-radiation. And whereas the DQE of a parallel-recording electron detector can exceed 0.5, measured values of collection efficiency for x-rays are typically no more than 0.003 (due largely to the limited solid angle of collection) and will be lower for light-element radiation because of absorption in the detector window. Therefore, the EELS signal should exceed that available from EDX, by a modest factor for heavy elements but by a factor of several thousand for light elements such as carbon.

But as discussed earlier, the sensitivity of a spectroscopic technique depends on signal/noise considerations rather than just on the signal. Here, EDX has the advantage since the background to the x-ray peaks is generally lower than in EELS. To measure the relative sensitivities of the two techniques, Leapman and Hunt (1991) used test specimens with small concentrations of F, Na, P, Cl, Ca and Fe deposited (from aqueous solution) on 10nm carbon films. Figure 7 shows the relative (EELS/EDX) values of signal/noise ratio, the noise component being obtained in each case directly from the MLS fitting procedure used for elemental quantification. EELS is seen to be advantageous in the case of very light elements, but also in the case of heavier elements if L-edges are used for the quantification. The corresponding x-ray L-emission peaks are in many cases unusable because of the relatively poor energy resolution of an EDX detector and consequent overlap problems.

The two techniques can also be compared on the basis of the accuracy of quantitative analysis. Ten percent accuracy of atomic ratios may be achievable from thin-film EDX spectroscopy, provided the system is calibrated (in terms of k-factor) for each element

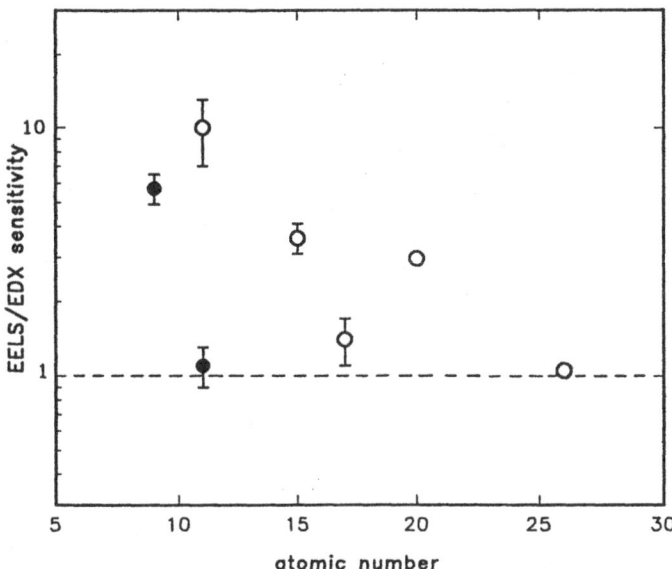

Figure 7. *Relative sensitivity of* EELS, *using parallel recording and first- or second-difference* MLS *processing of K-edges (solid circles) or L-edges (open circles), compared to* EDX *elemental analysis (using* UTW *detector and second-difference* MLS *processing of K-emission peaks). The increased sensitivity for Ca and Fe arises from the presence of white lines at the L-edge threshold, as seen in Figure 1.*

being analysed. However, this level of accuracy will be harder to achieve for light elements, where absorption of x-rays within the specimen (and at the detector) is not easy to quantify accurately. It is more difficult to specify the accuracy of EELS; the core-loss cross-sections are known in some cases to within 10%, but plural scattering may introduce systematic errors in thicker specimens.

In terms of ease of use, EDX has a clear advantage. Software for elemental quantification is well established and, once set up, the spectrometer system requires little maintenance or alignment. EELS requires more care and knowledge on the part of the operator, but in return the technique yields information beyond that of elemental composition, as we now illustrate.

11   Measurement of specimen thickness

A useful feature of energy-loss spectroscopy is that it provides a convenient way of estimating the local thickness of a TEM specimen. This information is needed to convert the areal densities provided by EELS or EDX elemental analysis into true concentrations, and for a variety of other procedures employed in electron microscopy.

The thickness information comes from measurement of the probability of inelastic scattering. The probability P_0 of *no* inelastic scattering is represented by the area I_0

under the zero- loss peak, relative to the total area I_t beneath the complete spectrum. This probability is also given by Poisson statistics, in terms of the total mean free path for inelastic scattering (all energy-loss processes) L. Setting $n = 0$ in Equation 1, we get $P_0 = I_0/I_t = \exp(-t/L)$, leading to:

$$t/L = \ln(I_t/I_0) \tag{18}$$

Although the value of L depends on the sample material, the incident-electron energy and the collection angle, it is of the order of 100nm for 100keV electrons, so a rough estimate of thickness is available immediately.

For thin specimens ($t/L < 2$), most of the intensity lies below 200eV energy loss and a single readout of the spectrometer (without gain change) is sufficient. For thicker specimens, the spectrum may have to be recorded over a wider range.

To get a true measurement of thickness, we need to know L for the material being examined. Ideally, the value of L is obtained from a calibration sample, whose thickness is determined by other means. However, measurements on elements and simple inorganic compounds (with constituents of comparable atomic number) have shown that L can be parameterised in terms of mean atomic number:

$$L = \frac{106\, F(E_0/E_m)}{\ln(2\beta E_0/E_m)} \tag{19}$$

$$F = \frac{(1 + E_0/1022)}{(1 + E_0/511)^2} \tag{20}$$

$$E_m = 7.6 Z^{0.36} \tag{21}$$

where L is in nm, E_0 in keV and β in mrad. Because of the logarithmic dependence on β, the latter need not be known to high accuracy. The accuracy of Equation 21 is believed to be 20% (Malis et al. 1988). Values of E_m have been tabulated for some common materials, allowing greater accuracy in these cases (Egerton 1992).

For better accuracy, one can also make use of a sum rule which relates the energy-loss intensity to optical properties via the Kramers-Kronig equations. It is necessary to know the quantity $Re[1/\varepsilon(0)]$, which in most materials can be taken as the reciprocal of the square of the optical refractive index, or as zero for a metal or semi-metal. Again, one records a low-loss spectrum, up to about 200eV. The data processing can take the form of a short computer program, run on the data-analysis system, yielding thicknesses which are believed to be accurate to about 10% (Egerton and Cheng 1987).

Another technique, based on the Bethe sum rule, provides an estimate of the mass-thickness of the specimen and is attractive for use with organic or biological specimens, since it requires no knowledge of the chemical composition (Crozier and Egerton 1989). However, it requires recording the spectrum over an extended range of energy loss, the removal of plural scattering by deconvolution and (in many cases) correction for the stray-scattering background of the spectrometer.

12   Information from EELS fine structure

It has long been recognised that the fine structure present in an energy-loss spectrum provides a wealth of information about the specimen. Fine structure in the low-loss region can provide a 'fingerprint' which has been used for identification of small regions of a TEM specimen. In alloys containing free-electron metals (*e.g.* Al), where the plasmon loss is sharp, small shifts in plasmon-peak position have been used to determine changes in local composition, on a 10nm scale (Williams and Edington 1976, Okamoto *et al.* 1992). Low-loss fine structure is prominent in organic specimens, where appearance of a sharp plasmon peak at around 6eV is often characteristic of double bonds (Isaacson 1972a,b).

12.1 · Near-edge fine structure (ELNES)

Recently, much attention has been directed to the interpretation and use of the near-edge fine structure (ELNES) of ionisation edges. This structure is directly analogous to the XANES structure observed in x-ray absorption edges, which has been studied by means of synchrotron radiation. Attempts to understand this structure are based on several seemingly different but basically equivalent approaches.

Band Structure Calculations

According to the 'Fermi Golden Rule' of quantum mechanics, the transition rate from an inner shell can be written in the form:

$$J_c(E) = C \, | \, M(E) \, |^2 \, N(E) \qquad (22)$$

where C is a constant, $M(E)$ is a transition-matrix element and $N(E)$ is the density of final (unoccupied) states. To the extent that the matrix term can be regarded as a slowly varying function of energy exchange E, the fine structure observed in $J_c(E)$ is a reflection of the density of states (DOS) above the Fermi level. One example is the detection of hole states just above the Fermi level in YBCO superconductors (Batson and Chisholm 1988).

The density of states can be calculated, by the augmented plane wave (APW) method for instance. Examples of the correlation between ELNES and calculated DOS are shown in Figure 8. It is now possible to get good agreement between experiment and theory, but only when the following details are taken into account.

1. The core-loss intensity is dominated by the effect of small-angle scattering (dipole regime), especially if a small collection aperture is used. The atomic transitions, like those in optics, therefore obey the dipole selection rule: the angular-momentum quantum number l must increase or decrease by one unit. For a K-edge ($l = 0$ initially), transitions to p-like states are therefore favoured. For an L_{23}-edge, transitions to s- or d-like states are possible, giving rise to a different fine structure. In general, $N(E)$ in Equation 22 should therefore be a symmetry-projected density of states, as available (for example) from the APW method.

2. Since the initial (core-level) states in the transition are localised close to atomic nuclei, $N(E)$ should be a local density of states. In a compound, $N(E)$ will be different at different atomic sites (cation and anion, for example), giving rise to dissimilar fine structures at the ionisation edges of the various constituent elements. Modern calculations can provide this information.

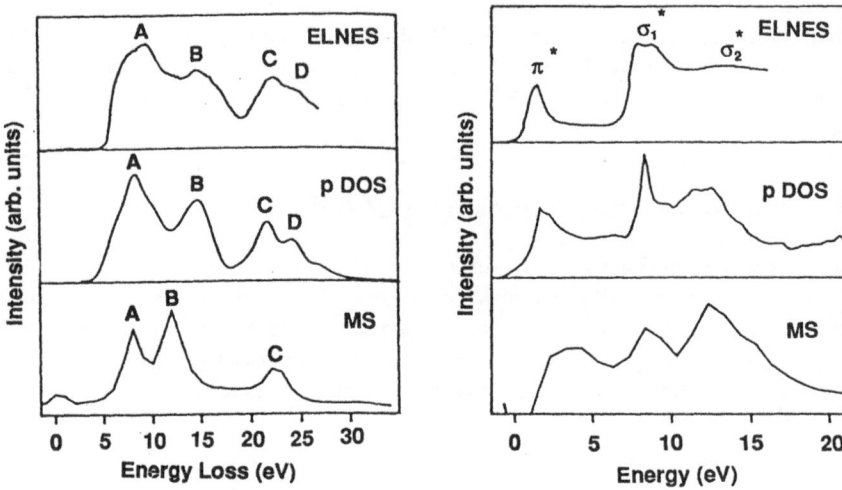

Figure 8. *K-shell* ELNES *in diamond (left) and graphite (right), compared with p-*
DOS *band-structure calculations and multiple-scattering* (MS) XANES *calculations (Weng*
et al. 1989).

3. For some materials, many-electron effects may modify the single-electron band-
structure analysis; for example, core excitons may be important in the case of
some insulators.

4. The observed structure will be broadened due to instrumental energy resolution,
the energy width of the core level (typically 0.1 to 0.5 eV) and lifetime broadening
of the final states.

Molecular Orbital (MO or LCAO) Theory

Because of the localised nature of the inner-shell transitions, it is tempting to interpret
the fine structure in terms of a linear combination of atomic orbitals (LCAO), determined
by the local symmetry of the atomic bonds. A simple example is that of graphite,
which consists basically of six membered carbon rings; the four carbon valence electrons
become sp^2 hybrids, which form three strong σ bonds within the atomic planes, and
a remaining p-electron perpendicular to the planes, which contributes to a delocalised
π orbital. The corresponding antibonding orbitals are denoted as σ^* and π^*; they are
the empty states into which core electrons can be excited, giving rise to distinct peaks
in the K-edge spectrum; see Figure 8. Diamond, on the other hand, is sp^3 hybridised
and the lower-energy π^* peak is absent. From the ELNES, we can therefore measure the
relative amounts of graphitic and diamond-like bonding, even in small localised regions
of a material (Berger *et al.* 1988).

In organic materials, the presence of delocalised or unsaturated bonding again gives rise to sharp π^* peaks at the edge threshold. In compounds containing carbon atoms with different effective charge, such as the nucleic acid bases, several peaks are observable as a result of chemical shifts (Isaacson 1972a,b). It is possible to observe this π^* structure disappearing during prolonged electron irradiation, reflecting radiation damage by bond scission (Isaacson 1972a,b).

Sharp threshold peaks known as white lines are found at L_2 and L_3 edges of transition metals (Figure 1), and are due to localised d-states, which in the solid form a narrow energy band. The relative height and separation of these lines has been found to provide an indication of electronic charge (oxidation state) (Rez 1992, Colliex *et al.* 1985, Okamoto *et al.* 1992).

XANES Theory

In the MO and band-structure pictures, the fine structure depends on the electron states, whose energy dependence is predicted by calculations performed in reciprocal space (electron momentum). It is also possible to do calculations in real space, by calculating the backscattering of the ejected inner-shell electron from neighbouring atoms in terms of its effect at the target atom. These calculations were first used to interpret x-ray absorption near-edge structure (XANES) and are a generalisation of the procedure used to analyse extended x-ray absorption fine structure (EXAFS), taking into account the possibility of multiple elastic scattering. Atoms surrounding the excited atom are divided into shells and the scattering from each shell is calculated (Rez 1992). This procedure is equivalent to calculating a local band structure (Colliex *et al.* 1985). Good agreement has been obtained between XANES predictions and measured ELNES in metal oxides (Rez 1992), minerals (Brydson *et al.* 1992) and in carbon; see Figure 8.

12.2 Coordination fingerprints

Understanding of the near-edge structure in terms of local bonding or band structure leads one to hope that information about the local environment of specific atoms might be obtainable by EELS, preferably without the need for elaborate calculations. In other words, one may be able to identify specific types of coordination from recognisable ELNES 'fingerprints'. The first attempt to relate ELNES with coordination is that of Tafto and Zhu (1982), who measured K-edges of Mg, Al and Si in various minerals and observed that these were more sharply peaked where the metal (or Si) atom was in tetrahedral rather than octahedral coordination; see Figure 9. Subsequent measurements have confirmed the possibility of coordination fingerprinting, with some reservations. For example, it is probably the symmetry rather than the coordination number of the surrounding atoms which is important, so if the bond angles are distorted, the ELNES structure changes (McComb *et al.* 1991). In the case of MgO, for example, the K-edge ELNES of magnesium and oxygen are appreciably different (reflecting the local p-DOS), even though the coordination is in each case octahedral (Colliex *et al.* 1985, McComb *et al.* 1991).

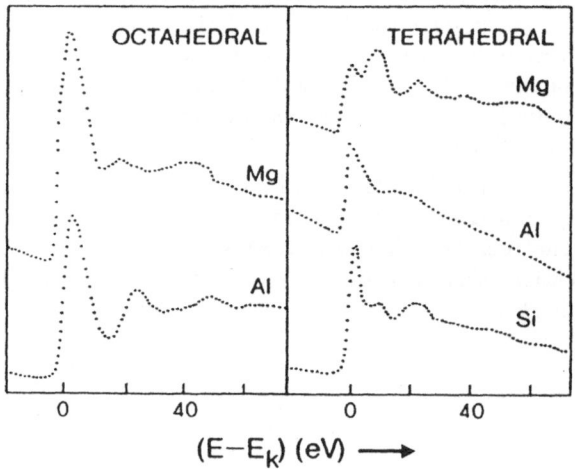

Figure 9. *K-edges measured in olivine, spinel and orthoclase (Tafto and Zhu 1982).*

12.3   Extended energy-loss fine structure (EXELFS)

Beyond 50eV from the ionisation threshold, fine-structure oscillations become weak but they continue for an energy range comparable to the threshold energy. The energy loss above the threshold value also represents the kinetic energy of the ejected inner-shell electron, and for a kinetic energy of a few hundred eV, the inelastic mean free path is of the order of 1nm. In other words, an outgoing electron is strongly scattered and the information it provides (via the transmitted energy-loss electron) is highly localised.

The oscillatory component of the core-loss intensity can be analysed according to the same equation as used in x-ray (EXAFS) studies, to yield a radial distribution function (RDF) whose peak positions and heights give the interatomic distances and coordination numbers. The EXAFS technique has been applied to many materials problems (Teo and Joy 1980) but the analogous EXELFS applications are fewer in number. Most of the data has been too noisy or covered an insufficient energy range to enable really accurate bond lengths to be determined. The use of parallel recording detectors and (possibly) higher-energy edges may alleviate these problems.

From the damping of EXELFS oscillations, it is possible to measure the amount of thermal or static (local) disorder in a material. This technique is being pursued to study short-range order in undercooled alloys, for example (Okamoto *et al.* 1992).

13   Energy-filtered images and diffraction patterns

Besides producing an energy-loss spectrum from a chosen area of the specimen, the electron spectrometer can form an image from a selected band of energy loss. It can be achieved in STEM mode, by scanning a small electron probe in a raster and displaying (as a function of time) the signal collected by the spectrometer slit or from a chosen region of the diode-array detector. An extension of this idea is spectrum-imaging, where a complete energy-loss spectrum is stored for each pixel for subsequent off-line analysis (Hunt and Williams 1991). Energy-selected images are also obtainable in CTEM mode, by utilising the imaging properties of a magnetic prism (Egerton 1986). In the latter case, the spectrometer is usually incorporated into the CTEM column, just below the objective lens, but imaging spectrometers can also be attached below the camera chamber (Krivanek *et al.* 1991).

One application is elemental mapping, where the spectrometer is set to an energy loss corresponding to a particular ionisation edge. Care has to be taken to correct for the pre-edge background and (for thicker specimens) to remove contrast arising from changes in mass thickness or diffracting conditions across the specimen. A straightforward extension of this idea would be to form images from different peaks in the near-edge structure, giving a 'chemical' map representing changes in bonding (*e.g.* different allotropes).

Another application is in CTEM imaging of relatively thick specimens, where zero-loss electrons are allowed through the energy-selecting slit so that image contrast and resolution are not degraded by chromatic aberration of CTEM imaging lenses. By similar filtering of diffraction patterns, the inelastic 'halo' can be removed to provide quantitative diffraction intensities, and to extract the radial distribution function of amorphous specimens, for example (Cockayne *et al.* 1991).

References

Ahn C C and Krivanek O L, 1983, *EELS Atlas*, (Gatan Inc, Warrendale)

Batson P E and Chisholm M F, 1988, *J Electr Microsc Tech* **8** 311

Berger S D, McKenzie D R and Martin P J, 1988, *Phil Mag Letters* **57** 285

Brydson R, Sauer H and Engel W, 1992, *in Transmission Electron Energy Loss Spectrometry in Materials Science, TMS EMPMD Monograph Series* eds Disko M M, Ahn C C and Fultz B, Ch. 6

Cheng S C, Crozier P A and Egerton R F, 1987, *J Microsc* **148** 285

Cockayne D, McKenzie D and Muller D, 1991, *Microsc Microanal Microstruct* **2** No. 2/3 359

Colliex C, 1984, *in Advances in Optical and Electron Microscopy* 65, eds Barer R and Coslett V E (Academic Press)

Colliex C, Manoubi T, Gasgnier M and Brown L M, 1985, *Scanning Electron Microscopy* **2** 489 (SEM Inc, Chicago)

Craven A J and Buggy T W, 1984, *J Microsc* **136** 227

Crozier P A and Egerton R F, 1989, *Ultramicroscopy* **27** 9

Disko M M, 1986, *in Microbeam Analysis* 429, eds Romig A D and Chambers W F (San Francisco Press)

Disko M M, Ahn C C and Fultz B, eds, 1992, *Transmission Electron Energy Loss Spectrometry in Materials Science*, (TMS EMPMD Monograph Series No. 2, The Minerals, Metals and Materials Society, Warrendale)

Egerton R F, 1986, *Electron Energy-Loss Spectroscopy in the Electron Microscope* (Plenum, New York)

Egerton R F and Cheng S C, 1987, *Ultramicroscopy* **21** 231

Egerton R F, 1992, in *Proc 50th Annual Meeting EMSA* p.1264

Geiger J, Nolting M and Schroeder B, 1970, in *Proc 7th International Electron Microscopy Congress* 111 (Grenoble)

Hofer F, 1991, *Microsc Microanal Microstruct* **2** No. 2/3 143

Hunt J A and Williams D B, 1991, *Ultramicroscopy* **38** 47

Ibach H and Mills D L, 1982, *Electron Energy Loss Spectroscopy and Surface Vibrations* Academic Press (New York)

Ibach H, 1990, *Electron Energy Loss Spectrometers* Springer Series in Optical Sciences No 64 (Berlin)

Isaacson M, 1972a, *J Chem Phys* **56** 1803

Isaacson M, 1972b, *J Chem Phys* **56** 1813

Isaacson M and Johnson D, 1975, *Ultramicroscopy* **1** 33

Joy D C and Maher D M, 1980, *Scanning Electron Microscopy* **1**, 25 (SEM Inc, Chicago)

Krivanek O L, Gubbens A J and Dellby N, 1991, *Microsc Microanal Microstruct* **2** No. 2/3 315

Leapman R D and Swyt C R, 1988, *Ultramicroscopy* **26** 393

Leapman R D and Hunt J A, 1991, *Microsc Microanal Microstruct* **2** No. 2/3 231

Leapman R D, 1992, in *Transmission Electron Energy Loss Spectrometry in Materials Science* TMS EMPMD Monograph Series, eds Disko M M, Ahn C C and Fultz B, Ch 3

Malis T and Titchmarsh J M, 1985, *in Electron Microscopy and Analysis* 1985 181 (Institute of Physics, Bristol)

Malis T, Cheng S C and Egerton R F, 1988, *J Electron Microsc Tech* **8** 193

McComb D W, Hansen P L and Brydson R, 1991, *Microsc Microanal Microstruct* **2** No. 2/3 561

Okamoto J K, Pearson D H, Ahn C and Fulz B, 1992 *in Transmission Electron Energy Loss Spectrometry in Materials Science, TMS EMPMD Monograph Series* eds Disko M M, Ahn C C and Fultz B Ch 8

Rez P, 1992, in *Transmission Electron Energy Loss Spectrometry in Materials Science, TMS EMPMD Monograph Series* eds Disko M M, Ahn C C and Fultz B Ch 5

Shuman H and Somlyo A P, 1987 *Ultramicroscopy* **21** 23

Steele J D, Titchmarsh J M, Chapman J N and Paterson J H, 1985, *Ultramicroscopy* **17** 273

Tafto J and Zhu J, 1982, *Ultramicroscopy* **9** 349

Teo B K and Joy D C, eds, 1980, *EXAFS Spectroscopy: Techniques and Applications* (Plenum, New York)

Titchmarsh J M and Malis T F, 1989, *Ultramicroscopy* **28** 277

Titchmarsh J, 1992, (these Proceedings)

Wang Z L and Cowley J M, 1988, *Surface Science* **193** 501

Weng X, Rez P and Ma H, 1989, *Phys. Rev. B* **40** 4175

Williams D B and Edington J W, 1976, *J Microsc* **108** 113

Yang Y-Y and Egerton R F, 1992, *Microscopy Research and Technique* **21** 361

Light Element Microanalysis and Imaging

D B Williams

Lehigh University
Bethlehem PA, USA

1 Introduction

Any definition of a 'light element' is somewhat arbitrary, but in the field of x-ray microanalysis, it is an element that cannot be detected by a Be-window energy-dispersive spectrometer (EDS), *i.e.* elements below sodium in the periodic table. For the materials scientist, therefore, Li, Be, B, C, N, and O and occasionally H and He are the light elements of interest and biologists are concerned with H, C, N, O, and F. Microanalysis of these elements is difficult, particularly when there is a need for high spatial resolution and/or high analytical sensitivity combined with quantification. For *all* the elements, the best approach, in the opinion of the author, is to combine microanalysis with digital imaging in the form of compositional images or maps (*e.g.* see Lyman 1992). Microanalysis in this form is also a microscopy technique, since high magnification elemental images are produced. This is a crucial advantage of mapping since it facilitates a one-to-one comparison between any composition variations in the specimen and the defect structure, or whatever other features are imaged by accompanying microscopy techniques.

The technique used for compositional mapping depends on the form of the specimen, either bulk (non-transparent to the radiation) or thin (transparent). The former has the advantage of relatively simple preparation from the parent sample and, consequently, relatively few artefacts. In addition, the sampling statistics are good since a large area of the specimen (many μm^2) is imaged and the minimum mass fraction is generally $\sim$1–100 ppm. The price to pay is relatively poor spatial resolution (typically not much less than a micrometer). Thin specimens generally offer high spatial resolution (a few nanometers) and good minimum detectable mass (a few hundred atoms). But thin specimen preparation and the need for a pristine surface are problems which often limit the quality of the microanalytical data.

Microanalysis in the materials sciences aims to relate composition variations to the properties of the material, often revealed through the defects imaged in the microscope. Within this framework, high resolution is not the only criterion for good microanalysis. For example, mechanical properties could just as easily be controlled by the chemistry of micron-sized inclusions (*e.g.* short-transverse toughness) as by nanometer-level Gibbsian segregation (*e.g.* temper embrittlement). So the best approach is to exercise all the necessary techniques to span the spectrum of low-resolution, high-sensitivity bulk microanalysis to high resolution, specimen-preparation limited thin-foil techniques.

2   Compositional imaging of bulk specimens

2.1   Electron probe microanalysis

More than thirty-five years ago Cosslett and Duncumb (1956) introduced the idea of x-ray dot mapping using wavelength-dispersive spectrometry (WDS) in the electron-probe microanalyzer (EPMA), as shown schematically in Figure 1a. This concept formed the basis for all subsequent scanning imaging processes. The mapping technique developed slowly but with the advent of x-ray energy-dispersive spectrometry (EDS) in the early 1970s, qualitative multi-element mapping became common. Now with digital beam control and the associated computer technology, quantitative EDS and WDS mapping are a reality (Newbury 1992). The technique remains limited by the serial nature of WDS and the need for complex time-consuming ZAF or $\theta(\rho z)$ correction routines applied to each pixel (Heinrich and Newbury 1991 and Goldstein *et al.* 1992). However, the full procedure (Figure 1b) is still essentially the same as that of Cosslett and Duncumb. Processing the signals during acquisition is slow and overnight scans are needed for quantitative images. While EDS mapping is common, the relatively poor signal to background in the EDS spectrum and associated peak overlap problems mean that WDS is the best technique when light elements, or relatively small amounts of material, are being sought. Figure 2 shows a typical quantitative EPMA x-ray map obtained using WDS. Light element mapping of this kind is also possible, but generally more difficult to quantify because of problems such as strong x-ray absorption.

The principal limitation of EPMA is the minimum spatial resolution of ~1 μm. There are fundamental physical limitations that restrict x-ray microanalysis to the micron level in bulk specimens. This dimension is controlled by the beam-specimen interaction volume, which at typical beam energies of ~20–30 keV in average materials (*e.g.* the transition metals) is a few micrometers. It is possible to reduce spatial resolution either by seeking more localised signals than x-rays (*e.g.* surface signals such as Auger electrons) or use other radiation sources such as ion beams that do not create such large interaction volumes as electrons. For the purposes of this paper we will consider only the latter in the form of high energy ion beams that, because of their momentum, permit the extension of surface analysis to destructive bulk analysis by the process of depth profiling.

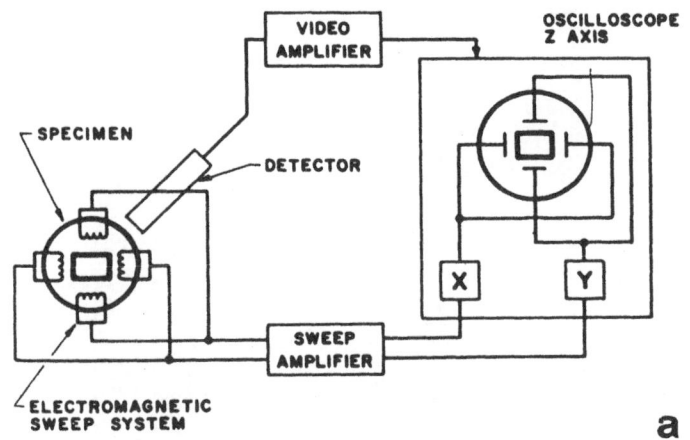

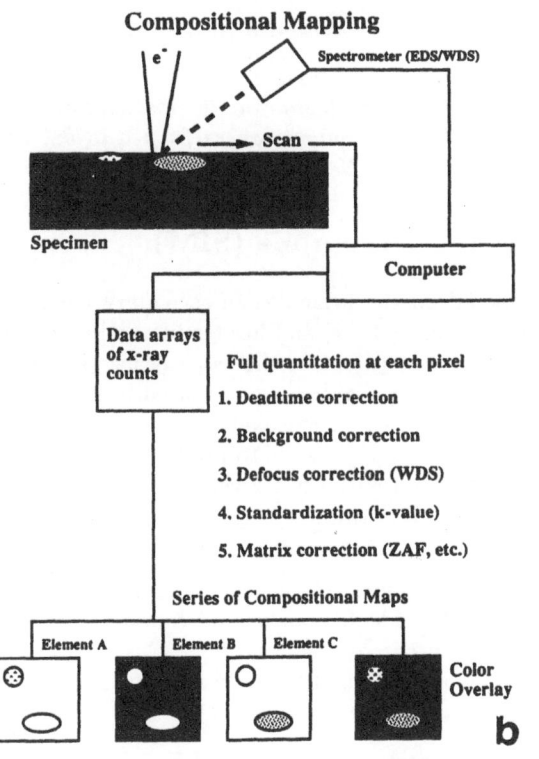

Figure 1. *(a) Schematic diagram of the method of mapping in an SEM (Cosslett, Duncumb 1956). The CRT and electron beam are scanned synchronously and the signal from any detector (e.g. x-ray) is used to modulate the CRT intensity. (b) The steps required for quantitative compositional mapping in an EPMA system(D E Newbury 1992)*

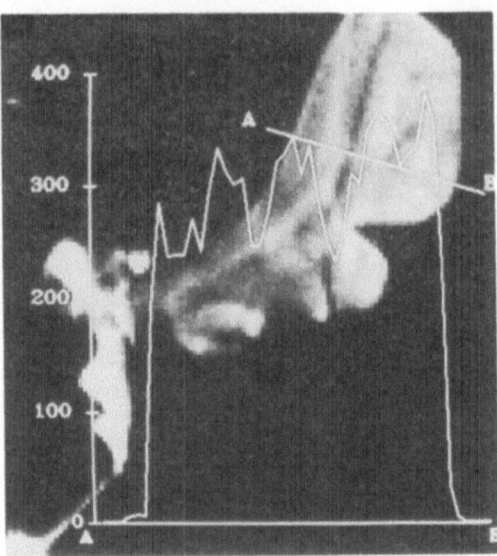

Figure 2. *Quantitative* EPMA *map showing the distribution of Zn at grain boundaries in polycrystalline Cu. The Zn content ranges from 0 to 0.4 wt.% along the line AB. Full scale 80 nm (courtesy D E Newbury 1992)*

2.2 Scanning ion microprobe (SIM)

Ion microscopy is based on the principles of secondary ion mass spectrometry (SIMS) (*e.g.* Czanderna and Hercules 1991) and has traditionally been the broad-beam, analog form which gives image resolutions of the same order as the EPMA (1 μm) (see Figure 3). New liquid metal (*e.g.* Ga) ion sources produce high brightness, small diameter beams that are capable of putting reasonable imaging currents into < 50 nm probes. Under these circumstances, it is possible to produce SIM digital images and the results presented here were obtained with the University of Chicago Scanning Ion Microprobe (UC SIM) described by Levi-Setti *et al.* (1991). The SIM images are formed by different ion species, with a spatial resolution of < 100 nm.

There are specific advantages and disadvantages of SIM and EPMA with regard to mapping of light elements. Ideally, one should combine the benefits of each, to derive a better understanding of the materials studied. A comparison of the two techniques has been given by Newbury *et al.* (1991) and relevant points are:

1. SIMS is sensitive to most elements and isotopes, in some cases to the ppm level. As usual, high spatial resolution and sensitivity are mutually exclusive. SIMS is very sensitive to light elements such as B, O and C, which are difficult to detect and quantify in EPMA or are undetectable (*e.g.* Li). But absolute quantification can be difficult with SIMS. It is not uncommon to have ±5% error in quantification data, even after extensive attempts to calibrate the measurements.

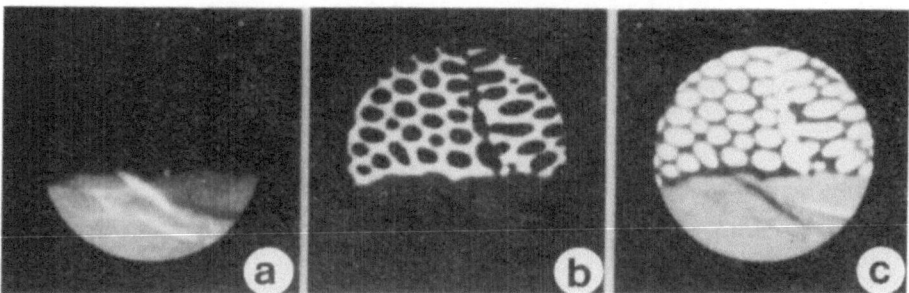

Figure 3. *Quantitative analog* SIMS *images showing the distribution of (a) Li, (b) Cu and (c) Al in a ternary alloy containing the coarse primary Li-rich T2 phase (Al$_6$Li$_3$Cu) and a eutectic mixture of the Cu-rich T1 phase (Al$_2$CuLi) and the Al-rich solid- solution phase. Full scale 150 mm (courtesy D E Newbury and D S Bright)*

2. The SIMS signals are usually significantly higher than those obtained with EPMA. The signal-to-noise for SIMS is typically $10^3 - 10^5$, compared with the equivalent peak to background of $10^2 - 10^3$ for EPMA. For comparison, the time required to accumulate a statistically significant 512×512 map of a single element is of the order of minutes with SIM, but can be several hours with EPMA.

3. Polished specimens often contain topographic relief at interfaces. This relief may disturb the absorption correction in EPMA. With care, one can study such samples with SIM because the ion signal is not as strongly influenced by topography.

4. The lateral resolution in SIM images can approach the probe size if there is a large concentration of the species of interest, or if the species has a large ionisation probability. The spatial resolution of SIM maps acquired with the UC SIM is routinely < 100 nm. The ion signal originates from the top two monolayers of the sample. Consequently, only a small microvolume ($\sim 50^3$ nm^3) is sampled for each image element. EPMA, on the other hand, samples the material composition to a depth of several micrometers, (depending on the beam energy and the average specimen composition) and laterally within $\sim 1\mu$m^2.

5. In a SIM, the sample can be sputtered *in situ* with the ion beam, prior to analysis. This permits recognition and elimination of polishing-induced artefacts. However, fragile structures can be difficult to study because SIMS is destructive.

The most significant drawback to SIMS is that quantitative concentration information cannot be readily extracted from the raw SIMS data. The secondary ion signal is affected by well known matrix effects, and secondary ionisation probabilities for a compound are not known *a priori*. Surface chemical effects, such as the enhancement of SIMS signals due to the presence of oxygen or alkalis, complicate the quantification which is possible if calibration standards similar in composition to the material under study are available. But correct calibration standards often cannot be chosen until the

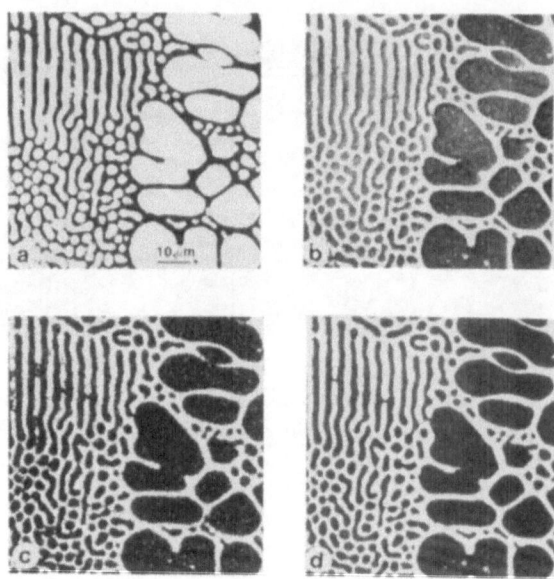

Figure 4. *Digital* SIM *images showing the distribution of various light element ion species (a)* Al$^+$, *(b)* Li$^+$, *(c)* F$^-$ *and (d)* O$^-$ *in the eutectic region of the same Al-Li-Cu alloy as in Figure 3, showing image resolution in the order of 0.1μm. The Al segregates to the solid solution part of the eutectic mixture while the Li is mainly in the T1 phase. F and O segregate preferentially to the Li-rich T1 (courtesy K K Soni and R Levi-Setti)*

phases are identified. Nevertheless, *relative* concentrations and gradients across a surface can be very accurately determined with SIMS ($\sim$ 1%). These relative concentration data provide valuable clues about material composition. Figure 4 shows SIM maps of various elements in an Al-Li-Cu alloy. Note the ease with which Li can be mapped, since the Li^{7+} ion is one of the most easily ionised species. The maps are not quantitative, but with due care and the generation of suitable working curves, quantification of Li can be achieved (Chabala *et al.* 1991).

3 Compositional imaging of thin specimens

Thin specimen microanalysis is carried out in the scanning/transmission electron microscope (S)TEM, often termed an analytical electron microscope (AEM). Possible signals for analysis include x-rays, Auger electrons and energy-loss electrons. The x-ray signal is not very useful for light elements. Even though current EDS systems can detect Be Kα x-rays (110 eV), the x-ray signal is very weak because of (a) the poor fluorescence yield ($< 10^{-4}$ for Be) (b) the reduced interaction volume in a thin specimen and (c) the increased absorption of the low energy characteristic x-rays in both the specimen and the detector. WDS is not yet an option in the TEM although compact spectrometers for AEMs are under development (Goldstein *et al.* 1989). The Auger signal is a possible source of light element surface analysis, but a pristine surface is required and this means that only UHV AEMs are useful. Some progress in interfacing Auger systems to AEMs is being made by several investigators, (*e.g.* Hembree, 1990 and Kruit, this volume). Currently the best option for light element thin specimen microanalysis is to detect and image the energy-loss electrons using electron energy-loss spectrometry (EELS). The EELS spectrum contains a wealth of information beyond the simple elemental ionisation edges, (which are especially useful for light element detection). The low-loss spectrum contains thickness, dielectric constant, interband transition and bonding data. The core-loss edges also contain fine structure that reflects both the local atomic bonding and the atomic environment surrounding the ionised atom.

Any kind of microanalysis in the AEM usually involves selecting a region of interest, positioning the beam on that region and gathering a spectrum for sufficient time to permit either immediate qualitative analysis or subsequent quantitative analysis. However, a single point analysis, or indeed several analyses, provides very poor sampling of the chemistry of the chosen region. A biased selection of the points of analysis and a pre-selection of the suspected elements present in the region are almost inevitable. In most analyses, there is only one opportunity to collect the data. So the analyst confines the search to certain elements and, particularly in the case of EELS, selects the appropriate portion of the spectrum to collect. If subsequent analysis of the spectrum indicates the presence of an unforeseen element, or the quantification process is unsatisfactory because of insufficient counts, then it is often difficult or impossible to find the exact analysis region again. The experiment has to be re-done on a different specimen or a different area of the same specimen. Even if the original region is found, the analysis point may have been damaged by the prior electron exposure, covered by contamination or oxidised during storage. All these problems can be overcome using the technique called *spectrum imaging* (Jeanguillaume and Colliex 1989, Hunt and Williams 1991).

3.1   EELS spectrum imaging

Spectrum imaging is the acquisition of a full EELS spectrum at every pixel in a STEM image of the specimen. With this process a complete record of all the detected beam-specimen interactions is stored at each point in the image. As a result of this approach no information is lost and all the data can be processed in batches after the acquisition, so (S)TEM time is only spent gathering, not processing data. It is possible to return to the stored data at any time to check for unforeseen features in the spectrum. A parallel-collection (PEELS) spectrometer is essential for spectrum imaging because a serial (SEELS) system is too slow: *e.g.* A 128×128 image has 16,384 pixels. A SEELS system recording a 60 s spectrum at each pixel requires over 11 days to collect a spectrum image. A PEELS system recording the spectrum in 0.2 s requires only 54 mins. A large data storage capability (10 Mb - 1 Gb) is essential because a 128×128 (pixel) $\times 1024$ (channel) spectrum image contains 16.8 Mb of data. A $512 \times 512 \times 1024$ spectrum image contains 269 Mb. A field-emission gun (FEG) STEM to maximise the count rate at each pixel makes spectrum imaging much more efficient because an FEG is 10^3 times brighter than the best thermionic source and spectrum acquisition times are reduced proportionately. Suitable software is required to retrieve and analyse the requisite fraction of this enormous amount of data in a reasonable time.

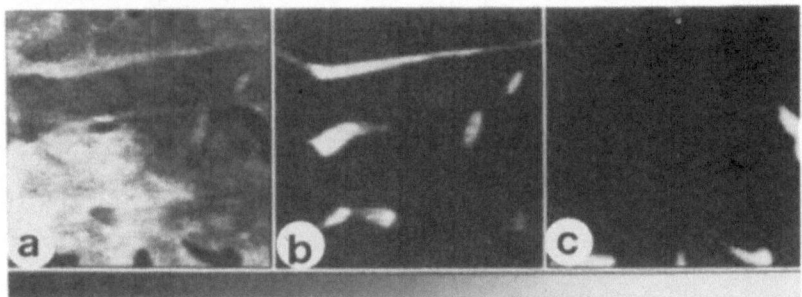

Figure 5. *Grey scale quantitative elemental maps using the Cu M, Be K and Co M edges. In Figure 5a the Cu is primarily in the matrix phase, compared to the Be (Figure 5b) which is located in the CuBe intermetallic lamellae. The Co ternary additions (Figure 5c) are segregated to other precipitates. Full scale is 1μm (courtesy J A Hunt)*

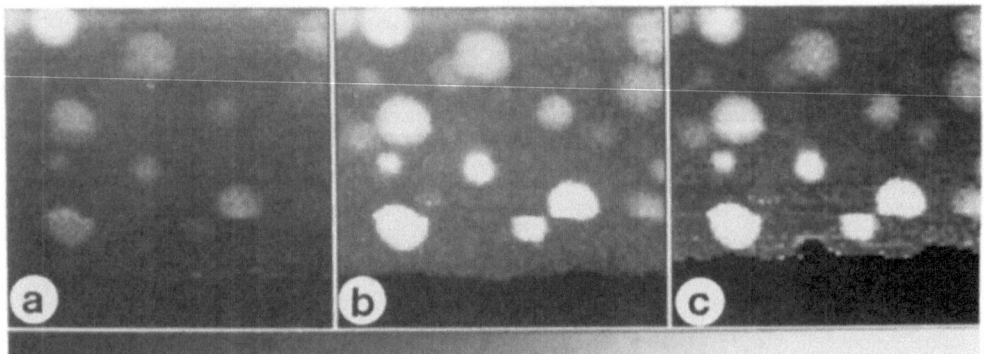

Figure 6. *The advantages of post-acquisition data processing in images of an Al-Li alloy. Figure 6a The Li distribution calculated from the Li K edge using the conventional power-law background removal shows a variation in Li content with foil thickness away from the edge at the bottom of the image. Figure 6b The Li distribution calculated after first difference background removal and multiple least squares fitting of standard spectra. The Li distribution is insensitive to thickness variations. Figure 6c The Li distribution calculated from the shift in the plasmon peak position. Full scale is 500 nm and a grey scale look up table is included across the base of each figure (courtesy J A Hunt)*

Once a spectrum image has been acquired and stored there are many ways to view the information. The simplest approaches are conventional methods such as a map of a specific ionisation edge intensity or a map of a specific plasmon shift. Alternatively, the image can be projected onto different image axes; *e.g.* projection along the energy-loss axis gives a total (unfiltered) image, or the total spectrum summed from the whole specimen. Spectra at certain pixels in the image can be summed, such as those from a feature in the image that has a shape not amenable to analysis in spot or line mode. A specific edge can be imaged to determine the distribution of an element when its localisation is not known *a priori*. Similarly, a specific edge can be sought when its presence was only suspected *a posteriori*. Finally, if the analysis routines prove unsatisfactory, then other routines can be applied to the original data without the need for further data acquisition.

Some advantages of spectrum imaging of light elements are illustrated in Figures 5 and 6. Figure 5 shows three maps of the Cu, Be and Co distribution in a thin foil of a Cu-Be-Co alloy. The segregation of Be to the lamellar precipitates is clearly shown. Quantitative mapping of Be on this dimensional scale would be impossible with x-rays. Figure 6 shows the distribution of Li in an Al-10 at% Li alloy, calculated using different quantification routines. The Li is segregated to the spherical Al_3Li precipitates.

In summary, spectrum imaging permits imaging of the distribution of any feature in *any* spectrum (which gives EELS an advantage over EDS. Qualitative imaging is straightforward, but quantitative imaging is only possible using sophisticated software and post-acquisition batch processing. It is possible to carry out searches for unsuspected elements, removing the possibility of having to reacquire the spectrum. Direct comparison of different data reduction schemes is easily performed.

Acknowledgements

This work draws on several years of collaborative research and discussions with many colleagues, whom I wish to thank:- J. A. Hunt, K. K. Soni, C. E. Lyman, J. I. Goldstein, R. Levi-Setti and D. E. Newbury. Funding for this work came from the National Science Foundation through grants DMR 8905459 and DMR 9111839.

References

Chabala J M, Levi-Setti R, Soni K K, Williams D B and Newbury D E, 1991, *Applied Surface Science* **51** 185–192

Cosslett V E and Duncumb P, 1956, *Nature* **177** 1172–1173

Czanderna A W and Hercules D M (eds), 1991, *Ion Spectroscopies for Surface Analysis* (Plenum Press, New York)

Goldstein J I, Newbury D E, Echlin P, Joy D C, Romig A D Jr., Lyman, C E, Fiori C and Lifshin E, 1992, *Scanning Electron Microscopy and X-ray Microanalysis* (2nd Edition) (Plenum Press, New York)

Goldstein J I, Lyman C E and Williams D B, 1989, *Ultramicroscopy* **28** 162–165

Heinrich K F J, Newbury D E (eds), 1991, *Electron Probe Quantitation* (Plenum Press, N.Y.)

Hembree G G, Liu, C H and Venables J A, 1990, *Proceedins XII International Congress for Electron Microscopy* (San Francisco Press, San Francisco CA) 382–383

Hunt J A and Williams D B, 1991, *Ultramicroscopy* **38** 47–73

Jeanguillaume C and Colliex C, 1989, *Ultramicroscopy* **28** 252–257

Levi-Setti R, Hallégot P, Girod C, Chabala J M, Li J, Sodonis A and Wolbach W, 1991, *Surface Science* **246** 94–106

Lyman C E, 1992, *Microscopy: The Key Research Tool (The Bulletin of the Electron Microscopy Society of America)* **22** 1–10

Newbury D E, 1992, *Microscopy: The Key Research Tool (The Bulletin of the Electron Microscopy Society of America)* **22** 11–21

Newbury D E, Marinenko R B, Myklebust R L and Bright D S, 1991, in *Images of Materials*, —edited by Williams D B, Pelton A R, and Gronsky R, (OUP, New York) 290–308

A Comparison of Quantification Methods and Analytical Techniques

A G Fitzgerald

University of Dundee
Scotland

1 Introduction

Electron energy loss spectroscopy (EELS), energy-dispersive X-ray microanalysis (EDX), Auger electron spectroscopy (AES) and X-ray photoelectron spectroscopy (XPS) have been developed over the past twenty years for the determination of the chemical composition of thin films and surfaces of materials. However, very few comparisons have been made of the results of analyses made using different techniques. Also, in surface analysis by AES and XPS, software to enable rapid comparisons and detailed investigation of matrix effects has not been available until recently (Fitzgerald *et al.* 1992a,b).

Difficulties arise in attempting to make these comparisons. For example, most composition analysis by EDX is currently carried out using X-ray detectors which do not detect oxygen. This can be a major contaminant in surface analysis and can result in poor agreement between surface compositions obtained by EDX and thin-film electron spectroscopy. Another problem arises in the surface cleaning required in preparing films for analysis by XPS and AES. The argon ion etching used in surface cleaning can in certain cases alter the surface stoichiometry.

In the work presented here applications of software for rapid surface composition analysis by AES and XPS are described with examples of the benefits of using this software to compare the effects of different parameter calculations on matrix factors.

Films of amorphous silicon alloys a-SiC$_x$:H and a-SiN$_x$: compound semiconductors and Cu/Ag alloys have been used to compare quantification methods. Homogenous amorphous silicon alloys can be deposited by plasma enhanced chemical vapour deposition (PECVD), with a wide range of compositions, and are ideal for quantitative analyses.

2   Experimental

The Auger electron spectra and X-ray photoelectron spectra were acquired using a VG Microscopes HB100 Multilab SEM, equipped with a CLAM 100 concentric hemispherical analyser. Electron energy loss spectra from thin films were obtained using a Gatan parallel electron energy loss spectrometer (PEELS) fitted to a JEOL 100C STEM. X-ray microanalysis spectra were acquired by a Link Scientific detector, attached to a JEOL T300 SEM.

3   Computer Quantification in Surface Analysis

Quantitative surface analysis by XPS or AES of a matrix containing a number of elements is obtained by measuring the X-ray photoelectron or Auger electron intensities. These are measured respectively for the matrix and for standards that contain elements present in the matrix. In AES, the ratio of the fractional atomic contents (Seah 1980), is given by

$$\frac{X_A}{X_B} = \frac{[1 + r_A^{\mathrm{std}}(E_A^b)][1 + r_m(E_B^b)]}{[1 + r_B^{\mathrm{std}}(E_B^b)][1 + r_m(E_A^b)]} \frac{\lambda_m(E_B)\lambda_A^{\mathrm{std}}(E_A)}{\lambda_m(E_A)\lambda_B^{\mathrm{std}}(E_B)} \left(\frac{a_B^{\mathrm{std}}}{a_A^{\mathrm{std}}}\right)^3 \frac{I_A/I_A^{\mathrm{std}}}{I_B/I_B^{\mathrm{std}}} \tag{1}$$

which can be written

$$\frac{X_A}{X_B} = F_{AB} \frac{I_A/I_A^{\mathrm{std}}}{I_A/I_B^{\mathrm{std}}} \tag{2}$$

where X_A and X_B are the fractional atomic contents in the sample of atoms A and B respectively, $\lambda_m(E_A), \lambda_m(E_B), \lambda_A^{\mathrm{std}}(E_A)$ and $\lambda_B^{\mathrm{std}}(E_B)$ are the attenuation lengths (AL) of electrons of energy E_A and E_B in the matrix and standards respectively, $r_A^{\mathrm{std}}(E_B^b)$ and $r_B^{\mathrm{std}}(E_A^b)$ are the electron backscatter factors in the standards, $r_m(E_A^b)$ and $r_m(E_B^b)$ are the electron backscatter factors in the matrix, and a_A^{std} and a_B^{std} are the respective atomic sizes in the standards for A and B. I_A and I_B are the electron intensities measured for the AES peaks from elements A and B in the sample, F_{AB} is the AES matrix factor, and I_A^{std} and I_B^{std} are the electron intensities measured by AES in the standards. The equivalent matrix factor for XPS can be obtained by setting the backscatter factors, r to zero.

Since parameters such as attenuation length and electron backscatter factor require knowledge of the unknown composition, determination of accurate surface composition involves an iterative process which can be conveniently installed on a PC. The programs AQUA for Auger Quantification and QUAX for Quantitative XPS have been developed at the University of Dundee to enable the rapid determination of surface composition by AES and XPS using iterative procedures from raw spectral data. This software includes a number of different calculation methods which are available for the matrix dependent parameters, attenuation length, (Seah 1980, Penn 1976, Tanuma *et al.* 1990) backscatter factor (Ishimura *et al.* 1983, Reuter 1971) and photoionisation cross-section (Scofield 1976, Yeh *et al.* 1985).

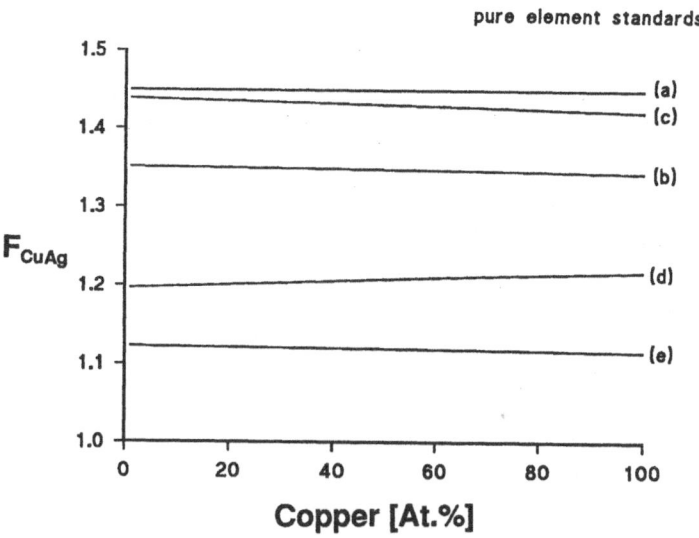

Figure 1. *Components of the* AES *matrix factor* F_{CuAg} *for the copper-silver system with pure element standards, (a) atomic size correction only; (b) atomic size plus backscattering correction; (c) full matrix factor using the TPP-2 formula; (d) full matrix factor using the Penn (1976)* IMFP *formula; (e) full matrix factor using Seah and Dench (1979) AL formula. Apart from (a) the Ichimura et al backscatter factor was used, incident beam energy 5keV. Courtesy Fitzgerald et al. (1992a) and John Wiley & Sons Ltd.*

4 Comparison of AES Matrix Factor Calculations

A useful feature of the AQUA/QUAX system is its ability to investigate the effects of various matrix correction attempts on surface quantification results. Using the "Plot Matrix Factors" option in AQUA, Figure 1 is obtained, showing how F_{CuAg} for Cu-Ag alloy with no AL correction compares with F_{CuAg} calculated using the various AL corrections. It is immediately apparent that the Tanuma AL correction (Tanuma *et al.* 1990) (denoted TPP-2) causes F_{CuAg} to increase by comparison with the curve shown in Figure 1b, indicating that this part of the correction is greater than unity. The Seah and Dench (1979) and Penn (1976) corrections, however, have magnitudes less than unity.

The determination of differences between matrix factors in terms of concentration difference is not possible from Figure 1 because any concentration difference vanishes as the sample tends towards a pure element. A matrix factor variation can be converted to a composition variation by using the relation (Fitgerald *et al.* 1992a,b).

$$\Delta = X_B \left(\frac{X_A(F_{AB}/F'_{AB}) - X_A}{X_B(F_{AB}/F'_{AB}) + X_A} \right) \tag{3}$$

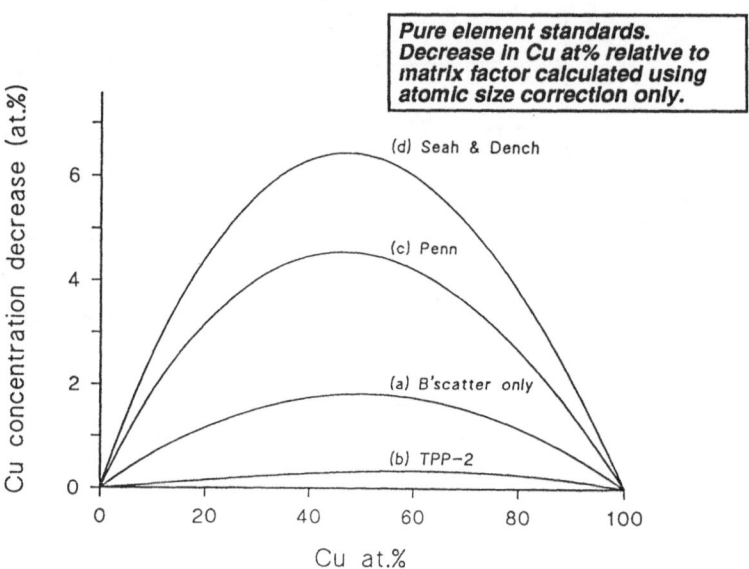

Figure 2. *Copper concentration observed after application of the following corrections in addition to the atomic size correction (a) backscattering only; and backscattering with (b) the TPP-2 AL correction formula, (c) the Penn AL formula; (d) the Seah and Dench AL formula. Courtesy Fitzgerald et al. (1992a) and John Wiley & Sons Ltd.*

where X_A, X'_A, X_B and X'_B are the fractional compositions obtained using two different matrix corrections and $\Delta = X_A - X'_A$.

When Equation 3 is applied to the data of Figure 1 the effect on concentration of applying three different attenuation-length corrections is shown in Figure 2. The consequence of applying these different corrections is to decrease the apparent copper concentration by a maximum of about 6 at.%.

From an examination of Figure 2, the results obtained using the TPP-2 AL calculation method and the Seah and Dench method might be expected to be in better agreement than results using the TPP-2 method and the Penn equation. However, reference to the actual matrix factor equation (Equation 1) shows it is the AL ratio that is important, not the AL magnitudes. When the attenuation lengths predicted for the two elements versus energy are plotted using each calculation method in turn it is found that the Penn and the Seah and Dench methods result in a longer AL in silver than in copper, whereas the TPP-2 method predicts the converse. It is this fact which causes the disagreement between the matrix factors shown in Figure 1 and the plotted concentration in Figure 2.

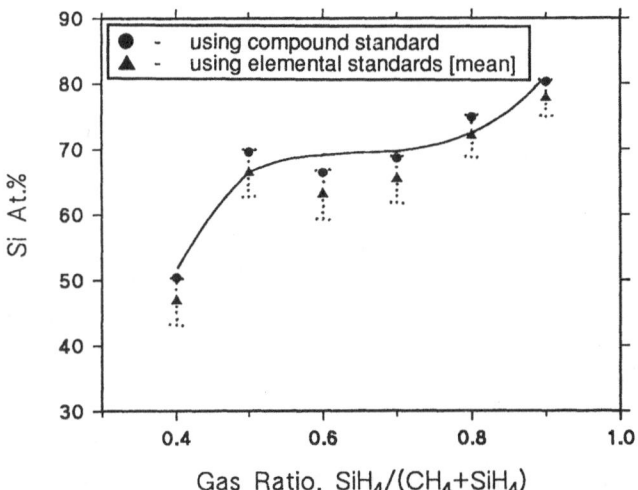

Figure 3. *Variation in surface composition for a range of a-SiC$_x$:H films relative to an SiC standard. The films were deposited by PECVD with a range of silane-methane gas mixtures. The mean values obtained relative to element standards are also shown, with the error bars corresponding to the spread of these values. Courtesy Fitzgerald et al. (1992b) and Elsevier Science Publishers.*

5 Comparison of XPS Matrix Factor Calculations

By use of the QUAX software the variation of the XPS matrix factor with composition for an amorphous silicon carbide film a-SiC$_x$:H can be studied for different matrix dependent parameters; for example, for different attenuation lengths obtained by a number of methods.

The use of a compound standard minimises the importance of the matrix factor. In Figure 3 a comparison of quantification of six a-SiC$_x$:H films prepared from different silane-methane mixtures, using both a compound standard and elemental standards is shown. The maximum divergence for quantification using a SiC compound standard was less than 1 at.% when using different attenuation-length calculations confirming the reduction of matrix effects in this case. Also shown are the mean compositions obtained with element standards using three different attenuation-length calculations.

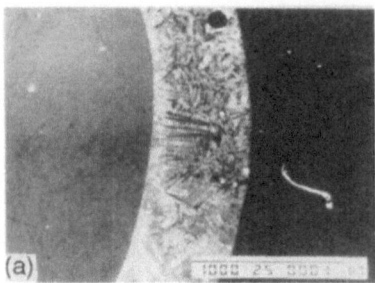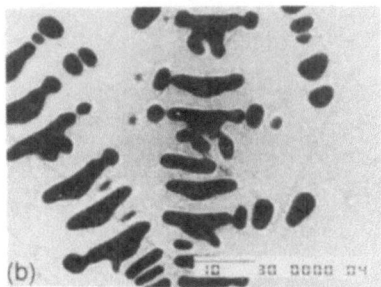

Figure 4. *Backscattered electron image (atomic number contrast) from a Cu/Ag sample showing (a) a region of intermediate composition between copper (right) and silver (left), scale mark 1000μm and (b) a region containing both a fine and a coarse eutectic structure, with silver-rich component, (light contrast) and copper-rich component, (dark contrast). Scale mark 10μm.*

6 Analyses of the Cu-Ag Interface by AES and EDX

The interface region between the two metals in a copper-silver bimetallic sample observed by backscatter electron imaging in the SEM contains two distinct phases characteristic of a eutectic structure (Figure 4). X-ray microanalysis shows that the eutectic is composed of copper-rich and a silver-rich phases, with an overall eutectic composition (39 at.% Cu). The crystal structure and composition of the constituent phases of the eutectic material have been confirmed by electron diffraction studies of a thinned foil of the eutectic composition.

When the surface composition of the ion beam cleaned eutectic material is examined by high resolution AES, a constant surface composition (53 at.%Ag and 47 at.%Cu) is obtained on quantification, using AQUA. This uniform surface composition can be explained by observation of the surface structure of the alloy after etching (Figure 5a). Preferential sputtering of one component of the eutectic material is evident in the micrograph with the formation of ridges and etched valleys. Figure 5b shows the same area imaged in the backscattered electron mode. The dark contrast (copper-rich) in the backscattered image corresponds to the top of the ridges which appear bright in the secondary-electron image. The backscattered electron image shows clearly that the ends of the surface ridges exhibit dark contrast, and are therefore copper-rich. During ion-beam etching the copper regions shield areas of silver-rich material from the ion beam. The resulting complex surface structure prevents an accurate analysis of the surface composition by Auger electron spectroscopy. Scattered electrons produced from silver-rich regions well below the surface of a ridge excite Auger electrons at the sidewall of a ridge. This surface is likely to be silver-rich at a depth of the order of 0.2μm since this is the dimension of the eutectic phase texture. A schematic diagram of the sites for the origin of Auger electrons emitted from the surface structure of the eutectic,

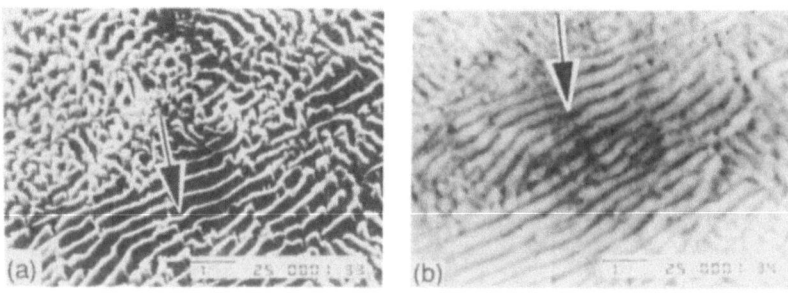

Figure 5. *Area of an ion etched surface of the copper-silver eutectic with the sample tilted 45^0 to the incident beam. (a) secondary electron image, (b) backscattered electron image. The arrow indicates equivalent areas in the two micrographs. Scale mark $1\mu m$.*

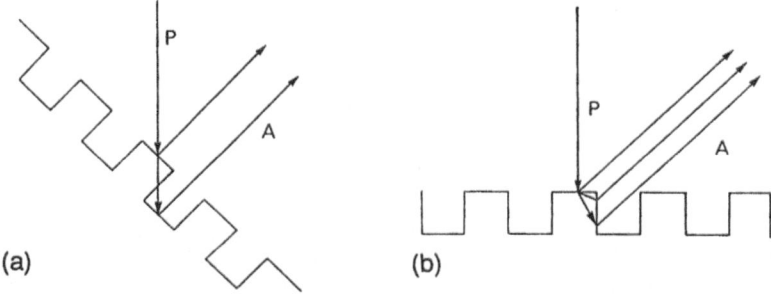

Figure 6. *Diagram showing the sites of origin of Auger electrons (A) generated by the primary beam (P) at ridges, ridge walls and furrows in an ion etched eutectic surface with a uniform distribution of the two constituent phases giving rise to an approximately constant copper-silver surface composition, (a) 45° incidence and (b) normal incidence. Only Auger electrons directed towards the detector are shown.*

obtained by ion beam etching, when normal and tilted to the electron beam, is shown in Figure 6. At a 45^0 angle of incidence for a regular eutectic structure, similar in form and size to the structure observed in Figure 5, an approximately constant composition across the eutectic will be detected. Although the form of this fine eutectic structure obtained by ion beam cleaning explains the surface composition obtained using AQUA, the effect of ion beam cleaning on surface composition is more complex. Applying AQUA to AES from large silver-rich areas of the eutectic gave a surface composition of silver of approximately 60 at.%. For large copper-rich regions the surface composition was found to have an approximately 40 at.% silver composition.

7   Comparison of Compositions by XPS and EELS

In Table 1 the compositions for a range of a-SiN$_x$ films obtained by XPS and EELS are compared. The NH$_3$/SiH$_4$ gas volume ratio (R), used in film preparation by PECVD, has been varied over a wide range. Films of a-SiN$_x$ are known to contain up to 15 at.% hydrogen. Since none of the analytical techniques can detect hydrogen, atomic concentrations are expressed as a percentage of (N + Si).

	Composition at%			
	XPS		EELS	
R	Si	N	Si	N
1.0	53.4	46.6	61.4	38.6
2.0	46.6	53.4	49.8	50.2
4.0	46.0	54.0	48.8	51.2
6.0	45.4	54.6	46.1	53.9
8.0	45.1	54.9	45.1	54.9
12.0	39.5	60.5	36.2	63.8

Table 1. *Comparison of compositions of a-SiN$_x$ films for a range of gas volume ratios, R (= NH$_3$/SiH$_4$).*

8   Thin Films on Substrates by XPS, AES and EDX

The quantitative X-ray microanalysis of thin films on substrates is a difficult analysis problem. Electron penetration into the substrate and any subsequent electron backscattering give rise to a complex X-ray spectrum which contains X-ray peaks from both film and substrate. Conventional ZAF calculations cannot be used. Instead Monte Carlo simulations can be applied to emulate the electron-sample interactions in thin films on substrates. The determination of the composition of a thin binary film on a substrate has been tested with a range of vacuum deposited films (Table 2).

	Composition at.%		
Element	Monte Carlo	AES	XPS
As	45	a	49
Se	55	a	51
Zn	42	50	43
Se	58	50	57
As	64	57	53
Te	36	43	47

Table 2. *Monte Carlo and surface analysis results from some vacuum deposited films. a - AES not possible due to specimen charging*

9 Conclusions

The AQUA/QUAX software for determination of surface composition by AES and XPS can be used to make rapid calculations of surface composition and provide a comparison of different matrix dependent parameter calculations.

Good agreement has been obtained between film compositions obtained by Monte Carlo simulations and by surface electron spectroscopies. Good agreement was also obtained in quantitative analysis by surface and thin film electron spectroscopic techniques.

By correlating the surface structure of an ion beam etched Cu-Ag eutectic with the Auger signal from this structure one source of error explaining the lack of agreement between the eutectic compositions obtained by AES and EDX has been found.

Acknowledgements

I would like to thank my colleagues B E Storey, P A Moir, H L L Watton, A D Gillies and P G LeComber for their help in this work. I am grateful to the Science and Engineering Research Council for Research Grant GR/F 92978.

References

Fitzgerald A G, Moir P A and Storey B E, 1992a, *Surface and Interface Analysis* **19** 200–294

Fitzgerald A G, Moir P A and Storey B E, 1992b, *J Elect Spectr and Relat Phenom* **59** 127–159

Ishimura S, Shimizu R and Langeron J P, 1983, *Surface Science* **124** L49-L54

Penn D R, 1976, *J Elect Spectr and Relat Phenom* **9** 29-40

Reuter W, 1971, *Proc 6th Inter Conf on X-ray Optics and Microanalysis*, eds Shinoda G, Kohara K and Ichinokawa I, 121-130 (University of Tokyo Press, Tokyo)

Scofield J H, 1976, *J Elect Spectr and Relat Phenom* **8** 129-137

Seah M P, 1980, *Surface and Interface Analysis* **2** 222-239

Seah M P and Dench W A, 1979, *Surface and Interface Analysis* **1** 2-11

Tanuma S, Powell C J and Penn D R, 1990, *J Elect Spectr and Relat Phenom* **52** 285- 291

Yeh J J and Lindau I, 1985, *Atomic and Nuclear Data Tables* **32** 1-155

Data Analysis and Processing

P Trebbia

Université de Reims Champagne-Ardenne
France

1 Introduction

In this Chapter,we shall refer to the result of the measurement (or of a measurement series) of a given variable (*e.g.* the number I of detected particles) as a function of various physical parameters (time, energy, temperature, coordinates in direct or reciprocal space, *etc.*) as data. These data are at least two-dimensional (*e.g.* a spectrum) but they may have a much higher dimension. For example, a time sequence of energy filtered images is a 5-dimensional data set (I versus x, y, E, t). Because there is no intrinsic difference between 2-dimensional or N-dimensional data, except for memory management, computing time and display convenience, in the following the particular dimension of the data set will not be specified. A measurement can thus be seen as a point in an N-dimensional space, with coordinates corresponding to the values attributed to each one of the physical variables. Due to the inherent imprecision of every measurement, these values have a random character (the probe is not exactly at position(x, y), the spectrometer is not exactly set at energy E, the number of detected particles per second is not exactly I). However, the relative precision of these values can be different by several orders of magnitude, so that one usually considers some of them (*e.g. I*) as random variables when compared to the other deterministic variables (x, y, t, E). Before starting any process, it is essential to understand the inherent character of the measurement.

1.1 Deterministic variables versus random variables

Deterministic data can be described by an explicit mathematical relationship. As a first step, this can be categorised (Bendat and Piersol 1986) as periodic or non-periodic data, periodic data being further classified as sinusoidal or complex periodic data, whereas non-periodic data contains both almost-periodic and transient data. The validity of such a classification can be confirmed by Fourier analysis in reciprocal space.

Let us suppose, for example, that we are measuring some physical phenomenon Y with respect to time t. Periodic data $Y(t)$ would give a Fourier spectrum with peaks well localised at given frequencies (the harmonics) that are all integral multiples of a fundamental frequency F. On the other hand, non-periodic data may have either a Fourier spectrum made of discrete peaks, like periodic data, but at frequencies that are not in rational proportions, or a continuous Fourier spectrum.

In any case, the Fourier Transform is obviously the tool to be used for any analysis and further processing (masking, filtering, *etc.*) of such deterministic data. Random variables are quite different. The same measurement, performed N times within exactly the same protocol, that is under the same experimental conditions, is likely to give N different values. Statistics have to be used to extract from such a collection of N values the general trends of the variable: mean value, variance and correlation functions. Let us consider for example, N measurements of the variable x as a function of time t.

$$\mu_x(t_1) = \lim_{N \to \infty} \frac{1}{N} \sum_{k=1}^{N} x_k(t_1)$$

$$\text{var}_x(t_1) = \lim_{N \to \infty} \frac{1}{N} \sum_{k=1}^{N} [x_k(t_1) - \mu_x(t_1)]^2 \tag{1}$$

$$R_{xx}(\tau, t_1) = \lim_{N \to \infty} \frac{1}{N} \sum_{k=1}^{N} x_k(t_1) x_k(t_1 + \tau)$$

are respectively defined as the mean value $\mu_x(t_1)$, the variance $\text{var}_x(t_1)$ and the auto-correlation function $R_{xx}(\tau, t_1)$ at a specific time t_1. If both $\mu_x(t_1)$ and $R_{xx}(\tau, t_1)$ do not vary as t_1 varies, then the variable $x(t)$ is said to be stationary in the wider sense. For such variables,

$$\mu_x = \mu_x(t_1) \qquad \text{var}_x = \text{var}_x(t_1) \qquad R_{xx}(\tau) = R_{xx}(\tau, t_1) \tag{2}$$

If, moreover, the same statistics can be obtained by time averaging over a specific measurement k, then the random variable $x(t)$ is said to be stationary and ergodic. In that case,

$$\mu_x = \mu_x(k) = \lim_{T \to \infty} \frac{1}{T} \int_0^T x_k(t) dt$$

$$\text{var}_x = \text{var}_x(k) = \lim_{T \to \infty} \frac{1}{T} \int_0^T [x_k(t) - \mu_x(t)]^2 dt \tag{3}$$

$$R_{xx}(\tau) = R_{xx}(\tau, k) = \lim_{T \to \infty} \frac{1}{T} \int_0^T x_k(t) x_k(t + \tau) dt$$

As will be emphasised below, a careful study of these three main statistical functions (mean, variance and correlation) will generally be sufficient to analyse and process data with a random character. Note, however, that if another random variable, y, has to be compared with x, one must introduce two other functions, called covariance and cross-correlation, defined by:

$$\text{cov}_{xy}(t_1) = \lim_{N \to \infty} \frac{1}{N} \sum_{k=1}^{N} [x_k(t_1) - \mu_x(t_1)][y_k(t_1) - \mu_y(t_1)]$$

$$R_{xy}(\tau, t_1) = \lim_{N \to \infty} \frac{1}{N} \sum_{k=1}^{N} x_k(t_1) y_k(t_1 + \tau) \tag{4}$$

1.2 The components of a measurement

Every time an experiment is performed, the resulting measurement has a random character: the value given by this measurement cannot be predicted exactly. One may arbitrarily split it into two components, the signal and the noise:

$$\text{Measurement} = \text{Signal} + \text{Noise}$$

where signal could be defined, for a stationary variable, as the mean value of the data obtained over N independent experiments carried out within exactly the same conditions, and noise could be defined as a purely random variable with zero mean.

$$\text{Signal} = \text{mean(Measurement)} \qquad \text{mean(Noise)} = 0$$

As defined above, signal can then be interpreted as a pure deterministic variable, following, for example, the well-established laws of physics, whereas noise is a purely random variable whose characteristics depend on the specific probability law which rules it (Poisson law, binomial law, multinomial law, *etc.*) (Bevington 1969). Among all the possible probability laws, the so-called Normal (or Gaussian) law is important. The central limit theorem (Bendat and Piersol 1986) states that if a random phenomenon is the result of the addition of many independent and random causes, the probability laws ruling each one of these being unknown, then this random phenomenon is normally distributed. It is worth pointing out, as defined above, that the signal generally consists of a 'useful signal' (what the experiment is supposed to reveal) and a 'useless signal', also called background. In an electron micrograph for example (Trebbia and Piersol 1990, Trebbia and Mory 1990), the protein signal (useful signal) that the biologist wants to see is superimposed over an approximately uniform background due to the resin embedding the specimen; in an EDX spectrum, the peak related to a given atomic excitation line, which proves the presence of this atomic species in the specimen, is superposed over a background due to bremsstrahlung.

$$\text{measurement} = \text{useful signal} + \text{background} + \text{noise}$$

Before processing the data to extract the 'useful signal' from the background (see Section 4) and recover the genuine object of study, with the use of deterministic laws describing the transfer function of the instrument, it is necessary to analyse the data to obtain an estimation of the relative importance of these three terms. Data analysis and data processing are very different by essence and must not be confused.

1.3 Analysis versus processing

Analysis is the act of understanding as much as possible about the content of a measurement (signal to noise ratio, contrast, information source, information level, *etc.*) without modifying the data. On the other hand, *processing* is the act of transforming the data (convolution, smoothing), and rejecting part of the content (filtering) in order to keep only what one defines as 'useful'. It is quite clear that an analysis can be conducted in a rather objective way, that is with no *a priori* convictions, whereas data processing requires that some external information is injected into the problem

(Trebbia and Bonnet 1990). For example, making a Fourier transform of a noisy image of a crystal lattice is an analytical approach to the problem of trying to differentiate between the signal and noise components. The decision to perform the reverse Fourier transform of only selected areas, around a few periodic peaks visible in the reciprocal space, relies however on an additional hypothesis, namely that 'the useful signal is lying inside the filter limits and the noise frequencies are outside these limits'.

From this example, it is clear that the information injected in the measurement may be partially biased, possibly leading to biased results (the 'what I see is what I want to get' syndrome) (Trebbia and Manoubi 1989):

a) the noise is not fully removed since it is likely to have some frequency components in the same range as those of the signal,

b) there is no hope of discovering a new, unexpected structure, with frequency components lying outside the filter window.

Such a bias is easily demonstrated (Van Dyck *et al.* 1988) by Fourier transforming an image showing nothing but pure noise, selecting periodic areas in the reciprocal space, and back transforming the filtered data, an image of a periodic lattice is built. However, all the information contained in this image comes from the hypotheses needed by the process (filter position and filter bandwidth).

It is therefore obvious that data analysis, without *a priori* assumptions, except for those which have been specifically identified as inducing no bias for the problem under study, is a mandatory step.

1.4   Information and entropy

There is a very simple means of estimating quantitatively both the information conveyed by a data set (an image, a spectrum, *etc.*) and the information changes due to any kind of data processing. Following the basic principles given by Shannon (Shannon *et al.* 1949, Jaynes 1957 a,b, Jaynes *et al.* 1988, and Kullback 1959) it is well known that the information Info(E) associated with an event E is defined as the cologarithm of the probability $P_b(E)$ of that event occurring *before it happened*:

$$\text{Info}(E) = -\log[P_b(E)] \tag{5}$$

Information is a dimensionless positive number (its unit depends on the base used for the logarithm: either *nat* for natural or *bit* for base 2), taking a *zero* value only for an event that we know *a priori* (*i.e.* before it happened) it will certainly occur.

The entropy $S(P)$ of the probability law $P(E)$ ruling a random variable E that may take N possible values $E_1, E_2, \ldots E_N$ is defined as the mathematical expectation of the information conveyed by these N possible events:

$$S(P) = -\sum_{k=1}^{k=N} P_b(E_k)\log[P_b(E_k)] \tag{6}$$

With these definitions, the information analysis of a data set is very simple. Let us suppose for example that an image $I(x,y)$ has been recorded, showing some structural details visible by a contrast variation throughout the image. We have merely to compute the histogram H_a of that image, to find the mean and variance of that histogram, and to compare H_a with a Gaussian histogram H_b built with the same mean and variance. The choice of a Gaussian histogram as a reference is justified by the central limit theorem. These two histograms H_a and H_b can be interpreted (Trebbia and Bonnet 1990, Trebbia and Mory 1990), after a proper normalisation, as the probabilities of occurrence of a particular grey level respectively before (H_b) and after (H_a) the experiment is done (*i.e.* the image is recorded). The relative entropy associated with such a change of probability law, that is, the information conveyed by the actual image when compared with a uniform but noisy image showing no structural contrast, is then defined as (Kullback 1959):

$$S(H_a/H_b) \; = \; \sum_{k=1}^{N} P_a(E_k)\log\left\{\frac{P_a(E_k)}{P_b(E_K)}\right\}$$

$$\text{with} \qquad P_i(E_k) \; = \; \frac{1}{N_p}H_i(E_k) \quad i = a, b \tag{7}$$

where N_p is the total number of pixels in the image (or the number of channels in a spectrum, *etc.*), N the total number of grey levels in that image and E_k the event 'that grey level k is present in the image'. In the same way, an estimate of the information change induced by a process can be carried out by comparison of the two histograms H_b and H_a, computed respectively before and after the process. An example of the use of these concepts for the control of different image processes (Gaussian convolution, median filtering, FAC filtering (see Section 4 below) can be found in (Trebbia and Mory 1990).

2 Analysis of the measurement

The ultimate goal of data processing is to split the 'useful signal' from the background and the noise. Since the treatment of the background can be done by deterministic equations, the first step of analysis is of course the evaluation of the so-called *signal to noise ratio* (SNR) which can be seen as a kind of 'quality factor' of the measurement.

2.1 Signal to Noise Ratio

Two definitions of SNR are generally given as:

$$\text{SNR} = \frac{\text{var}(signal)}{\text{var}(noise)} \quad \text{or} \quad \text{SNR} = \left\{\frac{\text{var}(signal)}{\text{var}(noise)}\right\}^{1/2}, \tag{8}$$

the first one being generally used in Optical Sciences. From these definitions, a single measurement (a single image, or a single spectrum, *etc*) is clearly insufficient for solution of the problem: what is the noise contribution to a pixel value of 567? As demonstrated by several authors (Frank 1980, Bonnet *et al.* 1991), at least two independent and stationary measurements M_1 and M_2, (*i.e.* performed in invariant experimental conditions) are necessary. They give two 'equations' from which it is possible to extract the two unknowns var *(signal)* and var *(noise)*.

The problem is solved by computing the correlation coefficient $R_{M_1 M_2}$ defined by:

$$R = R_{M_1 M_2} = \frac{\text{cov}_{M_1 M_2}}{\sqrt{\text{var}_{M_1} \text{var}_{M_2}}} = \frac{R_{M_1 M_2}(0) - \mu_{M_1} \mu_{M_2}}{\sqrt{\text{var}_{M_1} \text{var}_{M_2}}} \tag{9}$$

where $R_{M_1 M_2}(0)$ is the value of the central peak of the cross-correlation function, that is:

$$R_{M_1 M_2}(0) = R_{M_1 M_2}(\delta_x = 0, \delta_y = 0) \tag{10}$$

where δ_x and δ_y are spatial shifts playing the same role as τ in Equation 4. SNR is then deduced as:

$$SNR = \frac{\text{var}(signal)}{\text{var}(noise)} = \frac{R}{1 - R} \tag{11}$$

2.2   Some other applications of correlation techniques

It must be emphasised that auto-correlation and cross-correlation functions are also of valuable help in many other domains:

- in the evaluation of the experimental resolution (Frank 1980, Mory *et al.* 1991) (spatial resolution of an image, for example),

- in the comparison of data gathered from the same object of study but conveying different physical information (Jeanguillaume *et al.* 1992). For example, the York multi-spectral Auger microscope (MULSAM) which is able to deliver at least five different but complementary images of a single specimen (Barkshire *et al.* 1991, Kenny *et al.* 1991). The analysis of their respective correlation enables one to determine artifacts such as beam current fluctuations or angular variations in the secondary electron yield,

- in the determination of the relative positions of similar features in different data sets (Frank 1980, Van Dyck *et al.* 1988). Besides evident applications, the registration of individual data sets belonging to the same sequence of measurements is absolutely necessary, before any data processing, as it is the case in 3-D reconstruction from tomographic projections: a correct alignment (translation, rotation) of each projection has to be performed before running any reconstruction algorithm to get rid of possible misalignments.

2.3   Analysis of large data sets

Under specific circumstances, the data set to be analysed may be complex, either consisting of the number of individual measurements involved in the experiment (a time sequence or a series of several tens of images recorded at different energies, for example) or because one is collecting data from different information sources. Analytical electron microscopes are now able to deliver simultaneously data from several kinds of detectors (illumination current, bright field, dark field, EELS, Auger, EDX, EBIC, cathodolumi-nescence, secondary electrons, *etc.* (Jeanguillaume *et al.* 1992). Although the use of monovariate statistics, as listed above, is still possible, it is clearly not appropriate for a global analysis of all the available data. A comparison of all the possible pairs of images, obtained from a sophisticated analytical microscope, is obviously not the most efficient way to get an overall view of the measurement content.

Facing such a problem, one has to turn one's mind towards a new class of statistical tools, gathered under the general label of 'multivariate statistics'.

3 Multivariate statistics (MS)

This technique was developed at the beginning of the century (Pearson 1901) to help in the interpretation of data sets in the fields of Sociology and Econometry, its applications to spectroscopies and microscopies being only very recent (Van Heel and Frank 1981, Van Heel 1984, 1991, Van Heel *et al.* 1991, Garenström 1981, Prutton 1987, Hannequin and Bonnet 1988, Bonnet *et al.* 1992). Although it may appear as some 'magical spec-ulation' to non-specialists, MS is based on a solid mathematical background (Benzecri 1969, 1978), and relies on a very simple idea which is summarised below.

3.1 The basic principle of multivariate statistics

Let us suppose that one has performed several measurements pertaining to a single experiment: for example, a collection of N EDX spectra obtained at different locations on a given specimen (or N images of the same object at different doses, or a series of N energy filtered images of the same specimen, *etc.*). If there exists no elemen-tal concentration variation between the different sampling areas on the specimen, then all these EDX spectra should exhibit the same quantitative features (peak location, peak/background ratio, 'structural contrast' in the wider sense, *etc.*). In other words, they should be proportional to each other (the only varying parameter being the acqui-sition time) and, from a mathematical point of view, they are fully dependent. From any given spectrum of that series, one can build any other one (within statistical error) provided that one knows the arbitrary multiplicative coefficient (integrated intensity ratio) to be used as a scaling factor. The situation is identical to a collection of parallel vectors with different norms: all these vectors have the same support, that we shall refer to as a factorial axis, but different scaling factors containing information on the modulus. Each one of these scaling factors is the coordinate of the relevant vector on the factorial axis. On the other hand, if there exists any concentration variation between two probed locations, the two related spectra are not proportional and this situation is similar to two independent vectors generating a 2-D vectorial space.

Multivariate statistics takes benefit of such a property. Each measurement, an image made of P pixels, (or a spectrum made of P channels) is considered as a single vector (with P components), the whole data set, the N images, being a collection of N vectors. These N vectors are gathered in a single matrix (N by P) and MS merely analyses the characteristics of the vectorial space generated by this matrix. Looking for the eigenvectors (the so-called factorial axes) enables one to find whether some information is redundant (any pair of proportional images would pertain to the same eigenvector, thus reducing the dimension of the vectorial space). This also enables hierarchical classification of the non-proportionality between images (contrast variation, noise fluctuation, *etc.*) by looking at the eigenvalue associated with each eigenvector.

3.2 An example: the factorial analysis of correspondence

Several analysis methods belong to the class of multivariate statistics (Benzecri 1978, Lebart *et al.* 1982). One of them however, the factorial analysis of correspondence (FAC), is of particular interest because the metrics used for measuring the 'distance' between data, that is for measuring the similarity between individual data sets (between images or between spectra) can be interpreted in terms of variance. As already mentioned (Trebbia and Bonnet 1990), variance, contrast and information are closely connected concepts. The analysis of a measurement series by FAC retrieves and classifies, in hierarchical order, the different independent sources of information contained in the series. These sources are displayed as orthogonal factorial axes; the coordinates of each data set on these axes giving the weight of this information source in the data set.

Since signal, background and noise, as defined above, are three independent and orthogonal sources of information, one can easily understand that FAC is able to solve the problem stated in Section 1.2. This has been demonstrated for an energetic image series (Trebbia and Bonnet 1990, Trebbia and Mory 1990). Applications of MS to time series are under progress for analysing dynamical phenomena either as radiation damage (through EDX spectra (Jbara *et al.* 1992) or the diffusion of molecules through solid materials. An application of the MS technique to the very critical problem of automatic classification and alignment of data sets can also be found in (Frank *et al.* 1991, Van Heel *et al.* 1991).

4   Data processing

Once the data have been analysed, one has to decide whether they are worth processing or whether they should be rejected: whatever the sophistication of the above process, it will be unable to deliver more genuine information than that already included in the measurement, except for the external hypotheses required by the process itself (see Section 1.3). This can be summarised by the invariable rule: 'Good calculations can never be a substitute for good experiments'. The problem to solve is dependent on both the SNR of the data (is the noise an important component of the measurement?) and on the availability of a theoretical model describing either the background or the 'useful signal', or both.

4.1   Smoothing and/or filtering

Some possible artifacts induced by filtering operations have been already discussed in Section 1.3. Smoothing the data, to reduce the noise component and increase the SNR, is also a biased operation. In images, for example, smoothing is usually performed via a convolution product between the data and a squared matrix of normalised coefficients.

This operation results in a 'washed out and blurred' image: the SNR is indeed increased but at the expense of the spatial resolution. Perhaps it would be worthwhile to reduce the relative importance of the noise by increasing the counting time of the measurement. If the specimen cannot suffer higher exposure times, one has to ask oneself about the reliability of the data.

In any case, one must not forget that noise is also useful for checking many hidden instrumental artifacts. In a counting experiment (*e.g.* EDX, EELS, Auger spectroscopy), the data are supposed to obey a Poisson law. Any illumination instability would increase the expected variance, whereas a reduced variance of the data would cast suspicion on the stability of the main supply: a 50 Hz interference on a spectrometer is similar to a smoothing process on adjacent channels. The following principle can thus be stated: except in the most hopeless and difficult circumstances, never smooth the data systematically.

4.2 Signal estimation with an *a priori* deterministic model

After the noise level in the measurement has been checked to be natural, and provided a theoretical model is available to explain *a priori* the behaviour of the signal component, one may try to find the best estimates of the various parameters needed by this model with respect to the actual values of the experimental data. Assumption of the existence of such a model is, in fact, an external hypothesis (*i.e.* not included in the data) that must be tested in relation to the real experiment. Several procedures can be used, but it would be better to take into account, in this hypothesis test, the random character of each data and, therefore, to prefer the maximum likelihood method (ML) to other least squares methods (LS) (Taupin 1988). The basic idea of the Maximum Likelihood search is as following (Bevington 1969): Let us suppose there are N independent measurements ($\exp_i$) of a random variable, for example the content of N channels in an energy spectrum $I(E)$.

$$\exp_i = I_i(E) = I(E_i) \quad i = 1, \ldots N \tag{12}$$

with I being a random variable (number of detected particles) and E_i a deterministic variable (the energy value as given by the setting of the spectrometer). Let us suppose, also, that the model is a mathematical expression with P parameters. If the number of equations (N) is equal to the number of unknowns (P), then the determination of these parameters is purely deterministic and the degrees of freedom of that problem is zero. On the contrary, if N is greater than P, then the degrees of freedom $\nu = N - P$ allows one to take into account the random character of the data. Owing to the central limit theorem and because the values given by the model are interpreted as the mathematical expectation of the measurement, the probability of getting the experimental value $\exp_i$ is:

$$\mathcal{P}_i = \frac{1}{\sqrt{2\pi \operatorname{var}_i}} \exp \left[\frac{-(\exp_i - \operatorname{model}_i)^2}{2 \operatorname{var}_i} \right] \tag{13}$$

where the model gives, for energy E_i, the value model_i, and var_i is the variance of $I(E_i)$. Because the model is supposed to agree with all the independent values $\exp_i$, the overall probability of such an agreement is:

$$\mathcal{P} = \prod_{i=1}^{N} \mathcal{P}_i \tag{14}$$

The best estimates of the $\mathcal{P}$ parameters of the model are deduced from the conditions that the model must satisfy to maximise the overall probability (to find the Maximum Likelihood of the model). They are found by looking for the values of the parameters that minimise the quantity:

$$\xi^2 = \sum_{i=1}^{N} \frac{(\exp_i - \text{model}_i)^2}{\text{var}_i} \tag{15}$$

From Equation 15, the difference between the LS method and the ML method is clear. In both cases, a sum of squared differences between expected (model) and measured (experiment) values has to be minimised, but, in the ML method, each term of the sum is weighted by a factor taking into account the random character of the data.

It can be demonstrated that ξ^2 is a random variable ruled by the so-called χ^2 probability law, whose expected values are tabulated as a function of the degrees of freedom ν (Bevington 1969, Bendat et al. 1986). The hypothesis test on the validity of the model can then be conducted: at a given confidence level α (α being typically in the range 0.05 to 0.1), if the value of ξ^2 calculated with Equation 15 is smaller than the tabulated value of $\chi^2_{\alpha,\nu}$, the model is accepted; otherwise, it is rejected. Moreover, the precision on the optimised values of the model parameters can be deduced from the local variation of ξ^2 with respect to these (Bevington 1969). Applications of this ML method for the extraction of the 'useful signal' in EELS spectra can be found, for example, in (Trebbia 1988, Trebbia and Manoubi 1989, Manoubi et al. 1989).

4.3   Signal estimation without any *a priori* model

Sometimes, there is no deterministic model available to explain the behaviour of the data, or it is known *a priori* that any model is valid only over a restricted range of the variables (for example, in EELS spectra, a power law model for the background is only valid over several tens of eV (Trebbia and Manoubi 1989). In this case, one is usually tempted to try to fit the data with polynomial expressions of higher and higher degrees. This procedure has two serious drawbacks:

- the physical meaning of the model vanishes,

- any use of such a model outside the limited range where it has been tested leads to erroneous conclusions. Large and numerous oscillations are expected, the number being proportional to the degree of the polynomial. In other words, an extrapolation of the model might be hazardous.

Facing such a problem, multivariate statistics (Section 3) can be, here again, of valuable help:

- the eigenvalue associated with each factorial axis reveals the relative importance of that axis to the overall variance of the data, and gives, therefore, the relative weight of the relevant source of information (related to the axis) in the whole data set;

- by projecting each data point (each pixel of each image) onto each factorial axis, it is possible to see (as an image, a spectrum, *etc.*) each one of these axes and to assign a physical meaning to them;

- the coordinate of each data set (each image as a whole) on each factorial axis also helps in this assignation: the higher the absolute value of the coordinate, the higher the weight of the physical meaning of that data set in the physical meaning of the axis; coordinates of opposite signs revealing an anti-correlation between the two relevant sets for this particular axis. With all this information in hand, one is then ready to enter the process step. Like other processing methods, one has to inject some external information by selecting which ones among the available factorial axes are stated as conveying 'useful' information and 'useless' information. Filtering the data is thus quite simple: one has merely to eliminate the rejected axes from the linear combination describing each one of the data sets in order to obtain a filtered result.

When compared to the usual filtering processes as described in sections 1.3 and 4.1, FAC filtering has several important properties that must be emphasized:

- such a filtering operation does not modify the experimental resolution of the data set (the spatial resolution in images, for example),

- the filter conveys a physical meaning which is the sum of the information conveyed by the selected factorial axes. There is no *a priori* and general use filter, like a convolution product with a matrix of numbers that are set independently of the data to be processed.

The problem of splitting the signals from the noise and the background is then solved, as shown in (Trebbia and Mory 1990).

4.4 Other possible processes with FAC

Besides the application of a specific filter to the data, various kinds of processes can be conducted by multivariate statistics. A general review of these possibilities can be found in (Bonnet and Trebbia 1991) and are outlined here:

- the graph obtained by the location of each data set in the vectorial sub-spaces defined by the selected factorial axes can be modelled and interpolated or extrapolated with the underlying assumption that such a model is valid in the whole sub-space. In other words, it is possible to create fictitious data, that is, data that cannot be really recorded like images or spectra which would be obtained at a specific illumination (even at zero-dose, see (Jbara *et al.* 1992).

- the same graph can be used to define sub-volumes containing 'similar data', enabling automatic classification of the data.

- by neglecting all factorial axes with eigenvalues ranging below a critical value, it is possible to compress data without loss of significant information.

5 Conclusion

This short review is just an introduction to a wide variety of problems and solutions. Only a few of them have been sketched out in these pages. Beyond the technical details involved in data analysis and processing, three working rules are worth noting:

- experimental apparatus and computers are not magic black boxes. The sophistication of modern instruments must not prevent the operator thinking about which analyses and processes are best suited to a particular problem.

- sophisticated calculations never replace good experiments: it is absolutely impossible to extract from a measurement more information than it conveys by itself.

- one should always have a critical attitude of mind about which kind of information is 'useful' or 'useless'. Even noise may reveal a lot about instrumental artifacts.

References

Barkshire I R, Greenwood J C, Kenny P G, Prutton M, Roberts R H and El Gomati M M, 1991, *Surface and Interface Analysis* **17** 209

Bendat J S and Piersol A G, 1986, *Random Data: Analysis and Measurement Procedures* (John Wiley & Sons, New York)

Benzecri J P, 1969, *Methodologies of Pattern Recognition*, ed Watanabe S (Academic Press, New York)

Benzecri J P, 1978, *L'Analyse des Données* (Dunod, Paris)

Bevington P R, 1969, *Data Reduction and Error Analysis for the Physical Sciences* (McGraw Hill, New York)

Bonnet N, Lebonvallet S, El Hila H, Colloit G, Beorchia A, 1991, *J Phys III France* **1** 1349

Bonnet N, Simova E, Lebonvallet S and Kaplan H, 1992, *Ultramicroscopy* **40** 4

Bonnet N and Trebbia P, 1991, *Proc 10th Pfefferkorn Conference* (Cambridge) to be published in *Scanning Microscopy International* (Chicago)

Frank J, 1980, in *Computer Processing of Electron Microscope Images*, Topics in Current Physics **13** 187, ed Hawkes P W (Springer Verlag, Berlin)

Frank J, Penczek P and Liu W, 1991, *Proc 10th Pfefferkorn Conference* (Cambridge) to be published in *Scanning Microscopy International* (Chicago)

Garenström S W, 1981, *Appl Surf Sci* **7** 7

Hannequin P and Bonnet N, 1988, *Optik* **81** 6

Jaynes E T, 1957a, *Phys Rev* **106** 620

Jaynes E T, 1957b, *Phys Rev* **108** 171

Jaynes E T, Gull S F and Skilling J, 1988, *Maximum Entropy Bayesian Methods in Science and Engineering, Volume I: Foundations*, eds Erickson G J and Smith C R (Kluwer Academic Publishers, Boston)

Jbara O, Trebbia P, Bonnet N and Cazaux J, 1992, *Proc 10th European Congress on Electron Microscopy*, eds Rios A, Arias J M, Megias-Megias L and A-Lopez-Galindo *Electron Microscopy 92* **1** 443

Jeanguillaume C, Colliex C Ballongue P and Tence M , 1992 *Ultramicroscopy* **45** 205

Kenny P G, Prutton M, Roberts R H, Barkshire I R, Greenwood J C, Hadley M J and Tear S P, 1991, *Proc 10th Pfefferkorn Conference* (Cambridge) to be published in Scanning Microscopy International (Chicago)

Kullback S, 1959, *Information Theory and Statistics* (John Wiley & Sons, New York)

Lebart L, Morineau A, Fénelon J P, 1982, *Traitement des Données Statistiques* (Dunod, Paris)

Manoubi T, Tencé M and Colliex C, 1989, *Ultramicroscopy* **28** 49

Mory C, Kohl H, Tencé M and Colliex C, 1991, *Ultramicroscopy* **37** 191

Pearson K, 1901, *Phil Mag* **2** 559

Prutton M, El Gomati M M and Walker C G H, 1987, *Instr Phys Conf Ser* **90** 1

Shannon C E and Weaver W, 1949, *The Mathematical Theory of Communication* (University of Illinois Press, Urbana)

Taupin D, 1988, *Probabilities, Data Reduction and Error Analysis in the Physical Sciences* (Les Editions de Physique, Les Ulis)

Trebbia P, 1988, *Ultramicroscopy* **24** 399

Trebbia P and Manoubi T, 1989, *Ultramicroscopy* **28** 266

Trebbia P and Bonnet N, 1990, *Ultramicroscopy* **34** 165

Trebbia P and Mory C, 1990, *Ultramicroscopy* **34** 179

Van Heel M and Frank J, 1981, *Ultramicroscopy* **6** 187

Van Heel M, 1984, *Ultramicroscopy* **13** 165

Van Heel M, 1991, *J Mol Biol* **220** 877

Van Heel M Winkler H, Orlova E and Schatz M, 1991, *Proc 10th Pfefferkorn Conference (Cambridge)* to be published in *Scanning Microscopy International* (Chicago)

Van Dyck D and Coene W, 1988, *J Microsc Spectrosc Electron* **13** 463

Microscopy and Microanalysis of Insulating Materials

J Cazaux

Université de Reims
France

1 Introduction

Most materials in the human environment are insulators. Their investigation and analysis using incident charged particles (electrons in SEM, TEM, EPMA and AES and ions in SIMS) or neutral particles (x-ray photons in XPS) leads to difficulties and errors in quantification procedures. Starting from the physical properties that insulators have in common, these problems are described first. Next a critical analysis of the approach generally used for explaining the charging effects on bare insulators based on the total yield of the emitted particles is undertaken. An electrostatic–electrokinetic approach is then suggested to explain the influence of the experimental conditions and the nature of the surrounding media (such as the coating of the specimen). Some practical solutions are deduced and additional developments concerning the microscopic causes of charging, the time evolution of these effects as well as the beam induced currents are indicated.

2 Problems

By definition, insulators are materials which have the lowest electrical conductivity values γ. The range of conductivity is from around $10^{-8}\Omega^{-1}\mathrm{m}^{-1}$ (at the blurred frontier with semiconductors) up to $10^{-16}\Omega^{-1}\mathrm{m}^{-1}$ for excellent insulators. Using reasonable values for carrier mobilities, it can be deduced that the density of these carriers is less than $10^{8}\mathrm{cm}^{-3}$ and is associated with crystalline imperfections or impurities in the investigated insulators. If almost all the valence electrons are bound, they participate in the polarisation of the insulator submitted to the influence of an electric field E. This property can be characterised by the value of the dielectric constant.

When irradiated by incident particles, the atomic mechanisms involved in insulators are qualitatively the same as those of metals. These mechanisms concern the elastic scattering of incident particles, the sputtering of atoms (in SIMS) and the excitation of the valence and the core electrons of the atoms composing the insulator. These excitation processes are followed by de-excitation processes (x-ray and Auger emissions) and by the ejection of charged particles : secondary ion emission in SIMS and secondary electron emission in all the techniques (Cazaux and Lehuède 1992).

The main aspect of the investigation of an insulator with charged or neutral particles is that, in contrast to a metal, there are insufficient carriers to quickly restore the neutrality of the specimen. The consequences of this non-neutralisation of charge are :

1. The modification of the physical parameters of a given interaction because of acceleration or slowing down of the incident charged particles and their deflection (even when the specimen is coated by a thin conducting layer).

2. A change of the spectral distribution and total yield of the emitted charged particles such as a shift in energy of the secondary electron distribution, of the Auger lines or the photoelectron lines in XPS.

3. Possible modification of the specimen such as oxygen desorption in some oxides or the migration of the mobile ions in glasses.

The absence of conduction electrons in monocrystalline insulators may also explain the fact that these materials have, at least at the initial stage of the irradiation, a secondary electron yield of about one order of magnitude greater than that of metals (when irradiated under the same conditions). The absence of electron-electron interactions increases the attenuation length of the emitted secondary electrons. However, the situation changes rapidly because the electric field then established modifies the yield of the secondaries leading to complications in the analysis of a time dependent situation.

The trajectories (and interactions) of incident particles can be calculated using standard methods at the beginning of the irradiation when the insulator is not charged. However, the situation becomes more complex after some incident particles are trapped into the insulator and secondary electrons are emitted (leaving positive charges) and the electric field thus created modifies these interactions and trajectories which in turn modify the field first established, and so on (even in coated insulators).

3   Discussion

For a bare insulating surface (bounded by a vacuum) and submitted to electron irradiation, the most popular approach to determine the sign of the surface potential V_s is based on the evaluation of the number (and sign) of particles entering the specimen, I_0 , and the number (and sign) of particles exiting it : $I_r = \sigma I_0 = (\eta + \delta)I_0$ where η is the backscattered electron coefficient, σ equals total yield and δ equals the secondary electron yield (Reimer 1985). When σ is greater than the unity, a positive charging is expected *i.e.* when the primary beam energy E_0 is between E_{C1} and E_{C2} (see Figure 1). In the next step, a decrease of δ is introduced to explain some kind of

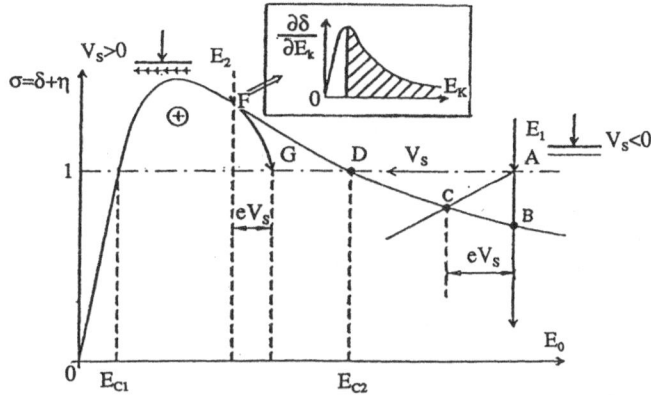

Figure 1. *Illustration of the conventional approach (for the charging of insulators by incident electrons) based on the total yield curve. Positive charging is expected when the primary beam energy E is between E_{c1} (50–200V) and E_{c2} (2.5–5 keV). The inset illustrates the decrease in the secondary electron emission as a function of a surface potential V increasing positively.*

stabilisation of the positive potential while the use of Ohm's Law $V_s = RI_s$ (where R is the leak resistance and I_s is the specimen current) is suggested to find V_s when the specimen is negatively charged, i.e when $E_0 < E_{C2}$. In fact, as we shall see below, this analysis fails for the positive region because the $E_{C1}-E_{C2}$ interval may change during the irradiation (because of the established electric field) while the induced current is non-ohmic for the negative charging region. The correct approach, valid at any time during the irradiation, is to use total charge conservation given by (see Figure 2 for the signs) :

$$I_0 = I_r + (\frac{dQ}{dt}) + I_s \tag{1}$$

where dQ is the charge trapped during the time interval dt. I_r and I_s are obviously dependent on the field created by the total charge, Q, $(= \int_0^t (dQ/dt)dt)$ and its distribution ($\int_V \rho dV = Q$) accumulated inside the insulator from the time $t = 0$ up to $t = t_0$ (Cazaux 1992). Consequently, for a permanent irradiation and even when the steady state is attained (at $t = \infty$, $dQ/dt = 0$), the equilibrium reached differs from that of the initial state.

For evaluation of the electric field, E, and potential, V, at any point of the space (inside the insulator and outside it) at any time, the need is to know the charge distribution $\rho(\mathbf{r})$ and its time evolution, $\dfrac{\partial \rho(\mathbf{r})}{\partial t}$. When the incident charged particles remain fixed (ion implantation) in the specimen, the problem is restricted to a search for solutions of the Poisson equation satisfying penethe boundary conditions and starting from a theoretical or experimental implantation profile.

The problem of irradiation by light and mobile particles such as electrons is more difficult. From elementary considerations, it can be established that all the incident electrons are not trapped, or if they are, the trapping time is very short because if

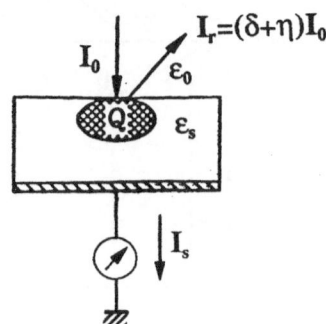

Figure 2. *Conventional positive signs are used for the charge conservation relation.*

this were not the case the field they would have created could reach a disruptive value (greater than $10^7 \mathrm{Vcm^{-1}}$) in less than one second (for $J_0 = 1\mu\mathrm{A/mm^2}$) and the specimen could be destroyed. Consequently for electron irradiation solution of the electrostatic problem requires a knowledge of the microscopic mechanisms of trapping, mechanisms which depend upon the composition and the crystalline structure of the specific specimen being irradiated (as well as the specific conditions of the irradiation).

To keep this general description of charging effects it is possible to avoid these difficulties by using reasonable models for the charge distribution inside irradiated insulators. Using the classical laws of electrostatics, it is easy to deduce the general form of the electric field and of the electric potential inside and outside the specimen in order to explain the main aspects observed during irradiation of insulators (Cazaux 1986a,b).

For uncoated insulators, the model distribution (Figure 3) can be considered to be composed of two cylinders having the same diameter, d_0 , (the incident beam spot size), the height of the first cylinder (positively charged) corresponds to the attenuation length of the secondaries. The height of the second cylinder corresponds to the range R of incident particles in the material and the sign of the charge distribution is the same as that of the primaries. This cylinder is only present when the insulator is coated by a thin conducting layer (at earth potential) while only the first cylinder is present in the case of irradiation by neutral particles (such as x-ray photons).

Of course such naïve models could be improved but, even in this simplified form, they can give useful information for the case of electron irradiation (which can be easily transposed to other types of radiation). Other models for the charge distribution are:

- A non-uniform electric field inside the dielectric which diverges (or converges).

- For a given charge distribution, the field characteristics change when the insulator is coated.

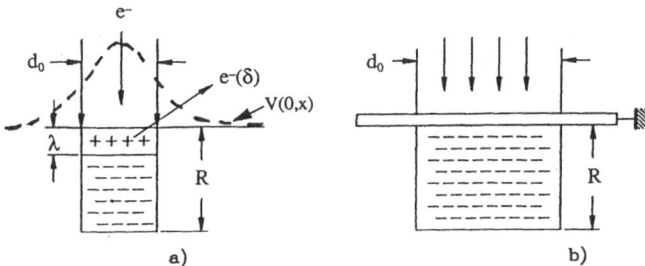

Figure 3. *Examples of the model for the charge distribution of irradiated insulators. Left: uncoated insulator. Right: coated insulator.*

- The potential at an uncoated surface is a function (and not a constant as is supposed in the total yield approach) and its shape depends upon the incident spot size diameter while its sign depends upon the sign of the algebraic sum of the positive and the negative charges trapped at the surface and in the bulk.

This type of analysis gives an explanation of the direction of migration of mobile ions in glasses (Figure 4) and the corresponding change of the emitted x-ray signal (Figure 5). It also explains the shift and distortion of the spectra of emitted particles; it is clear, for instance, that the trajectories of the secondary electrons into the vacuum are influenced by the electrostatic lens produced by the electric field built-up at the surface and around it. Moreover, because electrons are emitted at parts of the specimen where the potential function is not the same, their spectral distribution is distorted. This analysis also explains the distortion of thin (biological) specimens when investigated by transmission electron microscopy (due to the radial distribution of the electric field in this case (Cazaux 1986a).

For a bulk specimen it is also possible to explain the change of the observed spectra in terms of change of the experimental conditions such as incident probe size, defocused or not, change of width of the analyser field aperture or change of specimen environment or thickness (the boundary conditions lead to different fields if the specimen is in the form of an unsupported foil or if it is backed by the metal of the specimen holder or if it is coated by a thin conducting layer at earth potential). All these statements seem obvious (they are derived from classical electrostatics) but they may explain many conflicting results obtained on samples of the same nature but investigated using different equipment. In particular, to prevent the deflection of the incident charged beam into the vacuum, the insulator to be investigated is often coated by a thin conducting layer but it is also often forgotten that the electric field inside such a specimen may take very large values mainly at the metal/dielectric interface (Figure 6). During the irradiation, the electric field first established induces a slowing down of the next arriving incident electrons and a decrease of their penetration depth. Consequently the absorption correction factor to be used in EPMA analysis of these materials should be less than that used for a conducting specimen of the same composition.

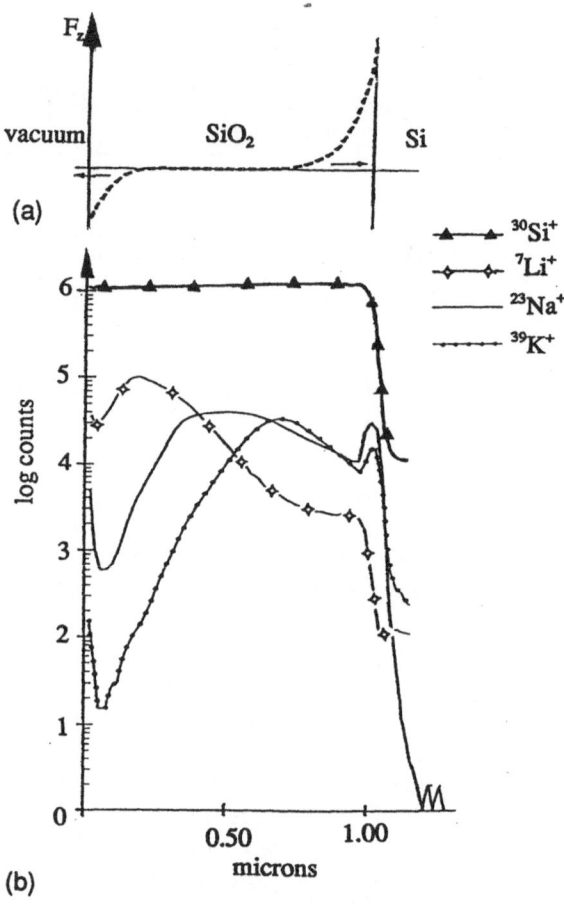

Figure 4. *(a) Positive ion profile obtained with* SIMS *on a vacuum-SiO$_2$-Si system showing a migration compatible with (b) the calculated electric field owing to electron irradiation (see the inset at the top) (Cazaux and Lehuède 1992).*

When an uncoated insulator is investigated using electron irradiation the induced electric field is mainly directed towards the bulk except close to the surface (Figure 4b). Auger analysis may lead to false values in this situation because of the migration of mobile ions (Figure 4a). However, during such an analysis a subtle change in the boundary condition may occur (such as carbon contamination or accumulation of metallic atoms at the surface). This change may modify the direction and amplitude of the electric field built up.

It should also be kept in mind that the parameter measured is often the potential (through the shift of the electron spectra) while the physically important parameter is the electric field (and its associated Coulomb force). For example, in the irradiation of thin oxide on a conducting material, when going from the surface to the interface, the change of the surface potential may be too low to measure while the local field

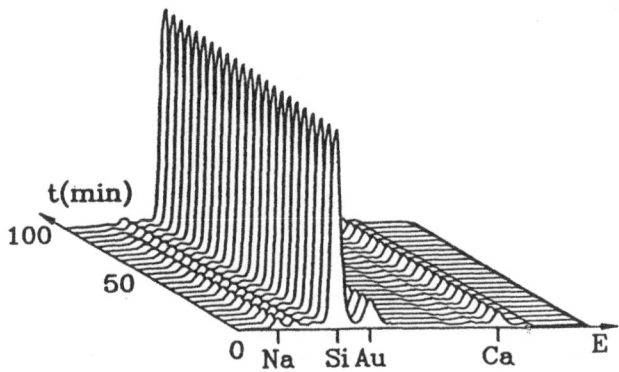

Figure 5. *Change in the x-ray emission spectrum as a function of time for glass coated by a thin gold layer and irradiated by 20keV electrons (note decrease of the Na signal).*

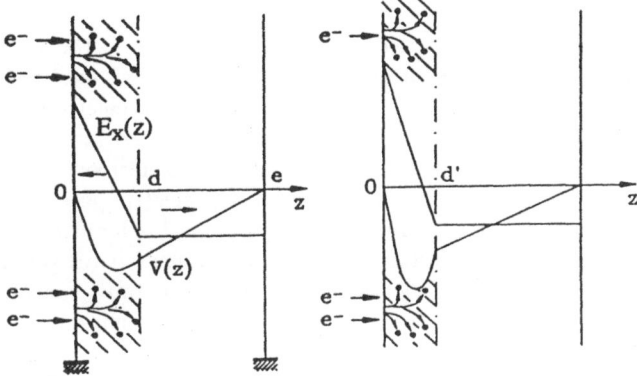

Figure 6. *Expected electric field and potential produced by a uniform charge distribution (up to depth, d) inside an insulator sandwiched by a thin conducting coating and the specimen holder. During irradiation the new incoming electrons are decelerated by the trapped earlier incoming electrons and their penetration depth decreases.*

may be sufficient to induce effects such as the migration of the species (for a thickness $t \simeq 100$Å, a voltage drop of 0.1V corresponds to a field of 10^5Vcm^{-1}).

If the electrostatic approach is considered in more detail for electron irradiation, it is necessary to obtain more information on the trapping mechanism for electrons. By analogy with the semiconductor situation, it is believed that incident electrons can be trapped on structural defects or impurities and the evolution of the charge density as a function of time is expected to follow (Cazaux and Le Gressus 1991a) :

$$\rho(t) = J_0 N \sigma \tau (1 - e^{-t/\tau}) \tag{2}$$

where N is the density of trapping centres, σ the capture cross-section, J_0 the incident beam density and τ the time constant of the detrapping process.

This result resembles the charging of a capacitor through a resistor (even if this analogy is not very satisfying) and the electric field and potential at a given point in space are expected to follow the same time dependence.

When applied to a perfectly monocrystalline specimen, Equation 2 leads to non-charging of the bulk ($\rho = 0$ when $N = 0$) and to surface charging (of density $Q_s(t)$) associated with the corresponding surface density of trapping defects N_s (N becomes N_s) (Cazaux *et al.* 1991b).

In fact, analysis of the experimental results can be complicated by additional effects such as contamination, surface neutralisation by the secondaries and flashover when a critical value is attained but, in predicting the time evolution of charging effects in a given insulator the main difficulty stems from the lack of reliable values of σ and τ. For example, does τ change as the function of the applied electric field?

For a complete investigation of charging effects, the specimen current must also be considered. As with a metallic specimen, the physical reason for the specimen current in insulators is the restoration of the neutrality of the irradiated sample. The direction of the specimen current depends upon the sign of the majority charges trapped in the solid. When this sign is positive neutralisation is due partly to the secondary electrons. Some in vacuum are reattracted and some inside the solid cannot escape into the vacuum: the two contributions lead to a decrease in δ and to a decrease in the incident electrons penetrating into the bulk in the case of electron irradiation. These electrons are also reattracted by the thin positively charged layer associated with the main part of the secondary electron emission. These two electronic contributions acting like space charges may induce several changes of the electric potential as a function of depth. In this situation the contribution from the specimen current is expected to be very low because of the poor conductivity of the specimen outside the irradiated region. An exception is specific arrangements such as partial metallic coatings of the surface (or grids) that facilitate the neutralisation by surface conductivity.

A quite different situation arises when the sign of majority charges is negative. In electron irradiation, the incident electrons injected into the insulator contribute to dQ/dt when they are trapped. However, the great majority of incident electrons are not trapped and they differ from the internal electrons existing prior the irradiation of the specimen. The density of the latter is below $10^8 \mathrm{cm}^{-3}$ and this low density explains the high electrical DC resistivity of insulators. The local density associated with the trapped incident electrons may be very large; for example, it has been shown (Cazaux *et al.* 1991b) that it was possible to inject current densities of around $1 \mathrm{\,mA/mm^2}$ or more into insulators having an electric resistivity of $10^{15} \Omega \mathrm{cm}$. Being injected by the incident beam, these additional trapped electrons are involved in a form of dynamical doping. After having lost most of their initial kinetic energy through the usual interactions with atoms of the solid, they are unable to interact with the valence electrons when their kinetic energy reaches values less than the band gap energy. In the conduction band their behaviour is similar to that of the secondaries. The only possible interaction is with phonons and thus with a large inelastic mean free path and a large scattering time (Cazaux *et al.* 1991b).

Being in a negatively charged region, these electrons, in excess, are pushed towards the earth potential where they produce the specimen current, I_s. The corresponding current is non-ohmic in the sense that its value is independent of the (low) conductivity of the material. A consequence of this analysis is to invalidate the popular use (Reimer 1985) of Ohm's Law, $V_s = RI_s$, between the surface potential V_s and I_s via a leakage resistance R. This is also true of resemblance between the charging of an insulator submitted to irradiation and the charging of a capacitor C through a so-called leakage resistance R.

4 Solutions

Some practical solutions for minimising charging effects have been reviewed by Werner and Warmholtz (1984) but a practical consequence of the electrostatic approach is that the unique solution for cancelling exactly the charging effects (*i.e.* the field and potential created by the trapped charges) is to neutralise exactly the charges creating these effects ($V = 0$ and $E = 0$ elsewhere when ρ and σ are null).

Such charge neutralisation may occur spontaneously, for instance when the charges are distributed at the surface of the insulator and when this surface is surrounded by a poor vacuum (the molecules of the surrounding atmosphere neutralising the excess charges, for example, in an environmental microscope).

Another example is the use of additional irradiations by charged particles (*i.e.* use of flood guns to counterbalance the excess charge resulting from primary irradiation). This is popular in surface analysis (Cazaux and Lehuède 1992). Using this solution, it must be kept in mind that this neutralisation is exact only when the penetration depth of neutralising charges is very close to the in-depth distribution of the charges responsible for the charging effects. The fact that the surface potential is stabilised at a value close to zero as well as the fact that the surface potential is null because of the metallic coating at earth potential does not ensure that the electric field inside the specimen is zero and does not prevent spurious effects from occurring (such as the migration of the mobile ions). Another elegant solution to charging problems consists in widely illuminating the specimen and the sample holder by photons having an energy greater than the band gap energy of the investigated insulator (UV radiations). The electron-hole pairs generated by photoabsorption induce a surface photoconductivity which allows excess charge to flow to earth (Figure 7). Finally, when successive signals (or spectra) are obtained during the irradiation of an insulator, the use of modern image processing methods may allow the evaluation of the signal or spectrum corresponding to the zero-dose limit (see Trebbia, this volume), *i.e.* prior to irradiation.

5 Conclusions

The main conclusion of this paper is that charging effects in insulators are complex phenomena that cannot be fully explained by using only an ideal total yield curve. The electrostatic approach explains at least qualitatively the main aspects of charging and the spurious effects that may occur during the investigation of insulating materials

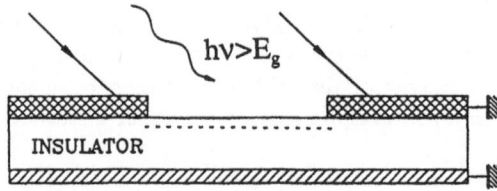

Figure 7. *Schematic drawing of a practical solution to reduce charging effects by ultraviolet irradiation.*

by incident electron beam techniques (and others). To go deeper into the subject a microscopic knowledge of the trapping mechanisms is required to enable quantification of these effects. This type of study is in its early stages and is difficult because of the dynamic processes occuring during irradiation. But it is also a very exciting field full of promising studies and applications such as secondary electron emission of insulators, electron beam nanolithography and non-ohmic induced current (Cazaux 1992).

References

Cazaux J, 1986a, *J Appl Phys* **59** 1418–1430
Cazaux J, 1986b, *J Micros Spect Electron* **11** 293–312
Cazaux J and Le Gressus C, 1991a, *Scanning Microscopy* **5** 17–27
Cazaux J, Kim K H, Jbara O and Salace G, 1991b, *J Appl Phys* **70** 960–965
Cazaux J, 1992, in *Ionisation of Solids by Heavy Particles*, ed Baragiola R A, (A NATO Advanced Workshop to be published by Plenum Press in 1993)
Cazaux J and Lehuède P, 1992, *J Elect Spectrosc Relat Phenom* **59** 49–71
Reimer L, 1985, in *Scanning Electron Microscopy*, published in *Optical Sciences* **45** 119–132 (Springer-Verlag)
Werner H W and Warmholtz N, 1984, *J Vac Sci Technol* **A2** 726–733

Electron Specimen Interactions

D C Joy

University of Tennessee and
Oak Ridge National Laboratory
USA

1 Introduction

The purpose of this chapter is to give a users guide to electron solid interactions. We start by laying out the fundamental principles of electron-solid interactions. Since this, by itself, is of limited value because such fundamental studies relate only to single electrons and single atoms in unrealisable conditions we then show how, by the application of simple Monte Carlo modelling techniques, this knowledge can be translated into useful and detailed predictions about the outcome of possible experiments in an electron microscope. Both the Monte Carlo programs, and a simple computational data base on electron-solid interactions, are available on an IBM PC compatible disk so that the data and these techniques will be readily accessible when required.

2 Scattering

An electron is a charged particle, so it can interact both with the positive charge on the ionic cores of atoms, and with the loosely bound electrons that surround atoms. These interactions can change both the direction of travel and/or the energy of the electron, and can result in the transfer of energy to the specimen and the production of various types of secondary emissions. Scattering events are divided into two distinct classes:

1. Elastic scattering events — those in which the direction of the electron trajectory is altered but the electron energy remains essentially unchanged;

2. Inelastic scattering events — those in which the electron transfers some energy to the sample. This results both in a reduction of the kinetic energy of the electron (from the conservation of energy) and to a change in the direction of its travel (because of the conservation of momentum).

The probability that a given event will occur is called the 'cross section', σ, of that event. If we consider the electron beam striking unit area then σ represents the fraction of that area that produces the event of interest. Thus the cross-section σ thus has units of cm^2/atom . In particle physics it is common to quote cross-sections in 'barns' where 1 barn=10^{-24}cm^2 and because this avoids the use of large exponents this convention will also be followed here. In practice, because electrons are travelling through the specimen, it is often more convenient to talk in terms of the average distance λ (cm) the electron travels between events of a particular type. λ is called the mean free path for the excitation of the event and is obtained from the cross-section σ by the relation:

$$\lambda = \frac{A}{\sigma N_0 \rho} \tag{1}$$

where A is the atomic weight of the target material, N_0 is Avogadro's number and ρ is the density (gm/cm^3).

2.1 Elastic Scattering

Starke's experiments (1898) showed that electrons were deviated through significant angles (much greater than 1 degree) when they interacted with atoms in a solid. Almost twenty years later Geiger and Marsden similarly found that α-particles could be backscattered from thin gold foils. Neither of these results was expected nor consistent with the uniform density models of the atom then current and it was not until Rutherford realised that these results had to imply that the nucleus was much more massive than the rest of the atom that progress was made. Rutherford derived a mathematical expression to explain the observations of high angle scattering events (for an excellent discussion on this important early work see Rhodes 1986). This 'Rutherford' cross-section model computes the effect of the positive charge on the nucleus on the negatively charged electron (Figure 1). In the simplest case we can neglect the effect of the orbiting electrons, and the incoming electron will be attracted to the nucleus by a Coulomb force $F = -dV/dr$ where

$$V = -\frac{Ze^2}{r} \tag{2}$$

Z is the atomic number and r is the instantaneous distance between the electron and the nucleus. Classical dynamics, incorporating Equation 2 gives the differential cross-section for scattering per unit solid angle as

$$\frac{d\sigma}{d\Omega} = \frac{4Z^2}{a_0^2 q^4} \tag{3}$$

where $a_0 = h^2/me^2$ is the Bohr radius for the atom (5.29×10^{-9} cm) and hq is the magnitude of the momentum gained by the nucleus and lost by the electron which is related to the scattering angle θ as

$$q = \frac{4\pi}{\lambda} \sin(\theta/2) \tag{4}$$

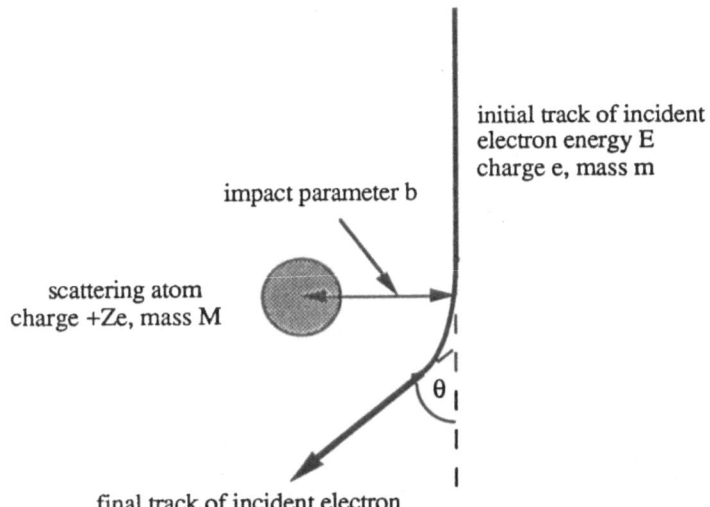

initial track of incident
electron energy E
charge e, mass m

impact parameter b

scattering atom
charge +Ze, mass M

θ

final track of incident electron

Figure 1. *Scattering of an electron by an atom in the Rutherford model.*

where λ is the electron wavelength. Neglecting relativistic effects, the energy ΔE lost by the electron of energy E and mass m and transferred to the nucleus (mass M) is

$$\Delta E = \frac{2m}{M}(1 - \cos\theta)E \tag{5}$$

Wave mechanical theory indicates that the influence of the orbiting electrons is, in fact, negligible for large angles of scattering ($\theta > 10°$ at $E = 100$ keV) so this simple Rutherford model can be used to describe backscattering effects. Putting $\theta = \pi$ in Equation 5 gives $\Delta E \approx 10$ eV for the maximum energy transferred in a 'head-on' collision with an atom such as silicon at 100 keV. Thus although this kind of scattering is normally described as elastic there can be an appreciable energy loss involved for large scattering angles. This energy transfer can result in permanent structural changes in the target, for example the displacement of atoms within the material (usually called radiation damage) or the ejection of material lying close to the surface (usually called sputtering).

For small angles of scattering ($\theta < 10$ degrees) the screening effect of the orbiting atomic electrons must be included. One way of doing this is to specify a screened Coulomb potential of the form

$$V = -\frac{Ze^2}{r}\exp\left(-\frac{r}{R}\right) \tag{6}$$

where the screening parameter $R = a_0/Z^{1/3}$. Then the differential cross-section becomes

$$\frac{d\sigma}{d\Omega} = \frac{4}{a_0^2}\frac{Z^2}{\left(q^2 + \frac{1}{R^2}\right)^2} \tag{7}$$

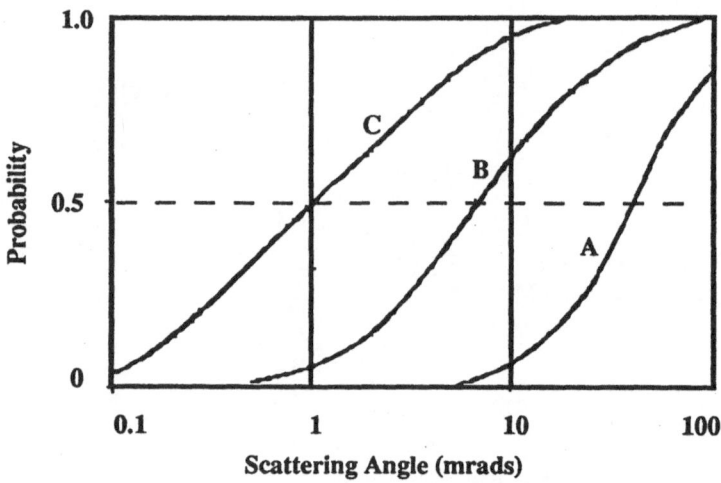

Figure 2. *The probability P(q) of scattering through an angle less than q for (a) elastic interactions in an amorphous material; (b) K shell excitations in carbon; (c) plasmon production.*

This differential cross-section can be integrated to give the integral cross-section $\sigma(\theta)$ for scattering through any angle less than θ. From Equation 6 we obtain

$$\sigma(\theta) = \frac{\sigma_T}{\left(1 + \dfrac{1}{q^2 R^2}\right)^2} \tag{8}$$

where $\sigma_T = \lambda^2 Z^{4/3}/\pi$ and is the total cross-section for scattering into all angles. Half of the total scattered intensity occurs for $qR < 1$ which is about 40 milliradians or 2 degrees for silicon or aluminium at 100 keV. Figure 2a plots this variation for silicon at 100 keV. Equation 8 describes approximately the angular dependence of elastic scattering in amorphous materials where diffraction effects arising from the wave nature of the incident electron are not too severe. σ_T is of the order of 10^{-18}cm^2 (10^6 barns) per atom, so the elastic mean free path is about 1000Å. In a crystalline solid electron diffraction is of major importance and the scattered intensity becomes sharply peaked about particular directions, the Bragg diffracted beams. For an ideal crystal with every atom at a lattice point the phase of the electron wave is such that all the atoms recoil simultaneously, this being equivalent to collision with a mass M equal to that of the entire crystal. The energy loss of Equation 5 thus becomes vanishingly small for all the Bragg beams, and the scattering is elastic in the strictest sense.

Equation 7 is often found in equivalent but differently expressed forms. For example Wentzel (1927) gives

$$\frac{d\sigma}{d\Omega} = \frac{Z^2 e^4}{16E^2} \frac{1}{\left[\sin^2\left(\dfrac{\theta}{2}\right) + \left(\dfrac{\theta_0^2}{4}\right)\right]^2} \tag{9}$$

where E is the energy (in keV) and e is the electron charge. θ_0 is the screening parameter, for which a convenient approximation (Cosslett and Thomas 1964) is $\theta_0 = 0.1167Z^{1/3}/E^{1/2}$ radians. As given in Equation 9 the cross-section is not relativistically corrected and is thus systematically in error for $E > 50$ keV or so. A correction can be made (Bethe and Ashkin 1953) by putting Equation 3 in the form

$$\frac{d\sigma}{d\Omega} = \frac{Z^2 \lambda_R^4}{64\pi^4 a_0^2} \frac{1}{\left[\sin^2\left(\frac{\theta}{2}\right) + \left(\frac{\theta_0^2}{4}\right)\right]^{-2}} \tag{10}$$

where λ_R is the relativistically corrected electron wavelength, given in centimeters by the relation

$$\lambda_R = \frac{3.87 \times 10^{-9}}{E^{1/2}(1 + 9.79 \times 10^{-4}E)^{1/2}} \tag{11}$$

where as before the energy E is in keV.

The screened Rutherford cross-section is widely used in calculations and simulations of electron scattering because it is quite accurate and is in a convenient analytic form which makes it easy to handle. However under some conditions, particularly at low beam energies ($E < 20$ keV) and for higher atomic number elements ($Z > 30$), where the scattering angles are large the Rutherford formula is only an approximation. When the cross-section is derived from the relativistic Dirac equation, taking into account spin-orbit coupling, then the 'Mott' scattering cross-section which results can differ significantly from the Rutherford values. The major problem in using the Mott cross-section is that it cannot be expressed analytically but must be computed for each element, energy, and scattering angle of interest. A complete tabulation of such values has been produced by Czyzewski *et al.* (1990) for all elements $4 < Z < 92$ and for energies between 20 eV and 20 keV. Figure 3 plots the angular differential form of the Mott cross-section for mercury at 2 keV from that data set. It can be seen that there are some interesting features in these cross-sections, particularly the oscillatory nature of the data in some angular regions. A plot (Figure 4) of the cumulative probability (*i.e.* the probability of an electron being scattered less than a certain angle) shows that there are differences in the scattering distribution between the Mott and Rutherford models which are quite marked at low angles. There are some important applications where the difference between the Mott and Rutherford models is significant. These are the cases where the interaction consists predominantly of one single, elastic, large-angle scattering event, such as in backscattering from thin films or the diffraction controlled backscattering that is observed in electron channelling patterns. In most other cases the standard Rutherford expression produces rather good data, and so can generally be used without a loss of accuracy.

As calculated above in Equation 7 the total elastic mean free path λ_E is of the order of tens of nanometers. If we imagine electrons passing through a target of thickness τ then, if τ is of the order of λ_E, on average each electron would experience just one elastic event in its transit. If τ is increased beyond λ_E however then each electron will likely experience several scatterings and is said to be undergoing *plural* elastic scattering. If the target is made so thick that each electron experiences perhaps 25 or more elastic events before leaving then the electron is said to be *multiply* scattered. The mathematical treatments of plural and multiple scattering calculate the probability

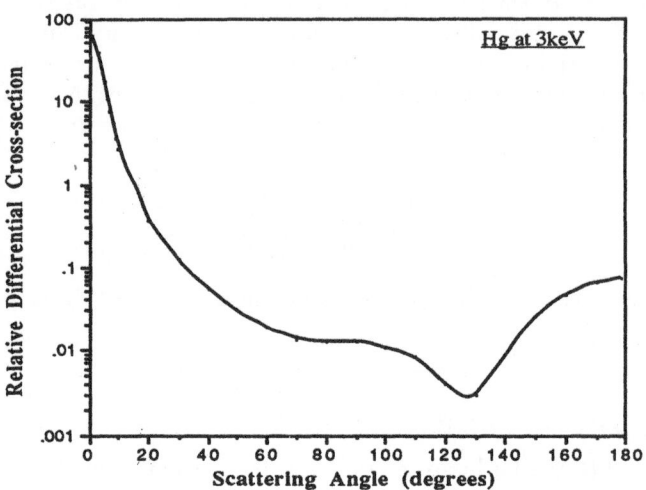

Figure 3. *Mott differential scattering cross-section for Hg at 3 keV.*

$P(\theta)$ that an electron will emerge at an angle between θ and $\theta + d\theta$ relative to its initial direction. In the case of plural scattering the result can only be expressed as an infinite series expansion (Bethe and Ashkin 1953), but for the multiple scattering limit we find (Bethe and Askin 1953)

$$P(\theta)d\theta = \frac{2\theta}{\langle \theta_m^2 \rangle} \exp\left(-\theta^2/\theta_m^2\right) d\theta \tag{12}$$

where

$$\langle \theta_m^2 \rangle = 3.9 \times 10^4 \frac{Z(Z+1)\rho\tau}{AE^2} \log\left(\frac{4\pi Z^{4/3} N_0 \rho\tau}{A} \left(\frac{h}{2\pi mv}\right)^2\right)$$

This relation shows that the mean scattering angle θ_m, varies as about $\tau^{1/2}$. At 100 keV for a thickness, $\tau = 1\mu$m in copper, θ_m is about 25 degrees.

2.2   Inelastic Scattering

Inelastic scattering events are those which cause a deflection of the incident electron and a transfer of energy. These events can be classified depending on whether the electrons of the sample are excited singly or collectively. Inelastic events are important because the energy transferred to the sample can be emitted in a form, such as secondary electrons, Auger electrons, x-rays *etc.*, which carries useful and quantifiable information.

Single Electron Excitations

Coulomb interactions between the fast incident electrons and the atomic electrons may cause the latter to change their quantum state, with an increase in internal energy.

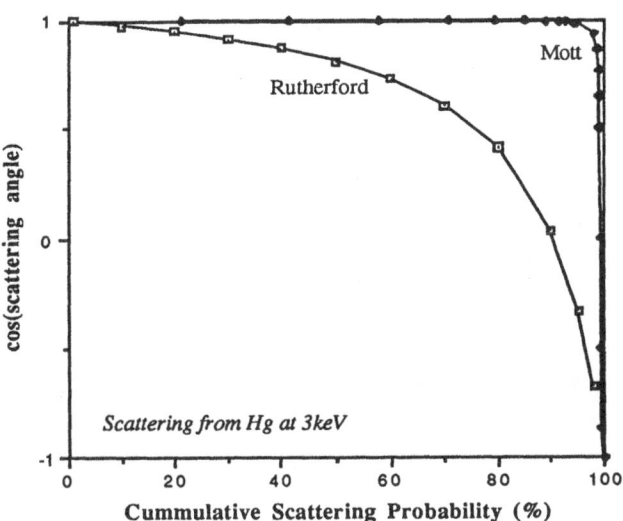

Figure 4. *Comparison of cumulative scattering distribution for Hg at 3 keV in the Mott and screened Rutherford models.*

There are three possibilities. (a) *Intraband transitions* of the outermost electrons surrounding the atom, which is possible in materials such as metals where the highest energy band is only partially filled. (b) *Interband transitions* of outer electrons where the initial and final states are separated by an energy gap such as a valence to conduction band transition in a semiconductor. Here de-excitation may involve emission of photons or secondary electrons. (c) *Core transitions* where a tightly bound electron is lifted to a partially filled level or to vacuum. Here the energy transfer may be many keV and the subsequent filling of the core-state vacancy may produce Auger electrons or x-rays.

For energy losses which are large compared with the binding energy of the valence electrons, the scattering can be treated as a Rutherford-type collision between two electrons. Putting $M = m$ in Equation 5 gives

$$\Delta E = E\theta^2 \quad \text{for} \quad \theta \ll \pi/2 \tag{13}$$

which implies that there is a unique angle of scattering for a given energy loss. For transitions of tightly bound electrons the Bethe expression for the differential cross-section is

$$\frac{\mathrm{d}\sigma}{\mathrm{d}\Omega} = \frac{A}{\theta^2 + \theta_E^2} \tag{14}$$

where

$$A = \frac{4a_0^2 f(E, \theta)}{(E/R)(\Delta E/R)} \quad \text{and} \quad \theta_E = \frac{\Delta E}{E}$$

$f(E, \theta)$ is known as the oscillator strength and is independent of θ for small scattering angles but decreases at large angles. By integrating Equation 14 the total cross-section is found to be

$$\sigma(\theta) = \pi A \log \left(1 + \frac{\theta^2}{\theta_E^2} \right) \tag{15}$$

which shows that the scattering is peaked in the forward direction about $\theta = 0$. Figure 2b shows that the scattered intensity is spread over angles larger than θ_E, which is typically an angle of the order of a few milliradians. Inelastic scattering therefore is associated with (relatively) small angle deviations of the beam while elastic scattering is generally high angle. Note that both elastic and inelastic scattering cross-sections increase with the atomic number of the target atom. For general forward scattering ($\theta < \pi/2$) the total elastic cross-section varies as about $Z^{4/3}$ while the total inelastic cross-section varies as $Z^{1/3}$. This indicates that heavier elements scatter electrons more efficiently than light elements, and also that inelastic scattering is more significant in low than high atomic number elements.

Three of the single electron excitations which are important in electron microscopy and microanalysis are:

1. *Secondary electrons*

    Secondary electrons (SE) are defined as electrons with energies between 0 and 50 eV which are ejected from a specimen when irradiated by a high energy, primary, beam. They were first studied by Austin and Starke in 1902, and a knowledge of their properties is essential to an understanding of scanning electron microscopy (SEM) which is mostly performed with SE. A cross-section for the production of a secondary electrons in metals and semiconductor by the interaction of a primary electron with a loosely-bound conduction band has been given by Streitwolf in the form:

$$\sigma_{SE} = \frac{e^4 k_F^3 A}{3\pi E n_c \rho N_0} \left[1 - \frac{1}{(50 - E_F)} \right] \tag{16}$$

    where k_F is the wave vector ($k = 1/\lambda$) corresponding to the Fermi energy E_F, and E_{SE} is the energy of the secondary electron ($0 < E_{SE} < 50$eV). Secondaries are also produced in insulators and other materials where there are few or no conduction band electrons and, in this case, the SE come from outer shell atomic electrons. In general production occurs in two stages, an initial SE is formed which then in turn generates other secondaries in a 'cascade' process. Because of the complexity of this procedure it is often easier to ignore the exact details of the interactions and adopt a phenomenological approach to secondary electrons, which will be described later.

2. *Fast secondary electrons*

    The SE familiar to the scanning microscopist are of low energy, but there are also collisions between the primary electrons and atomic electrons in which a large energy transfer occurs. These are often called Fast Secondary Electrons (FSE) to distinguish them from the more conventional SE. The FSE can be generated with energies up to that of the incident electron, but since (quantum mechanically) the primary and the FSE cannot be distinguished the cross section for the two electrons are added and the maximum possible energy loss is restricted to $\Delta E = 0.5E$. The cross-section can be written in the non-relativistic form (Moller 1931, Evans 1955):

$$\frac{d\sigma}{d\varepsilon} = \frac{\pi e^4}{E^2}\left[\frac{1}{\varepsilon^2} + \frac{10}{(1-\varepsilon)^2}\right] \tag{17}$$

where ε is $\Delta E/E$. The inelastic scattering which generates the FSE causes only a small deflection α of the incident beam where

$$\sin^2\alpha = \frac{2\varepsilon}{(2+t-t\varepsilon)} \tag{18}$$

and $t = E/511$. For typical energy transfers α is less than a degree, but the FSE leaves the impact point at an angle β to the original beam direction given by

$$\sin^2\beta = \frac{2(1-\varepsilon)}{(2+t\varepsilon)} \tag{19}$$

implying that β is usually in the range 70°–85°. Thus the FSE tend to move, initially, at right angles to the incident beam direction and this can have important consequences in some instances. This topic will be examined again later.

3. *Inner shell ionisations*

The primary electron can interact with a tightly bound, 'inner shell', atomic electron and eject it from its shell out to vacuum, thus ionising the atom. The minimum energy transfer required for the ionisation is about equal to the binding energy of the electron and is called the critical excitation energy E_c. The atom will subsequently de-excite by an outer shell electron taking the place of the ionised inner shell electron and emitting its energy as a fluorescent x-ray, or by a re-arrangement of the orbital electrons and the subsequent emission of an Auger electron carrying away the excess energy. The cross-section for the ionisation (Bethe 1930) is usually written in the form:

$$\sigma = \frac{\pi e^4 b_s n_s}{E E_c}\log\left(c_s\frac{E}{E_c}\right) \tag{20}$$

where n_s is the number of electrons in the subshell, and b_s and c_s are constants for the particular shell. Note that this is a total, not a differential, cross-section. The derivation of this equation makes several important assumptions. Firstly the energy E of the incident electron is sufficiently low that relativistic effects need not be considered. If this is not a valid approximation then Equation 20 must be modified, for example (Williams 1933)

$$\sigma = \frac{\pi e^4 b_s n_s}{(m_0 v^2/2)E_c}\log\left[c_s\left(\frac{m_0 v^2/2}{E_c}\right) - \log(1-\beta^2) - \beta^2\right] \tag{21}$$

where $\beta = v/c$ and m_0 is the rest mass of the electron. Secondly, E must be at least three to four times E_c, a condition which means that the Bethe cross-section is really not applicable to many situations of interest in the SEM or microprobe since there E_c is typically in the range 1–10 keV while E is only 10–20 keV.

To overcome these difficulties a wide range of other cross-section models have been devised, some from alternative theoretical formulations, and others as semi-empirical modifications to existing expressions. For high energy electron beams (100 keV and above) the analytical cross-sections for K shell ionisations formulated by Kolbensvedt (1967), and Gryzinski (1965), have been widely used by electron microscopists. The program EIS on the disk allows a comparison of the Bethe, Kolbensvedt, and Gryzinski cross-sections for different elements. Under some conditions these can differ rather significantly. An alternative approach (Bethe 1930, Inokuti 1971) uses a 'hydrogenic' model in which the atom of atomic number Z is treated as if it were a hydrogen atom with a binding energy equal to that of the K or L shell of the real atom. This method is attractive because the generalised oscillator strength (GOS) of the hydrogen atom (*e.g.* see Equation 14) can be expressed analytically and the desired ionisation cross-section can then by found numerically. Two short computer programs to perform these calculations are available, SIGMAK, for K-shells (Egerton 1979), and SIGMAL, for L shells (Egerton 1981). This approach is most often used for the computation of *partial* ionisation cross-sections — that is the cross-section for scattering into a given angular cone and specified energy range — but can be used to give an asymptotic approximation to the total (*i.e.* x-ray ionisation) cross-section as well. A routine to compute both SIGMAK and SIGMAL is part of the EIS program so that these variants of the cross-sections can also be related to the other values. Because the inelastic scattering is still forward peaked (see Equation 14) it is usually sufficient to select an acceptance angle (maximum scattering angle) which is only about $2\theta_E$. Finally, for the case where the beam energy is too low for the Bethe theory to be applicable (*i.e.* for overvoltage ratios below about 4) the semi-empirical model of Casnati *et al.* (1982) has been used with success and this cross-section is also included in the EIS package for comparison with the other models. Detailed first principles quantum mechanical calculations are also possible (*e.g.* Leapman *et al.* 1980, Luo and Joy 1991) which avoid the Bethe limitations. However such computations do not produce the data in an analytical form and consequently some curve-fitting and parametrisation is required to make the data conveniently accessible.

Interactions with many electrons

1. *Phonon scattering*

Because of thermal vibrations, the atoms of a crystal are never exactly at their lattice sites but vibrate. The primary electron can interact with these crystal oscillations (Hirsch *et al.* 1965) creating or annihilating phonons. The energy loss associated with this interaction is very small ($\simeq 0.02$eV) but the angular deflection is significant, giving rise to a diffuse background between the Bragg spots in a diffraction pattern. It is phonon scattering which ultimately sets an upper limit to the thickness of crystal through which an electron beam can pass.

2. *Plasmon Losses*

Plasmon oscillations result from the Coulomb forces between valence electrons in a solid. A local fluctuation Δn in the valence electron density n - caused for example by the passage of a high energy incident electron - produces a restoring force with an equation of motion (neglecting thermal motion and coupling between the valence electrons and the rest of the atom)

$$\frac{d^2(\Delta n)}{dt^2} + \frac{ne^2}{\varepsilon_0 m}\Delta n = 0 \tag{22}$$

which produce oscillations in the electron density with an angular frequency

$$\omega_p = \left(\frac{ne^2}{\varepsilon_0 m}\right)^{1/2} \tag{23}$$

The energy of these oscillations is quantised and can only change in steps of magnitude given by $\Delta E = h\omega_p$. In most materials $\omega_p = 10^{16}$ rad s^{-1} so $\Delta E = 10$ to 30eV. Since plasmons arise from electrons which are not bound to particular atoms the scattering event is not localised at some point. The scattering angle θ_E (see Equation 14) is of the order of 0.1 milliradians at 100 keV so plasmon scattering is sharply forward peaked. The total cross-section σ_P is

$$\sigma_P = \frac{\Delta E}{2a_0 n E}\log\left(\frac{\theta_c}{\theta_E}\right) \tag{24}$$

where θ_c is about 10 milliradians. This implies a mean free path of about 1000Å at E = 100 keV for most metals so a large proportion of the electrons transmitted through a sample in a **TEM** have undergone plasmon scattering.

3. *Bremsstrahlung*

An incident electron that passes through the Coulomb field of an atom is decelerated, and the kinetic energy that it loses is emitted as a photon of electromagnetic radiation by the electron. This 'bremsstrahlung' or 'braking radiation' forms a continuous distribution of photon energies from zero up to the incident beam energy. The production of bremsstrahlung is directly related to the direction of the incident electron and is strongly anisotropic. If the specimen is thin and the electron suffers only a few scattering events in passing though it then this anisotropy is observable, but in bulk samples where many scattering events occur the anisotropy is averaged out. Several cross-sections are available to describe the production of bremsstrahlung by energetic electrons. The simplest is a semi-empirical model derived by Kramers (1923) for solid specimens. This neglects the directional anisotropy but otherwise gives a good description of the effect, and has a cross-section which is of the form

$$\sigma_B = \frac{1.43 \times 10^{-21}Z^2}{4\pi}\left(\frac{E - E_B}{E E_B}\right) \tag{25}$$

where E_B is the energy of the bremsstrahlung. Note that σ_B is now in units of photon energy per unit energy interval/incident electron/ atom/cm^2/sr. We will make use of this later. A more detailed analysis was performed by Kirkpatrick and Wiedmann (1945) who made detailed numerical calculations based on Sommerfeld's (1931) quantum mechanical theory, and also by Koch and Motz (1959). The Kirkpatrick and Wiedmann results, which correctly account for the anisotropy of the radiation are available in a tabulated form which has subsequently be curve fitted to provide a compact representation. The final expressions are lengthy and will not be given here but Statham (1976) has given a **FORTRAN** routine to evaluate the expression.

3   Electron Energy Losses

As the electron travels through the specimen it loses energy because of the variety of inelastic scattering events that we have been discussing above. This means that for a given amount of incident energy an electron will only travel some finite and maximum distance in the target before transferring all of its energy to the specimen and coming to equilibrium with it. A knowledge of this distance is important because it sets a limit on the range of action of the electron beam. The rate at which the energy transfer occurs along the electron trajectory is also important because this determines how certain emissions of interest such as secondary electrons and x-rays, are generated.

If the electron loses energy $-dE$ in travelling a distance ds along its trajectory than $-dE/d(\rho s)$, where ρ is the density of the target, is called the electron stopping power. This is not a constant but varies with the energy of the electron. If we know the form of this variation then we can compute an electron range R from the equation

$$R = \int_E^0 \frac{1}{-\left(\dfrac{dE}{d(\rho s)}\right)} dE \tag{26}$$

The apparent problem involved in determining the stopping power is that it involves the sum of an energy loss which is essentially continuous, the slowing down of the electron due to the Coulomb field of the atoms which gives rise to bremsstrahlung, and a large number of discrete and statistically random events such as the production of plasmons, phonons, secondary electrons and x-rays. Every electron will suffer a different sequence of scattering events in type of event, number of events, and the distribution of these events. The solution to this dilemma was suggested by Bethe (1930). Instead of trying to account individually for every kind of inelastic event we make the assumption that the effect of both the electrostatic drag on the electron and the discrete energy losses occurring during inelastic scattering events can be combined and approximated by a model in which the incident electron is slowing down continuously. While these results will not be an accurate representation for any particular electron that we could track through the sample, if we average the data for the 10^9 or so electrons which strike our specimen each second in the electron microscope then the model will be correct.

The rate at which energy is lost by the incident electron is written in the form (Bethe, 1930)

$$\frac{dE}{d(\rho s)} = -785 \frac{Z}{AE} \log\left(\frac{1.166E}{J}\right) \quad \text{eV/Å} \tag{27}$$

where E is the energy of electron (in eV), and Z and A are respectively the atomic number and atomic weight of the target. J, which has units of eV, is called 'the mean ionisation potential' and represents the effective average energy-loss per interaction between the incident electron and the solid. This single parameter incorporates into

Element	J (eV)	Element	J (eV)	Element	J (eV)
H	21.8	Cr	257	Pd	470
Li	40.0	Mn	272	Ag	470
Be	63.7	Fe	286	Cd	469
B	76.0	Co	297	Sn	488
C	78.0	Ni	311	Gd	591
Na	149	Cu	322	Ta	718
Al	166	Zn	330	W	724
Si	173	Ga	334	Pt	790
Ca	191	Ge	350	Au	790
Sc	216	Zr	393	Pb	823
Ti	233	Nb	417	U	890
V	245	Mo	424		

Table 1. *Measured values of mean ionisation potential J*

its value all possible mechanisms for energy loss that the electron can encounter, so allowing the Bethe equation to provide a convenient and compact way of accounting for the variety of energy losses experienced.

Nuclear physics techniques have been used to measure J experimentally for a wide range of materials and compounds, and Table 1 gives values for some common elements (in solid form). There is a monotonic and almost linear increase of J with the atomic number of the element and Berger and Seltzer (1964) showed that this variation could be fitted with good accuracy by the relation

$$J = \left[9.76Z + \frac{58.5}{Z^{0.19}}\right] \quad \text{eV} \tag{28}$$

and a comparison of this expression with the data in Table 1, for example for silicon ($Z=14$) and gold ($Z=79$), shows that the fitted value and the experimental value are generally within a few electron volts of each other. However the fit is not perfect and where the highest accuracy is required the tabulated value should be used. Other expressions for J are in use, particularly for low Z elements, but for all practical purposes the Berger-Seltzer expression is good enough, especially bearing in mind that J occurs only in a logarithmic term in the stopping power expression. The EIS program contains a table of measured J values, like Table 1, and also the Berger-Seltzer approximation of Equation 28 so that the measured and predicted values of J can be compared (option 4 on the disk). For compounds the appropriate value of J can again be found using Equation 28 but replacing Z by Z_{av} the mean value of Z for the compound. So, for example, if a material has the composition AB_2, *i.e.* 33 atomic percent of A and 66 percent of B, then

$$Z_{av} = (1 * Z_A + 2 * Z_B)/3 \tag{29}$$

where Z_A and Z_B are the atomic weights of A and B respectively. This simple average in most cases produces a value for Z_{av}, and thus for J, which is of acceptable accuracy. In the case of complex materials however the composition may not be known. Table 2

Material	J (eV)	Material	J (eV)	Material	J (eV)
Nylon	63.9	Aluminium Oxide	145	Adipose Tissue	63
Teflon	99.1	Calcium Fluoride	166	Bone	107
Paraffin Wax	48.3	Lithium Fluoride	94	Muscle	75
PMMA	74.0	Silicon Dioxide	139	Skin	74
Polythene	57.4	Sodium Iodide	452	Air (1 atm.)	86
Polystyrene	68.7	'Pyrex' Glass	134	Blood	75
Plastic Scintillator	64.7	Photographic Emulsion	64.7	Liquid Water	75

Table 2. *Measured values of mean ionisation potential J*

therefore gives J values, for some compounds likely to be encountered in electron microscopy. Although this list is far from exhaustive is does give a useful guide as to probable values for generic types of materials. Using either the measured or computed value of J, Equation 27 can now be evaluated for a given electron energy E. Figure 5 attempts to plot the rate of energy loss, $(-dE/ds)$, or $\rho(dE/ds)$, for copper ($\rho = 7.6$ gm/cm^3) over the energy range 100eV to 100 keV. The predicted energy loss values show a steady increase from about 0.1eV per Å at 100 keV up to almost 10 eV per Å at beam energies of a few keV. These values are typical of those found in all materials, and are in good agreement with experimental data. However, at an energy of about 2 keV the curve reaches a maximum value before starting to fall rapidly as the energy is further reduced. At still lower energies the predicted stopping power will actually change sign. Experimentally it is also found that the actual stopping power is significantly higher in the low-energy range than the predicted values. Therefore while the Bethe stopping power equation as formulated in Equation 27 is convenient for energies such that $E \gg J$, it cannot be used at lower energies without encountering problems. In particular, since J is of the order of 0.3–0.6 keV for most materials the study of electrons travelling with initial energies of 1–10 keV, *i.e.* typical energies encountered in a scanning electron microscope, will be difficult since a significant part of the trajectory will occur in the energy range where the Bethe relation cannot be applied.

The inaccuracy of the Bethe expression at low energies, and its non-physical change of sign when $E \approx J$ are not inherently due to the failure of the original model but rather to the simplified form in which it is usually employed. The full expression derived by Bethe gives the stopping power as:

$$\frac{dE}{d(\rho s)} = -785 \frac{Z}{AE} \sum_{n,l} \frac{f_{n,l}}{Z} \log \left(\frac{2E}{A_{n,l}} \right) \qquad (30)$$

where for the l-th shell containing n electrons $f_{n,l}$ is the oscillator strength and $A_{n,l}/2$ is the binding energy. To derive Equation 27 from this expression it is necessary to make two approximations, firstly that the sum rule applies *i.e.* that $\sum f_{n,l} = Z$, and secondly that the binding energies are assumed to be all approximately equal in value and small compared to E. Equation 30 can, however, be evaluated directly for any given atom provided that, in order to avoid the non-physical consequences of the logarithmic term going negative, the summation over a given shell is terminated as soon as $E \leq A_{n,l}/2$,

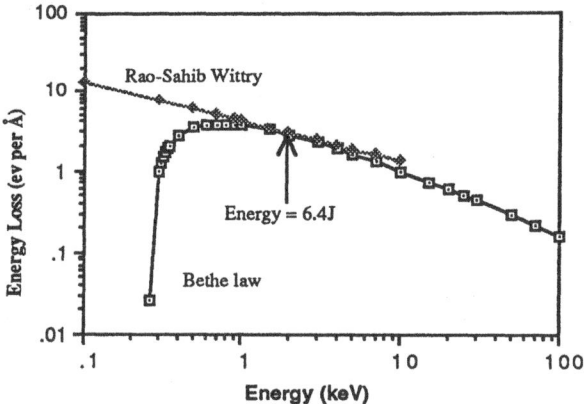

Figure 5. *Stopping power for copper (in eV/Å) from the Bethe equation, and the Rao-Sahib Wittry extrapolation.*

an assumption which corresponds to the physical reality that a given shell cannot be ionised by an electron whose kinetic energy is less than the critical energy of the shell. If these precautions are observed then Equation 30 is valid and accurate over a wide energy range, although somewhat clumsy to evaluate.

If, however, we prefer to use a simpler expression like Equation 27 we must make allowance for the fact that J is not a constant but changes with the energy of the incident electron because certain inelastic processes cannot occur below a specific threshold energy (*e.g.* inner shell ionisation). One approach to this is a method due to Rao-Sahib and Wittry (1974). This uses a parabolic extrapolation from the tangent to the Bethe curve at the energy $E = 6.4J$, where the curve has an inflection, down to $E = 0$. Thus for $E < 6.4J$ Equation 27 is replaced by the expression:

$$\frac{\mathrm{d}E}{\mathrm{d}(\rho s)} = -\frac{624Z}{\sqrt{EJA}} \quad (\mathrm{eV}/\text{Å}) \tag{31}$$

Figure 5 shows how this modification extends the range of the original Bethe expression of Equation 27 at low energies, and how it smoothly merges into this expression at higher energies. This expression provides a convenient extrapolation which is well behaved over the low energy range, and so this approximation has been widely used. It is not, however, neither physically realistic nor accurate. Figure 6 compares the rate of energy loss, $(-\mathrm{d}E/\mathrm{d}s)$, or $\rho(\mathrm{d}E/\mathrm{d}S)$, for copper ($\rho = 7.6\mathrm{gm/cm}^3$) over the energy range 100 eV to 100 keV obtained using the Bethe model and the Rao-Sahib Wittry extrapolation with the results of detailed computations by Tung *et al.* (1979). While the agreement between the composite Bethe-Rao-Sahib-Wittry profile and the Tung *et al.* data is satisfactory at energies down to a few keV once the energy falls below 1 keV neither the original Bethe curve nor the Rao-Sahib Wittry extrapolation is close to the Tung *et al.* value, the error being a factor of two to three times at an energy of a few hundred electron volts. The same sort of result is found for all other elements, with the errors becoming worse for high atomic numbers since these have a larger J

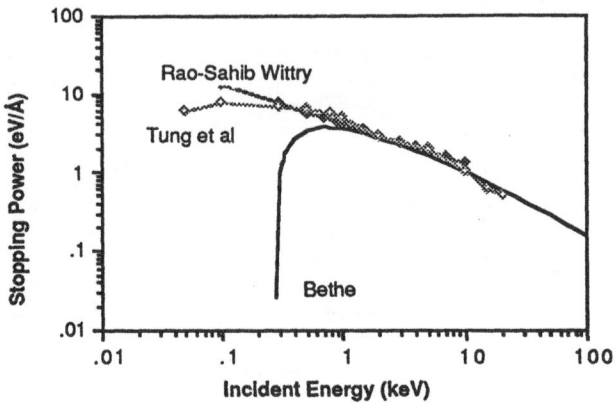

Figure 6. *Comparison of the Bethe stopping power curve, the Rao-Sahib-Wittry extrapolation, and the calculations of Tung et al.*

value and consequently the Bethe expression becomes unusable at energies as high as 4 or 5 keV. We can simultaneously escape the mathematical singularity of the original Bethe expression, eliminate the need to use the Rao-Sahib Wittry extrapolation, and improve the accuracy of the stopping power as compared to the Tung *et al.* data by using a modified version of the Bethe equation suggested by Joy and Luo (1989). We write the stopping power as:

$$\frac{\mathrm{d}E}{\mathrm{d}\rho s} = -785\frac{Z}{AE}\log\left(\frac{1.166(E + 0.85J)}{J}\right) \tag{32}$$

At high energies $E \gg J$ this expression converges to the original Bethe expression of Equation 27, but the addition of the extra term $0.85J$ in the numerator of the logarithmic term ensures that for all positive values of E the log term evaluates to a positive quantity. As shown in Figure 6 this modified expression is now a good fit to the Tung *et al.* data for all energies above 50 eV. It can be shown that this result is equally applicable to all other elements and compounds of interest, producing a stopping power value which agrees well with more detailed calculations for all energies above a few tens of electron volts. Although this is an empirical modification a little algebraic manipulation of Equation 32 will show that, in effect, this is Equation 27 but with J now being a function of E and is thus consistent with the reasoning outlined above. This version of the Bethe expression is used in option 3 of the program EIS to compute the stopping power of the chosen element.

Stopping powers can be determined experimentally by a technique relying on electron energy loss spectroscopy (Luo *et al.* 1991). The analysis of such data shows that the stopping powers computed from the Bethe expression, and as modified for low energy use in Equation 32, are in excellent agreement. For scanning electron microscopy and microprobe analysis these equations can therefore be used with confidence. At higher beam energies however a relativistic correction must be applied to the Bethe model, in the form

$$\frac{dE}{d(\rho s)} = \frac{1.535 \times 10^{-3} \, Z}{A \left(1 - \frac{m_0 c^2}{(E + m_0 c^2)} \right)} \left[\frac{T^2(T+2)m_0 c^2}{2J^2} + (1 - \beta^2) + \frac{T^2 m}{8} - \frac{(2T+1)\log 2}{(T+1)^2} \right]$$

(33)

where $m_0 c^2 = 511$ keV is the rest mass energy of the electron, $\beta = v/c$, and T is the kinetic energy in units of 511 keV. The EIS program computes both the standard Bethe expression and this relativistically corrected version so that their values can be compared. You will note that at high energies the relativistic version generally predicts a higher stopping power, and hence a smaller electron range, although this situation is often reversed at lower energies.

When Equations 27, 32 or 33 are used to evaluate Equation 26 then the result is known as the Bethe range R_B. This measures the total distance travelled by the electron along its trajectory and thus R_B does not bear any fixed relation to quantities such as the average depth into a specimen that any given electron may travel. The Monte Carlo models discussed below and provided on the disk, answer that sort of question in detail. However since there are occasions when a quick evaluation of the likely depth of penetration of the electron is required the expression

$$\text{range} = \frac{760 E^{1.66}}{\rho} \, \text{Å}$$

(34)

where E is in keV and the density is in gm/cm^3, is a useful rule of thumb. The range of data available in option 4 of the EIS program is a numerical integration of Equations 26 and 32 and give a good value for R_B, while the Monte Carlo program JustBS on the disk outputs information about the depth of penetration and the depth from which backscattering can occur.

4 Modelling electron-solid interactions

The sections above have described the fundamental effects which govern the passage of an electron through a solid. What we wish to do now is to make use of this theoretical background to give us detailed information about the kinds of electron-beam interactions that give us information in the microscope. We must therefore move from the scattering of one electron by one atom to the scattering of a beam of electrons (perhaps 10^{10} per second) by a target containing perhaps 10^{20} atoms in the irradiated volume. It is clear from the variety of interactions that any one electron can undergo that there is little merit in trying to represent the macroscopic details of the interaction of a beam with a specimen by some analytical expression. While simple functional relationships can be derived between certain macroscopic properties of the interaction, such as the backscattering or transmission coefficients, and the parameters describing the specimen and the electron beam, the enormous number of different ways in which a given electron could complete the sequence of interactions involved in a single trajectory precludes the construction of a detailed analytic model. Monte Carlo sampling techniques which give us random numbers as a means of predicting the magnitude of various events and as a

way of selecting between possible scattering options, were first introduced by Ulam and
Von Neumann during the Manhattan project (Rhodes 1986). With the advent of powerful personal computers this technique allows us to model electron solid interactions
accurately, conveniently, and flexibly,

4.1   Basic principles of Monte Carlo simulations

The Monte Carlo technique, as applied in this context, attempts to describe the trajectory which takes the electron through the solid. Although no individual trajectory
produced by the simulation will represent a 'real' trajectory, if the physics of the processes encountered by the electron are properly modelled then predictions based on a
large number of trajectories will accurately describe effects which can be experimentally
observed. (Note that we cannot apply these techniques in cases where either the wave
nature of the electron (*i.e.* diffraction effects) or the crystalline or atomistic nature of
the sample are dominant.) In order to make these calculations we need two basic pieces
of information, the angles through which the electron is deflected as it travels in the
specimen, and an estimate of how far (on average) the electron will travel given some
particular value of incident energy. The simulations described here make two significant
approximations in order to answer these questions:

1. We assume that only elastic scattering events are significant in determining the
   path taken by any given electron as it moves through the solid. As we have seen
   above elastic scattering, described by the screened Rutherford cross-section and
   produced by the coulombic attraction between the negatively charged electron
   and a positively charged atomic nucleus, results in angular deflections of from
   a few degrees up to 180°. The great majority of inelastic scattering events ,
   on the other hand, produce angular deflections which are typically 0.5° or less.
   Consequently elastic scattering events are likely to be the ones which dominate in
   determining the path taken by the trajectory, and ignoring the effects of inelastic
   scattering introduces only negligible error while greatly reducing the number of
   computations that are required.

2. The electron is assumed to lose energy continuously, at a rate determined by the
   stopping power equations discussed earlier, rather than as the result of discrete
   inelastic events. This simplification allows the net result of all possible inelastic
   scattering processes to be accounted for without having to worry about the exact
   details of the individual events.

   Neither of these assumptions is essential to the successful construction of a Monte
Carlo program, and indeed much work has been put into simulations which specifically
seek to avoid such radical simplifications. In practice, however, the benefits resulting
from the gain in accuracy achieved by a more rigorous approach are usually outweighed
by the substantial increase in computing time required. The procedures discussed
here provide an acceptable degree of accuracy (*i.e.* as accurate as a typical experiment
performed on an electron microscope) while at the same time remaining capable of
generating statistically valid data in a reasonable time period (*i.e.* a few minutes to a
couple of hours) on a personal computer.

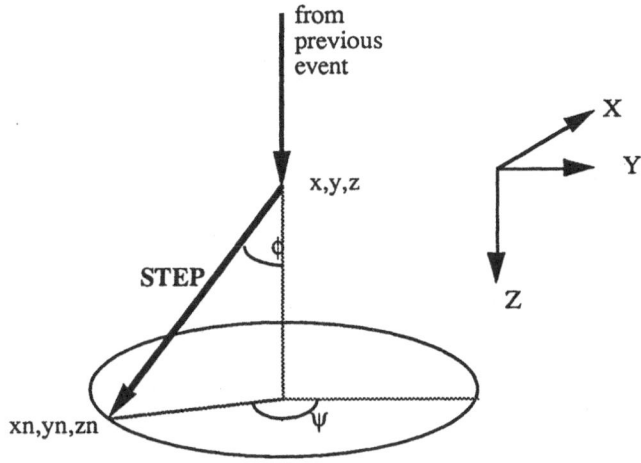

Figure 7. *Coordinate system for Monte Carlo simulation.*

The single scattering Monte Carlo model

Within the constraints discussed above the most accurate Monte Carlo simulation of the electron beam interaction is one which attempts to account for each elastic scattering event suffered by the electron as it travels through the sample (Newbury and Myklebust 1981). We assume (Figure 6) that the electron undergoes an elastic scattering event at some point represented by the coordinates (x, y, z), after having travelled from its previous scattering event. We wish to calculate the coordinates (xn, yn, zn) of the next point to which the electron is scattered. The parameters which describe the instantaneous situation of the electron are its energy E and the direction cosines cx, cy, cz of the trajectory segment that brought the electron from its previous scattering location to the point (x, y, z). These direction cosines are relative to a fixed set of axes attached to the specimen defined with the convention that the positive z-axis is normal to the specimen surface and directed into the specimen, the x-axis is parallel to the tilt axis, the x-y plane is the surface plane of a flat (*i.e.* untilted) sample, and the y-axis completes a right-handed set of axes. When the specimen is tilted the positive direction of the y-axis is down the surface of the specimen. To calculate the position of the new scattering point (xn, yn, zn) we need to know the distance between it and the point (x, y, z), and the elastic scattering angles ϕ and ψ.

We will assume that the incident electron will be scattered as described with our screened Rutherford cross-section, and that we can associate with this a total elastic scattering cross-section, and consequently an elastic mean free path λ, which represents the average distance that an electron will travel between encountering elastic scattering events. Experimentally the actual distance that an electron travels between successive scatterings will, of course, vary in a random fashion. In our Monte Carlo simulation this variability is introduced by saying that the distance (or step length) between the scattering events at x, y, z and xn, yn, zn is given by the relation

$$\text{step} = -\lambda \log(\text{RND}) \quad (\text{cm}) \qquad (35)$$

where RND is a equidistributed random number between 0 and 1 selected by the computer.

Figure 8 plots the variation of the step length, in units of λ , as a function of the random number chosen. Since the random numbers are uniformly distributed between 0 and 1 we see, for example, that there is a 10% chance of drawing a number such that RND < 0.1 in which case the step will be equal to or greater than 2.3λ, and equally there is a 10% chance of picking a number such that RND > 0.9 in which case the step length would be equal to or less than 0.1λ. The step lengths therefore vary over a wide range of values depending on the random number picked by the computer but, as can be verified by integrating Equation 35, the average step length will be λ.

In the scattering event at (x, y, z) which marks the start of the step, the electron is deflected through some angle ϕ relative to its previous direction (see Figure 6). The size of this deviation is determined by $d\sigma_E/d\Omega = \sigma'$, the angular differential form of the Rutherford cross-section (*i.e.* Equation 9) and in the program is found by solving the equation

$$\text{RND} = \int_0^\phi \frac{\sigma'}{\sigma_E} d\Omega \qquad (36)$$

where σ_E is the total Rutherford cross-section given above, and integration extends to a maximum value of ϕ. The right hand side of Equation 36 represents the probability of the electron being scattered through an angle less than ϕ. Since we do not know, for any given scattering event, what the probability actually is we pull a random number RND from the computer, equate this to the probability and run the equation backwards to determine the angle for which this value would be correct By evaluating Equation 36 an equation can be derived (Newbury *et al.* 1976) which relates the scattering angle ϕ to the random number RND:

$$\cos \phi = 1 - \frac{2\alpha \text{RND}}{(1 + \alpha - \text{RND})} \qquad (37)$$

where α is the screening coefficient given above in Equation 6. This equation generates a unique scattering angle in the range $0° < \phi < 180°$, producing an angular distribution which matches that obtained experimentally. Although all angles between $0°$ and $180°$ are possible the great majority of scattering events are predicted by Equation 37 to be low-angle (*i.e.* less than $10°$).

Figure 9 plots the probability of obtaining an angular scattering of greater than some minimum value ϕ for the case of a silicon target irradiated at 100 keV. Note that while there is only a 1 in 10,000 chance of an electron being scattered by an angle in excess of $110°$, more than 50% of all electrons are scattered through at least $1.5°$. The electron can scatter to any point on the base of the cone shown in Figure 7 so the azimuthal scattering angle ψ is given as

$$\psi = 2\pi.\text{RND} \qquad (38)$$

All of the information needed to specify the scattering step from (x, y, z) to (xn, yn, zn) is now available. Although the calculation is straightforward the algebra is cumbersome

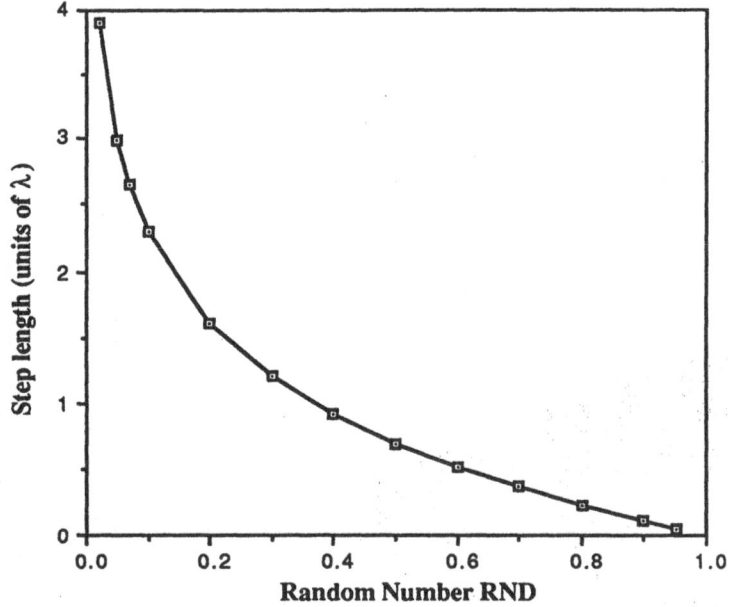

Figure 8. *Variation of step length with* RND.

because of the need to present the result relative to the initial fixed coordinate axes described above. Following Newbury *et al.* (1976) we get:

$$
\begin{aligned}
xn &= x + \text{step}.ca \quad (a) \\
yn &= y + \text{step}.cb \quad (b) \\
zn &= z + \text{step}.cc \quad (c)
\end{aligned}
\tag{39}
$$

where

$$
\begin{aligned}
ca &= (cx.\cos\phi) + (V1.V3) + (cy.V2.V4) \quad (a) \\
cb &= (cy.\cos\phi) + (V4.(cz.V1 - cx.V2)) \quad (b) \\
cc &= (cz.\cos\phi) + (V2.V3) - (cy.V1.V4) \quad (c)
\end{aligned}
\tag{40}
$$

and

$$
V1 = AN.\sin\phi, \quad V2 = AN.AM.\sin\phi, \quad V3 = \cos\psi, \quad V4 = \sin\psi \quad (a)
$$

$$
AN = -\frac{cx}{cz} \quad \text{and} \quad AM = \frac{1}{\sqrt{1 + AN^2}} \quad (b) \tag{41}
$$

Using this information the electron, given a starting energy, position and direction, can be tracked through the specimen one step at a time. As the electron travels it loses energy in the way predicted by the stopping power Equation 32. The actual energy ΔE, lost along the step from (x, y, z) to (xn, yn, zn) is then:

$$
\Delta E = \text{step}.(dE/dS) \tag{42}
$$

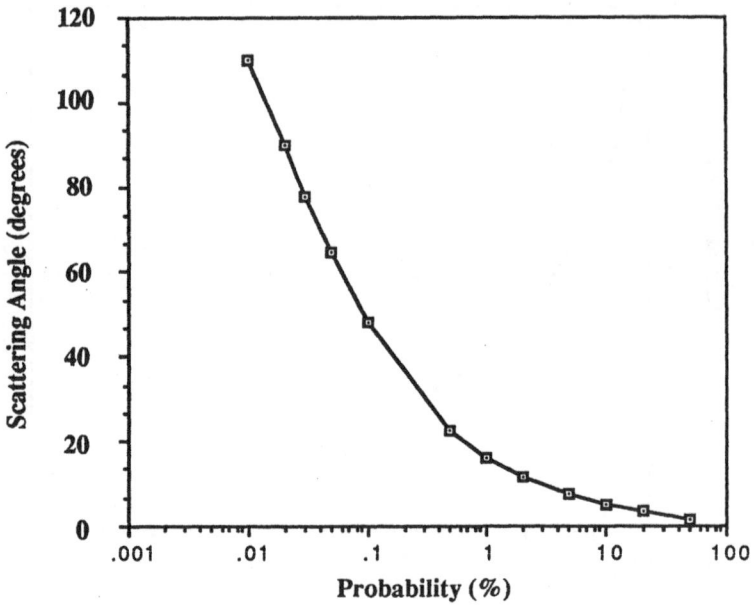

Figure 9. *Scattering in the screened Rutherford model.*

Typically we continue to track the electron until it either leaves the specimen (by transmission or backscattering) or falls so low in energy that we can ignore it. In the code discussed here the trajectory is terminated if the energy falls below 500 eV.

The sequence of operations needed to simulate the electron path through the specimen can now be written out schematically in an algorithmic form:

repeat
      Get starting energy E of electron
      Get starting coordinates x, y, z for the step
      Get direction cosines cx, cy, cz relative to initial axes
           Compute mean free path λ for energy E and given material
               Calculate the step length **step** from Equation 35
           Find the scattering angles ϕ, ψ from Equations 37, 38
           Compute final coordinates xn, yn, zn from Equations 39, 40, 41
               Compute finish energy $E\prime = E - \text{step}.(\mathrm{d}E/\mathrm{d}s)$
      Reset coordinates $x = xn, y = yn, z = zn$
      Reset direction cosines $cx = ca, cy = cb, cz = cc$
      Reset energy $E = E'$
until electron leaves sample or falls below some minimum energy

The program loops through this sequence of steps to generate one electron trajectory. The process is then repeated to generate a large number of trajectories. Since this process is technically a random sampling of data, the precision of the result will depend on the number of trials that are made. If a given event occurs with a probability p in each trajectory then the standard deviation of the results after N trajectories have been run is $1/\sqrt{N}p$. So if $p = 0.1$ we need to run 100,000 trajectories to get 1% precision. In practice it is often sufficient to run 5000-10000 trajectories, which with the programs provided here will take of the order of ten minutes on a typical 286 IBM PC or clone (provided the computer is equipped with an 80x87 maths coprocessor chip).

The plural scattering model

The single scattering model outlined above is accurate and ideally suited for the simulation of events in materials which have at least some electron transmission (*i.e.* TEM, STEM samples). If, however, the specimen is 'bulk' then a very large number of scattering steps are required before the average electron leaves the specimen or gets sufficiently low in energy for the run to be terminated, and so the program becomes very slow to run. For the study of SEM phenomena we can use instead a modified approach, the Plural Scattering Monte Carlo, which is much faster but slightly less accurate. Whereas the previous program tried to account for each elastic scattering event suffered by the electron, the plural scattering program tries to consider only the resultant effect of a number of scattering events. Note that this is not a rigorous mathematical expression of plural scattering of the type discussed earlier (Equation 12) but a much simpler kind of analysis.

The basic assumptions of the plural scattering Monte Carlo model are the same as those for the single scattering model, but the implementation is markedly different. The total length of the electron trajectory within the sample is taken to be the Bethe range R_B defined from Equation 26 and again using the stopping power equation. This Bethe range is then divided into, typically, fifty segments of equal length. This ensures that, unlike the single scattering case, there is a constant and relatively small number of computational steps associated with each trajectory. $E[n]$, the energy of the electron at the start of the nth step of the trajectory, is found by numerically solving the equation:

$$E[n] = E[n-1] - \int_{\text{step}} \left(\frac{\mathrm{d}E}{\mathrm{d}(\rho s)} \right) \mathrm{d}(\rho s) \tag{43}$$

where $(\mathrm{d}E/\mathrm{d}(\rho s))$ is again obtained from Equation 32. $E[1]$ is set equal to the incident beam energy E_0 and $E[51]$ is set equal to zero.

The azimuthal scattering angle ψ is given by the same expression as previously used and the axial scattering angle ϕ is again described by the screened Rutherford cross-section but using a different formulation of the equation. We write ϕ in the form

$$\cot \left(\frac{\phi}{2} \right) = \frac{2p}{b} \tag{44}$$

where p is the impact parameter (*i.e.* the projected distance of closest approach of the electron to the nucleus of atomic number Z), and b is $1.44 \times 10^{-2} Z/E$, where E is the instantaneous energy of the electron in keV. In each of the 50 steps making up one trajectory a large number of scattering events will occur (since the step length is now much larger than λ). Some of these deflections may add, and others may cancel so, following the original suggestion of Curgenven and Duncumb (1971), the net resultant scattering angle ϕ is written as:

$$\cot\left(\frac{\phi}{2}\right) = \frac{2p}{b}\sqrt{\text{RND}} \tag{45}$$

where RND is another random number between 0 and 1. In practice Equation 44 has been found to introduce a systematic error into the simulation because it does not allow for a sufficient amount of small angle scattering. The equation is therefore rewritten as:

$$\tan\left(\frac{\phi}{2}\right) = \frac{b}{2p}\left(\frac{1}{\sqrt{\text{RND}}-1}\right) \tag{46}$$

in which form ϕ approaches zero as RND goes to unity. The final problem is in determining a suitable value for the impact parameter p. The approach used here derives from Love *et al.* (1977) who rewrite Equation 45 in the form

$$\tan\left(\frac{\phi}{2}\right) = \tan\left(\frac{\phi_0}{2}\right)\left(\frac{E_0}{E}\right)\left(\frac{1}{\sqrt{\text{RND}}}-1\right) \tag{47}$$

where as before E_0 is the incident beam energy and

$$\tan\left(\frac{\phi_0}{2}\right) = \frac{0.0144Z}{2pE_0} \tag{48}$$

ϕ_0 thus represents the minimum scattering angle for the incident electron with energy E_0. As can be seen from the functional form of the Bethe equation (*i.e.* Equation 32) the variation of (E/E_0) is substantially independent of the atomic number Z (the variation coming only from the mean ionisation potential J which occurs inside the logarithmic term, and the random number RND will average to a mean value of 0.5 when a large number of trials is made). It therefore follows that the backscattering coefficient η should depend only upon $\cot(\phi_0/2)$. Experimentally this turns out to be a good approximation and we find that, for any element, the relation between $\tan(\phi_0/2)$ and the backscattering coefficient η can be written as a polynomial:

$$\tan(\phi_0/2) = 0.016697 + 0.55108\eta - 0.96777\eta^2 + 1.8846\eta^3 \tag{49}$$

To make Equation 48 a usable one in the program we need an estimate for η for our target. This can be done by using a relation due to Hunger and Küchler (1979) which gives the backscattering coefficient η of a material of atomic number Z at incident beam energy E as:

$$\eta(Z, E) = E^m C$$

where

$$m = 0.1382 - \frac{0.9211}{\sqrt{Z}}$$

and

$$C = 0.1904 - 0.2235(\log Z) + 0.1292(\log Z)^2 - 0.01491(\log Z)^3 \tag{50}$$

Given E and Z then Equation 49 gives a value of η and hence Equation 48 a value of $\tan(\phi_0/2)$. (It should be noted that if the target is not a single element but a homogeneous compound then the correct procedure is to find a value for η_{mix} — the backscattering coefficient of the compound - using the relation (Castaing 1960)

$$\eta_{mix} = \sum_i c_i \eta_i \qquad (51)$$

where the c_i are the concentrations of the elements, $\sum c_i = 1$, and the η_i are found from Equation 49, and then these values are used in Equation 50.

This procedure has the special advantage that the Monte Carlo simulation built around it can correctly predict the variation of the specimen backscattering coefficient with incident beam energy, an effect which is quite significant at energies below 5 keV. This is not normally possible with a model using a screened Rutherford, rather than a Mott, scattering cross-section. The Monte Carlo loop then follows closely to the procedure described above and can be represented in algorithmic form as:

for n=1 to 50
    begin
        Get starting energy $E[n]$ of electron
        Get starting coordinates x, y, z for the step
        Get direction cosines cx, cy, cz relative to initial axes
            Find the scattering angles ϕ, ψ from Equations 22 and 8
            Compute final coordinates xn, yn, zn from Equations 9,10, 11
            Check if the electron has been backscattered
                If **yes**, exit the loop and add 1 to backscatter total, otherwise:
        Reset coordinates $x = xn, y = yn, z = zn$
        Reset direction cosines $cx = ca, cy = cb, cz = cc$
    end

Despite the apparent simplicity of this approach the agreement between predictions made using the plural scattering model and the single scattering model is generally excellent and except for a few cases where the granularity (*i.e.* the size of the step length) of the plural scattering model is too high to permit a model to be realistic it is usually preferable to employ this technique when dealing with bulk samples since any slight drop in accuracy is outweighed by the gain in precision obtained from the much higher number of trajectories that can be run.

5 Applications of the models

In the final part of this chapter the theoretical material discussed at the start and as implemented by the Monte Carlo procedures just considered will be used to investigate quantitatively some aspects of image and data generation in both the SEM and TEM/STEM. In this way it is readily possible to see how different parameters affect the behaviour of electrons in a solid, and how this in turn affects such things as x-ray production and SE generation. This can be investigated by using the single scattering Monte Carlo program SS_MC on the disk to get a feel for the overall volume of the electron beam interaction, for example, by looking at the interaction of a high energy

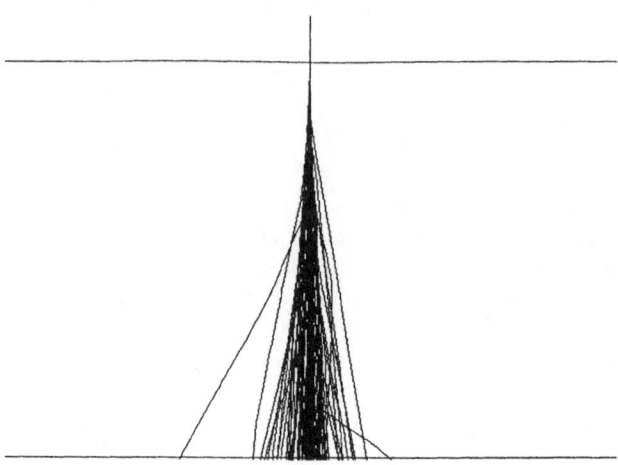

Figure 10. *Monte Carlo simulation of the interaction of a 100 keV electron beam with a 100 nm copper film*

electron beam (*e.g.* 100 keV) with thin (≈ 100 nm) films of carbon, silicon, copper and gold (Figure 10). Note how the beam, as would be expected from the Rutherford cross-section equation, remains mostly forward scattered. However, the relatively low number of scattering events causing deviations of a few degrees ultimately result in the beam broadening into a cone such that the volume of the interaction at the bottom surface is higher than that at the top. In this limit the beam interaction volume is effectively determined by the dimension of the sample in the direction of the beam as well as by the beam diameter and the physical parameters (the density and atomic number) of the foil. Thus under this condition, corresponding to that used for TEM and analytical TEM, the interaction volume is strongly specimen dependent and so can be controlled physically.

When the same simulation is run with a beam energy of 10 keV, even though many of the incident electrons are still transmitted through the foil, the interaction volume is much greater laterally because each electron is scattered many more times in its transit through the sample and the probability of a high angle event occurring is consequently increased. As a result a finite number of electrons are now backscattered. If the foil is made so thick that no electrons are transmitted (*i.e.* it becomes a 'bulk' rather than a thin sample) then the beam interaction volume becomes independent of the sample dimension and varies only with the beam energy and sample characteristics (Figure 11). The shape of the interaction volume varies with the material. If this is low atomic number material then the forward scattering behaviour still dominates and the volume hangs from the surface like a rain drop, but if the target is of high atomic number then a sufficient fraction of the scattering events are high enough in angle to expand the beam laterally so that the volume resembles the shape of an egg pressed

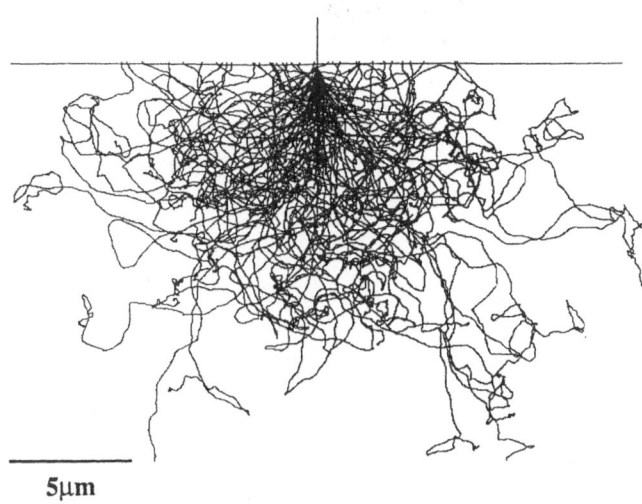

5μm

Figure 11. *Monte Carlo simulation of the interaction of a 10 keV electron beam with a bulk copper sample.*

against the surface.

In this plural scattering regime, corresponding to the situation usually encountered in the SEM or microprobe, the interaction volume can be controlled by the choice of beam energy. Using either SS_MC, the single scattering model program, or saving time by using the plural scattering model PS_MC, the beam interaction volume is found to vary quite rapidly with the accelerating voltage, being over 10,000 times larger at 10 keV than at 1 keV. However, at the lowest beam energies the interaction volume is once again comparable to that found at high beam energies in the thin foils, suggesting that there is always more than one route to examining a desired volume of material.

5.1   Backscattered Electrons

Study of backscattered electrons (BSE) resulted in the first evidence of the nature of the interaction between electrons and solids. Now BSE are important as a source of imaging contrast in the SEM, as an important contributory factor to the production of SE, and as a correction in x-ray and Auger microanalysis. Since BSE are incident electrons that have been scattered through more than 90° in the specimen then any modelling program can be used to investigate them. Using the PS_MC program the backscattering coefficient for carbon, silicon, copper, silver and gold at 10 keV can be computed. The data shows the monotonic rise with atomic number Z first observed by Starke (1898) and widely used as the basis for 'atomic number imaging' in the SEM. The fact that some of the electrons are backscattered also implies that the energy deposited in the sample by the incident beam is less than expected, hence the existence of BS electrons requires a correction to many analytical procedures. Procedure 4 in the EIS program also computes the expected backscattering coefficient for a given material and beam energy using the parametric model of Hunger and Kuchler (1979). It should be noted that the Monte Carlo models enable 'experiments' to be carried out that would be difficult to perform in a microscope, but which shed light on points of interest. For example, using the PS_MC program the effect of changing the density of the target, leaving all of the other parameters constant can be investigated. Intuitively the backscattering yield would be expected to change but as the computations reveal the yield is actually constant. Once this observation has been made it is then possible to apply hindsight to understand why it happens that way, and so to gradually turn a theoretical knowledge of beam interactions into a useful and practical body of data.

If the sample is tilted to the beam the number of backscattered electrons will increase, as can be verified by running PS_MC program for incident angles of 10°, 30°, 45° and 60°. It is found that the effect of tilting the beam with respect to the surface is effectively to swing the interaction volume closer to the surface without greatly changing its size or shape. As a consequence the backscattering yield will increase only slowly for low Z targets, since these exhibit mainly forward scattering, but more rapidly for high Z materials which have substantial lateral scattering.

It is clear that the BS electrons can have penetrated quite deeply into the material before escaping back to the surface. The program JustBS allows this effect to be studied. For a given target and set of conditions this will calculate the average depth from which BSE emerge, the greatest depth from which any BSE has reached the surface, and the diameter of the surface region from which these BSE are coming, which is essentially the resolution of the BSE image in the SEM. Typically the average depth of BS emission is found to be about one-fourth to one-third of the Bethe electron range for the material, and the resolution is of the same order of magnitude. Thus at high beam energies the BSE signal interacts with the sample on a scale of micrometers and is consequently low in resolution and surface specificity. At low beam energies however the BSE volume is reduced to only tens of nanometers in both depth and width and hence high resolution data from surfaces is definitely possible.

5.2   Secondary electrons

We saw earlier in this chapter that secondary electrons (SE) are an important example of a single electron inelastic event. More practically, secondary electrons are the most commonly used source of information in the SEM, and they have a profound influence on techniques as diverse as electron beam lithography and x-ray microanalysis. Fast Secondary Electrons (FSE), (see Section 1.1), are a special case of SE since their energy can be a high fraction of that of the incident beam. Although this makes their behaviour atypical, they do provide some insight into the characteristics of genuine SE. The program FSE_MC makes it possible to examine the behavior both of the incident electron beam and of the FSE. For example, choosing a beam energy of 20 keV, and a sample of copper foil 200 nm thick as a target, it is possible to examine firstly the trajectories of the incident electrons in the sample, then those of the FSE that are produced. The incident electrons are, as described earlier, essentially forward scattered so their interaction volume gradually broadens out as the number of scattering events increases. The FSE on the other hand are found to have a totally different distribution. When formed they are typically travelling almost normal to the incident beam direction. Since they also are mostly forward scattered, they form the horizontal limbs in a tree-like structure about the beam direction. The diameter of this FSE interaction is independent of the depth of penetration of the beam into the sample, so in cases where FSE effects are significant they may limit the spatial resolution that is attainable even at the limit of high beam energies and thin specimens.

A good example of this is found in electron beam lithography, where the incident beam 'exposes' a polymeric resist. When subsequently developed the polymer dissolves preferentially in areas where the beam has deposited energy. From the earlier discussion on interaction volumes it could be surmised that the finest line that could be fabricated would be attained at the highest beam energy and for the thinnest film of polymer. However experimentally it is found (Broers 1981) that regardless of the beam energy or how thin the film is made, the developed line never falls below about 100Å in width. The reason for this can be understood by thinking both about the FSE and about electron stopping power (*e.g.* Equation 32). The transfer of energy from electrons to the resist is governed by the Bethe stopping power. Since this contains a $1/E$ term lower energy electrons deposit their energy more efficiently than higher energy ones. The average energy of the FSE is only about 3–5% of the incident energy so these electrons will deposit their energy at a rate 20-30 times greater than that of the primaries. Thus even if the yield of FSE is only a few percent, they may account for half of the energy deposition. Secondly, because of the form of the FSE trajectories, this energy deposition will be directed laterally and the broadening will be almost independent of the thickness.

A similar effect occurs in x-ray production because again the cross-section for the production of the required effect varies as $1/E$. It is thus often more efficient to indirectly produce an x-ray by first generating an FSE and then using this to produce the ionisation. As the energy of the incident electrons goes higher, increasing the overvoltage ratio U, this type of mechanism becomes more important. Thus at 300 keV over 80% of all the oxygen K x-rays produced from a steel sample would be generated by the FSE component. This has two consequences; firstly the spatial resolution of the analysis for the oxygen will be much worse than the value predicted by the spreading of the primary electrons as can be confirmed by running the FSE_MC program; secondly a quantitative analysis of the oxygen will be difficult because the probability of x-ray production via the FSE will depend on factors such as the density and mean atomic number of the sample.

When considering SE in the SEM the fast secondary model is of limited value because the simple cross-section that is employed is inaccurate at low energies. In fact, a detailed model of the production of 'true' SE is very complex because of the need to consider the cascade multiplication of the SE as they travel through the solid. To avoid this sort of difficulty we can use an approach originally made independently, and almost simultaneously, by Salow (1940) and Bethe (1940). The assumption is that SE are generated at a rate N_{SE} proportional to the stopping power of the incident electron *i.e.*

$$N_{SE} = -\left(\frac{1}{\varepsilon}\right)\frac{\mathrm{d}E}{\mathrm{d}S} \tag{52}$$

where ε is a constant of proportionality which is, in effect, the average energy needed to create an SE. Although the validity of Equation 52 is not self evident in practice it has

been found that this is a good approximation — both for electrons and ions. A second
assumption is that the SE diffuse away from their creation point with a characteristic
attenuation length λ, so the probability $p(s)$ of a secondary travelling a distance s from
the source point is

$$p(s) = \exp\left(-\frac{s}{\lambda}\right) \tag{53}$$

which is known as the Straight Line Approximation (Dwyer and Matthew 1985). Although it represents a gross oversimplification it again has been found to be quite
accurate in its predictions. This representation of the generation and subsequent escape of SE is easily integrated into the Plural Scattering Monte Carlo model to give the
program SE_MC on the disk which given suitable values for the two unknown parameters ε and λ can then be used to compute how the SE yield varies with beam energy.
The task of finding ε and λ requires good experimental data on SE yields, and the
program contains a list of values that have been optimised for some common elements
and materials.

The physics behind this program can be applied in many different ways. For example, the majority of SEM images are recorded by using the SE topographic contrast
mode in which the yield of secondaries varies with the angle of incidence of the beam
to the surface. Using the program the secondary yield for copper at tilts of 0°, 30°,
45° and 60° for the usual energy range can be computed. At high beam energies the
SE yield does indeed vary with tilt, whereas at low energies the SE yield is found to be
almost independent of tilt. (A comparison of the values of λ and the electron range at
say 1 keV will show why this happens - the two are of the same order of magnitude and
thus essentially all of the generated SE escape and so there is no longer any increase
in yield to be gained from tilting the sample.) Hence the form of SEM images can be
expected to be different under low voltage conditions.

5.3   X-rays

The Monte Carlo methods discussed above are well suited to the study of x-ray microanalysis problems. An important first step is to distinguish between the interaction
volume of the electron beam and the corresponding generation volume of the x-rays
of interest. In the case of an electron transparent sample this is relatively straightforward. The energy of the electrons will generally be much higher than the critical
excitation energy of the line(s) of interest, so the x-ray generation volume will be equal
to the electron interaction volume if the FSE effects discussed above are ignored. The
program AEM_MC on the disk demonstrates how this information can be used. The program tracks the trajectories of electrons transmitted through the foil noting where each
electron leaves relative to the optic axis. Since x-ray production will be approximately
constant along the trajectory then the volume distribution of the x-rays can be determined. The variation of resolution with foil thickness, composition, beam energy, and
probe diameter can then be studied and used to analyse data in a given experimental
situation.

In the case of a bulk sample many of the electrons will fall below the critical excitation energy before they are backscattered or come to equilibrium, thus in general the
x-ray excitation volume is smaller than the interaction volume. The program PHIROZ

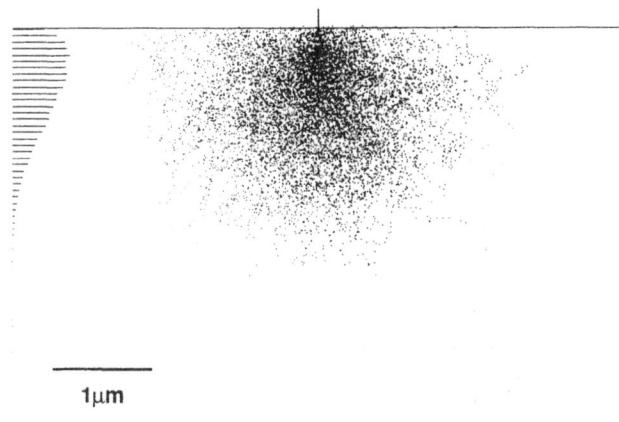

Figure 12. *X-ray generation volume obtained by using the program PHIROZ.*

computes the production of x-rays, generating a dot to represent the photons and cutting off when the electron energy falls too low (Figure 12). A comparison of the x-ray dot map and the normal trajectory plot then shows how the different volumes are related. It can readily be seen that the x-ray production is highly inhomogeneous, being concentrated on the axis and near the surface. Thus a simple statement of the relative diameters of the two volumes gives little guidance as to the likely spatial resolution that could be achieved in an analysis.

The name PHIROZ comes from '$\phi(\rho z)$', the name given by Castaing to the function which describes the variation with depth of x-ray production. The program displays this quantity as a histogram on the side of the display. For incident beam energies much higher than the critical x-ray excitation energy the $\phi(\rho z)$ curve reaches a maximum at some depth below the surface (because the cross-section will rise as the electron energy falls, *e.g.* Equation 20), while for energies close to the critical excitation value the $\phi(\rho z)$ curve will fall sharply to zero below the surface. Since both electrons travelling forward (incident electrons) and those being backscattered can contribute to the x-ray production it is clear that this is yet another example of a parameter of the beam interaction that cannot easily be deduced analytically. The $\phi(\rho z)$ curve is an important factor in the theory of quantitative microanalysis since its form determines how much absorption the x-rays experience as they leave the sample travelling towards the x-ray analyser.

It is instructive to compare the data with PHIROZ with the data from BREMS. This program does a very similar type of calculation but now for the continuum (bremsstrahlung) production. Since the form of the cross-section for continuum generation *e.g.* Equation 25, is different to that for characteristic x-ray production (Equation 20) then the depth variation of the signal is also different. BREMS displays the continuum $\phi(\rho z)$ curves for five different energies and it is found that the form of these is markedly different to those of the characteristic signal, as well as varying greatly with energy. This program also computes in the effect of x-ray absorption within the sample for a given x-ray spectrometer geometry to produce a simulated spectrum showing the energy profile of the continuum as it would be observed experimentally.

6   Conclusions

In summary, while a knowledge of the fundamentals of electron-solid scattering is desirable, this background is of limited value without the ability to put the information to practical use. Hence the reason for the emphasis here on computer software to make a variety of numerical data immediately accessible and on the simulation techniques which permit real questions to be modelled and analysed. Time spent in running the Monte Carlo programs under a wide variety of conditions and assumptions will help build up a realistic, quantitative, understanding of the details of electron solid interactions as they apply to microscopy and microanalysis.

A copy of the Monte Carlo programs and the EIS program described here can be obtained by writing to the author at EM Facility, Department of Zoology, University of Tennessee, Knoxville, TN 37996-0810, USA.

References

Austin L and Starke H, 1902, *Ann Phys, Leipzig* **9** 271
Berger M J, and Seltzer S M, 1964, *Studies in the Penetration of Charged Particles in Matter*, NAS-NRC Report 1133 (National Academy of Sciences, Washington DC) 205
Bethe H A, 1930, *Ann Phys, Leipzig* **5** 325
Bethe H A, 1940, *Phys Rev* **59** 940
Bethe H A and Ashkin J, 1953, in *Experimental Nuclear Physics*, **1** 166
    ed Segre E (Wiley, New York)
Broers A N, 1981, *J Electrochem Soc* **128** 166
Casnati E, Tartari A and Baraldi C, 1982, *J Physique* **B15** 155
Castaing R, 1960, *in Advances in Electronics and Electron Physics* **13** 317
Cosslett V E and Thomas R N, 1964, *Brit J Appl Phys* **16** 883
Curgenven L and Duncumb P, 1971, *Tube Investments Research Laboratories Report*, 303
Czyzewski Z, O'Neill D, Romig AD and Joy DC, 1990, *J Appl Phys* **68** 3066
Dwyer V M and Matthew J A D, 1985, *Surface Science* **152** 884
Egerton R F, 1979, *Ultramicroscopy* **4** 69
Egerton R F, 1981, *Proc 39th Ann EMSA*, ed Bailey G W, 198
    (Claitor's Publishing, Baton Rouge)
Evans R D, 1955, *The Atomic Nucleus*, 576 (McGraw-Hill, New York)
Gryzinski M, 1965, *Phys Rev* **138** A336
Hirsch P B, Howie A, Nicholson R B, Pashley D W and Whelan M J, 1965, in
    Electron Microscopy of Thin Crystals (Butterworths, London)
Hunger H-J and Küchler L, 1979, *Phys Status Solidi*(a) **56** K45
Inokuti M, 1971, *Rev Mod Phys* **43** 297
Joy D C and Luo S, 1989, *Scanning* **11** 176

Kirkpatrick P and Wiedmann L, 1945, *Phys Rev* **67** 321
Koch H W, and Motz JW, 1959, *Rev Mod Phys* **31** 920
Kolbensvedt H, 1967, *J Appl Phys* **38** 4785
Kramers H A, 1923, *Phil Mag* **46** 836
Leapman R D, Rez P and Mayers D F, 1980, *J Chem Phys* **72** 1232
Love G, Cox M G C and Scott V D, 1977, *J Phys D: Appl Phys* **10** 7
Luo S, and Joy D C, 1991, *Proc 49th Ann Meeting EMSA*, 325, ed Bailey G W
    (San Francisco Press, San Francisco)
Luo S, Zhang X and Joy D C, 1991, *Rad Effects and Defects in Solids*, **117** 235
Moller C, 1931, *Z Phys* **70** 786
Newbury D E and Myklebust R L, 1981, in *Analytical Electron Microscopy 1981*, 91
    ed Geiss R H (San Francisco Press, San Francisco)
Newbury D E, Yakowitz H and Myklebust R L, 1976, in *Use of Monte Carlo Calculations in*
    Electron Probe Microanalysis and Scanning Electron Microscopy, 151
    eds Heinrich K F J *et al.* (NBS Special Publication 460)
Rao-Sahib T S and Wittry D B, 1974, *J Appl Phys* **45** 5060
Rhodes R, 1986, *The Making of the Atomic Bomb*, 46 (Simon and Schuster, NY)
Salow H, 1940, *Phys Z* **41** 434
Starke H, 1898, *Über die Reflexion der Kathodenstrahlen, Ann Phys, Leipzig* **66** 49
Statham P, 1976, *X-ray Spectrometry* **5** 154
Sommerfeld A, 1931, *Ann Phys, Leipzig* **11** 257
Tung C J, Ashley J C and Ritchie R H, 1979, *Surf Sci* **81** 427
Wentzel G, 1927, *Z Phys* **40** 590
Williams E J, 1933 *Proc Roy Soc (London)* **A139** 163

Electron Probe X-ray Microanalysis

P Van Espen

University of Antwerp
Belgium

1 Introduction

The purpose of electron probe x-ray microanalysis is, in the strictest sense, to perform elemental analysis by measuring the x-rays excited in a microvolume under bombardment with energetic electrons, and in general to study the composition of materials at the microscopic level. Using this technique we can perform:

- spot analysis (local analysis, inclusions)

- line scan (composition over an interface)

- x-ray mapping (lateral distribution)

- in-depth analysis (multi-layers)

X-rays are generated in electron microprobes or scanning electron microscopes. For the measurement of the x-rays we use semi-conductor detectors (energy dispersive spectrometry, EDS) or crystal spectrometers (wavelength dispersive spectrometry, WDS).

In EPXMA accelerating voltages in the range 5–30 kV are used. Typical currents are in the nanoamp range for EDS and up to microamps for WDS. The probe diameter (0.1–1μm) depends on the probe current and is not critical since the interaction volume for x-ray production is always of the order of μm^3. In contrast to EPXMA, we speak of Analytical Electron Microscopy (AEM) when we use a transmission electron microscope (TEM, STEM) in combination with an x-ray detector (only EDS). This method is used to analyse (very) thin samples with high spatial resolution ($\sim$10nm).

EPXMA is capable of an absolute detection limit of the order of 10^{-13}g, and of analysing volumes of $\sim$1μm^3. Relative detection limits are in the range 0.1% –3%, with an accuracy and precision of 1% – 5% and a detectable element range from boron to uranium. The historical development of EPXMA is given in Table 1.

~1950	Static electron microprobe (Castaing)	
~1956	Scanning probe	Scanning electron microscope
~1968	EDS on microprobe (Fitzgerald, Keil, Heinrich)	
1970's	Electron probe microanalysers (EPMA) High quality x-ray analysis using 1 to 5 WDS and EDS	Scanning electron microscopes (SEM) High quality images (with optional EDS)
1990's	EPMA and SEM both for high quality images and x-ray analysis using WDS and EDS	

Table 1. *Historical development of* EPXMA

To appreciate the capabilities of EPXMA as an analytical tool, let us consider a steel sample containing 2% by weight of chromium. The weight of chromium in an analysed volume of $V = 1\mu m^3$ is

$$\frac{C_{Cr} \times \rho \times V}{100} = 0.02 \times 8 \times 10^{-12} = 16 \text{ pg}$$

and the number of chromium atoms analysed, n will be

$$n = \frac{g \times N_A}{A_{Cr}} = \frac{1.6 \times 10^{-13} \times 6.02 \times 10^{23}}{52.0} \approx 2 \times 10^9.$$

For any chemical analysis we require the results to be precise and accurate. Precision refers to the variation of results due to random (uncontrollable) factors in the experiment. The counting statistics in x-ray detection contribute to the lack of precision. Accuracy refers to how much the results deviate from the true value because of systematic errors and, in order to obtain accurate and precise results, we need to understand:

- the physics of the processes that are involved

- the instrument, how it works and its limitations

- the procedures that need to be used to quantify the results.

In the following, these three topics will be discussed. Rather than giving the most complete and complex theory, a number of numerical examples will be given which can easily be verified with a calculator or a spreadsheet program. An idea of the order of magnitude of the various phenomena is important in understanding and performing accurate and precise EPXM-analysis.

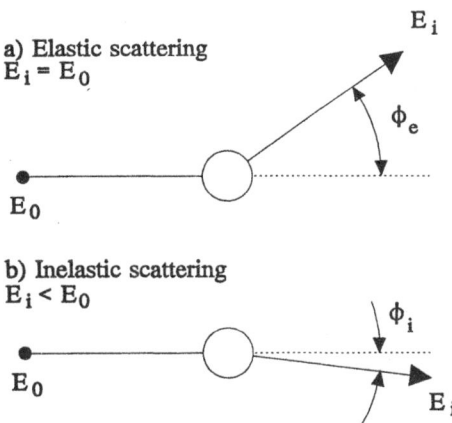

Figure 1. *Electron scattering (a) elastic (b) ineleastic*

2   X-ray production

2.1   The interaction of electron beam and specimen

To understand x-ray production we must consider the interaction between the electron beam and the specimen. We generally distinguish between elastic and inelastic events. During elastic scattering the beam-electron interacts with a specimen atom (nucleus) causing a change in direction (ϕ_e from $0°$ – $180°$, typically $5°$) and a very small energy loss (< 1 eV) (Figure 1a). If inelastic scattering occurs part of the energy of the electron is transferred to the sample (Figure 1b).

There are various inelastic scattering processes:

1. excitation of conduction band electrons ($\Delta E < 50$eV) giving secondary electrons,

2. excitation of the 'free electron gas' (plasmon excitation, $\Delta E \approx 15$ eV),

3. inner shell excitation to give emission of characteristic x-rays and Auger electrons ($\Delta E \approx 1$–20keV),

4. deceleration of electrons in the Coulombic field of the atom giving Bremsstrahlung production ($\Delta E \approx 1$–20 keV), and

5. lattice vibrations (phonon excitation, *e.g.* a rise in temperature of $10°$C for a 1nA beam current).

Elastic scattering is the dominant interaction. Out of hundreds of interactions only a few are inelastic, and out of all possible inelastic phenomena only the ionisation of the inner shells (K,L,M) produces characteristic x-rays that are useful in EPXMA. The

Bremsstrahlung production is responsible for the continuous background observed in the x-ray spectrum.

2.2   Scattering cross-sections

The probability that a particular interaction will occur is determined by its cross-section. For an interaction of type j, we can write:

$$\sigma_j = \frac{N_j}{n_t n_i} \tag{1}$$

where N_j is the number of interactions per cm^3, n_t is the number of targets (atoms) per cm^3, and n_i is the number of incident particles (electrons) per cm^2. The cross-section σ thus has the dimension of cm^2/atom and indicates the apparent size of the atom for a given interaction with an electron. From the cross-section we can derive the mean free path, $i.e.$ the average distance an electron travels between two interactions of type j:

$$\lambda_j = \frac{A}{N_A \rho \sigma_j}. \tag{2}$$

The Rutherford scattering cross-section formula,

$$\sigma(\phi_e > \phi_0) = 1.62 \times 10^{-20} \frac{Z^2}{E^2} \cot^2 \frac{\phi_0}{2} \tag{3}$$

gives the probability that an electron with energy E (in keV) undergoes elastic scattering by an atom of atomic number Z with a scattering angle ϕ_e larger than some specified angle ϕ_0.

As examples, the elastic scattering cross-sections for aluminium and copper as calculated from Equation 3 are tabulated below. The cross-sections are given as cm^2/atom.

	$\sigma(\phi_e > 2°)$ $E = 10$ keV	$\sigma(\phi_e > 2°)$ $E = 20$ keV	$\sigma(\phi_e > 30°)$ $E = 20$ keV
Al	9.0×10^{-17}	2.3×10^{-17}	9.5×10^{-20}
Cu	45.0×10^{-17}	11.0×10^{-17}	47.0×10^{-20}

It can be seen that forward scattering is more likely ($\sigma(\phi_e > 2°) \gg \sigma(\phi_e > 30°)$), increases with atomic number Z ($e.g.$ atomic number contrast in backscattered electron images), and decreases for higher energies.

The mean free path λ_e between two elastic scattering events can also be determined from this data; an example calculation for 20 keV electrons in aluminium is given below.

$$\lambda_e = \frac{A_{Al}}{N_A \rho_{Al} \sigma_{Al}(\phi > 2°)} = \frac{26.98}{6.02 \times 10^{23} \times 2.70 \times 2.3 \times 10^{-17}} = 7.2 \text{nm}$$

Inelastic scattering events decrease the energy of the electron beam until it comes to rest. The energy loss per unit of distance travelled in the solid is given by the Bethe equation:

$$\frac{dE}{ds} = -7.85 \times 10^4 \frac{Z\rho}{AE} \ln(1.166E/J) \quad \text{[keV/cm]} \tag{4}$$

where E is the mean energy of the beam electron along the path, s is the distance along the trajectory (not straight down). J is the mean ionisation potential of the sample, *i.e.* the average energy loss per interaction; considering all possible interactions we have

$$J = (9.76Z + 58.5Z^{-0.19}) \times 10^{-3} \quad \text{[keV]} \tag{5}$$

The stopping power is related to the energy loss by

$$S = -\frac{1}{\rho}\frac{dE}{ds} \quad \text{[keVcm}^2\text{/g]} \tag{6}$$

To obtain a measure of these quantities, consider 20 keV electrons incident on aluminium ($E = 20$ keV, $Z_{Al} = 13$, $A_{Al} = 26.98$ g, $\rho_{Al} = 2.7\text{g/cm}^3$). The above equations then give: the average energy loss per interaction J=0.163 keV; the number of inelastic interactions (E_0/J) is $n_i \approx 123$; the energy loss $dE/ds = -25.3 \times 10^3$ keV/cm and the total loss over 1cm is 2.53keV; the stopping power $S = 9.37 \times 10^3$ keV cm^2/g.

2.3 Inner shell ionisation

The likelihood of inner shell ionisation is given approximately by the Bethe cross-section:

$$\sigma = 6.51 \times 10^{-20} \frac{n_s B_s}{U E_c^2} \ln(C_s U) \tag{7}$$

where n_s is the number of electrons in the shell (n_s=2 for the K shell), E_c is the ionisation energy of the shell (keV), U is the overvoltage ($U = E/E_c$ with E the energy of the electrons), and B_s and C_s are constants for a given shell. For the K shell B_s=0.9, C_s=0.65 and $4 < U < 25$. For example the Bethe cross-section for K shell ionisation of aluminium by 20 keV electrons can be calculated using $E_{c,Al} = 1.56$ keV and $U = 20/1.56 = 12.8$ to give

$$\sigma_{K,Al} = 6.51 \times 10^{-20} \frac{2 \times 0.9}{12.8 \times 1.56^2} \ln(0.65 \times 12.8) = 0.80 \times 10^{-20} \quad \text{cm}^2$$

Using this information the electron mean free path between two K shell ionisations can be determined as

$$\lambda_K = \frac{26.98}{6.02 \times 10^{23} \times 2.70 \times 0.80 \times 10^{-20}} = 2.1 \times 10^4 \quad \text{nm}$$

This value must be compared to the mean free path of 7.2 nm as calculated above for elastic scattering. The beam-electron thus frequently undergoes elastic scattering, changing the direction of the path. From time to time inelastic scattering occurs, slowing down the electron and sometimes producing an x-ray photon.

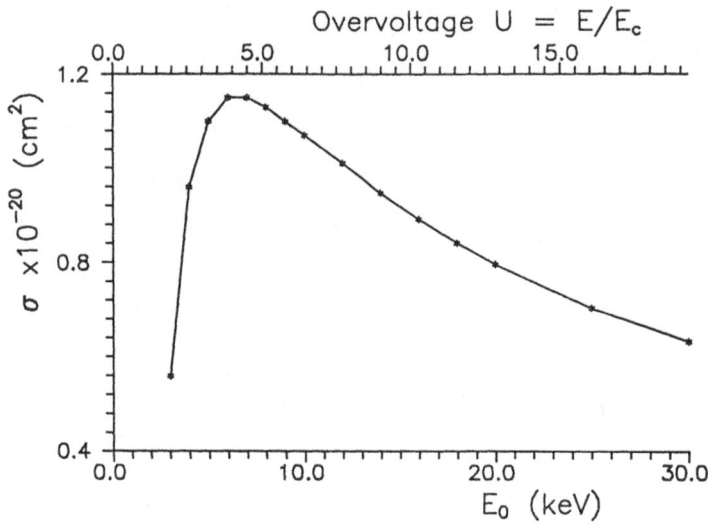

Figure 2. *K-shell ionisation cross-section as function of beam energy and overvoltage*

The K shell ionisation cross-section of aluminium as function of the beam energy and the overvoltage is shown in Figure 2. The ionisation cross-section mainly depends on the overvoltage U and reaches a maximum for $U \sim 4$.

2.4 Characteristic x-rays

In an atom, the electrons are arranged in shells and sub-shells: K, L_I, L_{II}, L_{III}, M_I $\ldots M_V$, $N_I \ldots N_{VII}$, corresponding to distinct energy levels E_c. If a beam electron has an energy higher than E_c, the orbital electron can be expelled (the inner shell ionisation process), and a vacancy is created in that shell. The atom is in a highly excited state and relaxes within approximately 10^{-12} seconds, moving an electron from a higher shell into the vacancy. The energy involved in this transition is released as an x-ray photon or as an Auger electron. E_c, the binding energy of the electron in the shell, is called the critical ionisation energy (or x-ray absorption edge energy).

The transitions which occur depend on the occupation of the levels and are governed by the quantum mechanical selection rules, *e.g.* the orbital quantum number l must change by ± 1. The K-L_1 transition is therefore forbidden ($\Delta l = 0$). Only the most intense lines are used in analytical work. The Kβ intensity is $\sim 13\%$ of that for Kα x-rays, the Lβ_1 and Lβ_2 are repectively $\sim 50\%$ and 20% of the Lα, while the Mα and Mβ_1 lines have almost the same intensity. These other weak lines must be considererd however because of the possible interference they cause with the other, stronger, lines used for analysis. The ratio of the K, L and M intensities (*e.g.* Lα to Kα) depend strongly on the overvoltage.

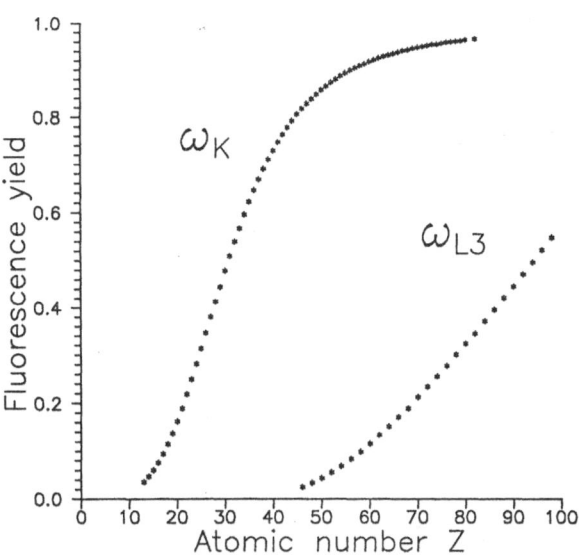

Figure 3. *Plot of fluorescence yield against atomic number for K and L shells*

Since the ionised atom can also relax by emitting an Auger electron, only a fraction of the vacancies produce photons. This fraction is called fluorescence yield, given by:

$$\omega_K = \frac{\text{number of K photons emitted}}{\text{number of K vacancies produced}}$$

A plot of fluorescence yield against atomic number for K, L and M shells is shown in Figure 3. The fluorescence yield is low for light elements, which is one of the main reasons why x-ray analysis methods are less sensitive for light element analysis.

2.5 Electron-solid interaction volume

From electron trajectory calculations using Monte-Carlo simulations we know that the interaction volume of the electrons has a pear-like shape. With increasing atomic number the dimensions of this interaction volume are reduced and the shape becomes more spherical. With increasing energy of the electrons, the size increases and deeper penetration occurs (Figure 4).

The electron interaction volume is however larger than the x-ray generation volume. As electrons slow down when they get deeper into the sample, their energy at a certain moment becomes less than the critical ionisation energy E_c, and x-rays are no longer produced. The range is approximately given by

$$R = \frac{0.0064}{\rho}(E_0^{1.68} - E_c^{1.68}) \quad \mu m \tag{8}$$

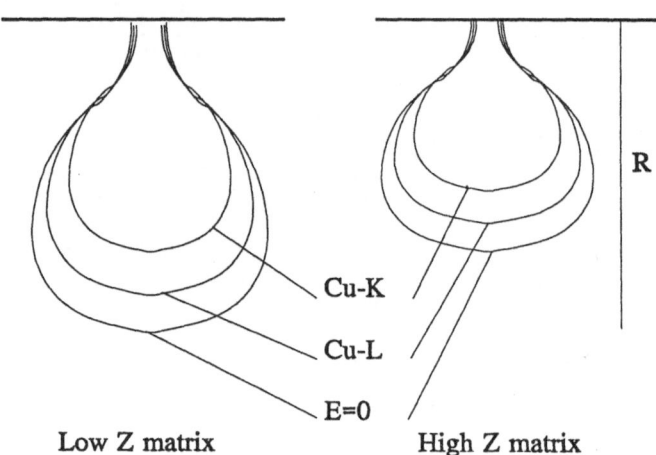

Figure 4. *Electron interaction volume in a solid for a number of emitted x-ray wavelengths.*

For 20 keV electrons penetrating into copper ($\rho_{Cu} = 8.93$ gm/cm³) the following ranges are observed :

$$
\begin{array}{lll}
\text{electron range} & (E_c = 0 \text{ keV}) & R = 0.110 \ \mu m \\
\text{Cu} - L_\alpha \text{ range} & (E_c = 0.933 \text{ keV}) & R = 0.109 \ \mu m \\
\text{Cu} - K_\alpha \text{ range} & (E_c = 8.980 \text{ keV}) & R = 0.081 \ \mu m
\end{array}
$$

Another way to descibe the interaction zone is by means of the phi-rho-z curve. The phi-rho-z curve , $\phi(\rho z)$, gives the production of x-rays as function of depth (see Figure 5 and the article by Joy in this Proceedings). X-ray production is maximum at a certain depth. This is because the width of the interaction layer increases and, as the electrons slow down, the ionisation probability reaches a maximum (at U $\sim$ 4).

The x-rays produced at a depth z can be absorbed in the material. The attenuation in a path length l is given by Beer's law:

$$
\frac{I}{I_0} = \exp(-\mu\rho l) \tag{9}
$$

where μ is called the mass attenuation coefficient, with units of cm²/g. If we include the take-off angle (Figure 6), the attenuation can be expressed as function of depth:

$$
\frac{I}{I_0} = \exp\left(-\frac{\mu\rho z}{\sin\psi}\right) \tag{10}
$$

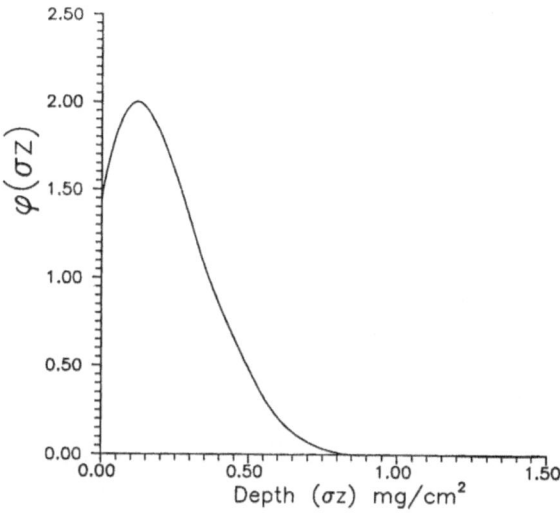

Figure 5. *X-ray production as a function of depth (phi-rho-z curve)*

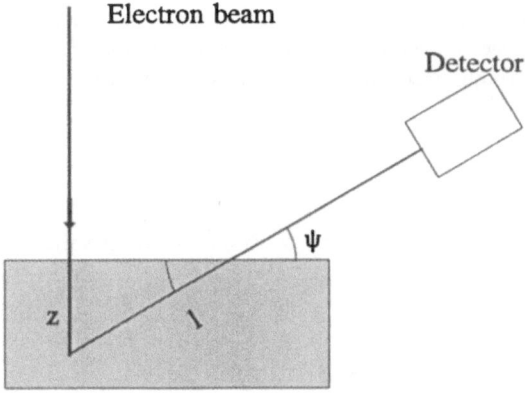

Figure 6. *Directions of electron incidence and x-ray take-off from a solid*

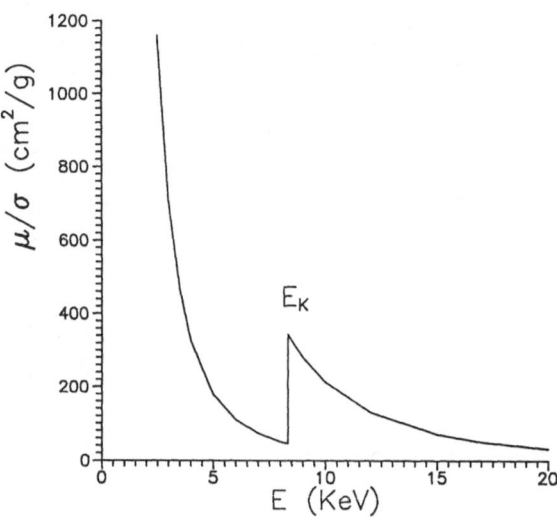

Figure 7. *Variation of mass absorption coefficient with x-ray energy for nickel, showing the K absortion edge*

For a given element, μ decreases as a function of the energy with jumps corresponding to the x-ray absorption edges (E_c), as shown in Figure 7. For a sample made up of n elements i, with weight fractions w_i, the mass attenuation coefficient is by:

$$\mu_M = \sum_{i=1}^{n} \mu_i w_i$$

Various tabulations and an expression to calculate the μ values exist. However, below 1 keV these values are not very accurate.

2.6   Number of x-rays produced

The ionisation cross-section σ gives the number of ionisations per electron per atom/cm². Since 1 cm³ contains ρ grams, ρ/A molecules and $\rho N_A/A$ atoms, a layer of thickness dx must contain $\rho dx N_A/A$ atoms. The number of ionisations in this layer is $\sigma N_A \rho dx/A$ and the number of x-ray photons produced (per electron) is $\omega \sigma N_A \rho dx/A$ For a bulk sample we integrate over the layers dx, taking into account the loss of energy of the beam-electron

$$I_c = \int_0^R \frac{\omega \sigma N_A \rho}{A} dx = \frac{\omega N_A \rho}{A} \int_{E_0}^{E_c} \frac{\sigma}{dE/dx} dE \qquad (11)$$

In reality less x-rays are produced because backscattered electrons are lost from the target. An approximate expression for this integral is

$$I_K = 0.116\frac{\omega}{A}(U - 1)^{1.67} \tag{12}$$

giving the number of K x-ray photons emitted into a solid angle of 4π per incident photon (neglecting backscattering). Using the above equation for a copper sample excited with a 30 kV, 2 nA electron beam, $A_{Cu} = 63.55$ g/mol, $\omega_{K,Cu} = 0.409$, $E_{c,Cu-K} = 8.98$ keV, we find $I_K = 0.003$ photons/electron into 4π steradians. If we have a Si(Li) detector with an area of 10 mm^2 at a distance of 40 mm from the sample, its solid angle is approximately given by $10/40^2 = 0.006$ steradians and hence the number of Cu-K photons detected per second (count rate) at 2nA is

$$\frac{0.006}{4\pi} \times \frac{2 \times 10^{-9}}{1.6 \times 10^{-19}} \times 0.003 = 18 \times 10^3 \quad \text{cps}$$

3 X-ray measurement

3.1 X-ray spectrometers

In EPMA we measure the energy (or wavelength) of the characteristic x-rays to identify the elements and the number (intensity) of these characteristic photons to quantify the concentration of each element. A dispersive device and a method of counting the x-ray photons are therefore required.

Two types of spectrometer can be used, namely crystal spectrometers based on Bragg reflection, called wavelength dispersive spectrometers (WDS) or Si(Li) or Ge(Li) solid state detectors called energy dispersive spectrometers (EDS).

The WD spectrometer works by allowing a small portion of the x-rays leaving the sample to impinge on the analyser crystal. If Bragg reflection occurs, the x-rays fall onto a detector. The electric pulse from this detector is amplified and counted.

Most often, curved crystal (focusing) spectrometers are used where the source (the place on the sample where the x-rays are generated), the crystal and the detector are positioned on a circle with radius R (the Rowland circle). All x-rays from the point source have the same angle of incidence and the diffracted x-rays will focus at the detector location (Figure 8). From the geometry it follows that $\sin\theta = L/2R$. For Bragg reflection it is required that $\lambda = 2nd\sin\theta$ so that when $n = 1$

$$\lambda = \frac{dL}{R} \tag{13}$$

Thus the wavelength of the detected x-rays is proportional to the source-crystal distance.

If the sample (source of x-rays) is not in focus, the efficiency of the geometrical arrangement will become lower, resulting in a lower observed x-ray intensity. The maximum sensitivity to defocusing is along the z direction (vertical axis). The position in the z-direction needs to be adjusted very accurately (within $\sim 2\mu m$) for satisfactory operation. To adjust the focus adjustment in electron microprobes the co-axial light

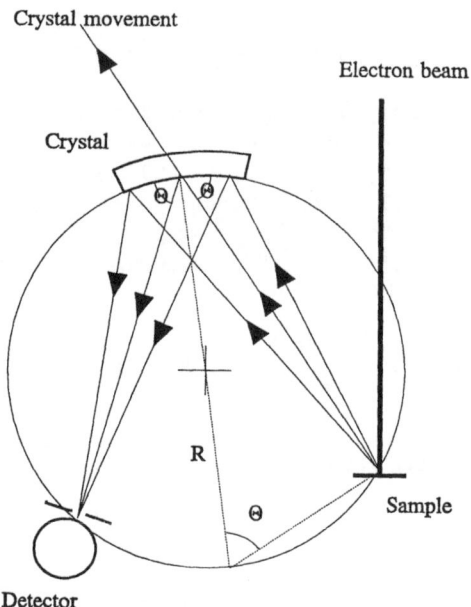

Figure 8. *Curved crystal (focusing) x-ray spectrometer*

microscope is pre-aligned with the x-ray optics. Bringing the sample into the focal plane of the optical microscope (z-movement of the sample) gives a focused condition for the x-rays as well.

The limited displacement of the crystal (L_{min} to L_{max}) requires crystals with different lattice spacing to be used in order to cover the full wavelength range required. Some examples are given in the following table.

Crystal (h, k, l)	$2d$ (Å)	Range (Å)	Range (keV)	Elements detected (Kα)
LiF (200)	4.027	0.9 – 3.7	3.3 – 14.0	K – Sr
PET (002)	8.742	1.9 – 7.9	1.6 – 6.5	Si – Fe
TAP (100)	25.757	5.6 – 23	0.5 – 2.2	F – P
STE (film)	98	21 – 89	0.14 – 0.6	B – O

For example, in a spectrometer with a lithium fluoride (LiF) crystal that can be displaced from $L_{min} = 80.5$ mm to $L_{max} = 254$ mm on a Rowland circle with radius $R = 140$ mm, the wavelength or energy range that can be detected ($\lambda = 12.3981/E$ [keV]) is given (using the above table for the values of $2d$) by

$$\lambda_{min} = \frac{dL_{min}}{R} = \frac{4.027}{2 \times 140} \times 60.5 = 0.870\text{Å} \quad (\sim 14.3\text{keV x-ray energy})$$

$$\lambda_{max} = \frac{dL_{max}}{R} = 0.014 \times 254 = 3.65\text{Å} \quad (\sim 3.4\text{keV x-ray energy})$$

Other important parameters of the spectrometer are the resolution and sensitivity.

The type of detector usually used in a WD spectrometer is a gas proportional counter, consisting of a tube filled with argon (or xenon, or a mixture of 90% argon 10% methane called P10). A tungsten wire at the centre of the tube is held at a potential of 1–2 kV. X-rays enter through a thin window, and are absorbed by the argon atoms (photoelectric effect) resulting in the production of energetic photoelectrons. Such electrons lose their energy by ionising other argon atoms, releasing more electrons. Because the wire is at a positive potential of 1–2 kV, the electrons initially released are accelerated to the wire and on their way they collide with other gas atoms, causing further ionisation and the release of electrons. The total number of electrons becomes much larger, but remains proportional to the number of initial electrons (hence we have a proportional counter). The gain factor or gas amplification factor, A is approximately 10^4.

The efficiency of a gas proportional counter (called the 'quantum efficiency') is reduced by the fact that x-rays of low energy are absorbed in the window, and highly energetic x-rays are not completely absorbed by the gas. At an x-ray wavelength of 1 Å(12.4 keV) the quantum efficiency of a Ar-CH$_4$ gas flow proportional counter is nearly zero, while a Xe proportional counter has an efficiency of 20%. The gas proportional counter is best for soft x-rays (up to 6 keV) provided that the window is thin enough.

For short wavelengths (high energy photons) a scintillation detector is often used. A thallium-activated sodium iodide single crystal, NaI(Tl), is normally used as the scintillator. X-rays are absorbed in the crystal producing photo-electrons, resulting in light emission (of wavelength 4100Å). This light falls on the photo-cathode of a photomultiplier tube, producing photoelectrons again that are accelerated to the first dynode of the tube. At impact, two or more secondary electrons are produced; these hit the second dynode, and so on. At the last dynode a pulse is produced proportional to the energy of the x-rays. Typically 8% of the light results in photoelectrons that reach the first dynode. The amplification factor of the photo-multiplier is very high, typically 10^6.

A preamplifier converts the total charge produced by each x-ray photon into a voltage pulse. This pulse is amplified and shaped by a main amplifier. Next, a single channel analyser is used to convert each pulse which exceeds a certain voltage level (the lower level discriminator threshold) into a rectangular pulse of fixed duration (5V TTL pulse). These pulses are counted for a preset amount of time. The pulse can also be sent to a rate-meter, stripchart recorder or a computer.

After the arrival of an x-ray photon, the detector and the associated electronics are dead for a very short time, called the dead-time. If a second photon arrives within this time interval it is not counted (dead-time losses). The dead-time depends on the recovery time of the detector itself and on the length of the pulses in the electronic circuit. At 1000 counts per second (cps), the average time between two events is 1 ms but the photons do not arrive in equal spaced time intervals, their arrival follows Poisson statistics. Thus even at this low count rate, two photons might arrive within 0.1 μs. Therefore whilst dead-time losses always occur, they are only significant at higher count rates. The relation between the true and the observed count rate is given by

$$I_{\text{true}} = \frac{I_{\text{obs}}}{1 - I_{\text{obs}} T_R} \tag{14}$$

For a dead-time T_R of 10^{-6}s, the dead-time losses at a range of true count rates are shown in the following table:

True rate (cps)	Observed rate	% loss
1000	999	0.1
10000	9901	1.0
100000	90909	9.1

The x-ray measurement is made for a fixed time T_P, during which the number of x-ray photons N_P occuring at the 2θ value corresponding to the maximum of the peak. This value needs to be corrected for the continuum background. We can measure this background at a 2θ value close to the peak, or take the average at two equal 2θ intervals from the peak. If we measure this background for a time T_B and obtain N_B counts, the nett x-ray intensity is simply given by

$$I = I_P - I_B = \frac{N_P}{T_P} - \frac{N_B}{T_B}. \tag{15}$$

X-ray measurements are affected by counting statistics (Poisson statistics) and uncertainty in the number of counts observed is therefore given by

$$\sigma_N = \sqrt{N}.$$

Using error propagation we can deduce that in a determination of the intensity

$$I = \frac{N}{T}$$

the uncertainty is given by

$$\sigma_I = \frac{1}{T}\sigma_N = \frac{\sqrt{N}}{T} = \frac{\sqrt{IT}}{T} = \left(\frac{I}{T}\right)^{0.5}$$

Applying this to the nett intensity gives :

$$\sigma_I^2 = \sigma_{I_P}^2 + \sigma_{I_B}^2$$
$$\sigma_I = \left(\frac{I_P}{T_P} + \frac{I_B}{T_B}\right)^{0.5}$$

Thus the relative counting error is given (as a percentage) by

$$\varepsilon = \frac{\sigma_I}{I} \times 100 = \frac{(I_P/T_P + I_B/T_B)^{0.5}}{I_P - I_B} \times 100 \tag{16}$$

If at the peak maximum $N_P = 1000$ counts, $T_P = 2$s, and at the background $N_B = 300$ counts, $T_B = 1$s then the nett intensity and the uncertainty in that intensity are

$$I = \frac{1000}{2} - \frac{300}{1} = 200\text{cps} \tag{17}$$

$$\sigma_I = \left(\frac{500}{2} + \frac{300}{1}\right)^{0.5} = 23\text{cps} \tag{18}$$

$$\varepsilon \approx 12\% \tag{19}$$

However, if both peak and background are measured for 20s the uncertainty becomes less:

$$\sigma_I = \left(\frac{500}{20} + \frac{300}{20}\right)^{0.5} \approx 6\text{cps} \tag{20}$$

$$\varepsilon \approx 3\% \tag{21}$$

This can be related to the detection limit as follows. If the average value of the background is 100 counts, its standard deviation σ equals 10 counts. In 95% of the cases the background will be in the range $\pm 2\sigma$, and is thus 100 ± 20 counts. If we observe a higher number (say 130 counts) it is unlikely that this comes from the background, and more likely that it is due to a peak (superimposed on the background).

The uncertainty in the nett number of counts $N = N_P - N_B$ is given by

$$\sigma = (\sigma_{N_P}^2 + \sigma_{N_B}^2)^{0.5} = (N_P + N_B)^{0.5}$$

so that near the detection limit, where $N_P \approx N_B$,

$$\sigma = \sqrt{2N_B}$$

The detection limit (in counts) at the 95 % confidence level is 2σ

$$DL_{\text{counts}} = 2 \times \sqrt{2N_B} \approx 3\sqrt{N_B}. \tag{22}$$

Since the concentration is related to the count rate we see that the detection limit is inversely proportional to the measuring time

$$DL_{\text{conc}} \sim \frac{3\sqrt{N_B}}{T_B} = 3\left(\frac{I_B}{T_B}\right)^{0.5} \tag{23}$$

Silicon drifted-lithium (Si(Li)) detectors are discussed by Titchmarsh in these Proceedings. A comparison of the Si(Li) detector with crystal spectrometers is given in the following table.

	WDS	**EDS**
Geometical efficiency	Variable ($< 2\%$)	$\sim 2\%$
Quantum efficiency	$\sim 30\%$	$\sim 100\%$
Detectable elements	$Z \geq 4$	$Z \geq 11$ (Be window) $Z \geq 4$ (ultra thin/windowless)
Complexity of instrumentation	very complex mechanics, simple electronics	mechanically simple, sophisticated electronics
Resolution	~ 10 eV	~ 150 eV
Max. count rate	50000 cps	~ 2000 cps - high resolution ~ 20000 cps - low resolution
Spectrum interpretation and evaluation	simple, few artefacts and interferences	difficult, many artefacts and spectral overlap, needs computer evaluation
Operating mode	serial (one element at a time)	parallel (all elements simultaneously)

4 Quantitative Analysis

Quantitative analysis requires stable experimental conditions, including a constant electron beam energy E_0, a constant electron dose (current × measuring time), and fixed detector type and geometry (take-off angle). A proportionality between the number of atoms i (or concentration C_i) and the number of characteristic x-rays detected, I_i, then exists. detected is

$$I_i = \text{'constant'} \times C_i \tag{24}$$

This proportionality 'constant' however also depends on the sample composition. This is called the matrix effect.

Concentrations can be calculated by a number of methods. They can be determined aproximately by the k-ratio, or by using calibration curves, or by applying matrix corrections to the k-ratio (using the standard ZAF method or the $\phi(\rho z)$ method), or by relying on fundamental parameters (the standardless ZAF method).

4.1 The k-ratio

The ratio of the x-ray intensity of element i in a specimen, I_i, to the intensity of the pure element $I_{(i)}$ is given by the so-called 'k-ratio'

$$k_i = \frac{I_i}{I_{(i)}} \tag{25}$$

The k-ratio can be used to determine an approximate concentration as follows. For an unknown sample we have $I_i = \text{constant} \times C_i$ whereas for a pure element we have $I_{(i)} = \text{constant} \times 1$ (where the concentrations are by weight). Thus

$$C_i \approx \frac{I_i}{I_{(i)}} = k_i \tag{26}$$

Note that this equation is only approximate since the 'constant' is not truly constant. An approximate concentration for tool steel obtained using this equation is given in the following table where we have taken $E_0 = 30$ keV, $\psi = 52.5°$.

Element	True concn. wt.%	$k_i \times 100$	% difference
C	0.82	0.17	-80
Cr	4.18	5.18	24
V	1.88	2.09	11
Mn	0.26	0.253	-3
Fe	81.8	80.8	-1

The approximation is very good for iron since the matrix (tool steel) is very similar to pure iron. The calculated carbon concentration is incorrect because the pure carbon matrix is quite different from a small amount of carbon in steel.

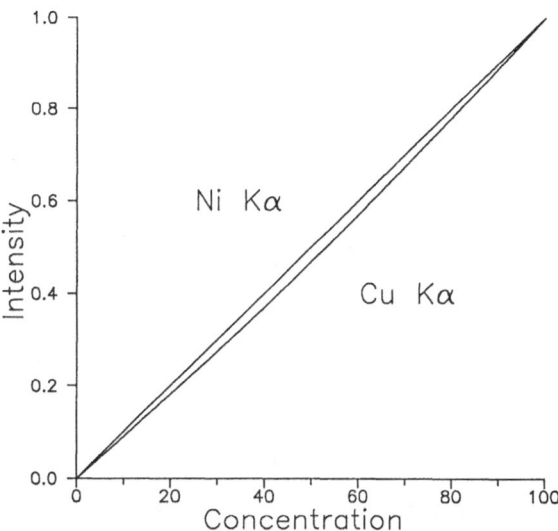

Figure 9. *Calibration curves showing the variation of the copper and nickel x-ray intensities with concentration in a copper-nickel alloy*

4.2   Calibration curves

To use the calibration curve method a series of standards are required which are, similar in composition to the unknown materials and cover the concentration range of interest. The calibration curve, such as that in Figure 9, is produced by plotting the x-ray intensity (or the k-ratio) of the element against the concentration. Very accurate results can be obtained using this technique. Observe the deviation from linearity. This indicates the size of the matrix correction.

4.3   The ZAF correction

Since the concentration and the x-ray intensity are proportional, there is a relation between the intensity and the concentration in an unknown sample (labelled u) and a standard (s). The concentration of an element i in an unknown sample is

$$C_i^{(u)} \approx C_i^{(s)} \frac{I_i^{(u)}}{I_i^{(s)}}$$

The ZAF method can be used to correct the observed x-ray intensities for differences in matrix composition between standard and unknown sample:

$$C_i^{(u)} = C_i^{(s)} \frac{I_i^{(u)} F^{(u)}}{I_i^{(s)} F^{(s)}}$$

The factor F involves the atomic number or Z correction, the absorption correction (A) and the fluorescence correction (F), so that

$$F = F_Z F_A F_F$$

which is the origin of the name 'ZAF correction'. Thus

$$C_i^{(u)} = C_i^{(s)} \frac{I_i^{(u)}}{I_i^{(s)}} \frac{(F_Z)_i^{(u)}}{(F_Z)_i^{(s)}} \frac{(F_A)_i^{(u)}}{(F_A)_i^{(s)}} \frac{(F_F)_i^{(u)}}{(F_F)_i^{(s)}} \tag{27}$$

or

$$C_i^{(u)} = C_i^{(s)} \frac{I_i^{(u)}}{I_i^{(s)}} Z_i A_i F_i \tag{28}$$

The terms Z_i, A_i and F_i are often referred to as the Z-, A- and F-corrections.

- Z_i, the atomic number correction, takes care of the difference in backscattering and stopping power,

- A_i, the absorption correction, corrects for the difference in absorption of the x-rays when travelling out of the sample,

- F_i, the fluorescence correction, corrects for the contribution from the x-rays that are not generated by the electron beam but by fluorescence induced by characteristic x-rays of other elements.

Note that the Z-, A- and F-corrections are ratios of the effect in the unknown sample and the standard. The more similar the standard and the unknown are, the closer the correction will be to 1, and the more accurate the concentration estimate.

To introduce the concept of k-ratio again we refer to the standards as pure elements:

$$C_i^{(u)} = k_i Z_i A_i F_i \tag{29}$$

This does not mean we have to use pure elements as standards, since we can calculate the pure element intensity from any multi-element standard, using

$$I_{(u)} = \frac{I_i^{(s)}}{C_i^{(s)}} \frac{(F_Z)_i^{(s)}}{(F_Z)_{(i)}} \frac{(F_A)_i^{(s)}}{(F_A)_{(i)}} \frac{(F_F)_i^{(s)}}{(F_F)_{(i)}} \tag{30}$$

We will now discuss the three factors in some detail. The relations shown are not necessarily the most accurate and up-to-date versions. They are chosen mainly because they are simple and serve the purpose of showing the magnitude of the various effects.

The absorption factor

X-rays produced at a given depth below the sample surface have to travel a certain distance through the specimen before they reach the detector. As they pass along this path they can suffer absorption. It is this decrease in x-ray intensity which is corrected for by the 'absorption correction'.

The x-ray intensity generated from a layer dz at a depth z below the surface is given by

$$dI = \phi(\rho z)d(\rho z)$$

Thus, if no x-rays are absorbed, the total generated intensity is

$$I_{\text{NOABS}} = \int_0^\infty \phi(\rho z)d(\rho z)$$

However because of absorption, the observed x-ray intensity from layer dz is less (Beer's law, equations 9 and 10):

$$dI = \phi(\rho z)\exp(-\mu z \csc \psi)d(\rho z)$$

Therefore the observed intensity from all layers is

$$I_{\text{ABS}} = \int_0^\infty \phi(\rho z)\exp(-\mu z \csc \psi)d(\rho z)$$

Clearly the absorption factor is

$$F_A = \frac{I_{\text{NOOBS}}}{I_{\text{ABS}}} \tag{31}$$

A semi-empirical expression for F_A due to Philibert, Duncumb and Heinrich (PDH) (Heinrich 1970) is

$$F_A = \left(1 + \frac{\chi}{\sigma}\right)\left(1 + \frac{h}{1+h}\frac{\chi}{\sigma}\right) \tag{32}$$

where, for an element i emitting x-rays of energy E_i,

$$\sigma = \frac{4.5 \times 10^5}{E_0^{1.65} - E_{c,i}^{1.65}}$$

$$h = 1.2 \sum_j C_j \frac{A_j}{Z_j^2}$$

$$\chi = \sum_j C_j \mu_j(E_i) \csc \psi.$$

In this model $\phi(0)$ is the value for x-ray generation at zero-depth. This model thus neglects the surface ionisation, which is excited by the primary beam and by the backscattered electrons on their return trajectories through the surface. This formula is often used in correction programs, though it fails in cases of strong absorption such as occur in the analysis of elements of low atomic number (when surface ionisation becomes important).

The absorption correction factor for Ni Kα radiation in a Fe-Ni alloy when the incident electron beam energy is $E_0 = 25$ keV and $\psi = 40°$ can be calculated using the values of the parameters given in the following table.

		Fe	Ni
$\mu(E_{NiK\alpha})$	cm²/g	370.2	60.0
Z		26	28
E_c	keV	7.111	8.332
A	g/mol	55.85	58.71
C	wt%	90.0	10.0

The cross section is

$$\sigma = \frac{4.5 \times 10^5}{25^{1.65} - 8.332^{1.65}} = 2654$$

For pure nickel taken as the standard we have

$$\chi = 1.0 \times 60.0 \times \csc(40°) = 93.3$$
$$h = 1.2 \left(1.0 \times \frac{58.71}{28^2}\right) = 0.0899$$

and hence

$$(F_A)_{(Ni)} = \left(1 + \frac{93.3}{2654}\right)\left(1 + \frac{0.0899}{1 + 0.0899}\frac{93.3}{2654}\right) = 1.038$$

For a sample with 90%Fe–10%Ni we have

$$\chi = (0.9 \times 370 + 0.1 \times 60.0)\csc(40) = 527.7$$
$$h = 1.2\left(0.9 \times \frac{58.71}{28^2} + 0.1 \times \frac{55.85}{26^2}\right) = 0.091$$
$$(F_A)_{Ni}^{(u)} = \left(1 + \frac{527.7}{2654}\right)\left(1 + \frac{0.091}{1 + 0.091}\frac{527.7}{2654}\right) = 1.219$$

and thus the absorption correction factor A_{Ni} is

$$A_{Ni} = \frac{(F_A)_{Ni}^{(u)}}{(F_A)_{(Ni)}} = \frac{1.219}{1.038} = 1.17$$

The extent that experimental parameters influence the magnitude of the absorption correction (for 90%Fe–10%ni) is demonstrated in the following table.

Acc. voltage E_0, keV	Take-off angle ϕ	Abs. corrn. A_{Ni}	
25	40	1.17	deep generation, short path
25	10	1.61	deep generation, long path
12.5	40	1.03	generation near surface, short path
12.5	10	1.15	generation near surface, long path

The absorption correction depends, for a given composition on the accelerating voltage E_0 (actually the overvoltage U) and the take-off angle. The correction is minimal for low overvoltages and high take-off angles.

The atomic number correction Z_i

The atomic number correction takes care of the difference in backscattering coefficient and stopping power and thus actually consists of two terms.

$$F_Z = F_R F_S \tag{33}$$

The backscatter correction is related to the backscatter coefficient η by the equation

$$F_R = \frac{I_{\text{NO BACKSCATTER}}}{I_{\text{BACKSCATTER}}} = \frac{1}{1 - \eta} \tag{34}$$

In ZAF corrections often the factor R is introduced with $R = 1 - \eta$. Thus for an element i the backscatter correction is given by:

$$(F_R)_i = \frac{1}{R_i} \tag{35}$$

The term R_i for an element i in a matrix is calculated from

$$R_i = \sum_j C_j R_{ij} \tag{36}$$

where R_{ij} can be interpreted as the effect of element j on the element under consideration i. An expression for R_{ij} has been given by Yakowitz:

$$R_{ij} = R'_1 - R'_2 \ln(R'_3 Z_j + 25) \tag{37}$$

with

$$
\begin{aligned}
R'_1 &= 8.73 \times 10^{-3} U^3 - 0.1669 U^2 + 0.9662 U + 0.4523 \\
R'_2 &= 2.703 \times 10^{-3} U^3 - 5.182 \times 10^{-2} U^2 + 0.302 U - 0.1836 \\
R'_3 &= (0.887 U^3 - 3.44 U^2 + 9.33 U - 6.43)/U^3
\end{aligned}
$$

and U the overvoltage, given by $U = E_0/E_{c,i}$.

The stopping power factor F_S is related to the expression

$$\left\{ \int_{E_c}^{E_0} \frac{Q}{S} dE \right\}^{-1} \tag{38}$$

where Q is the ionisation cross-section, considered as a constant, and S is the stopping power. An expression for the stopping power is

$$S = \text{const} \times \frac{Z}{A} \frac{1}{E} \ln\left(\frac{1.166 E}{J}\right)$$

Since this is a slow varying function of E, rather than integrating we take the value at an average energy

$$\bar{E} = \frac{E_0 + E_{c,i}}{2}$$

Thus the stopping power factor for an element i can be written as

$$F_S = S_i = \sum_j C_j S_{ij} \tag{39}$$

where

$$S_{ij} = \frac{Z_j}{A_j} \frac{1}{E_0 + E_{c,i}} \ln\left(\frac{583(E_0 + E_{c,i})}{J_j}\right)$$
$$J_j = 9.76 Z_j + 58.8 Z_j^{-0.19}$$

Combining the two factors gives the required atomic number factor

$$(F_Z)_i = (F_R)_i (F_S)_i = \frac{S_i}{R_i} \tag{40}$$

The above equations are used in the so called 'Duncumb Reed' correction (Duncumb and Reed 1968). The major influence comes from the difference in mean atomic number between standard and unknown.

As an example let us calculate the atomic number correction for the analysis of copper in aluminium when $E_0 = 25$ kV. Using the values $Z_{Al} = 13$, $A_{Al} = 26.98$ g/mol, $E_{c,Al} = 1.56$ keV, $Z_{Cu} = 29$, $A_{Cu} = 63.55$ g/mol, $E_{c,Cu} = 8.98$ keV the previous formulae give

$$J_{Cu} = 9.76 \times 29 + 58.8 \times 29^{-0.19} = 314$$

$$J_{Al} = 9.76 \times 13 + 58.8 \times 13^{-0.19} = 163$$

$$S_{Cu} = \frac{29}{63.55(25 + 8.96)} \ln \frac{583(25 + 8.98)}{314} = 0.0557$$

$$S_{Al} = \frac{13}{26.98(25 + 8.98)} \ln \frac{583(25 + 8.98)}{163} = 0.0681$$

$$R_{Cu} = 0.870 \quad R_{Al} = 0.947$$

Thus for a sample of composition 2% Cu - 98% Al:

$$S_{Cu} = 0.98 \times 0.0681 + 0.02 \times 0.0557 = 0.0678$$
$$R_{Cu} = 0.98 \times 0.947 + 0.02 \times 0.870 = 0.946$$

and hence

$$(F_Z)_{Cu}^{(s)} = \frac{0.0687}{0.946}$$

Since for pure Cu we have

$$(F_Z)_{(Cu)} = \frac{S_{Cu}}{R_{Cu}} = \frac{0.0557}{0.870}$$

the atomic number correction becomes

$$Z_{Cu} = \frac{(F_Z)_{Cu}^{(s)}}{(F_Z)_{(Cu)}} = \frac{0.870}{0.946} \times \frac{0.0687}{0.0557} = 0.920 \times 1.233 = 1.14$$

For the determination of heavy elements in light matrices the atomic number correction is greater than 1; for light elements in a heavy matrix the correction is less than 1. The correction decreases with increasing accelerating voltage.

The fluorescence correction

Characteristic x-rays of element j can excite element i if the x-ray energy is higher than the absorption edge energy, $E_j > E_{c,i}$. Thus more x-rays of element i are produced than from electron excitation alone. The fluorescence factor is the ratio of the x-ray intensity if no fluorescence occurs to the intensity when fluorescence occurs

$$F_F = \frac{I_{\text{NOFLUOR}}}{I_{\text{FLUOR}}} = \frac{I_i}{I_i + I^f} = \frac{1}{1 + (I^f/I_i)} \tag{41}$$

Since a number of elements j can contribute to the fluorescence of element i, the correction becomes

$$(F_F)_i = \frac{1}{1 + \sum_j (I_{ij}^f/I_i)} \tag{42}$$

I_{ij}^f is the x-ray intensity of element i due to the fluorescence from element j. The expressions developed by Reed are as follows

$$\frac{I_{ij}^f}{I_i} = C_j Y_0 Y_1 Y_2 Y_3 P_{ij} \tag{43}$$

where

$$Y_0 = 0.5 \frac{r_i - 1}{r_i} \omega_i \frac{A_i}{A_j}$$

$$Y_1 = \left(\frac{U_j - 1}{U_i - 1}\right)^{1.67}$$

$$Y_2 = \frac{\mu_i(E_j)}{\mu_M(E_j)}$$

$$Y_3 = \frac{\ln(1 + u)}{u} + \frac{\ln(1 + v)}{v}$$

$$u = \frac{\mu_M(E_i)}{\mu_M(E_j)} \csc \psi$$

$$v = \frac{3.3 \times 10^5}{(E_0^{1.65} - E_{c,i}^{1.65})} \frac{1}{\mu_M(E_j)}$$

The parameters appearing in these formulae are: r_i, the absorption jump ratio of element i, (for K-lines $(r_i - 1)/r_i = 0.88$, for L-lines $(r_i - 1)/r_i = 0.75$); ω_i is the fluorescence yield; $U_j = E_0/E_{cj}$ and $U_i = E_0/E_{ci}$; $\mu_M(E_j)$ the mass attenuation coefficient of the matrix for radiation of energy E_j. Typical values of the P_{ij} are

$P_{ij} = 1.00$ for the K x-rays of element j exciting the K level of element i
$P_{ij} = 1.00$ for the L x-rays of element j exciting the L level of element i
$P_{ij} = 4.76$ for the L x-rays of element j exciting the K level of element i
$P_{ij} = 0.24$ for the K x-rays of element j exciting the L level of element i

The fluorescence correction is then:

$$F_i = \frac{(F_F)_i^{(u)}}{(F_F)_i^{(s)}} \tag{44}$$

If pure element standards are used $(F_F)_{(i)} = 1$ since the x-rays for element i cannot excite i x-rays and the fluorescence correction becomes

$$F_i = (F_F)_i^{(u)} = \frac{1}{1 + \sum_j (I_{ij}^f / I_i)} \tag{45}$$

The fluorescence correction is only important for alloys of transition element such as chromium, iron and nickel.

The fluorescence of Fe by Ni in an Fe-Ni alloy (25keV, 40° take-off angle) can be calculated using the parameters given in the following table.

		Fe	Ni
$\mu(E_{\text{FeK}\alpha})$	cm²/g	71.1	92.3
$\mu(E_{\text{NiK}\alpha})$	cm²/g	370.2	60.0
ω		-	0.37
A	g/mol	55.85	58.71
E_c	keV	7.111	8.332
C	wt%	0.10	0.90

Only Ni-K can excite Fe (KK fluorescence) because $E_{\text{NiK}\alpha} = 7.478 > E_{c,\text{Fe}} = 7.111$ With Fe $= i$ and Ni $= j$

$$\sum_j \frac{I_{ij}^f}{I_i} = \frac{I_{\text{Fe,Ni}}^f}{I_{\text{Fe}}} = C_{\text{Ni}} Y_0 Y_1 Y_2 Y_3 \times 1$$

$$Y_0 = 0.5 \times 0.88 \times 0.37 \times \frac{55.85}{58.71} = 0.$$

$$Y_1 = \left(\frac{\frac{25.0}{8.332} - 1}{\frac{25.0}{7.111} - 1} \right)^{1.67} = 0.682$$

Mass attenuation coefficients for the matrix are given by

$$\mu_M(E) = C_{\text{Fe}} \mu_{\text{FE}}(E) + C_{\text{Ni}} \mu_{\text{Ni}}(E) \tag{46}$$

therefore

$$\mu_M(E_{\text{FeK}\alpha}) = 0.1 \times 71.1 + 0.9 \times 92.3 = 90.18$$

$$\mu_M(E_{\text{NiK}\alpha}) = 0.1 \times 370.2 + 0.9 \times 60.0 = 91.02$$

$$Y_2 = \frac{\mu_{\text{Fe}}(E_{\text{FeK}\alpha})}{\mu_M(E_{\text{NiK}\alpha})} = \frac{370.2}{91.02} = 4.07$$

$$u = \frac{\mu_M(E_{\text{FeK}\alpha})}{\mu_M(E_{\text{NiK}\alpha})} \csc(40) = \frac{90.18}{91.02} \times 1.56 = 1.54$$

$$v = \frac{3.3 \times 10^5}{25^{1.65} - 7.111^{1.65}} \frac{1}{\mu_M(E_{\text{NiK}\alpha})} = 20.47$$

$$Y_3 = \frac{\ln(1 + 1.54)}{1.54} + \frac{\ln(1 + 20.47)}{20.47} = 0.755$$

$$\frac{I_{Fe,Ni}^f}{I_{Fe}} = 0.9 \times 0.155 \times 0.682 \times 4.07 \times 0.755 = 0.292$$

$$F_F = \frac{1}{1 + 0.292} = 0.774$$

Thus 29% of the iron x-ray intensity is due to fluorescence.

The fluorescence correction is minimal at low operating voltages and low take-off angles, and increases with the concentration of the element causing the fluorescence:

C_{Ni} wt%	$\phi°$	E_0	$I_{Fe,Ni}^f/I_{Fe}$	F_{Fe}
90	40	12.5	0.185	0.844
90	10	12.5	0.108	0.902
90	40	25	0.292	0.774
90	10	25	0.187	0.843
50	10	12.5	0.042	0.960

Calculation of concentrations for the unknown

The calculation of the concentrations is made via an iterative scheme since the concentrations must be known approximately so that the corrections may be calculated. An example of this iterative type of calculation is given below for a Fe-Ni alloy of actual composition 75% Fe - 25% Ni.

- *Initialise*

    - Experimental conditions : E_0=30 keV, take-off angle 40°.
    - The experimentally determined k-ratios are : k_{Fe}=0.777, k_{Ni}=0.219
    - An initial guess of the concentration $(C_i=k_i)$ is C_{Fe}=0.777, C_{Ni}=0.219

- *Iteration #1:*

    - ZAF corrections based on initial C_i are $(ZAF)_{Fe}$=0.970, $(ZAF)_{Ni}$=1.144
    - New concentration estimates, $C_i=k_i(ZAF)_i$ are

$$C_{Fe} = 0.777 \times 0.970 = 0.754$$
$$C_{Ni} = 0.219 \times 1.144 = 0.251$$

- *Iteration #2:*

    - ZAF correction based on revised C_i are: $(ZAF)_{Fe}$=0.965, $(ZAF)_{Ni}$=1.140
    - New concentration estimates:

$$C_{Fe} = 0.777 \times 0.965 = 0.750$$
$$C_{Ni} = 0.219 \times 1.140 = 0.250$$

- *Iteration #3:*

  – ZAF correction based on revised C_i are: $(ZAF)_{Fe}=0.965$, $(ZAF)_{Ni}=1.139$
  – New concentration estimates:

$$C_{Fe} = 0.777 \times 0.965 = 0.750$$
$$C_{Ni} = 0.219 \times 1.139 = 0.250$$

Since no change occurs in the calculated concentrations obtained from the last two iterations, the process is stopped at this point.

Alternative matrix correction methods

There are many variants of the ZAF method, each using different equations to calculate the corrections. There is also an evolution from ZAF methods to so called phi-rho-z methods. The ZAF method is not really correct from a physics point of view beccause the atomic number, absorption and fluorescence effects are not multiplicative. Matrix corrections have been developed which are based on the evaluation of x-ray production as function of depth, $\phi(\rho z)$. These methods perform much better for light element analysis, but are computationally quite complex. The biggest problem in quantitative analysis, remains the uncertainty in the mass attenuation coefficients.

A discussion on these phi-rho-z methods is beyond the scope of this chapter. An excellent overview of this subject with contributions from the leading experts in the field cane be found in Heinrich and Newbury (1991).

Special requirements for the samples

To achieve the full potential of EPMA

- the samples must be flat and polished (to reduce absorption and backscattering errors)

- the phase size must be larger than the x-ray range

- the samples must be conducting or coated with a very thin layer of low Z material (*e.g.* carbon)

- surface contamination should be minimal especially for light element analysis

- standards need to be homogeneous at the micro level

- standards and unknown should be as similar as possible.

5   Conclusions

Electron probe x-ray microanalysis is a well established microanalytical technique. The objective of this chapter has been to discuss the areas of application of this technique, the physical processes involved in x-ray production and detection of x-rays, the operation of wavelength dispersive spectrometers most commonly used in EPMA and the procedures which are used to quantify the results. In addition the background to the ZAF correction method has been developed. Numerical examples have been used to give a feel for the magnitudes of the large number of parameters involved in the interaction of electrons and x-rays with solids and to demonstrate the effectiveness of matrix correction methods.

Many excellent discussions on electron probe x-ray microanalysis can be found in the literature. The maturity of the technique shows from the abundant availability of good text books. The text book by Goldstein *et al.* (1981) is very complete, covering all aspects discussed in this chapter, including electron-specimen interaction, ED-WD x-ray measurement and quantitative analysis. The book by Reed (1975) is a standard work by one of the 'fathers' of EPXMA. The book by Friel and Bardi gives a short and practical introduction to x-ray microanalysis. Heinrich and Newbury (1991) in their recent book deal with the latest developments in quantification, and include an excellent discussion of phi-rho-z methods. Finally the book by Bertin (1978) is a text on x-ray fluorescence, rather than EPXMA, covering aspects of WD-spectrometry in depth.

Acknowledgements

I wish to thank S Montoro and W Saenen, who contributed to the preparation of this chapter and worked on the examples given.

References

Bertin E P, 1978, *Introduction to x-ray spectrometric analysis* (Plenum Press, New York)

Duncumb P and Reed S J, 1968, *Quantitative Electron Probe Microanalysis* Eds. Costaing R, Deschamps P and Philibert J p.240 (Herman, Paris)

Friel J J and Bardi N C, *X-ray microanalysis and computer-aided imaging* (PGT, 1200 State Road, Princeton, NJ 08540)

Goldstein J I and Yakowitz H (eds.), 1975, *Practical Scanning Electron Microscopy, Electron and Ion Microprobe Analysis* p.379 (Plenum, New York)

Goldstein J I, Newbury D E, Echlin P, Joy D C, Fiori C and Lifshin E, 1981, *Scanning Electron Microscopy and X-ray Microanalysis* (Plenum Press, New York)

Heinrich K F J, 1970, *Present state of the classical theory of quantitative electron probe microanalysis* NBS Technical Note 521 (US Dept. of Commerce, Washington DC)

Heinrich K F J and Newbury D E (eds.), 1991, *Electron probe quantitation* (Plenum Press, New York)

Reed S J B, 1975 *Electron microprobe analysis* (Cambridge University Press, Cambridge)

Energy Dispersive X-Ray Analysis (EDX) in the TEM/STEM

J M Titchmarsh

Harwell Laboratory
Didcot, UK

1 Introduction

Energy dispersive x-ray analysis (EDX) is the principal method for performing chemical analysis in the transmission electron microscope (TEM). The analytical data are acquired in the form of digitised spectra, viewed during accumulation on a computer monitor, which display the number of x-rays, I, detected as a function of E, the x-ray energy. Figure 1 shows a typical spectrum, from a sample of Fe_3O_4; the major features are the characteristic peaks from the Fe and O in the sample. Chemical composition is derived by subsequent analysis of the magnitudes of these peaks. The dotted spectrum shows a 32-fold magnification of the solid spectrum, to amplify background details. This chapter describes (i) the physical processes which control the generation of such characteristic peaks and the continuum radiation on which the peaks are superimposed, (ii) the basis of the x-ray detection, energy measurement and display system, which together determine the peak widths and the continuum shape, (iii) the different methods for deriving quantitative information, and (iv) sensitivity limits and spatial resolution. The description will reflect the present status of the method, problems and pitfalls. The reader is referred to more comprehensive and specialised treatments (e.g. Russ 1984, Williams 1984, Zaluzec 1979) for greater detail.

TEM, in the present context, refers to any microscope which forms an image of the sample by transmission of a beam of electrons with a kinetic energy typically in the range 80–1000 keV. It includes any conventional TEM with a facility to focus the electron probe to a small crossover, TEM/STEM instruments which can produce either a conventional, projected image or a rastered image on a CRT, and dedicated scanning transmission microscopes (STEM) which can only form rastered images. In this chapter, the name Analytical Electron Microscope (AEM) implies any of the above, using any type of electron source, to which an EDX system has been interfaced.

A discussion of energy dispersive x-ray analysis requires use of a large number of mathematical symbols. The symbol notation which will be used throughout this chapter is given below.

Symbol notation

B	Continuum x-ray intensity		τ	Acquisition time
E	x-ray energy		e_j	Detector efficiency of line j
$E_{K,L}$	Energy of K, L shell x-ray		$\sigma_{K,L...}$	K, L... ionisation cross-section
$E_{1,2,..}$	Energy level of 1st, 2nd,...shell		$\sigma'_j(E,\theta)dE$	Continuum ionisation x-section
I_0^j, I^j	Characteristic x-ray intensity			under peak j over energy range
	of peak j			dE, emitted at angle θ
E_0	Energy of the fast electrons		h	Parameter for background fitting
R	Rydberg constant (13.6 eV)		Ω	Detector solid angle as a fraction
Z	Atomic number			of 4π steradians
ε	Energy equivalence of detector		ω_j	Fluorescent yield of j x-ray line
	electron-hole pair production		a_j	Partition function of j x-ray line
F	Fano factor		k_{AB}	Cliff-Lorimer factor for
θ	Angle between beam direction			elements A and B
	and detector		e	Electronic charge
T_0	Electron energy		z, z_0	Depth, thickness in beam direction
	$= 0.5mv^2$ (non-relativistic)		f	Geometric factor in absorption
	$= 0.5mc^2(1 - 1/(1 + E_0/mc^2)^2)$.			correction
	(relativistic)		ρ_j	Density of element j
$(\mu/\rho)_j^i$	Mass absorption coefficient		U_K	Overvoltage ratio $= T_0/I_K$
	of line i in element j		I_K	Ionisation energy of edge K
G_i	Channeling factor for line i		b_K, c_K	Bethe parameters for K lines
$I^{r,c}$	Characteristic intensity at a		N_0	Avagadro's number
	random, or channeling orientation		A_i	Atomic weight of element i
f_A	Fraction of impurity atoms		m	Electron rest mass
	on A atom sites		v, c	Velocity of electrons, light
L	Interplanar spacing in the		$n, n_{A,B...}$	Atomic density of matrix,
	beam direction			of element A, B ...
B_0	Electron source brightness		d	Probe diameter
i	Probe current		α	Probe convergence semi-angle
t, t_i	Foil thickness, thickness		δ	Close-packed inter-planar
	of ith layer			spacing

2   Generation of x-rays

2.1   The origin of characteristic x-rays

Electron states in the atom

For the simple case of the electron/proton pair of the hydrogen atom, analytic solutions of the Dirac equation exist which exactly define the allowed electron states. Every possible electron state is identified by a set of four quantum numbers which specify a unique combination of binding energy and momenta (orbital and spin). For more

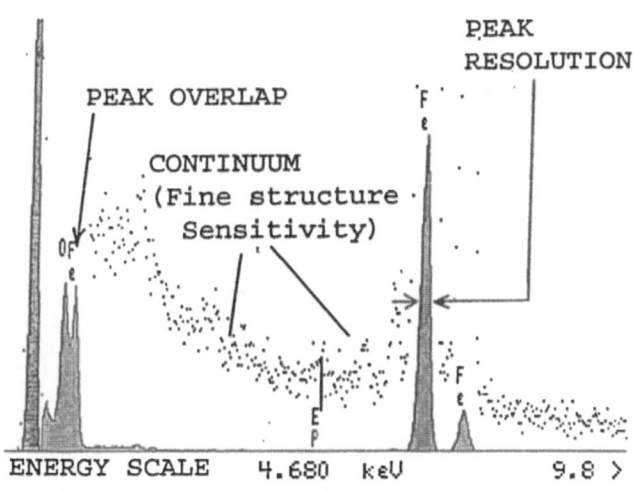

Figure 1. EDX *spectrum from Fe_3O_4: dots show continuum intensity at a 32-fold magnification.*

complex atoms than hydrogen, containing two or more electrons, interactions between the electrons complicate the analysis, and only numerical solutions of the electron states are available. However, the set of four quantum numbers can still be used to classify the permitted electron states, and no two electrons can have identical values for all four quantum numbers (Pauli exclusion principle). The quantum numbers are:

n = principal quantum number, a non-zero, positive integer defining the binding energy of the electron.

l = angular momentum quantum number, $0 < l < n - 1$, defining the orbital angular momentum.

s = spin quantum number, $s = 1/2$, defining the spin momentum.

j = inner quantum number, where $(l + s) \geq j \geq |l - s|$, reflecting the interaction of the spin and angular momenta.

Experimental investigation of electron states requires the imposition on the atom of an external, directional field (electric, magnetic, *etc.*) by which the measured electron characteristics, *e.g.* angular momentum vector, are a projection into the field direction. The measured electron properties are, therefore, different from, but directly related to those in the undisturbed atom. An understanding of the production of characteristic x-rays, following the disturbing influence of the electric field of an incident high energy electron, is obtained using measured values, l_j, s_j, m_j, of l, s and j, respectively, where

n	Shell	l	Orbit	s	j	m_j	Shell	No of electrons
1	K	0	s	1/2	1/2	±1/2	K	2
2	L	0	s	1/2	1/2	±1/2	L_I	2
2		1	p	1/2	1/2	±1/2	L_{II}	2
2		1	p	1/2	3/2	±3/2 ±1/2	L_{III}	4
3	M	0	s	1/2	1/2	±1/2	M_I	2
3		1	p	1/2	1/2	±1/2	M_{II}	2
3		1	p	1/2	3/2	±1/2 ±3/2	M_{III}	4
3		2	d	1/2	3/2	±1/2 ±3/2	M_{IV}	4
3		2	d	1/2	5/2	±1/2 ±3/2 ±5/2	M_V	6

Table 1. *Electronic Structure*

the allowed values are:

$$l_j = l, l-1, \ldots 0 \ldots -(l-1), -l$$
$$s_j = \frac{1}{2}, -\frac{1}{2}$$
$$m_j = j, j-1, -(j-1), -j$$

The electrons in any atom will try to fill the lowest energy states wherever possible, and the most tightly bound electrons are those with the lowest value of $n=1$ (K shell). However, when the exclusion principle is applied to the possible values of the quantum numbers, only two electrons are permitted to occupy such states. Additional electrons must successively fill states with $n=2$ (8 electrons–L shell), $n=3$ (18 electrons–M shell) *etc.*, which are less tightly bound. Table 1 lists the quantum numbers of the permitted electron states of the K, L and M shells, divided into their sub-shells, and with their orbit descriptors (s = sharp, p = principal, d = diffuse, f = fine) assigned historically from the appearances of the lines in optical spectra. Higher shells are excluded for brevity.

Ionisation and decay

An incident high-energy electron can lose kinetic energy during a collision with a bound electron in an atom of the AEM sample, ejecting the bound electron and leaving an ionised atom with a hole in the electron structure. The ionised atom almost instantly reverts to a lower-energy configuration by the transition of a less tightly-bound electron into the hole. The energy released by this decay, the difference in the binding energies of the two electron states, appears (i) as a characteristic x-ray, or (ii) is transferred to another bound electron, which is then ejected as an Auger electron. Auger electron

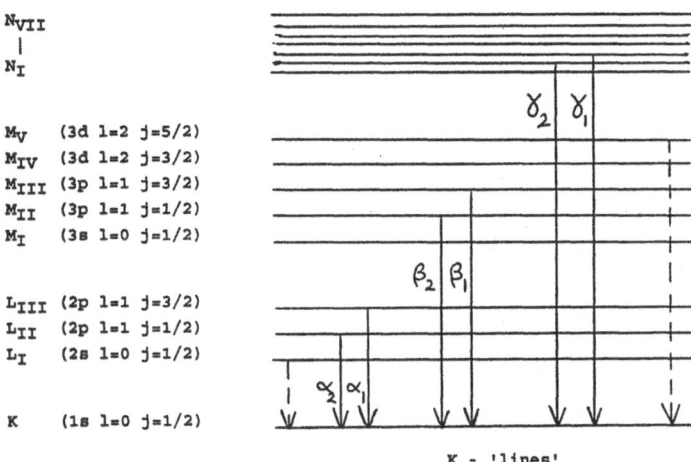

Figure 2. *Allowed and forbidden K-shell transitions.*

spectroscopy is described by Seah and Prutton in this Proceedings. The characteristic x-rays are emitted isotropically.

Selection rules

Transitions are not possible between all pairs of electron states: the selection rules which determine which electron states are allowed to participate in transitions are:

(i) $\Delta n > 0$, $\quad i.e.$ $L \rightarrow K$, $M \rightarrow K$, but not $L_{III} \rightarrow L_{II}$
(ii) $\Delta l = \pm 1$: $\quad \Delta j = -1, 0, +1$

Application of these rules to the states listed in Table 1 enables the prediction of the characteristic x-rays. Figure 2 shows examples of allowed (solid lines) transitions contributing to the x-ray 'K-series', as well as some forbidden (dashed) transitions. The L-, M-series, *etc.*, arising from transitions into ionised states with $n = 2, 3$ *etc.* are not shown in the figure. The Greek letters do not have any logical significance, being merely historical assignations. The approximate energies of the K and L x-rays can be derived from the following formulae:

$$E_K = E_n - E_1 = (-)\,R\,(Z-1)^2\,(1/n^2 - 1) \qquad n = 2, 3, \ldots \qquad (a)$$
$$E_L = E_n - E_2 = (-)\,R\,(Z-7.4)^2(1/n^2 - 1/2^2) \quad n = 3, 4, \ldots \qquad (b). \quad (1)$$

The values of 1 and 7.4 are subtracted from Z in a simple attempt to allow for the screening of the nucleus by the remaining K shell electron, and the remaining K and seven L shell electrons, respectively, for the K and L shell ionisations. The positions of the characteristic peaks in Figure 1 are determined, approximately, by these expressions.

2.2   Bremsstrahlung (continuum) x-ray emission

In addition to the emission of characteristic x-rays which have a very small energy
spread (typically the order of 1 eV), other x-rays are generated which have a continuous energy distribution between 0 and E_0. Any discussion of EDX analysis should
consider the continuum emission for several reasons: (i) quantification of the characteristic signal requires separation from the continuum; (ii) the continuum limits the
sensitivity and accuracy of EDX analysis; (iii) the relative intensities of a characteristic
peak and its background is a guide to the quality of the whole EDX system , and can
be used to indicate experimental artefacts, an excessively thick sample, and poor spec-
imen/detector geometry; (iv) the shape of the continuum can be used to determine the
absorption correction factors which are needed to determine detector efficiences at low
E (see Section 5.3).

According to classical physics, and in agreement with experimental observation,
any charged particle which is accelerated emits radiation. However, Newtonian mechanics and electromagnetism cannot explain satisfactorily the interaction of the fast
electron and the nuclear field. There is no inherent reason for a maximum x-ray frequency, or means of explaining relativistic effects. The classical treatments, such as
that of Kramers (1923), while predicting $B \propto Z^2$ at low E, in reasonable agreement with
observation, do not predict the observed dependence of B on the angle of emission. Relativistic quantum mechanics must be used to formulate and explain the observations,
and even then, uncertainties arise from the treatment of screening, retardation and exchange. Exact calculations of the energy and angular variations in B as a function of
E_0 and Z require complex numerical integrations which have only been performed for
a limited number of cases (Tseng and Pratt 1971, Pratt *et al.* 1977). Various approximate analytic solutions have been reported and compared with experiment (Chapman
et al. 1983, Gray *et al.* 1983). Such analytic expressions are useful in modelling the
continuum to determine absorption corrections, and for background removal from experimental spectra. The expression recommended by Chapman *et al.* (1983) known as
the Modified Bethe-Heitler solution (MBH), although algebraically complex, is suitable
for modelling with a modern personal computer. Figure 3 shows B predicted by the
MBH theory for variations in Z and in x-ray emission angle (equivalent to detector orientation in the electron microscope). Heavier elements generate much larger B, and there
is a much greater intensity emitted in the forward beam direction. Hence, detectors are
best positioned to collect x-rays emitted from the beam-entrance side of the sample, to
improve the characteristic/continuum ratio.

3   X-ray detectors

The AEM x-ray detector design is influenced by limited access to the sample due to the
proximity of the objective lens pole-pieces, aperture rod, anti-contamination devices,
and the sample holder. Hence, the solid-state energy-dispersive (semiconductor) detector is preferred to crystal spectrometers and proportional counters. Although, Ge
detectors were developed first, for high-energy radiation analysis, the Si detector was
more efficient for analysis of the low-energy x-rays generated in the EPMA and SEM.
Such detectors were readily adapted for use with the AEM. As the accelerating volt-

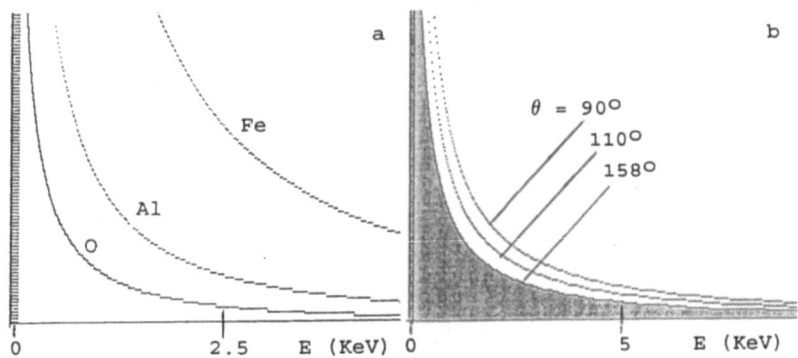

Figure 3. *Continuum x-ray intensity as a function of x-ray energy, $E_0 = 100\ keV$: (a) for O, Al, and Fe, detector take-off angle = $103.5°$; (b) for Fe, with detector take-off angles of $90°$, $110°$, and $158°$. (Beam direction = $0°$).*

age of the AEM increases, it is now becoming attractive to use Ge detectors for some applications. Statham (1991) has recently reviewed the performance of EDX detectors.

3.1   Electron-hole pair production

X-rays interact and lose energy in solids in several ways:

(i)   Rayleigh scattering, which is quasi-elastic, and is the basis of the wavelength dispersive spectrometry (WDS) method used in the EPMA (*e.g.* van Espen, this Proceedings);

(ii)   Compton scattering, involving a partial loss of energy by collision with a single electron, causing the photon to be deflected, and the emission of energetic electrons;

(iii)   Photoelectron emission, whereby the photon energy is completely absorbed by one electron, which is emitted with equivalent kinetic energy (allowing for any binding potential) into the surrounding solid, where it eventually thermalises;

(iv)   Electron-positron pair production, which can only occur for photons with energy E is greater than twice the rest mass of the electron (approximately 1023 keV), and so is outside the present scope of interest. Of these interactions, photoelectron production is by far the most relevant to the EDX detector.

A high-energy photoelectron in Si has a limited range (typically microns, or less) and it loses energy by exciting softer x-rays, plasmons and valence electrons, which in turn dissipate energy. One result of the thermalisation processes is the generation of electron-hole pairs: an x-ray of energy E_K, will produce, on average, E_K/ε electron-hole pairs, where $\varepsilon=3.8$ eV for Si, and 3.0 eV for Ge. The charge carriers eventually diffuse

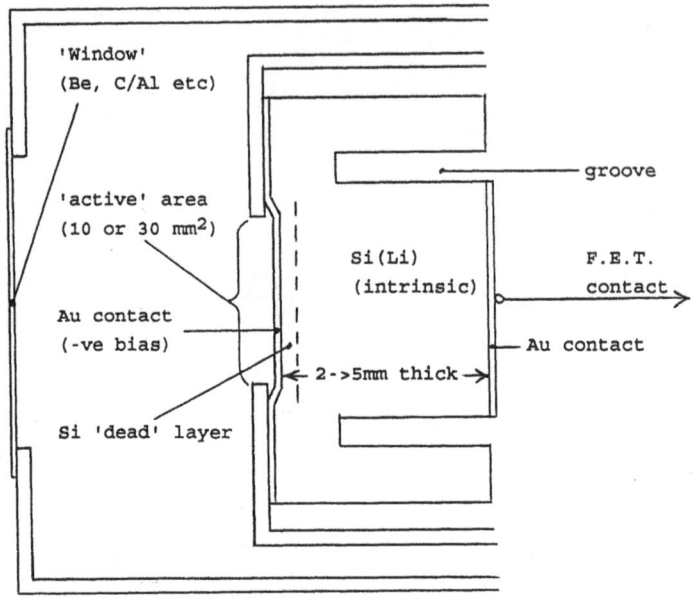

Figure 4. *Schematic construction of a Si detector.*

and recombine at traps within the lattice. The proportionality between E_K and the number of electron-hole pairs provides a means for measurement of the x-ray energy.

3.2 Detector structure

The structure of a typical Si detector is shown in Figure 4. The Si single crystal is typically 2–5 mm thick, and has traditionally been drifted with Li to fill any traps which could act as recombination sites for charge carriers. A gradual improvement in quality nowadays allows intrinsic Si, without Li drifting, to be used by some manufacturers. Au contact layers are evaporated onto opposite faces, and a negative bias of a few hundred volts is applied to the front contact, to produce a field of 100–200V/mm within the Si. The Si is cooled to about 90 K to reduce thermal generation of carriers. The active area of the detector, up to 30 mm^2, is defined by a circular aperture in a surrounding shield. Before 'detection' can occur, an x-ray must first pass through the 'window', the purpose of which is to maintain a good vacuum in the pre-pumped vicinity of the Si crystal, and to prevent the build-up on the Si of ice and contaminants from the microscope system . The window is commonly made from Be sheet, about 10 μm thick, but this significantly absorbs x-rays of low energy (<2 keV), and so there is an increasing trend towards the use of thinner window materials (*e.g.* mylar, diamond), or no window at all, when the AEM pressure is very low, to increase sensitivity at low E. Most x-rays with $E<20$ keV are absorbed within the Si, and produce a cascade of electron-hole pairs which are separated by the applied field, diffuse to the contacts and generate a charge pulse with a duration of about 1 ns.

3.3   Pulse amplification and processing

A cooled (150 K) FET preamplifier is placed immediately against the rear contact of the Si, the voltage output of which rises from zero to a maximum, typically about 1 V, in a series of steps of variable height. Each step occurs in about 1ns, with a height which is proportional to the energy of a detected x-ray. The random arrival of successive x-rays determines the time intervals between the steps. The output voltage is automatically restored to zero whenever the maximum is reached, either by a pulse of light, or by direct charge injection into the FET gate. The FET output is then processed by the main amplifier to convert each step into a pulse which rises quickly from a baseline voltage to a maximum, and then slowly decays over a duration of about 0.1 ms.

The pulses are passed through a discriminator to reject as noise all those below a selected threshhold, before being directed to the pulse-height measuring circuit. As the amplified pulse voltage rises, a sensing capacitor is quickly charged, but when the pulse voltage reaches its maximum, the capacitor is automatically disconnected from the pulse amplifier output. Uniform charge pulses, generated at a constant, clocked rate, are then directed into the capacitor. The number of clock cycles, n, required for the total discharge of the capacitor is, therefore, directly proportional to the x-ray energy. The value stored in the nth channel of the MCA is incremented by one, and the system then prepares for the next pulse.

3.4   Pulse pile-up rejection

In practice, the FET output is processed by two amplification chains — a 'slow' amplifier, the role of which is described above, and a 'fast' amplifier, which prevents the analysis of overlapping pulses. If a second pulse arrives before the first pulse output from the slow amplifier has reached a maximum, the fast chain amplifier alerts the system to the fact that the first pulse analysis will be corrupted, and both pulses are rejected. The analysis chain is inactivated while it returns to the quiescent state, and this recovery time is registered as 'dead time'. If the separation of two pulses is less than the discrimination time of the fast amplifier, the system erroneously analyses the combined pulses as a single pulse, such that a count is registered in the MCA at a position corresponding to the sum of the two x-ray energies. The 'sum peaks' and the dead time increase with the x-ray intensity incident on the detector. As the incident rate increases from zero, the rate of analysis increases accordingly, begins to saturate, and eventually falls to zero as the dead time approaches 100% and the analyser chain never reaches the quiescent state required to initiate analysis. The saturation rate can be adjusted by altering the bandwidth of the amplifiers, and provides a parameter by which different systems can be compared.

3.5   Detector resolution

Increasing the amplifier bandwidth increases noise, leading to broader spectral peaks. The peak full-width, half maximum (FWHM) is a measure of the resolution; it is determined by two factors: (i) the amplifier noise; (ii) the statistical variation in the number of electron-hole pairs generated by x-rays of identical energy. The natural width of the characteristic energy is too small, typically less than 1 eV, to contribute significantly. Amplifier noise is a complex function of electronic component values and temperatures, and is difficult to calculate with accuracy: even in the best modern systems the noise contribution, $(\text{FWHM})_0$, is at least 40 eV (Statham and Nashashibi 1988). The statistical variation is not completely random because the energy-loss processes of the photoelectrons follow similar patterns. Although the random variance for an average number of pairs, E_K/ε would also be E_K/ε, the actual variance is FE_K/ε, where F, the Fano factor, is a measure of the deviation of the pair generation from a purely random process. For example $F{\sim}0.11$ for Si. Hence, the standard error is $(FE_K/\varepsilon)^{1/2}$ electron-hole pairs, or $(FE_K\varepsilon)^{1/2}$ eV. The Gaussian FWHM (2.355 times the standard error) is 117 eV for Mn K x-rays, and this is the best available resolution, even with an ideal, noise-free amplifier. The actual experimental FWHM for an x-ray of energy E_K is a combination of the two components, added in quadrature:

$$\text{FWHM}^2 = \text{FWHM}_0^2 + 5.55\,(FE_K\varepsilon). \tag{2}$$

3.6   Detector artefacts

Photoelectron production in the detector inevitably causes Si K-shell ionisations, the decay of which generates Si K x-rays, which are then reabsorbed to produce electron-hole pairs. However, some Si K x-rays, particularly those generated near the surface, escape from the detector. The energy measured from the original high-energy x-ray is apparently reduced by that of the Si K energy. A small 'escape' peak is displayed at an energy of 1.74 keV below the true peaks. The largest escape peaks arise for x-rays with energy just above the Si–K ionisation edge (P–K, S–K, Mo–L etc.), because these are most strongly absorbed, and deposit their energy closest to the Si surface. Small corrections, e.g. ${\sim}2\%$ of P–K, ${\sim}0.1\%$ of Ge–K, are therefore required for accurate analysis, and false identification of escape peaks as characteristic peaks can be made by the unwary. In Ge detectors, both L and K escape peaks occur and, because Ge–K x-rays can escape from much greater depths within the detector, the K escape peaks are much larger than in Si.

The layer in the detector just below the Au contact is often referred to as the 'dead layer'. It contains recombination sites which remove part of the charge pulse when the electron-hole pairs are created near the entrance surface: this is known as 'incomplete charge collection' (ICC). A characteristic peak displays a low-energy tail or plateau, the intensity and shape of which is difficult to predict as it can vary from detector to detector, and as a function of time for any particular detector. For $Z{>}14$ the effect is a maximum for the same lines as escape peaks, i.e. x-rays of energy just above the Si K absorption edge (Craven et al. 1985). It can also be a severe problem for K x-rays from elements of low atomic number, which are always absorbed close to the detector surface. ICC can also produce a step in the continuum energy distribution at the Si absorption edge.

3.7   Germanium detectors v silicon

Ge has some theoretical advantages over Si as a detector material, but also some disadvantages, such as the escape peaks described above (Cox *et al.* 1988). It has a smaller band gap, and so produces more pairs than Si for a given x-ray energy (ε=3.0 for Ge, ε=3.8 for Si), which in turn improves statistical accuracy. This is partly offset by a poorer Fano factor (F=0.13 for Ge, F=0.11 for Si). Ge has a higher dielectric constant, leading to a higher detector capacitance and a poorer noise resolution. Overall, a theoretically predicted advantage in resolution for Ge has yet to be realised: Si and Ge are routinely similar at the present time. Ge absorbs x-rays more strongly, and allows analysis of K lines from elements of high Z when E_0 is large enough to exploit this. However, the ionisation cross sections fall with increasing Z, and statistical accuracy will be poorer than for the L lines. Ge is less easily damaged by stray, high-energy electrons which can sometimes enter the detector.

4   System artefacts

The overall performance of an EDX system on an AEM is influenced by the design of the microscope and the collimation of the detector. Charge pulses in the detector can be produced by both x-rays and electrons from a variety of sources, apart from those generated at the selected position of the electron probe on the sample. The sample can be bathed completely in hard x-rays generated in the condenser section of the microscope at the probe defining aperture. These can be scattered into the detector, or cause photoelectron production in the sample, which leads to further x-ray generation. High-energy electrons can be scattered through large angles to generate x-rays in pole-pieces, the sample holder, aperture rods etc, which can either enter the detector directly, pass through a poorly designed collimator, or further excite photoelectrons and x-rays (Nicholson *et al.* 1982).

Methods for assessing the integrity of AEM/EDX systems have recently been proposed (Williams and Steel 1987, Zemyan and Williams 1992), as a means of comparing system performance and detecting problems. They suggest the use of a standard sample, an evaporated, fine-grained Cr film, which is flat uniformly thick, metallic, conducting, self-supporting, and suitable for long-term storage. The threshold for displacement damage (approx 400 keV) makes it suitable for most AEMs. The parameters to be measured are: the Cr K FWHM and FWTM (full-width, tenth-maximum), to assess resolution and ICC: the peak/continuum ratio as a function of E_0, for comparison with theoretical values: the L/K peak ratios for thin-window detectors, as a function of time, to assess if a build-up of ice on the crystal is affecting the efficiency of low-energy performance. These suggestions appear to offer a clear way forward to EDX system assessment.

5   Spectrum processing

The processing of the spectrum to derive chemical concentration is achieved using software packages designed to separate peaks from the continuum, to identify and correct for escape and sum peaks, and to deconvolute overlapping peaks. Characteristic peak areas are converted into elemental values; these can then be further refined by correcting for absorption, secondary fluorescence within the sample, and crystallographic artefacts. These topics are described in this section. Several manufacturers of EDX systems offer comprehensive packages which fulfil analytical requirements to a greater or lesser extent, and with varying degrees of speed and accuracy.

5.1 Background removal

The simplest method is by linear interpolation, for a peak which is not overlapped and lying on a continuum which is not significantly curved. The method is inaccurate for small peaks in the region of a strong absorption edge in the continuum. A more sophisticated method is to generate and subtract a calculated continuum e.g. using MBH, which is scaled over peak-free regions of the experimental spectrum in the medium energy range ($4 < E < 15$ keV) where detection efficiency is ~ 1 (see Section 5.3). However, the method first requires an estimate of foil composition to construct a continuum, and ideally a knowledge of foil thickness, geometry and detector parameters (see later) to allow for absorption at the low-energy end of spectrum. The third, and most widely used method, digital filtering, provides a spectrum which is similar to the second differential of the data, a process which automatically removes almost all the continuum contribution, except in the low energy range where there can be considerable curvature.

Peak areas are ideally derived by comparing the experimental peaks, after continuum removal, with those acquired from single-element standards with the same AEM/detector, and stored in a reference library. Overlapping peaks are deconvoluted by systematically varying the proportions of two standard components until an optimum statistical match is achieved. When digital filtering is used, the standard spectra used for peak matching must also be filtered. An alternative method calculates Gaussian peaks with the appropriate resolution for all the elements of interest, and then scales the theoretical peaks to obtain the best match with experiment. The latter method does not allow for peak distortions from ICC. In practice, it is very tedious, and sometimes impossible, to obtain single-element standards using an AEM, and analysis can then be performed using a library of standards provided by the manufacturer of the EDX system. The highest accuracy will be obtained by using standards collected and processed on the same AEM as used for the analysis of the unknown sample.

5.2 Characteristic x-ray generation and collection rates

The atomic density of the jth element in the sample is:

$$n_j = c_j \, \rho_j \, N_0 / A_j$$

the number of j ionisations in time $\tau = n_j(i/e)\,\tau\,t\,\sigma_j$. The number of x-rays in a particular line actually collected by the detector is:

$$I^j = n_j(i/e)\,\tau\,t\,\sigma_j\,\Omega\,\omega_j\,a_j\,e_j \tag{3}$$

Chemical analysis implies that at least two elements, A and B, are present, so the counts ratio derived using Equation 3 is:

$$\frac{I^A}{I^B} = \frac{c_A k_A}{c_B k_B}$$

where we have introduced the parameters

$$k_j = \sigma_j \omega_j \, a_j \, e_j \, / \, A_j.$$

Hence

$$\frac{c_A}{c_B} = k_{AB} \left(I^A / I^B \right) \tag{4}$$

where we have defined $k_{AB} = k_B/k_A$. The use of the k parameters, which are specific to each detector through the efficiency factors, e_j, was first proposed by Cliff and Lorimer (1975). Equation 4 is the basis of AEM/EDX quantitative analysis.

5.3 Alternative analytical methods

Equation 4 implies that the ratio of any two elements can be derived from a spectrum provided that the k-factors are known. The k-factors can be acquired in several ways: (i) using 'standard' samples to measure experimental k-factors, or k-factor ratios; (ii) calculating k-factors from first principles, using measured or assumed e_j; (iii) using published k-factor ratios measured on similar systems; (iv) measuring k-factor ratios where feasible, and extrapolating to other elements using theoretical ratios, when particular elemental standards are not available.

Standards

In the EMPA, a large number of bulk, single-element standards can always be kept within the sample chamber, and accessed within a matter of minutes. Accurate peak profiles and proportionality constants can be measured to compensate for changes in beam current and detector resolution, on a daily basis, to optimise the analysis of the unknown sample. An equivalent procedure is not really feasible for thin foils in the the AEM. Sequential interchange of several standards required for a particular analysis would typically take hours. Additional difficulties arise from the requirement in all measurements to determine the local thickness t, density ρ, and orientation for absorption correction (Section 5.4), as well as probe currrent i, to allow normalisation of the proportionality constants.

Multi-element thin film standards are generally used in AEM to determine ratios of the k-factors defined in Equation 4, but there are still several sources of experimental error using this approach. It is difficult to ensure that the composition of a standard determined by a 'bulk' analysis method is homogeneous on the microscopic scale of AEM. When several elements are present in the 'standard', AEM analysis requires several minutes to ensure adequate counting statistics for high accuracy. During this time the sample can drift, damage, or contaminate. Samples may be covered by surface oxides or other films which are preferentially enriched with some elements. The measured *k*-factor ratios will then depend on the thickness t. If thicker regions are analysed to minimise the contribution of surface artefacts then corrections for self-absorption and secondary fluorescence become more important. Further errors in analysis of crystals can occur (Section 6).

It may be necessary to combine *k*-factor ratios measured from two or more standards to generate ratios for particular element combinations, and this degrades accuracy. For example: $k_{CA} = k_{AB}k_{BC}$, but the error in k_{CA} will be the sum of the errors of the two factors in the product.

Theoretical estimation of *k*-factors

From Equation 4, it follows that theoretical estimates of *k*-factors require values of σ_j, ω_j, a_j, A_j, and e_j, for every element of interest. Calculation of the σ_j based on the quantum mechanical treatment of Bethe (1930) have been developed into parametric forms, as described by Inokuti (1971) and Powell (1976). For K-shell ionisation, the non-relativistic parametric form can be written as

$$\sigma_K = 4\pi e^4 b_K \ln(c_K T_0/I_K) / T_0 I_K \tag{5}$$

With $\beta = v/c$, the equivalent relativistic form

$$\sigma_K = 4\pi e^4 b_K \left(\ln(c_K T_0/I_K) - \ln(1 - \beta^2) - \beta^2 \right) / T_0 I_K \tag{6}$$

is becoming more relevant as the energy of the electrons E_0 in AEMs increases. Equivalent formulae can be used for the L-shell (σ_L). Experimental verification is made by rewriting Equation 5 in the form

$$\sigma_K U_K I_K^2 = 4\pi e^4 b_K (\ln U_K + \ln c_K) \tag{7}$$

and plotting $\sigma_K U_K I_K^2$ linearly against $\ln U_K$, to derive values of b_K and c_K from the intercept and gradient. Such a graph is known as a 'Fano plot'.

Experimental estimates of b_K and c_K vary considerably, and have been reviewed by Powell (1976, 1989). The large experimental variation, particularly those derived from thin samples in the AEM, is hardly surprising because of the large uncertainties in many of the parameters in Equation 3, used to derive (σ_K) from the measured count rates. One recent suggestion by Paterson *et al.* (1989) for improving accuracy is to ratio the characteristic and continuum rates. The continuum intensity can be written as

$$B^j = (i/e)\tau (c_j \sigma N_0/A_j)\sigma' j (E, \theta)\Delta E \, t\Omega \, e_j \tag{8}$$

so that, from Equations 3 and 8 we obtain

$$\frac{I^j}{B^j} = \frac{\sigma_j \omega_j a_j}{\sigma' j(E, \theta) \Delta E}. \tag{9}$$

Many of the experimental variables are no longer present in Equation 9, but experimental errors are replaced by the uncertainty in the MBH, or alternative theory, for continuum cross-sections. This approach gave the following values:

$$\text{non-relativistic}: \quad b_K = 0.67 \qquad c_K = 0.79$$
$$\text{relativistic}: \quad b_K = 0.45 \qquad c_K = 1.11$$

Theoretical estimates of the fluorescent yields, ω_j, for K and L shell x-ray generation, and the partition functions, a_j, for the sub-shells, have been listed by Bambynek *et al.* (1972), Schofield (1974), Krause (1979) and others. These are based on detailed quantum mechanical calculations, a description of which is beyond the scope of this chapter. Parametric expressions based on these data, for the convenience of computer calculation, should only be used with caution. For example, ω_j can be derived from the expression (Burhop 1955):

$$\left(\frac{\omega}{1-\omega}\right)^{1/4} = A + BA + CZ^3 \tag{10}$$

where different constants A, B and C, are used for the K, L, and M shells, but this, like the other parametric expressions, has limited accuracy. The tabulated data from the above references provide greater accuracy. Estimates of ω for the L and higher shells are complicated by the occurrence of Coster-Kronig transitions. These are radiationless transitions within sub-shells such that, for example, a hole created by ionisation of the L_I sub-shell is filled by a transition of an electron from the L_{II} sub-shell, which is then filled by a transition from the M-shell. It is the latter transition which yields the characteristic x-ray. Each sub-shell has its own effective value of ω_j, which is partly related to ionisations of the other sub-shells. Estimates of the magnitude of these transitions are given by Krause (1979).

Detector efficiency, e_j

The detector efficiency, e_j in Equation 3, depends on several factors. Low-energy x-ray intensity is significantly attenuated before it reaches the detector crystal by absorption in the 'window', in any ice and contamination, in the Au contact layer, and in the Si 'dead layer' (see Figure 4). Calculation of the transmitted intensity, I, requires a knowledge of ρ_i and thickness, t_i, of each of these layers, and the value of the mass-absorption coefficient $(\mu/\rho)_i^j$ for the jth x-ray in the ith layer, for each characteristic x-ray of interest.

$$I^j = I_0^j \prod_i \exp\left(-(\mu/\rho)_i^j \rho_i t_i\right) \tag{11}$$

where I_0^j is the intensity incident on the window.

For high x-ray energies, e_j is reduced by transmission of the x-rays through the Si crystal:

$$I^j = I_0^j \left\{1 - \exp\left(-(\mu/\rho)_{Si}^j \rho_{Si} t_{Si}\right)\right\} \tag{12}$$

Hence, to calculate e_j for standardless analysis it is necessary to measure, or assume values of t_i for the layers, and to consult tables of $(\mu/\rho)_i^j$, (*e.g.* Thinh and Leroux 1979, Heinrich 1987). Measurements of t_i are made by carefully scaling computed B to experimental spectra. Typical values are: 8 μm for a Be window, 10–20 nm for the Au contact, and 100–200 nm for the 'dead layer'. Ice and contamination layers are likely to vary from detector to detector, and with time for any particular detector. Windowless detectors must be periodically warmed to evaporate ice.

Experimental measurements, *e.g.* Paterson *et al.* (1989), for a windowless detector, indicate that e_j is close to unity for $2 < E < 10$ keV. Table 2 lists some of their measurements for various characteristic x-rays. The effect on e_j of absorption at low E,

Line	E(keV)	e_j	Line	E(keV)	e_j
Fe–L	0.70	0.53	Fe–K	6.40	1.0
Co–L	0.77	0.59	Co–K	6.93	1.0
Ni–L	0.85	0.66	Ni–K	7.48	1.0
Cu–L	0.93	0.71	Cu–K	8.04	1.0
Ge–L	1.19	0.82	Ge–K	9.88	1.0
Mo–L	2.29	0.95	Mo–K	17.48	0.81
Ag–L	2.98	0.97	Ag–K	22.16	0.58
Ag–L	3.44	1.00	Sn–K	25.27	0.45

Table 2. *Measurements for various characteristic x-rays for a windowless detector*

and the transmission right through the detector at high E, is well illustrated by these measurements.

The combination of the various parameters described above allows the calculation of k-factor ratios for analysis without standards. Comparisons of experimental k-ratios for different AEMs, *e.g.* Cliff and Lorimer (1975), Wood *et al.* (1981), Schreiber and Wims (1981), usually show agreement within a few per cent for the K shells for $14<Z<30$, but increasing differences for other shells, and for $E<1.5$ keV, where small errors in t_i for the different detectors can lead to large divergences. An appreciation of this can be gained from Figure 5, which shows, theoretically, the sensitive dependence of e_j with E, to relatively small changes in t_i. Hence, an experimental determination of k factors is ideally required for accurate quantitative analysis of low Z elements: theoretical calculations are likely to be more erroneous.

5.4   Analysis corrections

Self-absorption

Once an estimate of chemical composition has been derived from the analysis of a spectrum, it may be necessary to apply corrections, the most important of which is for the self-absorption within the sample itself. The correction factor can be determined with reference to Figure 6, which depicts the specimen/detector geometry in two dimensions. In a thin, homogeneous foil it is assumed that there is a uniform rate of generation of x-ray intensity along the path of the electron beam, of length z_0. The contribution from a small increment of the foil, of thickness dz and depth z, emitted towards the detector, is reduced by self-absorption along the path L in the direction of the detector. Integration over z shows that the intensity, I, escaping the sample is related to the generated intensity, I_0, by:

$$I^A = I_0^A \left\{ \frac{1 - \exp((\mu/\rho)_A^A \rho f z_0))}{(\mu/\rho)_A^A \rho f z_0} \right\} \tag{13}$$

where f is a geometric factor dependent on the angle of elevation of the detector, the tilt of the sample from the horizontal, and the azimuthal angle of the detector with

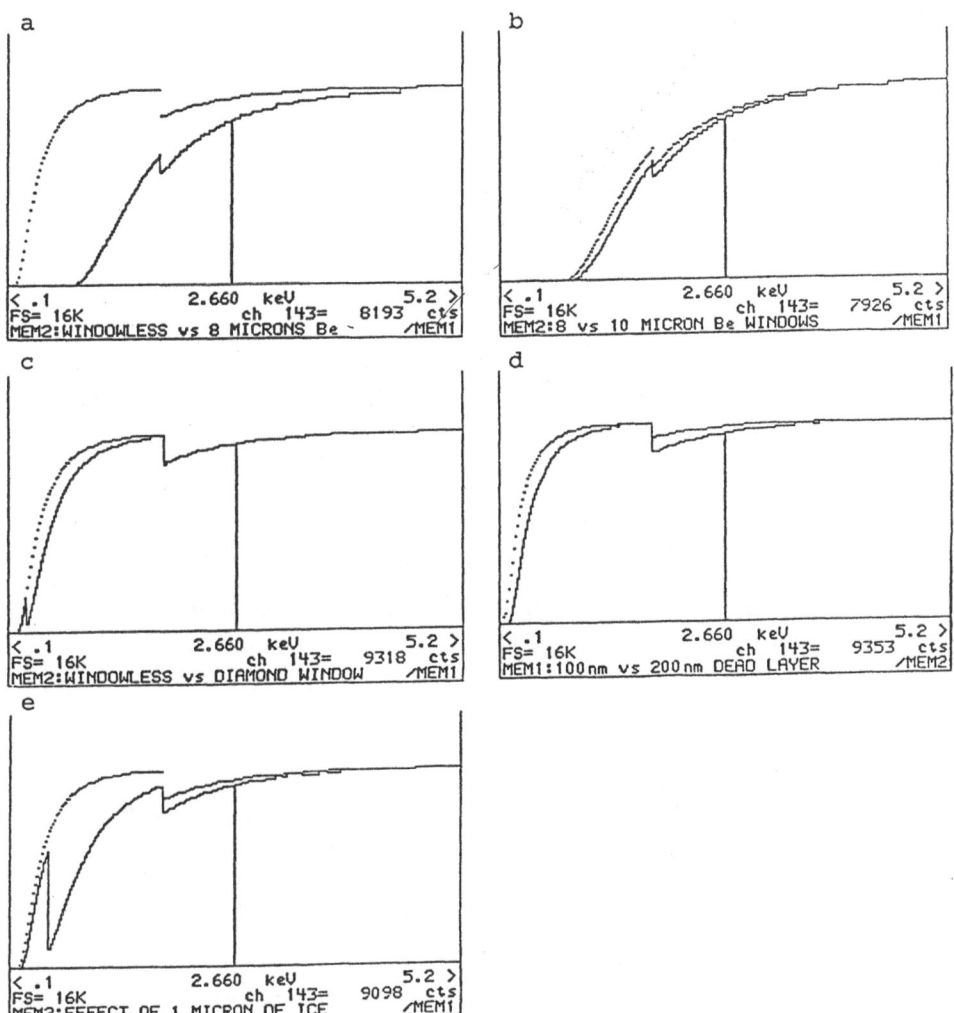

Figure 5. *Theoretical variations of detector efficiency with x-ray energy: comparison of (a) windowless vs 8 μm Be window, 200 nm Si dead layer; (b) 8 μm vs 10 μm Be windows, 200 nm Si dead layer; (c) windowless vs 100 nm diamond window 200 nm dead layer; (d) 100 nm vs 200 nm dead layer, windowless; (e) windowless with 200 nm deadlayer, with and without 1 μm of ice.*

respect to the sample tilt axis. This must be determined from the specific geometry of the analysis which, in practice, can be very tedious and uncertain for bent samples.

Equation 13 applies to the trivial case of sample consisting of a single element. The self-absorption in a binary alloy, composed of elements A and B, with fractional concentrations c_A and c_B, will now be considered. The mass absorption coefficient of each characteristic x-ray is the sum of the contributions of both elements, scaled by the

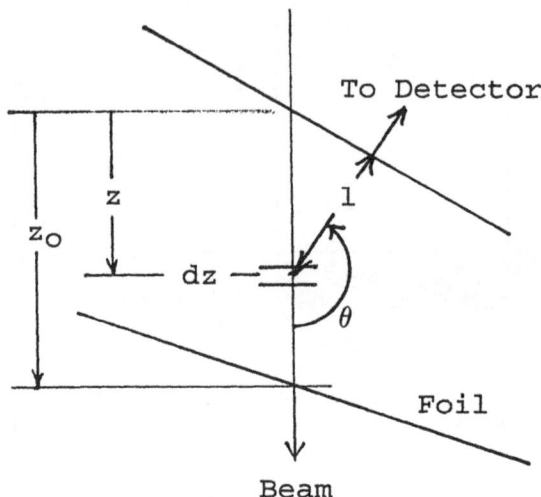

Figure 6. *Self-absorption path length, l, in the direction of the detector, for x-rays generated in element dz of foil, at a depth z.*

elemental concentrations. Hence:

$$(\mu/\rho)^A_{AB} = c_A(\mu/\rho)^A_B + c_B(\mu/\rho)^A_B \tag{14}$$

with an equivalent expression for $(\mu/\rho)^B_{AB}$, and the ratio of the measured x-ray intensities is related to the ratio actually generated in the sample, by:

$$\frac{I^A}{I^B} = \frac{I^A_0(\mu/\rho)^B_{AB}(1 - \exp((\mu/\rho)^A_{AB}\rho f z_0))}{I^B_0(\mu/\rho)^A_{AB}(1 - \exp((\mu/\rho)^B_{AB}\rho f z_0))} \tag{15}$$

The solution of Equation 4, corrected for self-absorption, using Equations 14 and 15 can only be achieved by an iterative method. A first estimate of composition is found from Equation 4, without absorption correction. This estimate is then used with Equations 14 and 15 to generate a better estimate, corrected for absorption, which is then fed into Equation 4 again, and the process repeated. The method can be extended to include as many elements as required, and the solution, performed within seconds on a computer, usually converges after two or three iterations to the limit of statistical accuracy.

Secondary fluorescence

Strong self-absorption implies that the absorbed x-rays produce photoelectrons by ionisation of atoms in the sample, which generates an additional contribution to the collected signal. Because characteristic x-rays are emitted isotropically, and can penetrate much further than electrons, this secondary fluorescence can arise over an extensive volume of the sample, and from the sample environment, *e.g.* holder, pole pieces etc. Fluorescence from absorption of the continuum will also occur. A correction formula has been derived by Nockolds *et al.* (1980), for the ideal geometry of a parallel-sided sample. As this is a gross assumption which is further limited in accuracy by measurements of sample thickness, and is a second-order correction anyway, there are few situations where the use of the correction can be justified. These will usually be in cases where the $E_{K,L...}$ of one element lies just above the ionisation edge, $I_{K,L...}$ of a second element.

Foil thickness determination

When correction for self-absorption is necessary, measurement of z_0 can be performed in several ways; none is very quick and the accuracy of all is questionable. Some investigators took advantage of the high residual hydrocarbon content present within the 'vacuum' of early AEMs, which caused the growth of a carbon contamination spot on both surfaces of the foil at the point of analysis. Subsequent tilting through a known angle separated the projected images of the two spots, allowing a measurement of z_0 (Lorimer *et al.* 1976). Errors arise from indistinct edges to the spot images and surface films, and can be as high as 200% (Stenton *et al.* 1981). AEMs are now much cleaner, and contamination is minimised to improve EELS, microdiffraction and HREM, so the method should only be used if no other is available.

In materials science, many poly-crystalline samples contain planar features such as stacking faults and interfacial boundaries which intersect both surfaces of the foil. Tilting the sample through a known angle, about a known axis, varies the projected image width of these features and enables the calculation of z_0 from the variation. However, such features do not routinely occur close to every selected analysis point. An alternative method for crystal samples requires tilting the foil to excite a known, two-beam, dynamical diffracting condition, and deriving the thickness from the spacings of the fringes in the convergent-beam diffraction discs (Kelly *et al.* 1975). This method requires tedious analysis, the fortuitous availability of a suitable vector close to the orientation selected for analysis, and is then only useful for foils of intermediate thickness (Allen 1981).

For reasons which are described below, EDX should usually be performed at a kinematic crystal orientation to avoid strong excitation of any diffraction vectors. A comparison of the incident electron intensity with that transmitted through the foil at a kinematic orientation then provides a means of measuring z_0. Measurement of i can be made using the exposure meter of the photorecording system , a Faraday cage or aperture rod connected to an electrometer, or an energy-loss spectrometer. However, a calibration is first required to relate the i to z_0 by some alternative experiment. When EELS is available, however, analysis of the low-loss region of the spectrum (Egerton, 1986) and a calculation of the mean free path for inelastic scattering (Malis *et al.* 1988) allows an estimate of z_0 within a few minutes, with an accuracy of 10–20%, which is sufficiently accurate for most purposes.

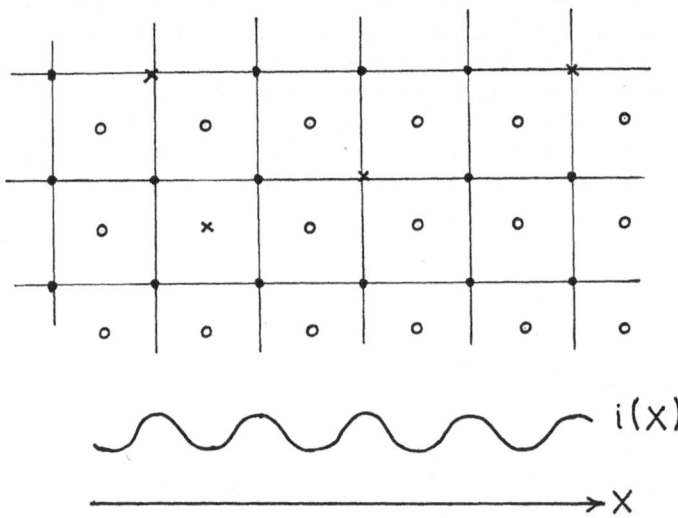

Figure 7. *Schematic projection of a bcc lattice in an ordered binary alloy; A atoms are located at the cube corners (•), and B atoms at the cube centres (○). Impurity atoms (×) can occupy either site. The spatial modulation of the electron beam is indicated in the lower part of the figure.*

6   Crystallographic effects in EDX

6.1   ALCHEMI

Many samples in materials science are crystallographic, and this can affect EDX analysis through the way in which the electron beam intensity is spatially distributed within the sample. Figure 7 shows the positions of the atoms, in a [100] projection, in an ordered, binary, bcc crystal AB, with atoms A at the cube corners, and atoms B at the cube centres. At any particular depth below the entrance surface, the electron beam is spatially modulated, or channeled, and propagates as Bloch waves (Hirsch *et al.* 1965), such that for the symmetry orientation in Figure 7, there might be a significant difference in the intensity of the beam passing close to the nuclei of the two types of atoms, as depicted in the lower part of Figure 7. Because K and L characteristic x-rays arise from ionisations of tightly bound orbits, close to the atomic nuclei, the effective i in Equation 3 will be different for A and B. Because the distribution of i varies sensitively with orientation close to zone axes and strongly excited Bragg reflections, x-ray generation (Hirsch *et al.* 1962), and hence k-factor ratios, become orientation-dependent by factors as much as 20–30% in some alloys. Analysis close to such an orientation should be avoided by first viewing the diffraction pattern before proceeding to acquire data. Variations will not occur in disordered alloys because non-equivalent atomic sites are occupied by equal fractions of different atoms.

This phenomenon has been used to advantage in the technique known by the acronym **ALCHEMI** (Atomic Location by Channeling Enhanced Microanalysis, Spence and Tafto 1983). A channeling factor, G_i, for element i, is defined from x-ray intensities, I^r and I^c, at a random, and a channeling orientation, for a constant i and point on the sample, as:

$$G_i = (I^c - I^r)/I^r \qquad (16)$$

Suppose that, for the crystal shown in Figure 7, substitutional impurity atoms of element C are unequally distributed between sites A and B, with fractions f_A and $(1 - f_A)$. Then:

$$G_C = f_A G_A + (1 - f_A)G_B \qquad (17)$$

from which f_A can be derived. However, the above analysis cannot always be applied. In Figure 7 the widths of the peaks in the current distribution are shown to be much broader than the inner orbital widths, so that i is effectively uniform across the projected area of an atom for which a particular ionisation occurs. In practice, this is not so, particularly for L and higher electron shells, which have larger projected areas (*i.e.* cross-sections). Complex correction factors to the G_is for this 'delocalisation' are then required to retain the accuracy of analysis (Pennycook 1988). An alternative solution to the problem, using multivariate statistics, has recently been proposed by Rossouw *et al.* (1989).

6.2   Coherent bremsstrahlung

A second effect which is sometimes observed in the spectra from crystalline samples is manifest as additional peaks, relatively small in magnitude compared to the major peak intensities, and usually observed in the energy range $1 < E < 4$ keV. These have been called 'coherent bremsstrahlung' (**CB**) peaks because, like the continuum, they also arise as a consequence of fast electron acceleration. Such peaks are strongest when low-index, lattice planes are perpendicular to the electron beam. The fast electrons then experience successive acceleration and retardation as they pass through the planes, separation L nm, with a frequency of $2\pi v/L$. The **CB** radiation is emitted with the same frequency, and lies in the x-ray range of the electromagnetic spectrum. Because the fast electrons in the **AEM** are relativistic, and the above frequency refers to the fast electron frame of coordinates, the energy (frequency) of the x-ray detected in the laboratory frame of reference is modified to:

$$E(\text{keV}) = \left(\frac{1.24}{L}\right)\frac{\beta}{1 - \beta\cos\theta}. \qquad (18)$$

where θ is the angle between the beam direction and the detector and $\beta = v/c$.

When $E_0 = 100$ keV, and $\theta = 110°$ (*i.e.* a 20° detector 'take-off' angle), $E = 0.537/L$, and because typically $0.2 < L < 0.5$ nm, **CB** peaks often lie in the 1–2 keV range. Detailed analysis predicts that multiples of the basic energy should occur, and this is confirmed by observation (Spence *et al.* 1983).

Although **CB** peaks are strongly excited when the beam is parallel to a low-index zone axis, they persist over large angular ranges, and gradually change in energy as the crystal is rotated. The dependence of **CB** peak energy on sample tilt, and on E_0, enables their identification. They affect analysis by (i) being falsely identified as small characteristic peaks, (ii) preventing the accurate analysis of minor elements by peak overlaps, and (iii) by interfering with continuum fitting and removal. An example of their detrimental effect is in the analysis of minor alloying elements, Si, Mo, and impurities, P, S, in steels (Titchmarsh, 1987)

6.3   Anomalous self-absorption in crystals

A third crystallographic effect can arise when there is strong self-absorption present. When an x-ray channeling direction in the sample is oriented towards the detector there can be anomalously low self-absorption; this can result in relative changes in detected intensity ratios of several per cent as the sample is tilted by only 1–2° (Hashimoto *et al.* 1990). However, this only occurs when the detector subtends a small solid angle at the sample, as for the high take-off-angle detectors on a few AEMs, and can generally be disregarded for the vast majority of modern detectors with Ω=0.01, or greater.

7   EDX resolution, beam broadening and sensitivity

The best spatial resolution of EDX analysis in the AEM is theoretically defined by the electron beam diameter. In practice, the spatial resolution is determined by a combination of foil thickness, the required statistical accuracy in an acceptable analysis time, the particular element combinations in the sample, the type of electron source on the AEM, and the probe-forming parameters employed.

7.1   Probe diameter

The electron source, and the formation of probes for high resolution analysis is described in detail by Kruit in this Proceedings. The electron current, i, in a probe of diameter d, formed by demagnification of a source of brightness B_0, using a semi-angle of convergence α, is:

$$i \ = \ B_0 \left(\pi \alpha d/2\right)^2 \tag{19}$$

For a fixed B_0, and a minimum i for any given specific analytical requirement, d can only be reduced if α is increased in proportion. However, any increase in α is limited by spherical aberration, which increases d as α^3. Hence, there is a minimum useful probe size. The B_0 of a heated W source is, typically, about one tenth of a heated LaB_6 source, which is, in turn, about one hundredth weaker than a field emission source (FEG). The FEG, therefore, gives at least an order of magnitude improvement in resolution compared with other types of source, for a given statistical accuracy and analysis time.

In the limit, the theoretical radial distribution of current, $i(r)$, in a very narrow probe, produced by demagnifying the electron source, can be computed using wave optics (Munro 1977, Mory 1985). The distributions vary with E_0, focus, and lens aberration coefficients. Although in practice it is now possible to generate an electron probe in an AEM with a FEG having a FWHM of less than 0.4 nm, this can only be achieved at the expense of reducing i to $\sim 10^{-11}$A which, although sometimes useful for imaging and EELS, is inadequate for routine EDX. Probe diameters of about 1 nm or greater, containing $i > 10^{-10}$A, are typically required for EDX with $\tau = 100-200$ seconds.

Probe 'size', and 'diameter' are terms which have been very poorly defined in the EDX literature. In only a very few instances have experimental measurements been made to determine $i(r)$. Because all the electrons in the probe can excite x-rays, and because $i(r)$ can have an extended 'tail' many times the FWHM, a statement of probe 'diameter', d, only has meaning when accompanied by an indication of the fraction of the i within d (Titchmarsh 1987). This is particularly important for the interpretation of data obtained from samples containing inhomogeneities on a scale similar to, or smaller than $i(r)$, and for attaining the best spatial resolution. Measurement of $i(r)$ is usually made from changes in an image, EDX, or EELS signal as the probe is scanned across an abrupt edge of the sample, which must be aligned exactly parallel to the electron beam direction. Any measured profile of chemical concentration cannot be narrower than the convolution of the real distribution with $i(r)$.

7.2 Beam broadening

In a typical sample in the AEM (*e.g.* when $t = 100$ nm) only a small fraction of the fast electrons excite a characteristic x-ray because the $\sigma_{K,L}$ are so small. However, the electrons are likely to undergo elastic scattering, possibly through large angles, more than once in such a thickness, and then cause ionisation at a large distance from the probe axis. This 'beam broadening' limits the accuracy of the EDX analysis when the sample is inhomogeneous. A simple estimate (Goldstein *et al.* 1977) is often used to calculate the diameter, b, on the exit surface of the foil within which 90% of the electrons emerge, where:

$$b = 6.25 \times 10^5 (Z/E_0) t^{3/2} (\rho/A)^{1/2} \tag{20}$$

with units of cm (b and t), eV (E_0), and gm cm^{-3} (ρ). A 'resolution' is then estimated by combining b and d in quadrature. However, b is larger than the diameter of the cylinder within which 90% of the x-rays are produced, and is of limited practical application. At high spatial resolution, elastic scattering affects the analysis of small embedded particles, diffusion profiles, and interfacial segregation, mainly by increasing the proportion of the signal from the surrounding matrix, which increases the background under the peaks of interest and reduces sensitivity. If spatial resolution is defined as the FWHM of the change in a signal, as the probe is traversed across an inhomogeneity, beam broadening scarcely affects resolution. Only when the tails of the signal are incorporated into the definition of the resolution does it become significant. In practical analysis, 'resolution' is irrelevant; it is the accuracy of the data interpretation which is important, and this usually requires a model of the chemical inhomogeneity and the use of complex calculations to deconvolute the interactions of a measured radial probe current distribution, the beam-broadening, and the sample geometry. Monte Carlo calculations (*e.g.* Newbury and Myklebust 1981, Titchmarsh 1988) probably offer the most useful method for such deconvolutions, although these incorporate approximations such as neglecting the crystallinity of the sample which might lead to an overestimate of the beam broadening (Furdanowicz *et al.* 1991)

7.3   Sensitivity

The most important measure of EDX sensitivity is the detection limit for the fraction, f, of a dilute element, A, in a matrix of element B (*e.g.* Joy and Maher, 1977). The Signal-to-Noise Ratio, SNR, of a peak, I, is defined as:

$$\text{SNR} = I/\sqrt{\text{var}(I))} \qquad (21)$$

Because var(I) varies with the uncertainty in fitting the continuum under the peak, var$(I)=hB$, where h is a function of the background fitting method, and $h\geq 1$. Both I and var(I) are output by the spectrum processing software, and so the experimental SNR can be calculated for any selected peak in a spectrum. The value of the SNR determines the confidence with which the peak is said to be 'detected', and this is related to Gaussian statistics. Thus,'detection' with 99% confidence, for example, implies: SNR$\geq$3, or:

$$I/\sqrt{hB} = 3 \qquad (22)$$

Equations 3, 8, and 22 can be combined to yield:

$$f = \frac{n_A}{n_B} \sim \frac{n_A}{n} = \left(\frac{3}{\sigma_A \omega_A a_A}\right) \left\{\frac{h\sigma_j'(E,\theta)\,dE}{ni\tau e_A\Omega}\right\}^{1/2} \qquad (23)$$

Thus, in practice, the sensitivity improves as the square root of i, τ, and other parameters which increase the total x-ray generation and collection rates. The probe current should be maximised to give the maximum counting rate that the detector can analyse, subject to the 'dead-time' limitations described in Section 3.4. For many combinations of A and B, a value of $f<0.001$ is feasible.

A more interesting problem for AEM, because of its capability for analysing very small volumes of the sample, is to calculate the fraction, f_{min}, of an atomic plane of impurity atoms which can be detected when segregated at an interface such as a grain boundary (Garrett-Reed, 1985; Titchmarsh and Vatter, 1989). Equation 23 must be modified to incorporate the spatial distributions of the current, i, and the solute, and yields the expression:

$$f_{min} = \frac{(2\pi)^{1/2} x_0 f}{\delta} = \frac{3x_0}{\delta \sigma_A \omega_A a_A} \left\{\frac{2\pi h\sigma_A'}{ni t\tau \varepsilon_A \Omega}\right\}^{1/2} \qquad (24)$$

In Equation 24, x_0 is the standard deviation of the current distribution, assumed to be Gaussian, and δ is the interplanar spacing of the matrix. In Figure 8 is shown the variation of P segregation concentration measured by positioning the electron probe at various distances from a grain boundary in a ferritic steel. The width of the profile reflects the width of the probe current distribution, because the segregation width is only about one or two atom planes thick. The FWHM of this profile is used to derive x_0 in Equation 24, and, hence, to determine a value for the sensitivity limit, f_{min}, and the amount of segregation present. The f_{min} was typically less than 0.05 of an atom layer, for an analysis time of 100 seconds, and about 0.02 atom layers with an analysis time of 600 seconds. Comparison of segregation concentrations in the same materials, measured by EDX and Auger analysis, (Vatter and Titchmarsh, 1989), suggested similar sensitivities for both methods.

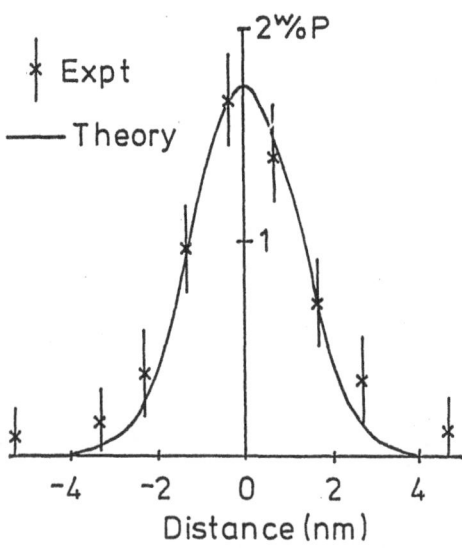

Figure 8. *Variation of measured P segregation concentration close to a grain boundary in a ferritic steel. The theoretical curve is a Gaussian distribution with a standard deviation equivalent to that of the measured probe current.*

References

Allen S M, 1981, *Phil Mag* **A43** 325

Bambynek W, Crasemann B, Fink R W, Freund H-U, Mark H, Swift C D, Price R E and Rao P V, 1972, *Rev Modern Physics* **44** 716

Bethe H, 1930 *Ann Phys* **5** 325

Burhop E H S, 1955, *J Phys Radium* **16** 625

Chapman J N, Gray C C, Robertson B W and Nicholson W A P, 1983, *X-ray Spectrom* **12** 153

Cliff G and Lorimer G W, 1975, *J Microsc* **103** 203

Craven A J, Adam P F and How R, 1985, *Inst Phys Conf Ser* **78** 189

Cox C E, Lowe B G and Sareen R, 1988, *IEEE Trans Nucl Sci* **NS-35** 28

Egerton R F, 1986, in *The Electron Microscope* (Plenum, New York/London)

Furdanowicz W A, Garrett-Reed A J and Vander Sande, J B, 1991, *Inst Phys Conf Ser* **119** 437, ed Humphreys F J (IoP, Bristol)

Garrett-Reed A J, 1985, in *SEM* **1985-I** 21, ed O Johari (SEM Inc, O'Hare, Chicago)

Goldstein J I, Costley J L, Lorimer G W, and Reed S J B, 1977, in *SEM* **1977-1** 315, ed O Johari (IITRI, O'Hare, Chicago)

Gray C C, Chapman J N, Nicholson W A P, Robertson B A and Ferrier R P, 1983, *X-ray Spectrom* **12** 163

Hashimoto I, Wakai E and Yamaguchi H,, 1990, *Ultramic* **32** 121

Heinrich K F J, 1987, *Proc 11th ICXOM* p 67 eds J Brown and R H Packwood, (Univ W Ontario, London/Ontario

Hirsch P B, Howie A and Whelan M J, 1962 *Phil Mag* **7** 2095

Hirsch P B, Howie A, Nicholson R B, Pashley D W and Whelan M J, 1965, in
    Electron Microscopy of Thin Crystals (Butterworths, London)
Inokuti, M, 1971, *Rev Mod Phys* **43** 297
Joy D C and Maher D, 1977, in *SEM* **1977-I** 325, ed O Johari (SEM Inc, O'Hare, Chicago)
Kelly P M, Jostsons A, Blake R G and Napier J G, 1975, *Phys Stat Sol* **A31** 771
Kramers H A, 1923, *Phil Mag* **46** 836
Krause M O, 1979, it J Phys Chem Ref Data **8** No 2 307
Lorimer G W, Cliff G and Clark J N, 1976, in *Developments in Electron Microscopy and*
    Analysis p 153, ed Venables J A (Acad Press, London)
Malis T, Cheng S C, and Egerton R F, 1988, *J Electron Microsc Tech* **8** 193
Mory C, 1985, PhD Thesis, Universite de Paris-Sud
Munro E, 1977, *Proc. VIIIth IXCOM* Paper No 19, eds R Ogilvie and D Wittry
    (NBS, Washington, DC)
Newbury D E and Myklebust R L, 1981, in *AEM 1981* p 91 (San Francisco Press)
Nicholson W A P, Gray C C, Chapman J N and Robertson B W, 1982, *J Microscopy*
    125 25
Nockolds C, Nasir M J, Cliff G and Lorimer G W, 1980, *Inst Phys Conf Ser* **52** 417,
    ed T Mulvey (IoP, Bristol)
Paterson J H, Chapman J N, Nicholson W A P and Titchmarsh J M, 1989,
    J Microscopy **154** 1
Pennycook S,1988, *Ultramic* **26** 239
Powell, C J, 1976, *Rev Mod Phys* **48** 33
Powell, C J, 1989, *Ultramic* **28** 24
Pratt R H, Tseng H K, Lee C M, Kissel L, MacCallum C and Riley M, 1977,
    Atomic Data and Nuclear Data Tables **20** 175
Rossouw C J, Turner P S, White T J and O'Connor A J, 1989, *Phil Mag Letters* **60** 225
Russ J C,1984, *Fundamentals of Energy Dispersive X-ray Analysis* (Butterworths, London)
Schreiber P T and Wims A M, 1981, *Ultramic* **6** 323
Scofield J H, 1974 *Phys Rev A* **9** 1041
Spence J C H and Tafto J, 1983, *J Microscopy* **130** 147
Spence J C H, Reese G, Yamamoto N and Kurizki G, 1983, *Phil Mag B* L39
Statham P J, 1991, *Inst Phys Conf Ser* **119** 425, ed F J Humphreys (IoP, Bristol)
Statham P J and Nashashibi T, 1988, in *Microbeam Analysis 1988* p 50, ed D E Newbury
    (San Francisco Press)
Stenton N, Notis M R, Goldstein J I and Williams D B, 1981, in *Quantitative Microanalysis*
    with High Spatial Resolution p 35 eds G W Lorimer, M H Jacobs and P Doig
    (The Metals Soc, London)
Thinh T P and Leroux J, 1979, *X-ray Spectroscopy* **8** 85
· Titchmarsh J M, 1987, *Proc 11th ICXOM* p 337, ed J Brown and R H Packwood,
    (Univ. W. Ontario, London/Ontario)
Titchmarsh J M, 1988, in *Microbeam Analysis 1988* p 65, ed D E Newbury
    (San Francisco Press)
Titchmarsh J M and Vatter I A, 1989, in *NATO ASI Series B* **203** 111, ed D Cherns
    (Plenum Press, London)
Tseng H K and Pratt R H, 1971, *Phys Rev A* **3** 100
Vatter I A and Titchmarsh J M,1989, *Ultramic* **28** 236
Williams D B, 1984, *Practical AEM in Material Sci* (Philips Electronic Instr Inc, New Jersey)
Williams D B and Steel E B, 1987, *AEM 1987* p 228, ed D C Joy (San Francisco Press)

Wood J E, Williams D B and Goldstein J I, 1981, in *Quantitative Microanalysis with High Spatial Resolution* p 24 eds G W Lorimer, M H Jacobs and P Doig (The Metals Soc, London)

Zaluzec N J, 1979, in *Intro to Analytical Electron Microscopy* ch 4, eds J J Hren, J I Goldstein and D C Joy (Plenum Press, New York)

Zemyan S M and Williams D B,1992, *J Microscopy* in press

Elemental Analysis and Imaging by Proton-Induced X-Ray Emission (PIXE)

J L Campbell

Department of Physics
University of Guelph, Canada

1   Introduction

The frequently cited analogy between PIXE and the more venerable technique of electron-probe micro-analysis or EPMA provides a good context for newcomers to PIXE. The two techniques are depicted schematically in Figure 1. In each case, a charged-particle beam creates vacancies in the inner atomic shells of elements in a specimen; these vacancies de-excite by various processes, one of the predominant means being via the emission of characteristic x-rays. Spectroscopy of the x-rays identifies the elements; then, with either standards or a fundamental parameter approach, the x-ray intensities are mapped into concentrations. In addition the beam may be focused to micron width and rastered over the specimen to provide images of elemental distributions in two dimensions. With either electrons or protons, given proper precautions, the analysis is non-destructive.

The source of protons for PIXE is a small Van de Graaff accelerator. The success of PIXE and other ion beam analysis techniques has revitalised technological development of both single-end and tandem accelerators providing proton energies up to 3 MeV. The beam is transported to the specimen chamber, in a vacuum of some 10^{-6} mb, using simple beam optical elements (electrostatic or magnetic dipole steering, magnetic quadrupole focusing, collimation) and the sample stage is designed to accommodate the maximum number of specimens. The x-ray spectroscopy is accomplished with energy-dispersive lithium-drifted silicon (Si(Li)) detectors, which reveal all detectable elements in a single spectrum measurement.

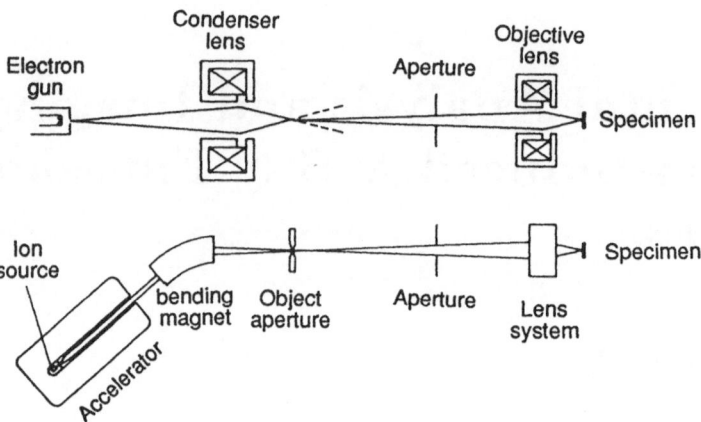

Figure 1. *Schematic of electron and proton microprobes (Johansson et al. 1988). Copyright © 1988 John Wiley and Sons Ltd.*

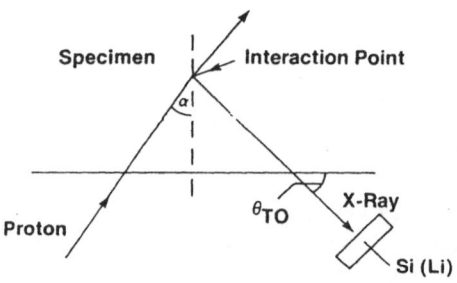

Figure 2. *Beam-specimen geometry in* PIXE.

2   Beam-specimen interactions

The first (partial) breakdown of the analogy with EPMA is the fortunate fact that for PIXE the interaction physics is simple and is accurately described in analytical terms. This is in contrast to EPMA, where multiple scattering of the electron beam is sufficiently complex as to necessitate approximate analytical treatments or Monte Carlo simulation of the beam-specimen interaction.

The x-ray production process may be explored with the generalised beam-specimen geometry of Figure 2. PIXE's simplicity derives from the almost linear path of an MeV-energy proton as it slows down via a multitude of small-energy-loss inelastic encounters with electrons in the specimen. The path length is typically a few microns to a few tens of microns. Semi-empirical expressions with tabulated parameters give the stopping power $S_M(E)$ for each element; this is proportional to the energy loss per unit path dE/dx, as a function of proton energy E. If the major element concentrations are known, then the matrix stopping power $S_M(E)$ is obtained as a concentration-weighted linear combination. With the energy profile along the path known, the K or L ionisation cross-section $\sigma_Z(E)$ in each segment $(x, x+dx)$ of the path is obtained. This is done via published fits to the extensive cross-section versus energy tables that have resulted from critical assessment of thousands of cross-section measurements in the atomic physics literature. Finally, the x-ray intensity reaching the detector from each segment is reduced by an exponential transmission factor through all overlying segments. This involves a matrix attenuation coefficient $(\mu/\rho)_M$ which is the concentration-weighted sum of the attenuation coefficients of the major elements for the x-ray of interest. Summing over all the x-ray yields from all segments of the proton path we have the expression

$$I(Z) = \left(\frac{N_{\mathrm{av}} \omega_Z b_Z t_Z \Omega \epsilon_Z^i}{4\pi A_Z} \right) N_p C_Z \int_{E_0}^{E_f} \sigma_Z(E) \exp \left[-\left(\frac{\mu}{\rho} \right)_M \frac{\cos \alpha}{\sin \theta_{T0}} \int_{E_0}^{E_f} \frac{dE}{S_M(E)} \right] \frac{dE}{S_M(E)} \tag{1}$$

for the K x-ray intensity of element Z(atomic mass A_Z) recorded by the Si(Li) detector. In this expression Ω is the solid angle subtended at the specimen, ϵ_Z^i is the intrinsic efficiency of the detector, ω_Z is the K shell fluorescence yield, b_Z the fraction of K x-rays that fall in the main Kα line, t_Z is the transmission of that x-ray through any absorbers interposed (see below) between specimen and detector, N_p is the number of protons and N_{av} is Avogadro's number. The integral, computed numerically, is simply the summation of intensities from each path segment.

Equation 1 implies that the energy E_f of protons emerging from the specimen must be known. In practice most PIXE is done with specimens so thin that $E_f \cong E_0$, as in air particulate deposits and biological tissue sections, or specimens so thick that $E_f = 0$, as in mineral grains, alloys and archaeological artifacts. (However one can also treat the intermediate case, and even cases of multiple thin layers, such as semiconductor structures). In the 'thin-target' case the yield is simply

$$I_Z = \frac{N_p m_a(Z) \sigma_Z(E_0) \omega_Z b_Z t_Z \Omega \epsilon_Z^i}{4\pi A_Z} \tag{2}$$

where $m_a(Z)$ is the areal density of element Z. Comparing these equations shows that, as in EPMA, the thick-target case resembles the thin-target case with multiplicative matrix corrections. If the major element concentrations are known then the matrix corrections are immediately calculable; if they are not known they may be determined by iterative solution of Equation 1, as described later.

The bremsstrahlung emitted by the decelerating protons is of negligible intensity, in marked contrast to that of the decelerating electrons in EPMA; the reason is the inverse proportionality of bremsstrahlung intensity to square of projectile mass. However the secondary electrons ejected in proton-atom collisions emit secondary bremsstrahlung and this is the determinant of PIXE's limits of detection (LODS). Figure 3 compares EPMA and PIXE bremsstrahlung backgrounds for a given specimen, demonstrating PIXE's superior detection limits. In classical terms the maximum energy of the secondary bremsstrahlung is the maximum energy transferable from a proton to an essentially free electron, viz $4(m_e/m_p)E_0$; for 3 MeV protons this is 6 keV. The background at higher x-ray energies is determined by a few particular light elements such as Na, Al

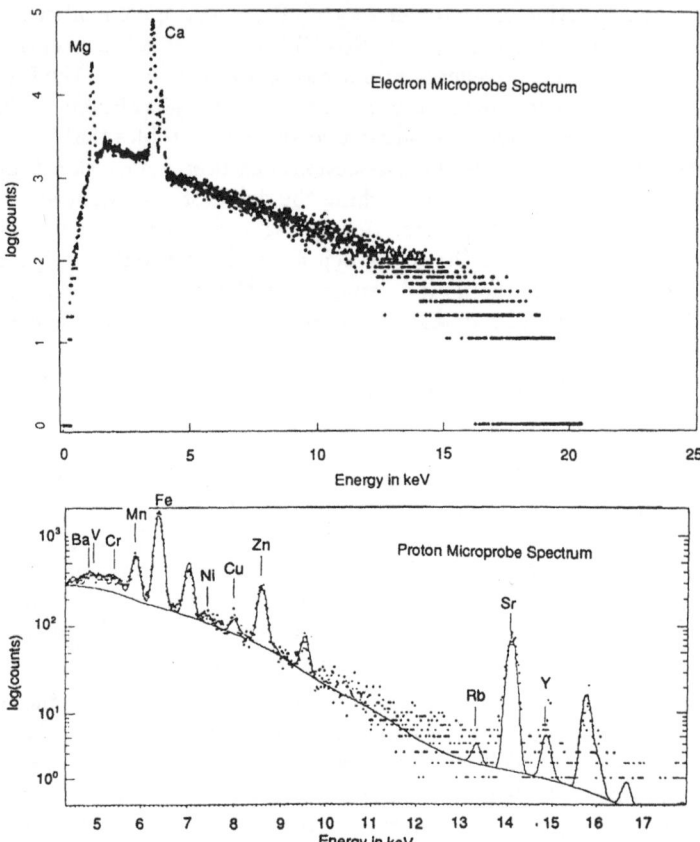

Figure 3. EPMA *and* μ-PIXE *spectra of a dolomite specimen (Watt et al. 1987). By permission of Elsevier Science Publishers.*

and F which undergo (p, γ) nuclear reactions, and the emitted gamma ray background can worsen the LODS considerably in PIXE. Finally the specimen undergoes heating, and so attention to limiting the beam current is necessary.

3   Energy dispersive x-ray spectroscopy

Trace element analysis demands the maximum geometric efficiency for x-ray collection and so wavelength-dispersive detection using crystal spectrometers is not an option. A large-area energy-dispersive detector, which records the x-rays of all elements present in a single measurement, has to be situated as near the specimen as possible (1–2 cm). The typical 30mm^2 collecting area of Si(Li) x-ray detectors is adequate, and their energy resolution is excellent, providing clear separation of the Kα x-rays of elements differing by one unit in atomic number. In our experience, the efficiency gained by using an 80mm^2

detector is worth the slight worsening of energy resolution that is incurred. Si(Li) x-ray detectors are discussed in some detail elsewhere in these proceedings. Their beryllium windows permit the detection of K x-rays of elements down to sodium; since scattered protons must be prevented from damaging the detector, the window sometimes has to be supplemented by a Mylar absorber. The detector is coupled through an amplifier to an ADC, usually housed on a board in the PC controlling the instrument. The individual peaks in the pulse height spectrum recorded by the ADC are essentially Gaussian in shape. The mean pulse size or peak position in the spectrum is proportional to x-ray energy, and the peak width increases linearly with the square root of x-ray energy, in accord with the statistics of electron-hole formation in the silicon crystal. The spectra are rendered complex by the fact that each element's K x-ray series contains about 6 lines and its L series over 20, but the relative intensities within these series are well-known, which facilitates dealing with inter-element overlaps.

For a given proton energy the x-ray yield per unit concentration falls rapidly with the Z of the element concerned (reflecting the behaviour of the ionisation cross-sections). This Z dependence of the yield can be reduced somewhat by inserting a thin absorber (aluminium is most frequently used) between specimen and detector. Most often in PIXE interest focuses on high-Z trace elements in a lower-Z matrix. An example is silver at around 100 ppm.wt. level in the mineral sphalerite (ZnS); in this case the intensity ratio of zinc K x-rays to silver K x-rays would be about 5×10^5, and thus in a 10 minute analysis with a detector counting rate of $5000s^{-1}$, only six silver K x-rays would be detected, versus three million zinc x-rays; the silver K x-ray peaks would not be discernible. Insertion of an aluminum filter of 0.75 mm thickness reduces the undesired zinc intensity by a factor 3000 while transmitting over half of the silver K x-rays. The silver K x-rays now constitute a much greater fraction of the throughput, and the beam current may be increased to restore the overall data acquisition rate; the resulting spectrum (Figure 4) therefore contains sufficient silver data to provide analysis with detection limits of a few ppm.

High counting rates necessitate corrections for electronic dead time and also cause spurious spectral peaks due to pulse pile-up. An elegant solution to both of these problems is to have a fast electrostatic deflector element remove the beam from the specimen whenever an x-ray is being processed and return it when the electronic system is again ready. This also minimises specimen damage.

4 Quantitative analysis

Although Equations 1 and 2 offer an absolute approach, the data base is not perfect; determination of detector efficiency (which varies among devices of the same nominal dimensions) is complex and tedious, and absolute measurement of integrated beam charge when using nanoamp proton currents is a matter for experts. For these reasons analysis is invariably conducted relative to standards.

With thin specimens, where there are no matrix corrections, a suite of commercial thin-film standards evaporated on Mylar backings provides a direct measure of the quantity

$$k(Z) = \sigma_Z(E_0)\omega_Z b_Z t_Z \Omega \epsilon_Z^i N_{av}/4\pi A_Z \tag{3}$$

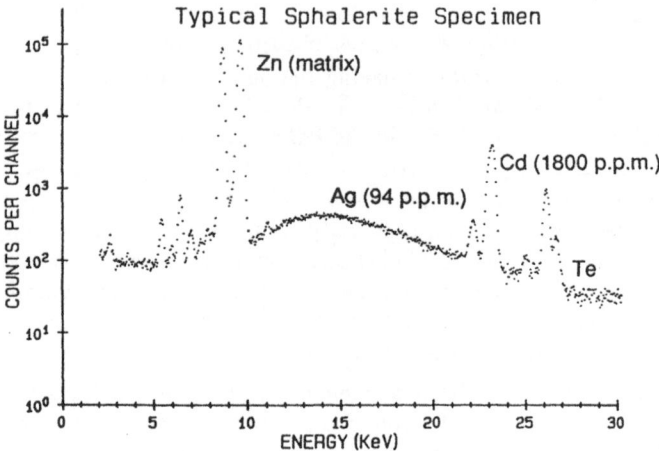

Figure 4. μ -PIXE *spectrum of a sphalerite (ZnS) mineral containing 94 ppm of silver.*

and the elements in the specimen are then handled by the equation

$$I(Z) = N_p k(Z) m_a(Z) \tag{4}$$

With thick specimens the constituent elements not only emit their own characteristic x-rays but affect (via proton slowing and x-ray attenuation) the intensities emitted by all other constituents of the matrix. Thus the measured intensity for element Z is

$$I(Z) = f(C_Z, ME, IE) \tag{5}$$

where ME and IE represent matrix and instrumental effects respectively. Following Newbury and Trace (1986) one has four possibilities. At the empirical extreme, working curves relating $I(Z)$ to C_Z may be obtained if one has standards with the identical matrix to the specimens and containing the same trace elements; however unless one's interests are very narrowly defined this is too restrictive. Some flexibility is gained by determining the ratios $I(Z)/C_Z$ relative to some major constituent whose concentration is already known, thus getting a degree of matrix effect cancellation; however multi-element standards are still needed. Fortunately the data base is known with sufficient accuracy that the matrix effects ME can be calculated accurately from fundamental parameters, either in absolute terms or as a ratio between standard and specimen; this removes the need for the standard to closely mimic the specimen and allows adoption of some broadly representative but minimal set of standards. For the reasons given earlier we reject taking this argument to the extreme of a standardless analysis.

Pursuing the third avenue and adopting the micro-Coulomb (μC) as the practical unit of integrated charge, we define $I_1(Z)$ as the theoretical x-ray intensity per μC per steradian and per unit of concentration (usually ppm); computation of $I_1(Z)$ is a straightforward exercise incorporated in various PIXE software packages. Then the measured intensity for Q μC of beam charge is

$$I(Z) = Q \Omega \epsilon_Z^i t_Z C_Z I_1(Z) \tag{6}$$

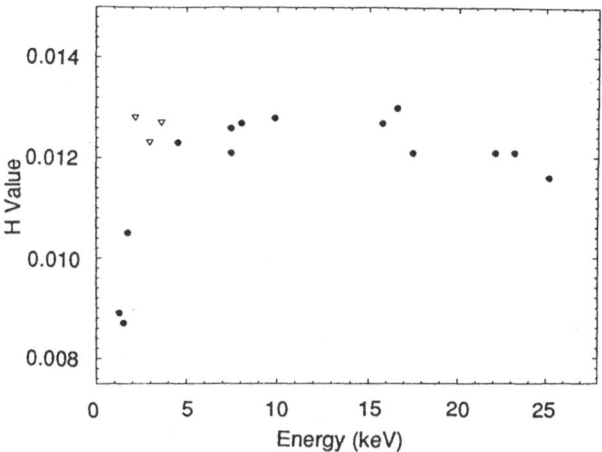

Figure 5. *Measured values of the H parameter (see Equation 7) for a particular μ-PIXE system.*

Note that a direct measurement of Q demands that the specimen be conducting; insulating specimens must be carbon-coated. However Q can also be measured indirectly, *e.g.* via the intensity of beam particles scattered into a surface barrier detector by a graphite vane rotating at a few Hz frequency through the beam; this implies a conversion factor f_Q between this instrument's response and Q. Even in direct measurement of Q there may need to be a corrective factor for practical reasons, for example to allow for charge leakage or the incomplete suppression of secondary electron emission. Determination of these factors can be averted by combining f_Q and solid angle into a single 'instrumental factor' H that characterises a given PIXE system: thus

$$I(Z) = H\epsilon_Z^i t_Z C_Z I_1(Z) \tag{7}$$

In practice, measured x-ray intensities from pure element standards, using a very carefully characterised Si(Li) detector, indicate that H is indeed constant (Figure 5) over a wide range of x-ray energies. But small errors in the data base (which are probable for elements of low atomic number), together with errors in the physical characteristics of the detector or in the description of its lineshape, can cause departure of H from constancy for x-ray energies below 2 keV and above 20 keV. In the example shown in the figure, the fall-off at high x-ray energy suggests that the nominal detector thickness of 3 mm is a slight over-estimate, resulting in an over-estimate of intrinsic efficiency and a corresponding slight under-estimate of H. It makes sense then to adopt the overall $H(E)$ relationship (H vs. x-ray energy E) as a full calibration that caters automatically for these uncertainties (as opposed to undertaking laborious studies of the detector properties). Note that this is equivalent to the k-ratio method of EPMA with the benefit that H is the same for most elements concerned.

The choice of standards depends upon the context. In alloy analysis in our own laboratory we use pure metal single-element standards. In analysis of trace and minor elements in mineral grains we use a set of synthetic sulphide mineral standards containing known trace levels of several mineralogically important elements.

Solution of Equation 7 is straightforward when the major elements and their concentrations are known (*e.g.* from prior EPMA analysis or from stoichiometry), since then the matrix stopping power and attenuation terms within the integral are immediately calculable. If the major element concentrations are not known, an iterative solution of Equation 7 is needed, with the matrix corrections being recomputed at each iteration; since the major elements sum to 100% concentration, one 'invisible element' (a low Z element whose x-rays are not detectable, such as oxygen) may be included in this procedure. In either case there is the chance of secondary fluorescence of particular elements by the intense proton-induced x-rays of the major elements. This contribution to the total x-ray intensity is computed from the same data base, supplemented by photo-electric absorption cross-sections. Many published analyses of standard reference materials and comparisons with EPMA analyses show that these procedures are accurate to a few percent.

5   The atomic physics data base

The standardisation method recommended depends heavily upon the existence of a reliable data base. The accuracy of this data base has been reviewed in detail elsewhere (Campbell *et al.* 1990), hence only brief remarks are needed here. Proton stopping powers $S(E)$ are known to ±3% over the relevant energy range. Reference tables of $\sigma_Z(E)$ for the K shell are accurate to approximately ±2%; ω_K and b_K are known at the ±1% level above $Z = 15$. The L-shell case is not yet satisfactory; inadequately known L-shell vacancy parameters such as Coster-Kronig yields result in uncertainties in the three subshell ionisation cross-sections deduced from experimental studies of proton-induced x-ray yields, and the relative subshell cross-sections seem to deviate from theory, which is not the case for the K shell.

The various semi-empirical schemes that have developed in the EPMA arena for quick provision of attenuation coefficients differ considerably, which has serious implications for PIXE given the proton penetration of tens of microns. Our preference is to store on disk the theoretical attenuation coefficients for all K and L x-ray lines in all elements; these numbers are obtained from the NIST computer code XCOM (Berger and Hubbell 1987), whose authors provide useful estimates of accuracy. Fortunately, the simplicity of PIXE's intensity-concentration relationship permits an easy calculation of the intensity error resulting from a given attenuation coefficient error. The results, shown in Figure 6, are almost matrix-independent, and they provide a rapid estimate of analytical error which, as would be expected, increases with the attenuation coefficient of the matrix.

In general the quantity $I_1(Z)$ can be computed to better than 5% accuracy for K x-rays, and significant cancellation of errors (*e.g.* in ionisation cross-sections) between specimen and standard results in the data base contribution to analytical error being markedly less than ±5%. Spectrum fitting, counting statistics, specimen homogeneity and beam-specimen alignment are likely to be the more important contributors to error.

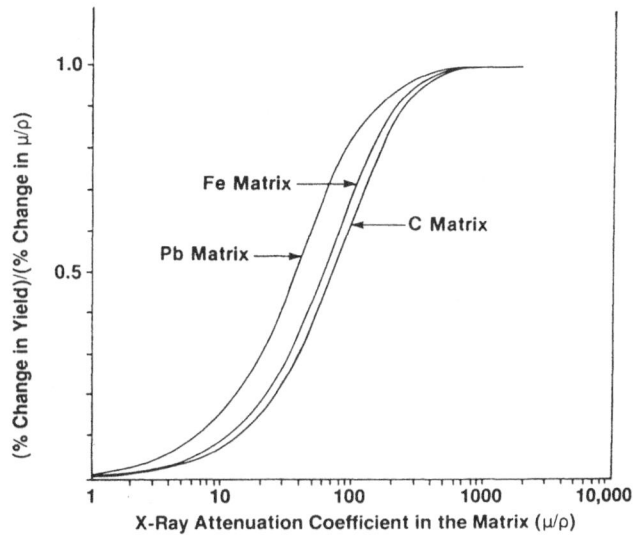

Figure 6. *Sensitivity of x-ray yields from 3 MeV protons to changes in attenuation coefficient in the specimen (Johansson et al. 1988). (By permission of Elsevier Science Publishers)*

6   Examples of PIXE

PIXE with proton beams of a few mm diameter competes with conventional X-ray fluorescence analysis and has found some important niches, usually involving thin specimens. But its most profound impact is via microbeams (see below) so we limit ourselves here to the single example of elemental analysis of atmospheric particulate collected on filter materials through which air is passed in sampling devices of various kinds. Figure 7 demonstrates the wide range of elements detected in an aerosol sample, the detection limits for transition metals being less than 1 nanogram per cubic metre of sampled air.

Conventional high-volume samplers collect rather large quantities of material on thick filters, which themselves have significant and variable metal content. In contrast PIXE needs only 0.1–1 milligram of material, which can be collected on a very thin filter (*e.g.* teflon, Nuclepore) which has a small trace element content. This can be analyzed by PIXE with zero preparation, providing a broad range of elements in a single non-destructive analysis.

This small-sample capability has facilitated sampling at remote sites where the air is reasonably pure, and where the particulate composition both provides information on long-range dispersal and may be correlated to visibility measurements—as in the U.S. IMPROVE network of some 60 samplers in national parks and wilderness sites (Eldred *et al.* 1989). A global network of IMPROVE type samplers, with PIXE as the principal analysis technique, is now being formed. We are using this approach to study long range transport of sulphur in fine particulate in Canada.

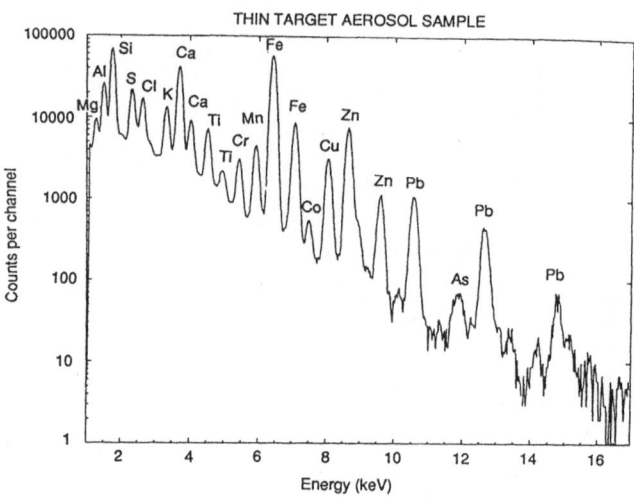

Figure 7. PIXE *spectrum of air particulate collected on a thin membrane filter.*

New sampler types have been invented to exploit PIXE's small-sample capability. The streaker sampler periodically moves its collecting substrate so that a linear or circular track of small deposits is created, providing time-dependent sampling. This can be used to cover long periods of time in remote locations or to sample concentrated urban air pollution with very short time resolution (minutes to hours). Other short-term sampling devices find application in industrial hygiene applications, *e.g.* in sampling the aerosol in welding shops, mines *etc.* The distribution of elements among particles of different sizes is important. Cascade impactors separate the particulate into typically six size ranges, resulting in even smaller samples. PIXE becomes very powerful in this context. In these small-sample cases, neither neutron activation analysis nor atomic absorption is feasible.

7   Specimen analysis software

As Figure 7 shows, PIXE spectra can be very complex. In extracting the areas of the principal x-ray lines of each element present, one faces the problems of estimating the continuous background and of separating overlapping peaks. Invariably the approach is that of non-linear least-squares fitting, where the parameters of a model spectrum are adjusted until an optimum fit to the measured spectrum, as indicated by the chi-squared criterion, is achieved.

For each element the relative x-ray line intensities are known from the data base and can be corrected for differential effects such as matrix attenuation, absorber transmission and detector relative efficiency. Hence only the height of the principal line (*e.g.* Kα) of each element is adjusted in the fitting process, and the minor lines then track the main line in the appropriate intensity ratio. With a perfect detector the individual

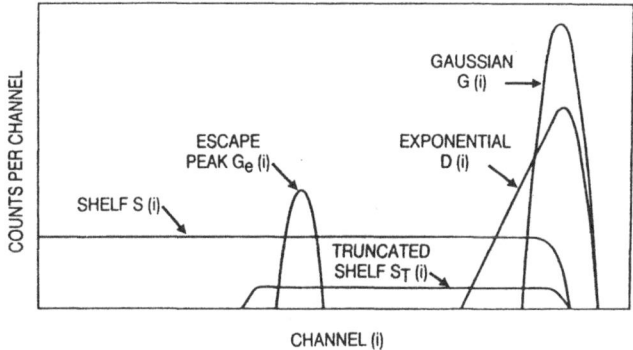

Figure 8. *Features of the lineshape (response function) of a Si(Li) x-ray detector.*

lines would have Gaussian shape but in practice detector imperfections and x-ray scattering en route to the detector cause degraded events which appear at lower energies in the spectrum. These can be modelled by suitable combinations of exponential tails and flat shelf-like features, as shown in Figure 8, and the parameters can be found by fitting the spectra of single elements; these parameters depend smoothly upon x-ray energy and thus need not be determined for every element. Incidentally, these tailing features increase the overall background and worsen limits of detection (LODS) for low-concentration elements whose x-rays lie upon the tails of high-concentration elements; the yttrium peak sitting on the scatter tail of zirconium in the zircon spectrum of Figure 9 is a good example.

The background continuum may also be modelled, and various quite complex analytical expressions have been used for this purpose; the underlying physics is too complex for there to be an empirical expression that is universally appropriate. This introduces several additional parameters which must also be fitted. Some workers prefer to strip out the background by a mathematical operation. Channel by channel convolution with the top-hat filter that is popular in EPMA completely removes a linear background. PIXE backgrounds are not linear but do vary slowly and smoothly; it follows that this approach will be effective with a spectrum in which the background is close to linear in any local region of width comparable to the filter size, which is typically 100–300 eV.

Various spectrum fitting codes are available. A recent comparison (Jaksic *et al.* 1991) of the results generated by two of the most widely used codes for various sample types showed good agreement despite the use of different background approaches, data bases *etc.* A potential source of error in any code is the superposition of a weak x-ray line on the low energy tail of a neighbouring intense line (again see the Y peak on the Zr tail in Figure 9), which reinforces the need for accurate analytical description of the non-Gaussian features of lineshape.

These PIXE routines, such as our own GUPIX program (Maxwell *et al.* 1989), run typically on microcomputers of the IBM-compatible 386 or 486 variety, and combine the tasks of spectrum fitting and conversion of peak intensities to concentrations. Full

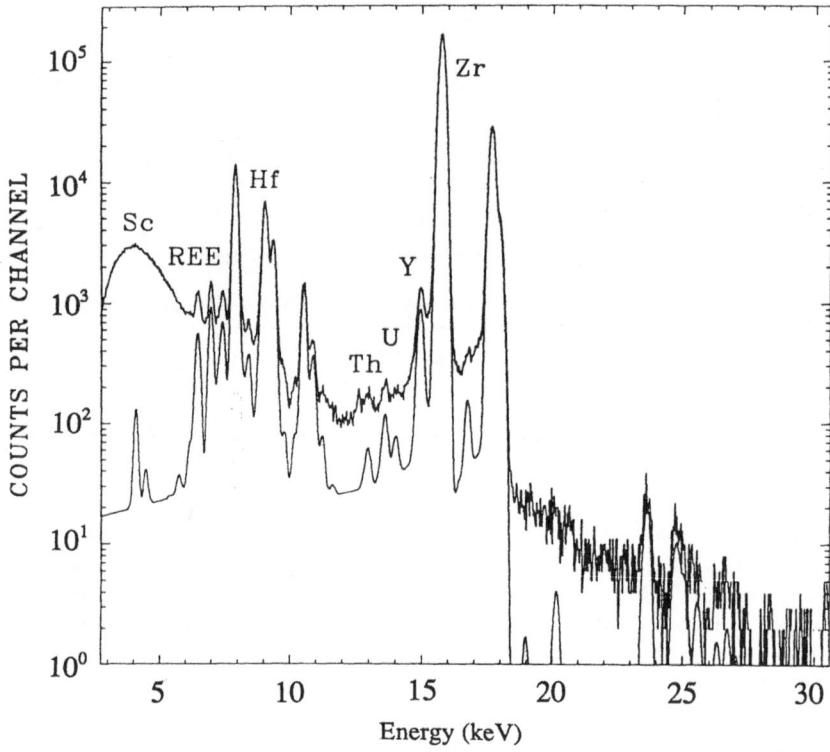

Figure 9. *μ*-PIXE *spectrum of a zircon.*

treatment of a complex spectrum takes about 2 minutes. In the GUPIX case the program size is about 0.7 MB and it can handle 30 elements with 600 x-ray lines. The user inputs the information pertaining to his own x-ray detector (dimensions, lineshape parameters, *etc.*) and standardisation (H values).

8 Micro-PIXE

It was at the UKAEA's Harwell Laboratory that Cookson *et al.* (1982) demonstrated that a proton beam could be focused to micron dimensions, with sufficient beam current to conduct PIXE and other types of ion beam analysis. This development was effectively the birth of a proton microprobe which offered x-ray emission analysis with the same spatial resolution as the electron microprobe and also, in many cases, markedly superior limits of detection. Subsequent technical progress is sketched in Section 9. Two examples illustrate the early recognition of this potential. In mineralogy EPMA had long been the technique of choice for measuring precious metal content in grains of sulfide minerals, the typical limits of detection being 300–500 ppm. Cabri *et al.* (see Johansson 1988) demonstrated with the Heidelberg proton microprobe that the corresponding μ-PIXE limits of detection were about 5 ppm, an observation that triggered the rapid development of μ-PIXE as a geochemical tool. Watt and Grime (1987) developed a raster-scanning capability on the Oxford microprobe and generated elemental images of the so-called Alzheimer's plaques, which were attracting interest because of the EPMA observation of aluminium as a constituent; the images revealed not only aluminium and silicon but also vanadium, chromium, manganese and titanium at lower concentrations. Whatever the explanation of the presence of these elements in the specimens, the power of μ-PIXE for imaging trace element distributions across a biological specimen was convincingly demonstrated.

Quantitative analysis with a stationary beam spot follows the formalism of Section 4, and again the preference in general is to employ standards. However various PIXE workers have reported inhomogeneity problems with standards. Many standard materials suitable for bulk analysis techniques are inhomogeneous at the micron level, which poses problems for both PIXE and EPMA. An excellent discussion of the varying homogeneity of various geological standard reference materials is given by Ryan *et al.* (1990). This is one reason for our own preference for synthetic mineral standards, although even there we have encountered occasional inhomogeneity problems. Another way to cope with inhomogeneous standards is to raster the microbeam over a an area of approximately 1 mm^2, in the hope of averaging over sufficient material.

9 Scanning proton microprobe: technology and instrumentation

The solenoidal lenses that focus low energy electrons create insufficient magnetic field to focus MeV energy protons, and so resort is made to the more complex quadrupole focusing systems. Since a quadrupole magnet focuses an axially transmitted ion beam in one plane but defocuses in the other, a pair of quadrupoles constitutes the minimum configuration needed to create a demagnified image of the object; this is a small aperture, usually defined by two pairs of vernier-adjustable tantalum jaws, one defining the horizontal and the other the vertical width. Most of the beam extracted from the accelerator is stopped by these jaws; the current transmitted is a few nano-amperes and the aperture structure is water-cooled to cope with the micro-ampere current that it intercepts. The aperture must have very smooth edges to avoid beam scattering which would create a 'halo' around the emerging axial beam.

The quadrupole lenses must be constructed and aligned with high accuracy. Construction imperfections result in the contamination of the field by higher multipoles. Two approaches have been taken to minimise these parasitic aberrations, which cause defocusing of the beam. The first (Grime and Watt 1987) employs computer-controlled machining of a single piece of iron to create a very high precision magnet of large bore, which consequently has high power consumption and requires water-cooling. These devices have produced beams as narrow as 0.3 μm diameter with sufficient current (100 pA) to conduct μ-PIXE analysis; no parasitic aberrations have been detected and the limits on decreasing beam size appear to be residual vibrations and thermal expansion effects due to temperature changes. The alternative approach involves a conveniently small air-cooled doublet constructed as a single package (Martin Goloskie 1988). Intrin-

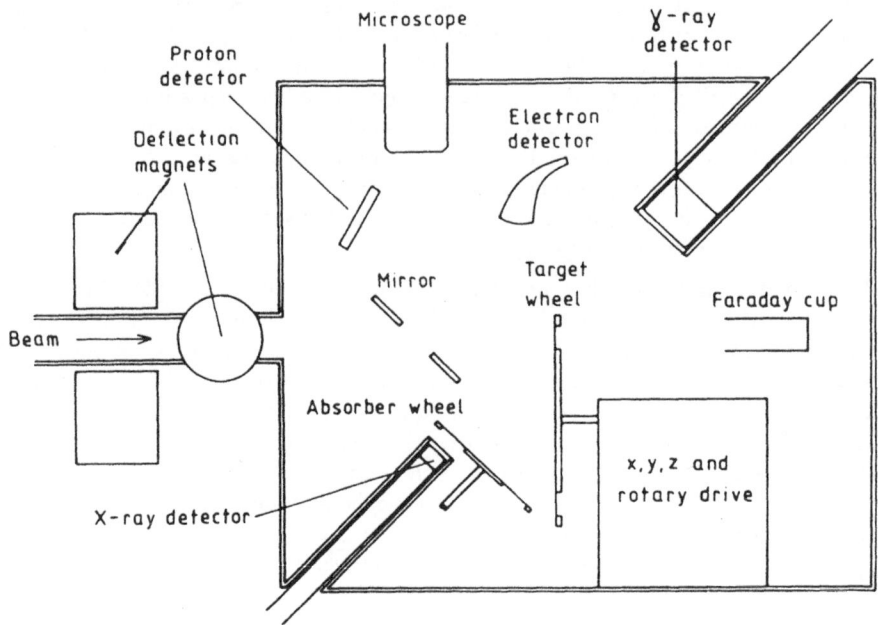

Figure 10. *Schematic of the instrumentation comprising a typical µ-PIXE specimen chamber. (Copyright © 1988 John Wiley & Sons Ltd., reprinted with their permission)*

sically, this smaller non-unitary device is less tolerant of machining and construction imprecisions; a set of eight electrodes is therefore incorporated in a configuration that permits one to generate an electric quadrupole field to compensate any alignment error, and also to cancel by electric dipole fields any magnetic sextupole aberrations (which may change as the device ages). An elegantly configured power supply makes the adjustment of the various electric and magnetic fields a straightforward task.

There are of course intrinsic aberrations in addition. Poor accelerator voltage stability will cause variable proton energy and hence chromatic aberrations. Spherical aberrations result when protons traverse the quadrupole field at an angle to the axis; this necessitates limiting the entry angles by a pre-lens aperture, which in turn is a limiting factor on current. The aperture also blocks off the outer field regions where higher order multipoles are most significant.

The specimen chamber is highly instrumented, as suggested schematically by Figure 10. An xyz specimen stage is driven by computer-controlled stepping motors or inchworms. A high-power optical microscope with a CCTV camera views the specimen and often the beam impact point is visible. A negatively-biassed Faraday cage returns secondary electrons to the specimen to ensure accurate measurement of the total beam charge via a charge digitiser. A close-in Si(Li) x-ray detector is often supplemented by detectors for scattered protons and nuclear gamma rays.

In the scanning version of the microprobe, the beam passes through a scanning element comprising either ferrite core magnets or mutually perpendicular pairs of planar electrodes which raster the beam across a defined area of the specimen. With a wide-bore quadrupole the scanning element may be placed ahead of the magnet; with the small magnetic doublets it has to be sited after the magnets since the beam would be defocused in the peripheral field regions. When the Si(Li) detector records an x-ray event, the digitised energy is stored along with the x, y co-ordinates of the beam spot. The 'list' of events resides on disk and may be played back during and after the measurement to construct images of elemental distributions in two dimensions.

For special applications the beam may be directed through a very thin window into the laboratory atmosphere or into a large helium-filled chamber. This enables the analysis of items too large to fit within the conventional specimen chamber or too delicate to withstand the heating that occurs in vacuum. The main relevance is to works of art, documents and archaeological artifacts.

10 Ancillary analytical techniques

The scanning proton microprobe (SPM) is usually equipped with a channel electron multiplier to detect secondary electrons; these are so copious that the channeltron can be several cm distant from the specimen. Secondary electron images can be useful both in diagnostic tests of the beam profile and in establishing the topography of very thin specimens such as biological tissue slices.

Beam particles backscattered from the specimen may be detected, usually at 170°, in an annular silicon surface barrier detector. This is especially useful with helium ion beams, where the interaction is simple Rutherford elastic scattering where the kinematics are precisely defined. With a thin specimen and a defined scattering angle, the scattered proton energy is a well-defined function of the scatterer's atomic number, and the intensities of the resulting peaks provide concentrations of major elements; with a thick specimen the energy loss of the scattered ion in the specimen draws the peak out into a shelf that runs from a defined energy down to zero. Since the elemental resolution is much poorer than in PIXE, the direct use of helium Rutherford back scattering (RBS) has been limited to a few special applications. However, proton scattering has proven to be a powerful complement to scanning μ-PIXE analysis of thin specimens.

This is illustrated in Figure 11, which is the energy spectrum of backscattered protons from a thin specimen of plant material . The flattening of the various peaks indicates the small degree of energy loss of the protons due to traversing the specimen material. Although the scattering cross-section is not Rutherfordian, a sufficiently large data base of proton reactions exists that there is a spectrum simulation program available (Jaksic 1991) which provides an excellent fit to the energy-distribution of the scattered protons. This enables a determination of the major biologically important elements of low atomic number that are not detected by PIXE. Equally important, it provides an accurate measure of the specimen thickness. Since a scanned specimen may have variable thickness, the scattering data provide a means of normalising PIXE x-ray intensities observed in scanning and the result is a significantly sharper image of the distribution of these elements.

A recent addition to the proton microprobe arsenal is scanning transmission ion microscopy or STIM. The bright field version of STIM measures the proton beam intensity

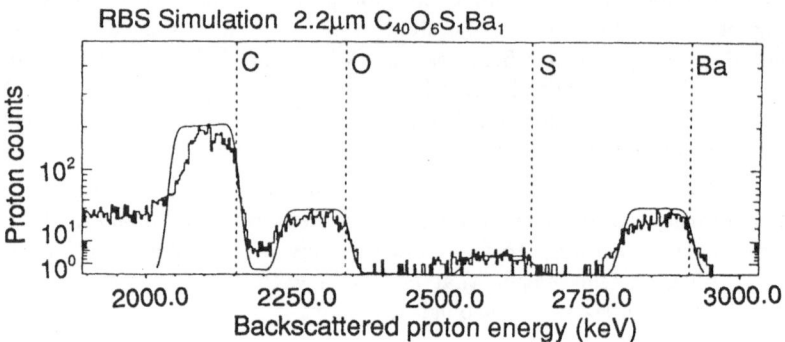

Figure 11. RBS *spectrum of a thin plant tissue specimen (Watt et al. 1991). (By permission of Elsevier Science Publishers.)*

transmitted without significant lateral scattering through a specimen of less than 50 μm thickness. This can be done with high efficiency by placing a particle detector behind the specimen; the benefit of recording all transmitted ions is that the two beam-defining slits may be closed down, reducing both the direct image size and the aberration contribution, which results in a spatial resolution of as low as 50 μm. Windows placed on the spectrum regions corresponding to different energy loss then provide two-dimensional images, the variations in which reflect either differing thickness or differing elemental content. An image of the mean energy loss is analagous to a high resolution x-ray microradiograph.

11 Examples of quantitative μ-PIXE analysis

Since the number of successful applications of μ-PIXE is now large, only a few specific examples are chosen here.

11.1 Mineralogy

The first two examples concern the quantitative analysis of individual small mineral grains of dimensions tens of microns; relatively broad beams (10–20 μm) are preferable to ensure homogeneous sampling and minimise heating damage. The current is typically a few nanoamps and 100–200 grains can be analysed in a day. Prior to the recovery of metals from ore bodies it is necessary to identify the particular mineral species that contain the metals of interest; then the appropriate physical processes may be selected to concentrate these species. Silver, gold and platinum group elements (PGEs) are extracted from massive sulphide deposits, and EPMA has long been the preferred technique for *in situ* analysis of individual grains to ascertain which species carry the metals of interest in solid solution. For silver, palladium, rhodium *etc.* in sulphides such as sphalerite, pyrite, chalcopyrite *etc.*, EPMA provides LODS of 300–500 ppm. The

μ-PIXE work at Heidelberg in the early 1980s demonstrated LODS of about 5 ppm for silver. In short order μ- PIXE showed that the known major silver carrier galena (PbS) was supplemented by chalcopyrite, sphalerite and pyrrhotite as minor carriers. Figure 4 shows a typical PIXE spectrum. Since concentrations of a few ppm of these metals are viable in economic terms it is important to have a technique which extends EPMA's concentration range down to the ppm level and which can be deployed on the same specimens in non-destructive manner. The overlap in concentration ranges of the two techniques has facilitated extensive cross-checking, and excellent agreement between them has been demonstrated. Another advantage of μ-PIXE is its ability to reveal the presence of small concentrations of elements deleterious to the refining process. The presence in sphalerite of small concentrations of selenium and arsenic, below the LODS of EPMA, is a case in point.

Another example of commercial relevance in geochemistry is the use of μ-PIXE to correlate the nickel content of garnets with formation temperature (Griffin *et al.* 1989). The typical Ni range is 10–100 ppm, well within μ-PIXE's capability (LOD 5 ppm) but below that of EPMA (LOD about 300 ppm). This establishes a geo-thermometer whereby the nickel content of potentially diamond bearing formations may indicate if the temperature of formation was indeed high enough for the creation of diamonds.

Oscillatory zoning, which results from mineral growth under conditions far from equilibrium, has been studied by using backscatter electron and EMPA imaging or major element distributions through the zones. μ-PIXE supplements these by imaging also the trace elements, whose variation may be decoupled from that of the major elements. Once identified in a two-dimensional image, zoning may be probed more precisely by a line scan across appropriate zones. Line scans are especially useful for probing boundary layers, as illustrated by Figure 12, for a titanate grain from a syenitic igneous intrustion; Nd, Th and Pb appear to be on the surface of the grain, suggesting that they are remnants of a fluid, enriched in incompatible elements, which migrated along grain boundaries.

11.2 Biology and medicine

Materials analysed in this section include bone, teeth, tissue, eyes and individual blood cells. Figure 13 shows a two-dimensional μ-PIXE image of an individual osteon from a worker exposed to lead and twice reported as showing clinical symptoms of lead poisoning. Bone sections were freeze-dried and microtomed to produce specimens for μ-PIXE (Johansson 1988). Localised deposition of lead is well-defined.

With thin specimens in the scanning proton microprobe the combination of analysis techniques sketched in Section 10 is very powerful. Secondary electrons provide specimen topography; scattered protons determine specimen thickness and the concentrations of the lighter major elements. PIXE provides minor and trace elements, the concentrations being normalised via the thickness data. Spatial resolutions below 0.5 μm have been reported and LODS in biological tissue are typically 1 ppm. The Oxford group refer to the integrated set of simultaneous analyses as scanning nuclear microscopy (SNM), and their work (Landsberg *et al.* 1991) on senile plaques demonstrates both the potential of the SNM and the practical difficulties in the study of elemental distributions in the particular case of biological tissue.

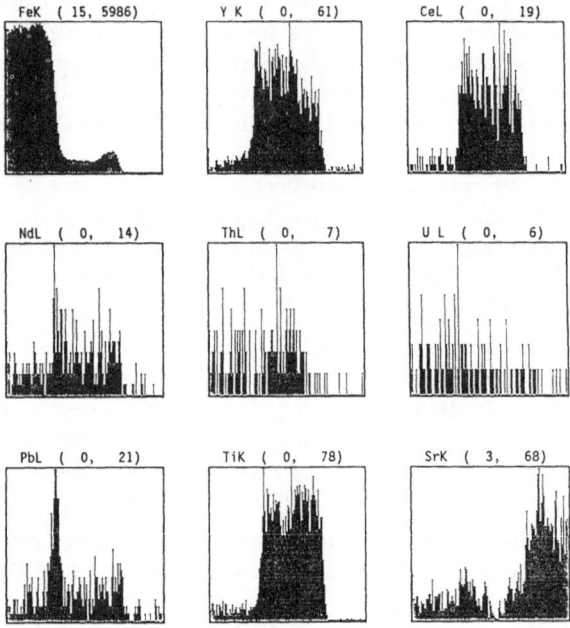

Figure 12. *μ-PIXE line scan across a titanate mineral grain.*

The suggestion that over 70% of mature-core senile plaques from Alzheimer's disease patients contain aluminum and silicon has stimulated debate on the possible role of these elements. The advent of STIM has made it possible to locate plaque cores without prior staining. This is because the senile plaques are composed mainly of amyloid protein, which is denser than the surrounding tissue and causes higher proton energy loss. It has been shown (Breese *et al.* 1992) by scanning stained tissue that the STIM images do specifically identify amyloid distributions; this provides the foundation for the first microanaylsis (by PIXE/RBS) of plaques that have undergone no chemical pretreatment. This combination of proton microprobe techniques has the potential to resolve the question of whether Al and Si are implicated in Alzheimer's disease.

11.3   Materials science

Applications here are diverse, but are of course dominated by those that meet PIXE's necessary precondition of a low Z matrix with the trace elements of interest having higher Z. An example of this genre is supplied by work (Hinrichsen *et al.* 1990) at the Montreal proton microprobe on trace impurity distributions in the polyethylene insulation of high-voltage underground cables. These cables deteriorate with age, and internal tree-like structures are implicated in premature failure. Water-trees appear to grow from voids or impurities, and electrical trees are breakdown channels of length 50–200 μm. It is important to understand the role, if any, of contaminants in nucleating

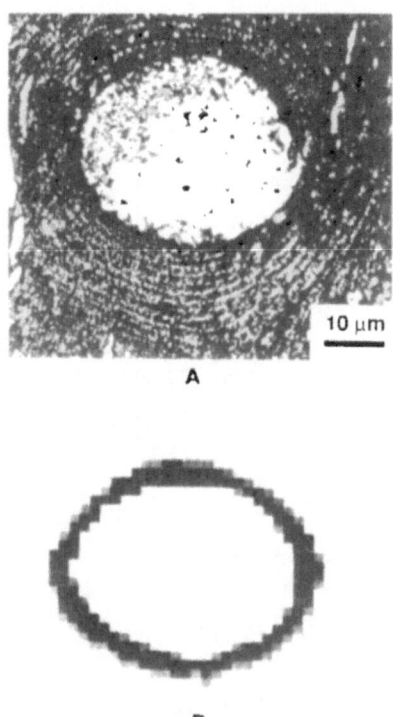

Figure 13. *Photomicrograph and μ-PIXE map of lead distribution in an individual osteon (Johansson and Campbell, 1988). By permission of Elsevier Science Publishers.*

these trees; contaminants may be introduced deliberately in the polymerisation process, may arise from inclusion of dust particles, or may result from diffusion of elements from the inner and outer conductors during elevated temperature incidents caused by voltage surges. Figure 14 shows radial concentration profiles of various elements in two electrical trees adjacent to a water tree that extends radially outwards from the inner conductor. The fact that sulphur and silicon, which are introduced in the polymerisation, have uniform concentration profiles, lends confidence to the remaining data, which indicate dramatic concentration gradients for various other elements, *viz* Cl, Na, Mg, Ca and K. In the first tree the strong decrease at 600 μm suggests that these elements have migrated radially inwards from the outer shield. Two-dimensional μ-PIXE imaging is now being used to better understand of the role of trace elements in the tree structures.

In materials science, specimens of interest often have a layered structure. Helium ion RBS is usually the technique of choice for layered specimens but there are examples where PIXE is optimally suited. They rest on the fact that the x-ray intensity from a buried layer may be very sensitive to proton slowing and to attenuation of substrate x-rays in the covering layers. A good example of this is the work of Demortier *et al.* (1990) on the determination of the thicknesses of nickel and zinc coatings on iron substrates.

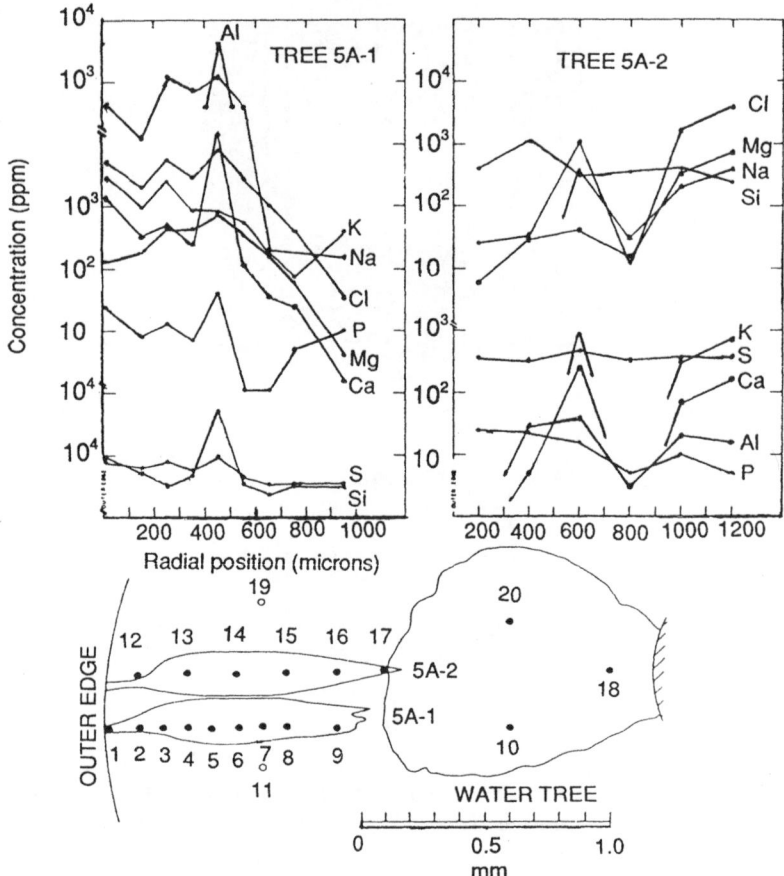

Figure 14. *Radial concentration profiles measured with μ-PIXE for two electrical trees which are adjacent to a water tree in the insulation of a field-aged high voltage cable (Hinrichsen et al. 1990). (By permission of Elsevier Science Publishers.)*

Helium ion RBS is an excellent method for coatings up to about 15 μm thickness; PIXE extends this range up to 30 μm. The quantity measured is either the $K\beta/K\alpha$ intensity ratio of the substrate x-rays, or the intensity ratio of the $K\alpha$ x-rays of surface layer and substrate. Figure 15 illustrates the marked dependence of these ratios upon the coating thickness, which translates to an accurate measurement of thickness.

A more complex example is found in films of the semiconductor cadmium mercury telluride (CMT), which are of interest for solar energy conversion. A potential method for inexpensive manufacture of large film areas is electro-deposition. We have used PIXE to characterise CMT films electro-deposited on glass which is first coated with an indium-tin-oxide layer. The band gap and the performance of the resulting device are dependent upon the stoichiometry of the CMT film. PIXE spectra were taken of the glass substrate,

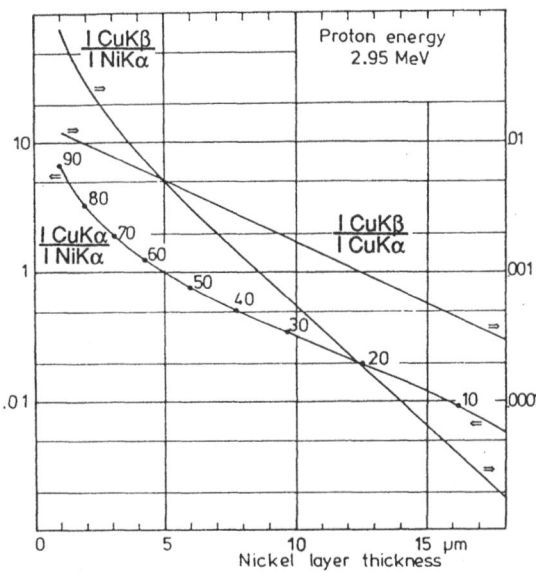

Figure 15. *Dependence of x-ray intensity ratios upon the thickness of a nickel layer on a copper substrate (Demortier et al. 1990). (By permission of Elsevier Science Publishers)*

glass plus ITO and glass/ITO/CMT, and processed by a version of the GUPIX code which caters for multiple layers. The analysis determined successively the thicknesses of the deposited layers via their attenuating effect on the barium L x-ray intensity from the glass. Once the thickness of each was known, the elemental composition was determined. Monotonic changes in band gap had been observed through the sequence of depositing twelve films from a given electrolyte. The PIXE analysis produced the data of Figure 16, showing that the process used was unable to guarantee a constant stoichiometry, and suggesting the need for rather complex means of maintaining the concentrations of various components of the electrolyte solution.

11.4 Environmental Science

The growing role of PIXE in air particulate analysis has already been described. There exist many fairly routine applications of PIXE's capability in the analysis of water residues and estuarine sludge, including, for example, the study of metal diffusion and sedimentation in the Venice Lagoon. These invariably use the broad-beam form of PIXE on a pelletised specimen of solid material or a thin target produced by chemical extraction techniques. The proton microprobe however makes possible some interesting variants. We have studied polymer-impregnated thin specimens of soil taken from an area on which sewage sludge was applied annually. Photo-micrographs indicated the position of vertical earth worm channels and the extent to which their surroundings

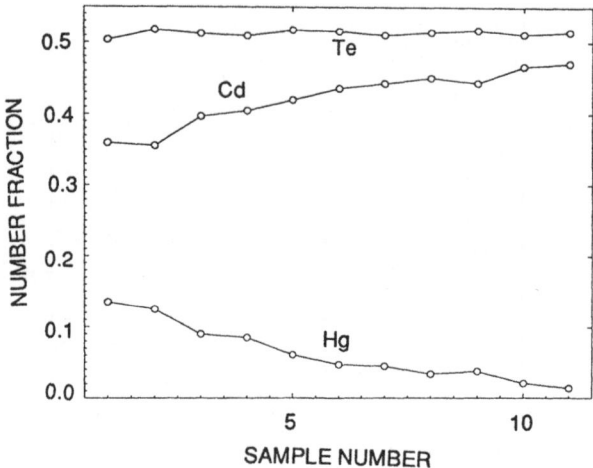

Figure 16. *Concentrations in a sequence of successively deposited cadmium mercury telluride films.*

were structurally modified; by using μ-PIXE mapping we located fecal pellets on the channel walls via their high phosphorous concentration. μ-PIXE analyses of the pellets at points from the wall into the surrounding soil showed that the primary transport mechanism of heavy metals from the sludge was via earthworms feeding on the surface and excreting at depth. The μ-PIXE data will provide a basis for modelling the metal transport from the sludge to the underlying aquifer.

Returning to the study of aerosols, the advent of the proton microprobe extends PIXE's role from analysis of collected particulate material to imaging and analysis of individual particles. In the case of fly ash particles below 10 μm diameter, the bulk concentrations are less important in terms of hazard than the fine details of individual particles. This is because the biological availability of toxic elements depends on the size of the particle, its matrix composition and the distribution of toxic elements within it or on its surface. Figure 17 presents μ-PIXE and RBS spectra recorded (Jaksic *et al.* 1991) with a sub-micron proton beam incident at the centre of a fly-ash particle. The centre was first located by μ-PIXE imaging of the element distributions. From the RBS spectrum the matrix composition is determined and the thickness is also measured. With the known thickness the PIXE data may be processed to find the concentrations of trace elements. This example is chosen to illustrate the merit of the simultaneous use of different ion beam analysis techniques in the proton microprobe.

11.5 Archaeometry

The non-destructive, in-situ nature of PIXE is well suited to the study of archaeological material (archaeometry). Large or delicate objects that cannot be placed in the evacu-ated specimen chamber can be analyzed by extracting the beam through a thin Kapton

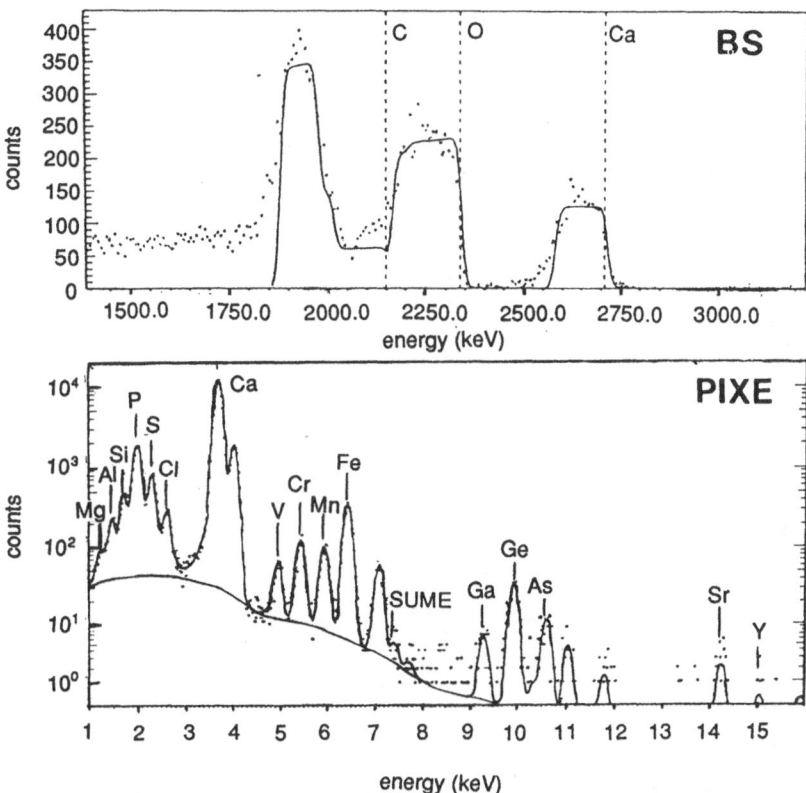

Figure 17. *Backscatter and* PIXE *spectra of a fly-ash particle (Jaksic et al. 1991). (By permission of Elsevier Science Publishers)*

window into the laboratory milieu; a flow of helium gas between specimen and x-ray detector copes with the problem of X-ray attenuation in the air. PIXE has been used to study individual lines of ink in a medieval bible, pigments in paintings suspected of being forgeries, solder smears used in construction of classical jewellery, and so on; indeed the Louvre museum has installed its own Van de Graaff accelerator, complete with PIXE and various ion beam analysis facilities, to pursue this avenue of research.

Work on artefacts in particular is radically different from mainstream PIXE work on biological, geochemical and environmental samples where the interest is usually in higher Z trace elements in a matrix of lower-Z major elements. In the artefact case, the lowest-Z matrix is usually silica, but frequently one contends with high-Z matrices such as bronzes, gold alloys, copper melting slags, and ceramics. This returns us to the issue of x-ray spectrum 'tailoring' mentioned in Section 3. Taking the bronze case as an example, the nickel content of ancient bronzes is a primary indicator for ore source discrimination.

If a simple silicon absorber is used to attenuate the predominant Cu K x-rays, this will enhance the relative intensities of K x-rays of trace elements having $Z > 29$. But

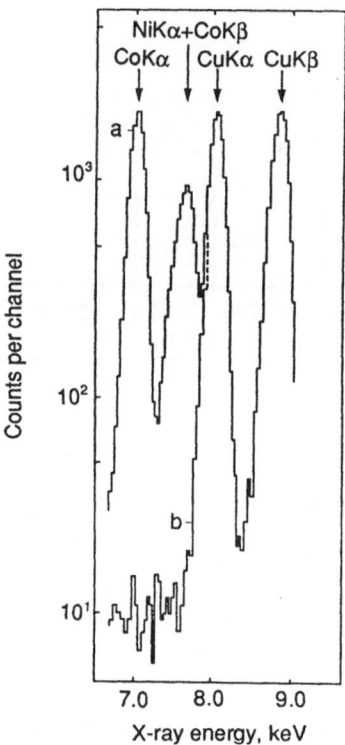

Figure 18. PIXE *spectra of a bronze artefact (Swann and Fleming, 1990), (a) with a selective filter of* 15.6mg/cm² *Co +* 7.5mg/cm² *V, and (b) with a simple attenuator of* 123 *mg/cm² Si. (By permission of Elsevier Publishers)*

the Ni K x-rays will be more heavily absorbed, and as a result the LOD for nickel is about 2%, which is inadequate for the task. The solution pioneered in the extensive archaeometric work of Swann and Fleming (1990) is the use of critical absorbers. Figure 18 shows bronze spectra recorded with 2 MeV protons. The silicon absorber is replaced by a selective filter of cobalt (which critically absorbs the copper K x-rays) and vanadium, which then suppresses the fluorescent cobalt K x-rays. This filter preferentially transmits the nickel K x-rays, while reducing the copper intensity by the same factor as before. The result is a 95 ppm limit of detection for nickel.

This type of work indicates that PIXE is by no means a 'black box' technique in some areas, but rather will provide rewards proportionate to the ingenuity of the analyst.

11.6   Other applications

PIXE and μ-PIXE are finding extensive use in a variety of very different areas. The reader wishing a broader survey is directed first to the books by Johansson and Campbell (1988) and Watt and Grime (1987). Additional useful references are from the following proceedings of conferences on PIXE and on proton microprobes: Johansson (1977, 1981); Martin (1986); van Rinsveld *et al.* (1987); Vis (1990); Grime and Watt (1987); Legge and Jamieson (1990); and Lindh (1992).

Acknowledgements

The author thanks the Natural Sciences and Engineering Research Council of Canada for its financial support and the many users of the Guelph Scanning Proton Microprobe for their stimulus. He also acknowledges the kindness of various authors in permitting reproduction of figures.

References

Berger M J and Hubbell J H, 1987, *National Bureau of Standards Report 87-3597* (U.S. Department of Commerce, Gaithersburg, Maryland)

Breese M B H, Landsberg J P, King P J C, Grime G W and Watt F, 1992, *Nucl Instrum Meth* **B64** 505

Campbell J L, Teesdale W J and Wang J X, 1990, *Nucl Instrum Methods* **B50** 189

Cookson J A, Ferguson A T G and Pilling F D, 1982, *J Radional Chem* **12** 39

Demortier G, Mathot S and Van Oystaeyen B, 1990, *Nucl Instrum Methods* **B49** 46

Eldred R A, Cahill T A, Wilkinson L K, Feeney P J and Malm W C, 1989, *Proc 82nd Meeting of the Air and Waste Management Assoc*

Griffin W L, Cousens D R, Ryan C G, Sie S H and Suter G F, 1989, *Contr Mineral Petrol* **103** 199

Grime G W and Watt F (eds), 1987, *Nucl Instrum Methods* **B30** 3

Grime G W, Dawson M, Marsh M, McArthur I C and Watt F, 1991, *Nucl Instrum Methods* **B54** 52

Hinrichsen P F, Kajrys G, Houdayer A, Jeremie A, Belhadfa A, Crine J P and Campbell J L, 1990, *Nucl Instrum Methods* **B45** 532

Jaksic M, Grime G W, Henderson J and Watt F, 1991, *Nucl Instrum Methods* **B54** 491

Johansson S A E (ed), 1977, *Nucl Instrum Methods* **142**

Johansson S A E (ed), 1981, *Nucl Instrum Methods* **181**

Johansson S A E and Campbell J L, 1988, *PIXE - a novel technique for elemental analysis* (John Wiley and Sons Ltd, Chichester)

Legge G J F and Jamieson D N (eds), 1991, *Nucl Instrum Methods* **B54**

Lindh U, 1992, *Nucl Instrum Meth* (in press)

Martin B (ed.), 1984, *Nucl Instrum Methods* **B3**

Martin F W and Goloskie R, 1988, *Nucl Instrum Methods* **B30** 242

Maxwell J A, Campbell J L and Teesdale W J, 1989, *Nucl Instrum Methods* **B43** 218

Newbury D E and Trace J, 1986, *Microprobe Techniques* **4** 103

Ryan C G, Cousens D R, Sie S H, Griffin W L and Suter G F, 1990, *Nucl Instrum Methods* **B47** 55

Swann C P and Fleming S J, 1990, *Nucl Instrum Methods* **B49** 65

Van Rinsveld H, Bauman S, Nelson J W and Winchester J W, eds, 1987, *Nucl Instrum Methods* **B22**

Vis R D (ed), 1990, *Nucl Instrum Methods* **B49**

Watt F, Grime G W, Brook A J, Gadd G M, Perry C C, Pearce R B, Turnau K and
    Watkinson S C, 1991, *Nucl Instrum Methods* **B54** 123
Watt F and Grime G W (eds), 1987, *Principles and applications of high-energy
    ion microbeams* (IOP Publishing Ltd, Bristol)

Ion-Beam Analytical Techniques

Rutherford Backscattering, Elastic Recoil and Nuclear Reaction Analysis

Ewa A Maydell

University of Strathclyde
Glasgow, Scotland

1 Introduction

In this chapter I shall address surface analytical methods based on ion-beam bombardment of a target material. These techniques are called Rutherford backscattering (RBS), elastic recoil depth analysis (ERDA) and nuclear reaction analysis (NRA).

Scattering of ions by the Coulomb field of an atom was first observed by Rutherford in 1911. For many years the effect was of purely academic interest, but commencing in the late sixties it gradually became a basis of a powerful tool (RBS) for surface and depth analysis of micron and submicron thick layers.

The energy spectrum of ions elastically backscattered from a thin surface layer of target material provides information on concentration and depth distribution of its component elements. Additional information on lattice damage and atomic location of impurities can be obtained from ion channelling experiments. ERDA employs the forward elastically recoiled atoms, while NRA exploits the inelastic processes that can occur.

2 Rutherford backscattering and channelling

Rutherford backscattering is a nondestructive analytical technique; it gives a quantitative measure of concentration of a heavy impurity in a lighter (host) target material plus atomic ratios for compound layers as well as depth distributions and, in combination

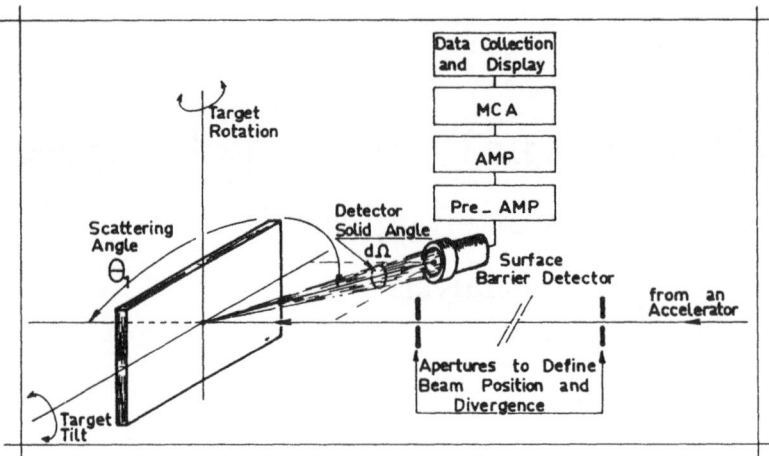

Figure 1. *Experimental arrangement for* RBS *measurement.*

with channelling, qualitative assessments of both crystalline quality, and substitution-ality of implanted and annealed dopants in single crystals. The sensitivity of RBS for a given element depends on the atomic number of the target atom. For high-Z elements, the detection limit can be as low as a fraction of one percent.

In this section the scattering of ions from a solid target is described and the analytical technique based on Rutherford scattering from random and single crystalline targets are outlined and illustrated with a number of examples, including the measurement of impurity depth distribution, lattice location of impurity atoms and the composition of thin layers.

2.1   Measurement of an RBS spectrum

In outline the method is simple. The sample is normally mounted on a goniometer in a vacuum target chamber ($\sim$1 x 10^{-6}torr) and irradiated with mono-energetic light ions (H^+ or He^+). For example He^+ ions in the range 1–3 MeV can be produced by a Van de Graaff accelerator. The beam is normally collimated to a half-angle of approximately 1/40 of a degree (to facilitate channelling experiments) and to a cross-section of about 1mm^2, defined by fixed apertures. A micro-beam of ions can be produced with a cross-section $\sim$200 to 500 μm^2 defined by a quadrupole based ion-optical system.

A small proportion of the ions ($\sim$1 x 10^{-5}) are elastically scattered backwards; hence the name Rutherford backscattering. A typical experimental arrangement is shown, schematically, in Figure 1. Only ions scattered through angle θ, within a solid angle $d\Omega$, are detected by the energy-sensitive surface barrier detector. This coupled with pulse-shaping and signal processing electronics, constitutes the spectrometer for measurement of an energy spectrum of backscattered ions.

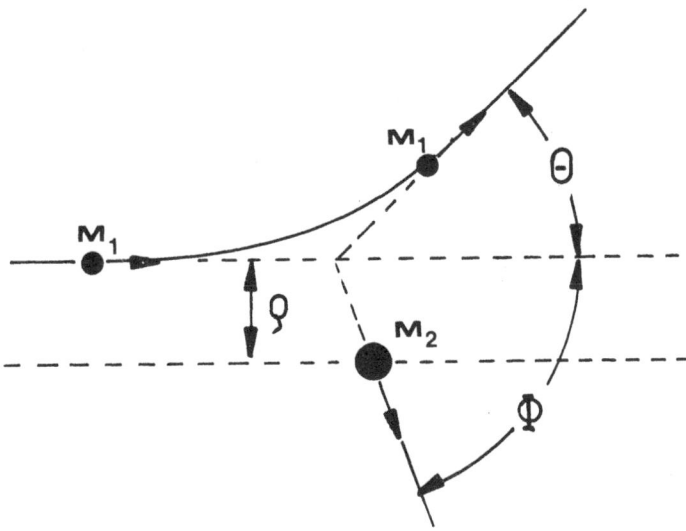

Figure 2. *Scattering of projectile (M_1, Z_1) from a target atom (M_2, Z_2) in laboratory coordinates. The figure defines the scattering angle θ, the recoil angle Φ and impact parameter ρ.*

2.2 Rutherford scattering of ions

The interaction of energetic ions with a solid target is a combination of elastic scattering of the projectile ions from the Coulomb potential of target-atom nuclei and their inelastic scattering from electrons. The elastic component defines the kinetic energy of the backscattered projectiles, resulting from 'hard sphere' collisions. The inelastic component results in the projectile ions losing their kinetic energy gradually, both prior and subsequent to the Coulomb scattering. We shall now examine the Rutherford backscattering process in detail, starting with ion scattering from an isolated single target atom, and progressing to ion-beam scattering from a thick target.

Scattering of an ion from a single atom

A positively charged energetic particle (say He^+) experiences a repulsive force on approaching another positively charged entity such as an atomic nucleus. This force causes the ion to be deflected; the angle of deflection being the greatest for ions that make the closest approach to the nucleus. Such a scattering event can be described in terms of an impact parameter, ρ, which is effectively the distance of closest approach of the projectile to the target atom.

The scattering process is referred to as 'elastic' when the kinetic energy and momentum are conserved and may then be described in terms of classical mechanics. Figure 2 shows schematically the scattering of a projectile from an isolated target atom in the laboratory coordinates.

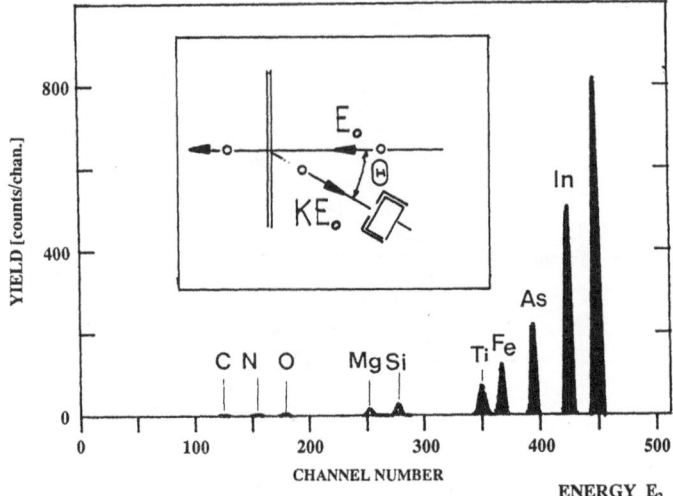

Figure 3. *Computer simulated* RBS *spectrum for 1.5 MeV He$^+$ ions backscattered through an angle of 160° from a thin-layer target comprising of $1 \times 10^{15} cm^{-2}$ atoms each of carbon, oxygen, nitrogen, magnesium, silicon, titanium, iron, arsenic, indium and gold.*

The probability of the projectile ion being scattered through the angle θ into a detector with an acceptance solid angle $d\Omega$ is given by the Rutherford formula:

$$\sigma(E_0, \theta)\, d\Omega \;\; = \;\; \left(\frac{Z_1 Z_2 e^2}{4E_0}\right)^2 \frac{1}{\sin^4 \theta/2}\, d\Omega \tag{1}$$

where E_0 is the projectile energy, Z_1 and Z_2 are the atomic numbers of projectile and the target atoms respectively, and e is electronic charge. Strictly, Equation 1 is an approximation for the scattering cross-section which is valid only for the case when $M_1/M_2 \ll 1$. For all other cases corrections must be made to account for the centre-of-mass motion.

The energy of the scattered projectile is different from its initial energy E_0, because of kinetic energy transfer to the target atom. The amount of the kinetic energy lost by the projectile in the collision depends on its impact parameter ρ, and the related scattering angle θ. The energy of the projectile after scattering, E_1, is given by:

$$E_1 \;=\; E_0 \left(\frac{M_1 \cos\theta + \sqrt{M_2^2 - M_1^2 \sin^2\theta}}{M_1 + M_2}\right)^2 \;=\; KE_0 \tag{2}$$

The coefficient K is the kinematic factor and has values between 0 and 1. For a given scattering angle θ, the value of K depends only on the masses of the colliding species. For example, in a 'head on' collision a He$^+$ ion may transfer a maximum of 64% of its kinetic energy to an oxygen atom but only 8% to a gold atom. The M_2 dependence of the collisional kinetic-energy transfer from a light projectile (H$^+$ or He$^+$) to target atoms and the dependence of the scattering cross-section on the atomic

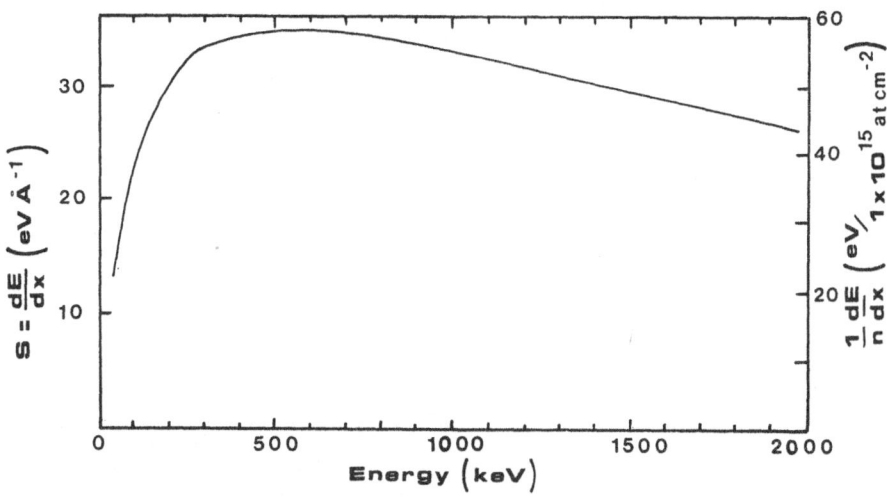

Figure 4. *Energy dependence of stopping power in aluminium for He^+ ions.*

number of the target element form the basis of an RBS measurement of composition for a target material. We note that for $\theta \to 0° K \to 0$ and $\sigma \to \infty$, while for $\theta \to 180°$, $K \to (M_2 - M_1)^2/(M_1 + M_2)^2$ and $\sigma \to$ its maximum value, $(Z_1 Z_2 e^2/4E_0)^2$.

Scattering from a thin multi-element target

We now consider He^+ irradiation of a very thin-layer target consisting of equal numbers of atoms of various elements. The target is so thin that the projectile will either be scattered from a target atom or pass through the film unscattered. Figure 3 shows the backscattered energy-spectrum for such a thin layer, simulated by computer. The simulated spectrum demonstrates (i) that the sensitivity of RBS to various elements increases with Z_2^2, and (ii) that the mass resolution decreases with increasing M_2 of the target atom.

We can express the number of projectiles $Y(N_T)$ scattered from N_T target atoms through the angle θ detected by a detector encompassing a solid angle $d\Omega$ by:

$$Y(N_T)\, d\Omega \;=\; N_p N_T \sigma\, (Z_1, Z_2, \theta, E_0)\, d\Omega \tag{3}$$

where N_p is the number of projectiles.

Scattering from a thick target

An energetic projectile is more likely to scatter through a small angle than a large angle θ (Equation 1), *i.e.* predominantly forward. This results in only small changes in kinetic energy of the projectile. At the same time the projectile, moving through the solid, loses its kinetic energy in inelastic processes involving ionisation and excitation of target atoms.

The rate at which a projectile loses its kinetic energy while penetrating a solid depends on its kinetic energy, the projectile and the target material. The projectile energy loss per unit path length is called the stopping power, S:

$$S(E) = \frac{dE}{dx} \equiv S(E, M_1, M_2, Z_1, Z_2) \tag{4}$$

(Note: some authors define $S(E)$ as $n_T^{-1}(dE/dx)$, where n_T is a number of target atoms in a unit volume and the stopping power so defined denotes the energy loss when the projectile traverses a layer containing a given number of atoms per unit area, *e.g.* $1 \times 10^{16} \text{cm}^{-2}$).

In Figure 4 the measured stopping power for He^+ ions in aluminium is shown as a function of the projectile energy. We see that $S(E)$ increases with projectile energy from 0 to $\sim$500 keV, when it starts to decrease. Stopping power data are known to $\sim$20% accuracy; polynomial fits are available for fast computer calculations (Ziegler 1977).

We now consider a 'thick' target (Figure 5) irradiated with mono-energetic projectile ions. By 'thick' target we mean one that causes a projectile to lose all its kinetic energy (become absorbed) within the thickness of the material. A projectile with an initial energy E_0 may either scatter from a target atom at the surface and change its kinetic energy to KE_0, or it may penetrate the target material to a depth x_0. At depth x_0 the projectile energy is reduced to E_1, where

$$E_1 = E_0 - \Delta E_{\text{in}} \tag{5}$$

and

$$\Delta E_{\text{in}} = \int_0^{x_0} S(E)\, dx \tag{6}$$

Scattering through an angle θ causes a further change in the projectile kinetic energy, to $E_1' = KE_1$. The projectile is then detected with energy E_2, given by:

$$E_2(x_0) = E_1' - \Delta E_{\text{out}} \tag{7}$$

where

$$\Delta E_{\text{out}} = \int_{\frac{x_0}{\cos\theta}}^0 S(E)\, dx \tag{8}$$

Projectiles that scatter from atoms within a thin layer Δx are detected with energies E_2 within an energy window ΔE_2 which relates to Δx by:

$$\Delta E_2 = \Delta x\, [S] \tag{9}$$

where $[S]$ called a stopping power parameter, is given by:

$$[S] = S(E_1) + \frac{K S(E_1)}{S(KE_1)} \tag{10}$$

Equation 10 is an approximation; it assumes $S(E)$ is constant within the layer Δx. To obtain a depth scale, an energy-depth conversion is needed; this is obtained by an iterative process, starting from the surface signal, where both, E_{in} and E_{out} are known

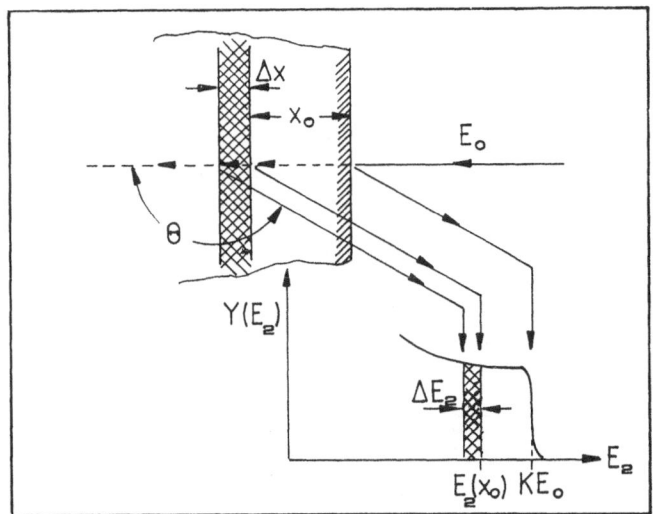

Figure 5. *Scattering of projectiles from a thick target; principle of generation of the* RBS *spectrum.*

accurately. The energy loss in consecutive layers is calculated either numerically or analytically to provide a depth scale corresponding to measured energies E_2.

The shape of the RBS spectrum for a thick target is defined by energy dependencies of both the scattering cross-section (Equation 1) and the stopping power (Figure 4). The energy spread of the primary ion-beam, the finite energy resolution of an energy-sensitive detector, and signal processing electronics all introduce a 'rounding' of the sharp edges of the spectrum and a broadening of the RBS signals for very thin (δ-like) layers.

Multi-element targets

When a target material contains two or more different elements then the projectile will interact with atoms of different kinds in proportion to their concentrations. The stopping power $S(A_a B_b)$ for a target material $A_a B_b$ is then derived using the Bragg rule as a weighted sum of the stopping powers for the individual pure elements.

$$S(A_a B_b) = \frac{aS(A) + bS(B)}{a + b} \tag{11}$$

where $S(A)$ and $S(B)$ are stopping powers for pure elements A and B.

Figure 6 shows the measured RBS spectrum for silicon implanted with bismuth. The bismuth RBS signal is observed in the energy window ΔE_2 from 1.4 to 1.5 MeV. The total (integrated) RBS signal, Y_{Bi}, in this window, depends on the total number of bismuth atoms N_{Bi} in the area irradiated with He^+ ions. The relationship between these is:

$$Y_{Bi} d\Omega = N_p N_{Bi} \sigma (Z_1, Z_2, E_0, \theta) d\Omega \tag{12}$$

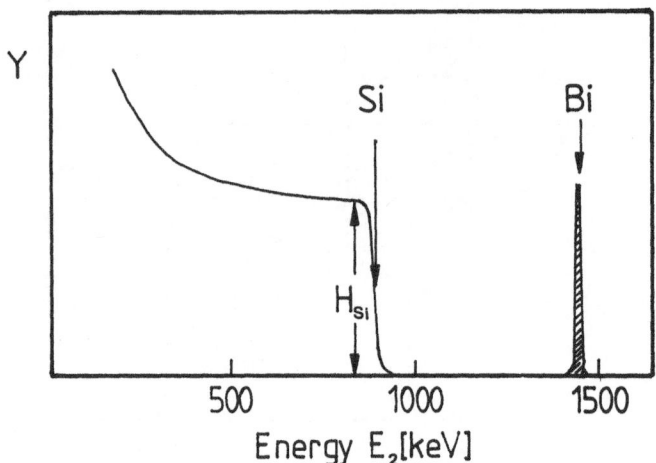

Figure 6. RBS *spectrum for bismuth-implanted silicon measured with 1.5 MeV He*$^+$
ions scattered through 160°.

Equation 12 correlates the backscattering yield for the impurity with its concentration. To obtain the bismuth concentration we need to integrate the bismuth signal and
calculate N_{Bi} using the instrumental parameters: θ, N_p, E_0 and $d\Omega$. Their values are
not always known accurately, and usually N_{Bi} is calculated with respect to the host
matrix. The intensity (height) of the RBS spectrum for silicon, marked H_{Si} in Figure 7,
i.e. the number of projectiles detected within the energy window of the MCA is given
by:

$$H_{Si}d\Omega = N_p N_{Si} \sigma (Z_1, Z_2, E_0, \theta) d\Omega \tag{13}$$

where $N_{Si} = n_{Si}\Delta x$ and n_{Si} is the atom density number (cm^{-3}). Using Equation 9 we
can write for Equation 13:

$$H_{Si}d\Omega = N_p n_{Si} \sigma (Z_1, Z_2, E_0, \theta) d\Omega \, \delta E/[S] \tag{14}$$

where δE is the channel width of the MCA. Dividing Equation 12 by 14 we obtain:

$$Y_{Bi}/Y_{Si} = (N_{Bi}/n_{Si}) [(\sigma(Bi)/\sigma(Si)] [S(Si)]/\delta E \tag{15}$$

Equation 15 does not contain any instrumental parameters (with exception of δE,
the channel width of the MCA) and enables us to calculate the concentration of the
implanted (or otherwise introduced) heavier impurity in a lighter matrix. It can also be
used to measure a thickness (in atoms cm^{-2}) of a thin surface layer. For Equation 15 to
give an accurate result, the impurity must be present in low enough concentration (1%)
not to affect the stopping power parameter $[S(Si)]$, and, in the case of a thin surface
layer, this must not be thick enough to reduce the primary ion energy (the value of the
scattering cross-section, for the matrix, affects H_{Si}).

Figure 7 shows an RBS spectrum measured for titanium silicide on a silicon substrate using 1.5 MeV He$^+$ ions scattered through an angle 160°. The titanium feature

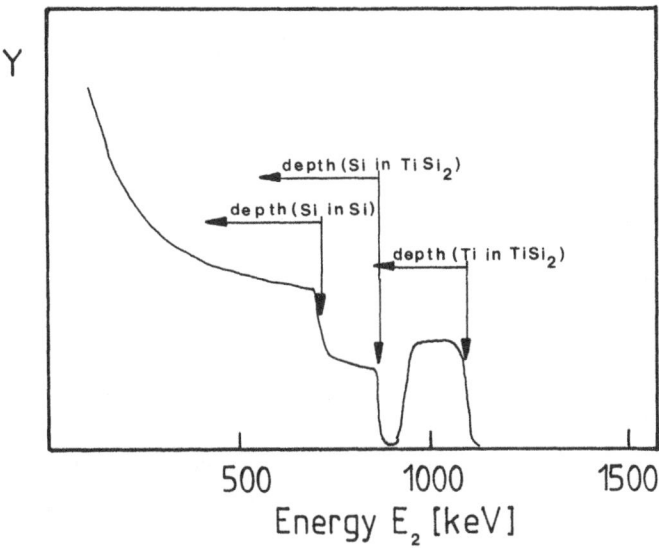

Figure 7. *Measured* RBS *spectrum of* ~250 nm *of TiSi₂ on silicon substrate using 1.5 MeV He⁺ ions scattered through 160°.*

in the spectrum results from He⁺ ions backscattered from titanium atoms in TiSi₂, while the silicon signal shows two features — a higher energy (step-like) feature which corresponds to He⁺ scattered from silicon atoms in TiSi₂, and a lower energy feature which corresponds to He⁺ scattered from silicon atoms in the silicon substrate. Note that the two depth scales, based on respectively the titanium and the silicon signals, are different. The stopping power parameter $[S]$ from Equation 10, contains K_{Si} for the energy depth-conversion of the silicon feature in the spectrum and K_{Ti} for the titanium feature.

The RBS spectrum for a uniform multi-element layer provides a measure of atomic concentrations of the elements present. Using Equation 13 for the silicon and titanium signals respectively, and dividing one by the other we obtain:

$$\frac{H_{Si}}{H_{Ti}} = \frac{n_{Si}}{n_{Ti}} \frac{\sigma(Si)}{\sigma(Ti)} \frac{[S(Ti \text{ in } TiSi)]}{[S(Si \text{ in } TiSi)]} \tag{16}$$

As a first approximation for the ratio n_{Si}/n_{Ti} we can take $(H_{Si}/H_{Ti})[\sigma(Ti)/\sigma(Si)]$, but for an accurate measurement we require more precise values for the stopping power parameter, $[S]$ for Si and Ti in TiSi₂.

Clearly, to account accurately for changes in the primary ion energy as the projectile traverses the solid, and for changes of the stopping power and scattering cross-section, a computer is of considerable help. In addition, computer simulations of spectra are used widely for RBS analysis.

There are two ways of analysing a measured RBS spectrum using computer simulation. The first is to deconvolute the estimated broadening introduced by the energy-spread of the primary beam and by the finite resolution of the detection and signal-

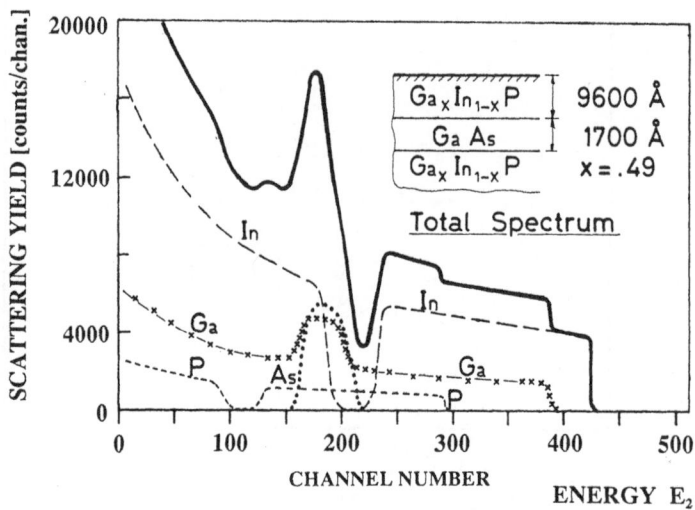

Figure 8. *Computer simulated* RBS *spectrum for a layered structure shown in the inset. The elemental* RBS *signals obtained for constituent elements are also shown.*

processing electronics, to obtain a 'de-broadened' RBS spectrum, and then compare this with simulated spectra for assumed target materials, adjusting thicknesses and atomic concentrations of trial layers until agreement is obtained between the simulated and deconvoluted spectra. The second is to use known algorithms plus realistic energy-spreads and stopping-power data to synthesise a spectrum for an assumed target material, compare this with the experimental result and then adjust parameters (atomic concentrations and layer thicknesses) to obtain agreement between the measured and simulated results. The second method is the more flexible and tends to be favoured; an example, for a complicated layered structure is shown in Figure 8.

2.3   Channelling of ions

In our foregoing discussion the solid from which the primary ion-beam is scattered consists of randomly distributed atoms. No assumptions were made or were needed about exact crystallographic structure of the target material. This simplification is valid for microcrystalline materials with no preferred crystal orientation, for which the primary ion-beam probes simultaneously a large number of crystal grains; ie, the beam spot diameter on the target is much larger than the dimensions of any individual crystallite. However, an important application of Rutherford backscattering is the characterisation of single-crystal materials modified by ion implantation with subsequent thermal annealing. In such a case the target material is a single crystal or a defective single crystal, and the crystal structure plays an important role in the scattering of the primary ions.

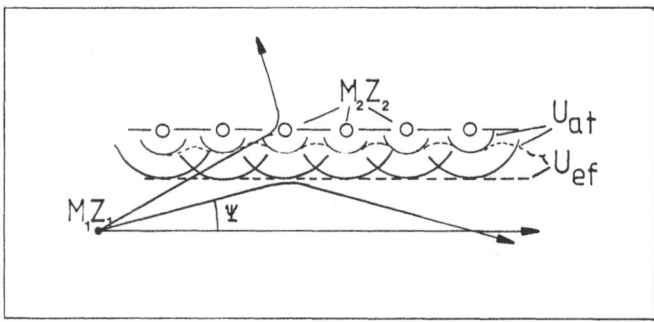

Figure 9. *Schematic of an atomic row and channelling conditions.*

2.4 Scattering of ions from a single crystal

We now consider a row of target atoms equally spaced along a straight line as in Figure 9. An approaching projectile ion experiences an effective electric field that results from a superposition of the electric fields generated by the individual atoms. From a large distance away the field appears uniform and a projectile ion approaching at glancing angle will be gently deflected from the atomic row. A projectile approaching at a more acute angle will penetrate the uniform field and experience the repulsive field of an individual atom. It will then be scattered in the usual way.

As a consequence of the gentle deflection experienced by a projectile ion impinging an atomic row at glancing angle, when a projectile scatters from a single crystal (a parallel arrangement of atomic rows) it will be propagated along the channel between the rows by being gently repulsed from the row to either side; the projectile may thus reach large depths in a single-crystal target, without experiencing wide angle scattering. The effect is called 'channelling'. On the other hand a projectile-ion that approaches an atomic row at large angle of incidence will be scattered in the normal way, referred to as 'random' scattering. Channelling along a crystallographic axis is called 'axial channelling'. A similar uniform (planar) potential is generated by atoms arranged in planes, and can result in channelling of ions between crystallographic planes; this is termed 'planar' channelling.

A critical angle for axial channelling: *i.e.* the maximum angle made by the projectile's trajectory to the atomic row (a densely packed crystalline direction) that allows channelling, ψ_c, depends on the projectile's energy, E, its atomic number, Z_1, atomic number of the target atom Z_2 and its lattice spacing, d:

$$\psi_c \propto \sqrt{Z_1 Z_2 e^2 / E d} \qquad (17)$$

Thermal vibrations of the lattice cause the critical angle to decrease, an effect which is important for single crystalline metal targets.

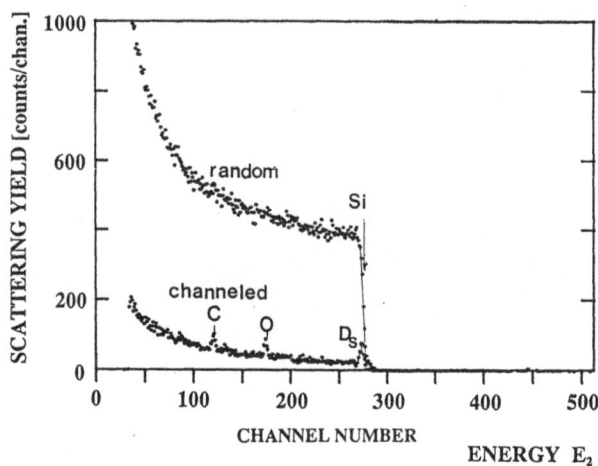

Figure 10. RBS *spectrum measured with 1.5 MeV He*+ *ions scattered through 160° for a (100) oriented silicon single-crystal target in 'random' and in 'channelling' directions.*

Projectiles channelled in a single crystal will interact only weakly with target atoms. The stopping power for such projectiles is much lower than for those that travel in a random direction. A channelled projectile may become 'de-channelled' if it encounters a lattice imperfection (*e.g.* interstitial atom 'blocking' the easy passage along a given channel; or a vacancy in an atomic row which causes distortion of the uniform electric field). Every channelled projectile will eventually become de-channelled in a perfect single crystal, as a result of its energy slowly reducing with depth causing an increase in the critical angle (Equation 17).

Channelling in a high quality single-crystal of silicon can lead to a thirty-fold reduction in the backscattered intensity.

To perform channelling experiments, the primary beam must be well collimated and the target surface highly polished so that all the projectile-ions impinge the surface at the same angle. The collimation is achieved by passing the ion-beam through a set of small apertures (~1 mm diameter) set along a 2m collimator base which allows only ions with trajectories diverging within, say, 3–5 arc seconds to pass. The single crystal is mounted on a two-axis, or three-axis goniometer, usually with automated sample rotation, for alignment of a given sample axis or crystallographic plane with the projectile ion-beam.

Figure 10 compares RBS spectra measured with 'channelling' and 'random' incidence of the projectile He+ beam for single-crystal silicon. The intensity of the channelled spectrum is reduced by approximately thirty times that of the random spectrum. This reduced backscattered intensity permits the detection of surface impurities, such as oxygen or carbon in monolayer concentrations. The detection of light impurities such as these is not possible in a random direction because of the high-intensity background signal that comes from a thick substrate. A small 'surface' peak, visible in all the channelling RBS spectra is marked D_s; it is normally observed in channelling experi-

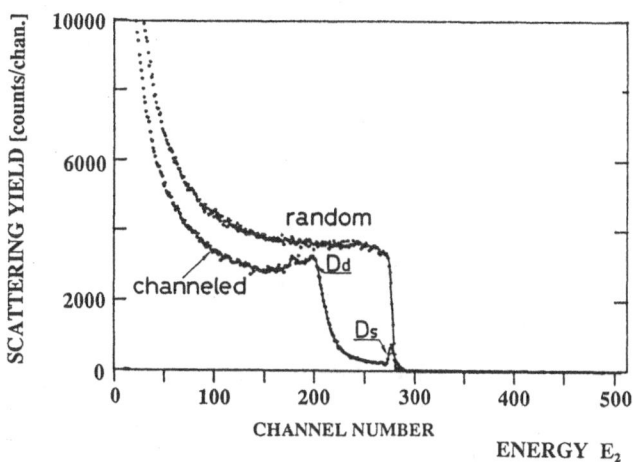

Figure 11. *Channelling and random* RBS *spectra measured for (100) single-crystal silicon implanted with* $3 \times 10^{16} cm^{-2}$ *200 keV oxygen ions.*

ments and is related to scattering of projectile-ions that impinge the surface near (small impact parameter) to atoms that terminate atomic rows in the surface plane.

Figure 11 compares random and channelled RBS spectra for single-crystal silicon implanted with 200 keV oxygen ions. The channelled spectrum shows a broad peak at an energy corresponding to the depth of radiation damage caused by the implantation. This channelling measurement indicates a defective single-crystal, at a depth corresponding to the maximum concentration of radiation defects. It is interesting to note that there are a number of primary ions, which after passing through a disordered layer, maintain their original direction (and their alignment with the silicon-crystal lattice) and are channelled at depths beyond the damaged layer. This is an observation consistent with the predominantly forward nature of Rutherford scattering.

Figures 12 and 13 are examples of application of channelling RBS made to characterise ion-implanted silicon single-crystals and to study their thermal annealing. Inspection of a channelled spectrum (Figure 12) shows that the backscattered yield in the high-energy part of silicon spectrum is identical to that in an RBS spectrum for random direction. This reveals the presence of an amorphous layer, approximately 300 nm thick, caused by implantation of 5×10^{15}As atoms cm^{-2} at an energy 150 keV. A step observed in the channelled spectrum, at which the intensity decreases by a factor of approximately 2, corresponds to an amorphous to single-crystal interface. Below this interface there is single-crystalline material, in which some of the projectiles can still channel. The arsenic feature in the spectrum is identical for both, the channelled and the random spectra, *i.e.* there is as much randomness in arsenic-atom arrangement as for the silicon atoms.

The silicon channelled spectrum, Figure 13, measured for the same sample as Figure 12 after thermal annealing, is identical to that for non-implanted single-crystal

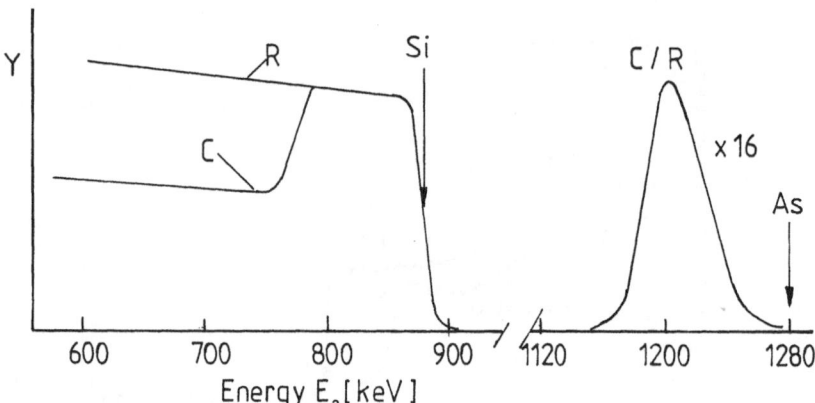

Figure 12. RBS *spectra measured for arsenic implanted silicon single crystal:* **R** *random direction,* **C** *channelling direction. The arsenic signal is identical in both directions.*

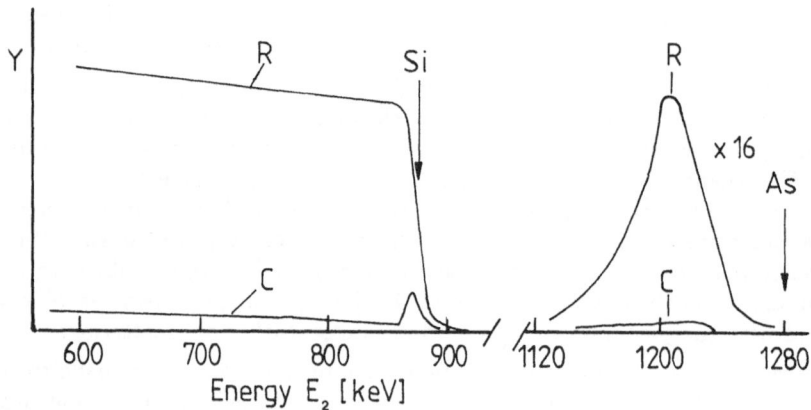

Figure 13. *Channelling and random* RBS *spectra measured for sample, shown in Figure 12 after rapid thermal annealing (scanned electron beam).*

silicon. This indicates that the thermal annealing restored the silicon to a perfect ordered lattice. The measured arsenic spectrum in the channelling direction has a very low intensity. This shows that arsenic atoms, in the restored silicon lattice, occupy silicon atom sites and are therefore 'invisible' to channelled projectile ions. The 'random' spectrum, for arsenic, has the expected high intensity, and a comparison of 'random' spectra for arsenic (Figures 12 and 13) reveals that the (rapid in this case) thermal annealing caused some redistribution of the implanted impurity; the maximum intensity

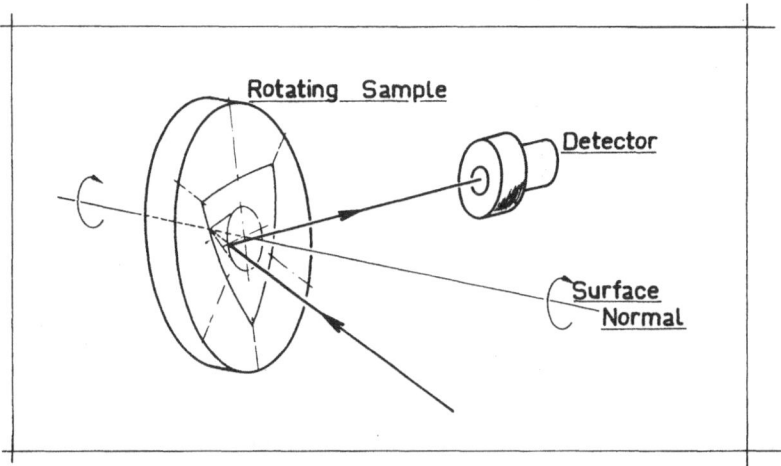

Figure 14. *Rotating-sample arrangement for measurement of (reproducible) 'random' spectra.*

is lower in Figure 13 than in Figure 12 while a 'tail' also develops to the low-energy side of the arsenic peak, showing that some arsenic has diffused deeper into the silicon. A ratio of channelled and random scattering yields, termed the 'reduced channelling yield' (χ_{min}), gives an indication of the crystalline quality for the annealed material. This ratio for silicon (Figure 13) is ~0.03, and is typical for good quality (100) single-crystal silicon. For arsenic we measure the ratio of channelling to random intensities as equal to 0.05. It can be shown that the non-substitutional fraction of impurity atoms, n_{ns}, is given by the expression:

$$n_{ns} = \frac{\chi_{min}(As) - \chi_{min}(Si)}{1 - \chi_{min}(Si)} \tag{18}$$

which gives 2% for the fraction of non-substitutional arsenic in our example. If the non-substitutional fraction of impurity defined by Equation 18 is low, this suggests high electrical activity of the implanted dopant. However, only a measurement of carrier concentration and its mobility constitutes a 'proof' of activation of a dopant.

It is also of note that prolonged exposure of doped single crystals to projectile-ion bombardment can cause dopant displacements with apparent non-substitutionality (Maydell *et al.* 1985). It is therefore important that exposures are reduced to the minimum needed; also that measurement of a channelled spectrum is made before the random spectrum.

In a high quality single-crystal material there is always some (high-index) order for some arbitrary direction. This makes it extremely difficult to avoid channelling when trying to obtain a 'true' random reference spectrum. A common practice is to incline the single-crystal sample at approximately 7° to the primary ion-beam in order to measure a so-called 'random' spectrum; this often causes unpredictable anomalous features to appear in the spectrum, which is evidence that some channelling (axial or planar) has occurred. A procedure that can be applied when rotation of the sample is possible during data acquisition, as shown schematically in Figure 14, is to measure the

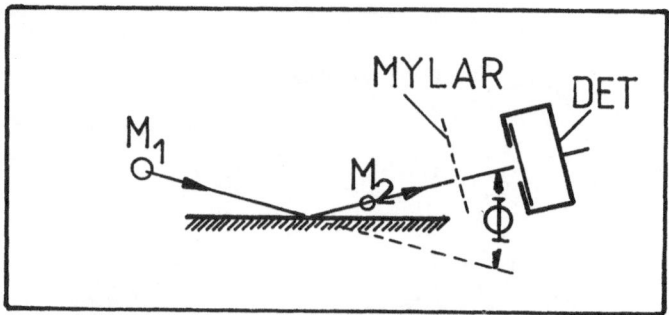

Figure 15. *Schematic of* ERDA *experiment.*

RBS spectrum with the sample rotating about an axis inclined by a few degrees to the direction of the projectile beam. The RBS spectra so produced are highly reproducible (Hemment *et al.* 1983); although it remains unclear whether such spectra are in fact truly 'random'.

3   Elastic recoil depth analysis

Elastic recoil depth analysis is a 'reverse' of Rutherford backscattering. It complements the RBS in detection and measurement of depth distribution of light impurities in heavier target host materials.

Primary ions are scattered 'backwards' only when they impact target atoms of larger mass. Otherwise, both the projectile and the target atoms move predominantly 'forward' after impact. This simple consequence of kinematics is exploited for the detection and depth analysis of light impurities in heavy targets, by energy analyzing the elastically recoiled target atoms. The analytical technique, is called Elastic Recoil Depth Analysis or ERDA. For a full description of the experimental method see, for example, Sofield *et al.* (1982).

The sample to be analysed is irradiated with a beam of heavy ions (usually chlorine or bromine) of energy 30–80 MeV. The impact geometry is near 'glancing' incidence (see Figure 15). The particle detector is positioned typically at 30° to the direction of the primary ion-beam and the atoms ejected in its direction are energy analysed. To prevent the detector from receiving a high flux of reflected primary ions, a 'beam stop', is introduced in front of the detector, usually a thin mylar foil, which allows the light recoiled target atoms to pass while suppressing the primary heavy ions.

The energy E_3 of a target atom (mass M_2) recoiling forward at an angle Φ to the direction of the primary ion (mass M_1), is given by:

$$E_3 \; = \; E_0 \, \frac{4M_1 M_2}{(M_1 \, + \, M_2)^2} \; \cos^2 \Phi \qquad (19)$$

The heavier the target atom, the higher the recoil energy it acquires on impact with a heavy projectile. Thus, in contrast to RBS measurement where light-element signals

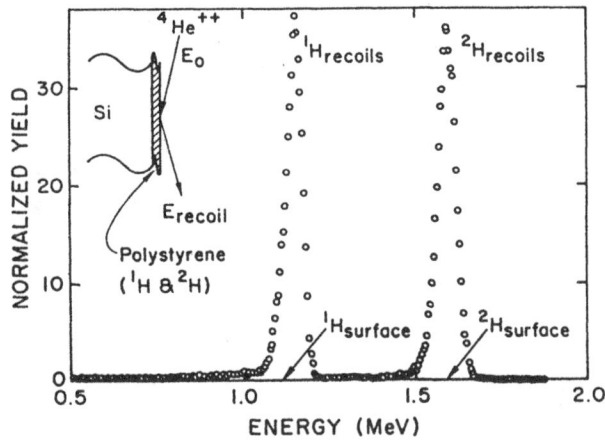

Figure 16. ERDA *measurement for hydrogen and deuterium recoiled species under He+ ion bombardment (Feldman and Mayer 1986).*

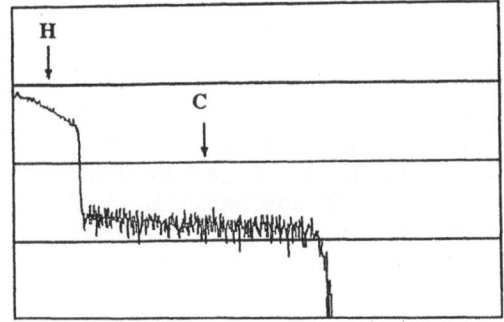

Channel: 1

Figure 17. ERDA *spectrum measured for hydrogen in a diamond-like carbon film using 30 MeV Cl+ ions. (El-Hossary et al. 1988).*

are superimposed on a background arising from the heavier substrate, in ERDA the heavier-substrate signal is superimposed on that of the light element. Therefore it is advantageous in some cases to chose a projectile ion to suit a specific analytical problem. For example, detection of hydrogen and deuterium by ERDA can be 'background free' if He+ (or α^-) particles are used for the primary beam. (See Figures 16, 17 and 18)

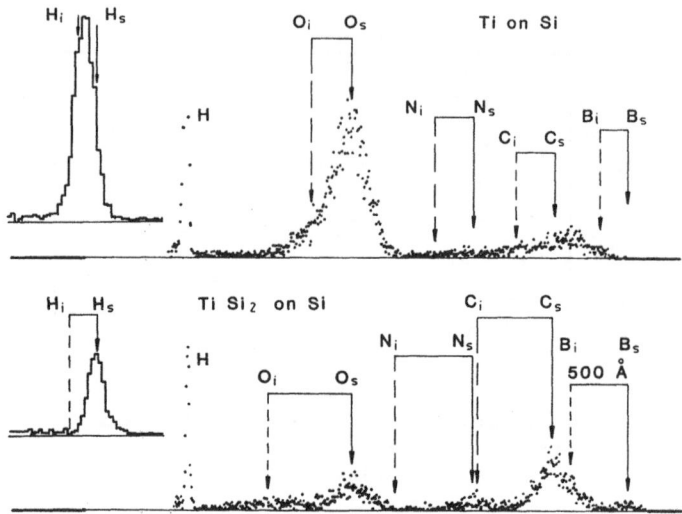

Figure 18. ERDA *spectrum showing hydrogen, boron, carbon and oxygen measured using 30 MeV Cl+ ions for: (a) 23.5 nm titanium and (b) 50 nm titanium disilicide (from Sofield et al. 1985).*

4 Nuclear reaction analysis

Analytical techniques based on elastic scattering of energetic ions have been used for materials analysis for more than twenty-five years. The inelastic processes that occur in these interactions can form a serious obstacle to the interpretation of RBS particle energy spectra when the backscattering cross-section is non-Rutherford. However, nuclear reactions, in which the nature of the target atom changes and/or the kinetic energy is not conserved, have only recently been used in material analysis. These measurements can complement RBS, and are, in some cases, more sensitive. The method competes as an analytical technique with secondary-ion mass spectrometry (see the contributions by Werner and Wittmaack in this volume).

The energies of the light primary ions (H+ or He+) used for Rutherford backscattering are such that these ions can penetrate the electron shells of target atoms, but eventually they scatter from the repulsive Coulomb field associated with the positive charge of the nucleus. Increasing the projectile energy can cause the ion to penetrate to and even beyond the Coulomb barrier of the nucleus, approaching the nucleus to a distance comparable with its radius. For an atom of an element of atomic number Z and atomic mass A, the 'nuclear radius' is $\sim 1.4 \times 10^{-13} A^{1/3}$cm. In other words for a projectile to overcome the Coulomb barrier of the target-atom nucleus its energy must be $Z_1 Z_2 e^2 / R = Z_1 Z_2 / A_2^{1/3}$ MeV. Once in the field of the nucleus the projectile may be 'captured' and form with it a so-called 'compound nucleus'. The probability of this happening depends on the specific nuclear structure.

The compound nucleus will be in an excited state, having partially absorbed the kinetic energy of the projectile. Excited states of nuclei are not stable; they relax by

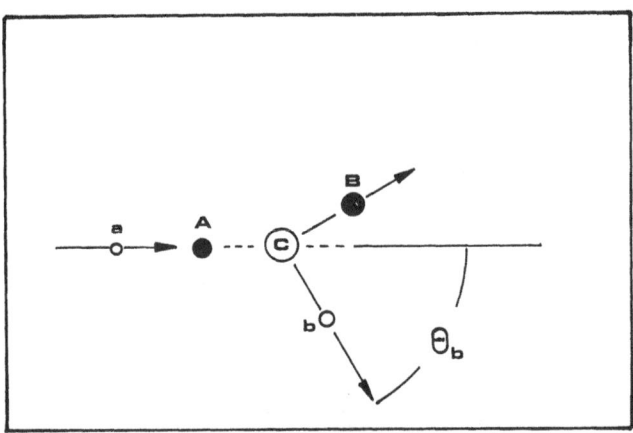

Figure 19. *Schematic representation of nuclear reaction.*

emitting either a secondary particle or γ radiation, and the atom produced will be in either a ground or an excited state. The latter may further relax by emission of an electron, a positron, or γ radiation. Thus the nuclear reaction, depicted in Figure 19, can be represented schematically by

$$a + A = B + b + Q \tag{20}$$

where a is the projectile ion, A the target atom, B a reaction product atom, b an emitted (light) particle and Q is the energy balance. Q may be positive or negative.

When Q is negative, the reaction can only take place if that amount of energy is available in the reacting system, and the reaction has then a specific (kinetic) energy threshold. The energy E_b of the emitted particle will depend on the geometry of detection and on Q, through:

$$(E_b)^{\frac{1}{2}} = C \pm (C^2 + D)^{\frac{1}{2}} \tag{21}$$

where

$$C = \frac{(M_a M_b E_0)^{\frac{1}{2}} \cos \theta_b}{M_b + M_B} \quad \text{and} \quad D = \frac{M_b Q + E_0(M_b - M_a)}{M_b + M_B};$$

M_a, M_b, M_A and M_B are the masses respectively of a, b, A and B; and θ_b is the angle of emission of b with respect to the initial direction of a.

When b is identical with a, the only identifiable differences between the nuclear reaction and a scattering process are in the kinetic energy balance, Q, and the delay between absorption and emission of particle a. The lifetime of a compound nucleus is typically 10^{-14}–10^{-18}s whereas the time required for the projectile to traverse the nucleus is 10^{-21}–10^{-22}s.

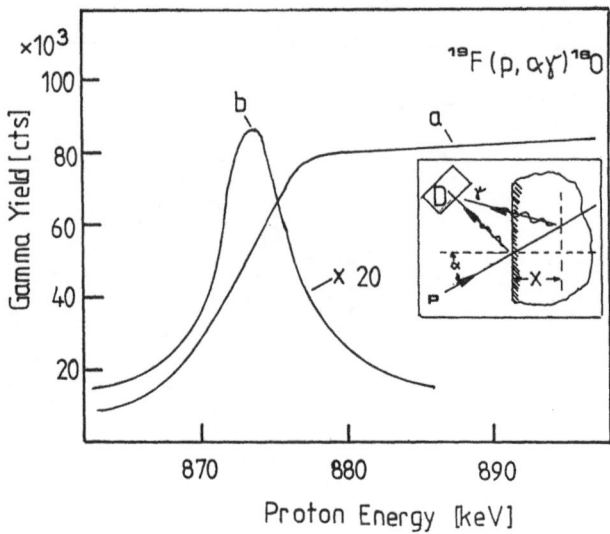

Figure 20. *Dependence of intensity of* γ *radiation on the proton energy:(a) AlF; (b) for* 2×10^{16} *Fcm^{-2} implanted into silicon using BF$_3$ ions.*

Once the compound nucleus is formed it can decay through various reaction channels. It is usual to denote the reaction described in Equation 20 by $A(a,b)B$. Some of the nuclear reactions that can occur during proton bombardment are then:

$$A(p,p)A \qquad \text{(Rutherford scattering)}$$
$$A(p,p')A \qquad \text{(inelastic scattering)}$$
$$A(p,\gamma)B \qquad \text{(prompt } \gamma \text{ emission)}$$
$$A(p,n)B_1 \qquad \text{(prompt neutron emission)}$$
$$A(p,\alpha)B_2 \qquad \text{(prompt } \alpha \text{ emission)}$$

The probability of a given reaction is described by the cross-section $\sigma(a,b)$. In the two-step model of the compound nucleus, the cross-section is given by

$$\sigma(a,b) = \sigma_c(a)\, x \text{ (relative probability of emission } b) \qquad (22)$$

In general the values of the cross-sections and probabilities of various emissions depend on the energy of the projectile (Feldman and Picraux 1977). The cross-sections for some reactions, while varying smoothly with projectile energy, show at some energies very high values, within a narrow energy range. These nuclear reactions are called resonant reactions.

Resonant reactions are exploited for detection, quantification and depth profiling of a range of light isotopes. An example of a nresonant reaction is $^{19}F(p,\alpha)\,^{16}O^*$, The resonance is observed at a proton energy of 872.5 ± 4.5 keV, and at this energy emission of α particles and γ radiation characteristic of $^{16}O^*$ is enhanced by a factor of ~ 10.

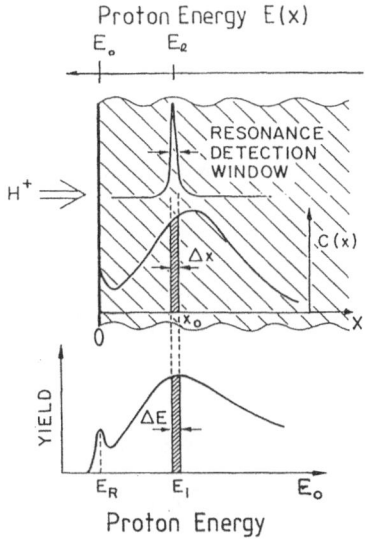

Figure 21. *The use of a resonant nuclear reaction for depth profiling.*

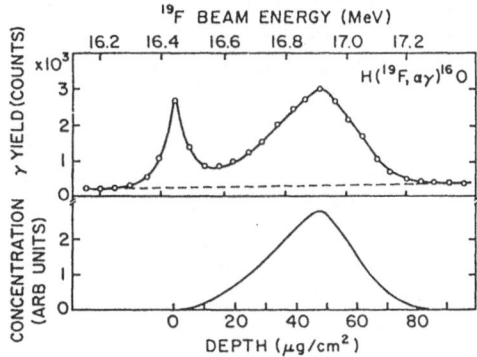

Figure 22. *Depth profile of hydrogen implanted into Al_2O_3. Upper figure shows experimental data; lower displays the hydrogen depth profile (Feldman and Mayer 1986).*

We consider now a target material containing fluorine uniformly dispersed in its matrix, (*e.g.* AlF), irradiated with a beam of protons of varying energy. Figure 20 shows the dependence on primary beam energy of characteristic γ ray emission from $^{16}O^*$, detected using the experimental arrangement shown in inset. Only a low intensity of γ radiation is observed for proton energies below resonance; it increases dramatically at resonance, and then for proton energies above resonance the γ yield slowly increases while the yield of α particles (not shown) promptly decreases. Detection of γ radiation is better suited for analytical purpose than detection of α particles. The α particles not only lose their energy in the target material and become completely absorbed if they originate at a considerable depth, but also have energies that are dependent on the primary proton energy (Equation 21). Resonant reactions exhibiting a single and narrow, strong resonance can be exploited for depth profiling in the case of suitable light elements. The principle is shown in Figure 21.

A measured depth profile for fluorine in a silicon substrate, using the resonant nuclear reaction $^{19}F(p, \alpha)\ ^{16}O^*$, is shown in Figure 20. Curve b in this figure represents the yield of γ radiation measured for silicon implanted with BF_3^+ ions, irradiated with protons. The yield first increases to a maximum corresponding to a peak in fluorine distribution and then decreases when the proton energy exceeds resonance. To obtain a depth distribution of fluorine we use the relationship between initial energy, of the proton beam, E_0, and the stopping power:

$$E_0 \ = \ E_R \ + \left(\frac{x}{\cos \alpha}\right) \left(\frac{dE}{dx}\right)\bigg|_{E_0} \tag{23}$$

where α is the angle of proton incidence. For more elaborate analyses of NRA data see Amsel and Maurel (1983).

Nuclear reactions are reversible; *i.e.* if a reaction $A(a, b)B$ is possible then the reaction $a(A, b)B$ is also possible. Since there is a considerable mass difference between the projectile A and target species a, the kinetic energy of the projectile must enable it to overcome the Coulomb barrier of the target species. The reverse reactions (with protons) are often used for quantification and depth profiling of hydrogen. Figure 22 shows the application of a high energy beam of ^{19}F ions for depth analysis of hydrogen implanted in Al_2O_3. Note the energy range of the fluorine ions.

References

Amsel G and Maurel B, 1983, *Nucl Instr and Meth* **218** 8

El-Hossary, Fabian D J and Sofield C J, 1988, *Thin Solid Films* **157** 29

Feldman L C and Mayer J W, 1986, *Fundamentals of Surface and Interface Analysis* (North-Holland, NY/Amsterdam/London).

Feldman L C and Picraux S T, 1977, in *Ion Beam Handbook for Material Analysis*, eds J W Mayer and E Rimini, (Academic Press, New York)

Hemment P F L, Maydell-Ondrusz E A, Stephens K G, 1983, *Nucl Instr Meth* **218** 103

Maydell E A, Wilson I H and Stephens K G, 1985, *MRS Symposia Proc*, eds F H Eisen, T W Sigmon and B R Appleton (North-Holland, New York/Amsterdam/Oxford) **47** 123

Sofield C J, Woods C J, Gowern N E B, Bridwell L B, Butcher J M, and Freeman J M, 1982, *Nucl Instr and Meth* **203** 509

Sofield C J, Harper R and Rosser P, 1985, *MRS Symposia Proc*, eds D K Bieglesen, J Rogonyi and C Shank, **35** 445

Ziegler J F, 1977, *Helium Stopping Powers and Ranges in All Elements* (Pergamon, NY)

Quantitative Analysis of Solids by SIMS and SNMS

K Wittmaack

ATOMIKA, Oberschleissheim, &
GSF, Neuherberg, Germany

1 Introduction

This chapter deals with some essential aspects of quantitative analysis by Secondary Ion Mass Spectrometry (SIMS) and Sputtered Neutral Mass Spectrometry (SNMS). The acronym SNMS can be misleading because neutrals can only be mass analysed after being ionised by a suitable technique. Nevertheless SNMS is the most common generic term for mass spectrometry of post-ionised sputtered particles (the term 'post'-ionisation is meant to indicate that the sputtering and the ionisation processes are decoupled in space and time). SNMS may be classified according to the method used for post-ionisation of the sputtered neutrals. Here discussion will be restricted to ionisation by electron impact. Laser-based SNMS is described in the chapter by Ledingham in this Proceedings.

The basic concept of compositional analysis by SIMS and SNMS is straightforward. The sample is bombarded with a beam of energetic 'primary' ions. As a result of primary ion impact atoms and molecules from the topmost layer(s) of the sample are sputter ejected into the vacuum outside the sample surface. Since all elements of the periodic table can be removed from a solid sample by sputtering, a mass analysis of the sputtered flux provides a direct measure of the composition of the eroded volume of the sample.

Usually, the vast majority of the particles in the sputtered flux are ejected as neutrals, particularly in the case of clean metals or semiconductors. In order to allow analysis of the sputtered flux by mass spectrometry, the ejected atoms and molecules must first be ionised. Two approaches can be adopted.

Ionisation during escape from the surface

In this case the charged sputtered particles are referred to as 'secondary ions', and the name Secondary Ion Mass Spectrometry is self explanatory. Positively and negatively charged secondary ions are commonly observed in the sputtered flux and both types of ions may be used for analysis. It has been known for some time that the ion fraction of the sputtered flux is very sensitive to the chemistry of the sample surface. Oxidized surfaces or surfaces covered with adsorbates such as water or hydrocarbons show high ion yields, notably in the positive SIMS spectrum, whereas the yields from clean metals and semiconductors are usually rather low. If one is interested in a highly sensitive analysis of the sample composition as a function of depth, it is necessary to choose experimental conditions that maximise the ion fraction in the sputtered flux.

Post-ionisation of sputtered neutrals by impact of electrons, ions or photons at macroscopic distances from the surface

In contrast to SIMS, SNMS makes use exclusively of positively charged ions. Whilst negative ions may be produced by electron attachment or charge transfer in collisions between neutrals and negative ions, these processes have not been used in routine analysis.

The key issues are the same in SIMS and SNMS. The aim is to make the signal I_i (counts/sec), measured for a certain isotope of element i, as high as possible. The problem may be illustrated with reference to Equation 1 which describes, in simplified form, the dependence of I_i on the degree of ionisation P_i of sputtered particles, and parameters of the sample and instrument,

$$I_i = c_i \, \gamma_i \, P_i \, Y \, \tau_i \, \eta_i \, i_0 \tag{1}$$

where c_i is the elemental concentration (atomic fraction), γ_i the isotopic abundance, Y the sputtering yield (atoms/ion) of the sample, τ_i the instrument transmission, η_i the detector efficiency, and i_0 the primary ion current (ions/s). Equation 1 might be taken on first examination to indicate that, for a given sample composition, *i.e.* for fixed value of $c_i\gamma_i$, we could increase I_i by making the last five factors as large as possible. However, if only a limited amount of material is available, maximising Y and i_0 does not result in a higher integral signal from the sputtered volume, but merely reduces the total time required for the measurement. Although fast analysis may often be desirable, the key parameter determining the signal intensity is the fractional (useful) ion yield, *i.e.* the product $P_i\tau_i\eta_i$, which defines the fraction of emitted secondary ions detected and analysed by the instrument. With modern secondary-ion detectors (*e.g.* channeltrons) detection efficiencies close to unity can be achieved routinely. Therefore, to achieve a high fractional ion yield we must concentrate on maximising P_i and τ_i.

In this chapter we shall deal mostly with the methods and mechanisms of ionisation in SIMS and SNMS. Whereas the basic procedures for maximising P_i are well established there are still many problems that need to be solved. Instrumental aspects are only discussed briefly, and since we are dealing with the analysis of sputtered particles, we start with a short overview of relevant aspects of the sputtering process. Reviews of various aspects of surface and depth analysis as well as on sputter depth profiling can be found elsewhere (Wittmaack 1991, 1992a).

2 Sputtering

Sputtering is the term used for the erosion of a sample by bombardment with heavy ions. The minimum energy required for sputter ejection of atoms (and molecules) is about 50 eV (Andersen and Bay 1981). In technical applications of the sputtering process ion energies range from about 0.5 to 10 keV. At these energies an ion entering a solid will lose its energy mostly in elastic collisions with target atoms. Before the ion comes finally to rest, it can set many target atoms into motion, either directly or via collisions of primary knock-on atoms with other target atoms. The collective event is called a 'collision cascade' (Sigmund 1969 and 1981). If the energy transferred to a target atom exceeds a threshold limit the struck atom is displaced permanently from its original lattice site, leaving a vacancy. Sputtering occurs if the recoiling atoms reach the solid-vacuum interface with an outward directed momentum and an energy exceeding the surface potential barrier.

The most common methods for determining the sputtering yield are based upon a post-bombardment measurement either of the change in mass of the sample or of the volume of the crater produced in the sample (Fine 1980). The latter method is usually employed in connection with depth profiling experiments because the quantity of interest, *i.e.* the erosion rate in the central part of the bombarded area, can be determined directly from the crater depth and the total bombardment time.

In the collision cascade region the theoretically predicted dependence of the (back-) sputtering yield on the bombardment and sample parameters (Sigmund 1969 and 1981) can be expressed as

$$Y(E, \theta, \mu) = \Lambda \alpha \left(\frac{dE}{dz} \right)_n \tag{2}$$

where z is the depth from the sample surface. Λ is a material parameter proportional to the depth of origin of sputtered particles L and inversely proportional to the surface binding energy U_0 (Sigmund 1969 and 1981):

$$\Lambda = \frac{L}{\pi^2 U_0} \tag{3}$$

The mean depth of origin of sputtered particles is less than 1 nm (Wittmaack 1991). For purely elastic collisions, the dimensionless parameter α depends only on the target-to-projectile mass ratio, μ, and the impact angle θ. The θ-dependence can be approximated by

$$\alpha(\theta) = \alpha(0) \cos^{-p} \theta \tag{4}$$

Transport theory predicts that the power p in Equation 4 depends on the mass ratio μ and decreases from almost 2 for $\mu < 1$, to less than 1 for $\mu = 10$ (Sigmund 1969). The dependence of sputtering yield on the primary-ion energy E and on the masses and atomic numbers of the incident ion and the target atoms is determined by the nuclear stopping power $(dE/dz)_n$ encountered by the incident primary ion as it enters the sample.

Crystalline targets that essentially retain their regular structure under ion impact exhibit energy and angular dependencies of the sputtering yield that differ significantly

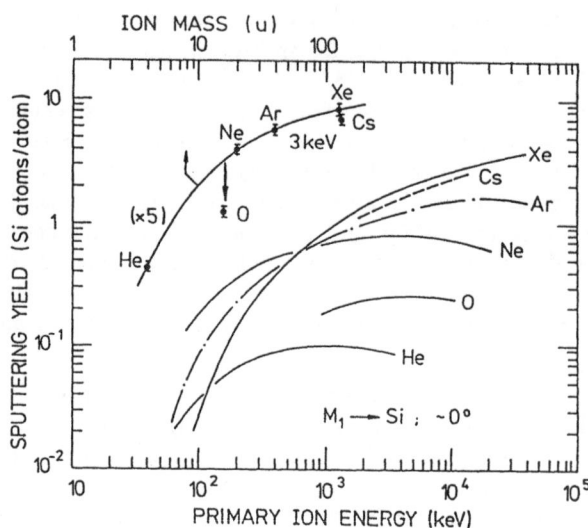

Figure 1. *Sputtering yield of silicon (measured as silicon atoms per atom of primary-ion species) versus the energy (bottom scale) for various primary ions at normal incidence. The data at the top left show the variation of yield with ion-mass (upper scale) for a fixed energy of 3 keV/atom.*

from the behaviour of amorphous materials (or those amorphized by ion bombardment). These differences are usually explained on the basis of transparency and channelling (Roosendaal 1981). A primary ion approaching the sample parallel to a low-index direction 'sees' only the outermost layer, whereas atoms beneath are hidden in the shadow of the top layer. Thus the probability for collisions between an incident ion and subsurface target atoms is significantly reduced and a sizeable fraction of the incoming beam is caused to propagate along lattice rows. Hence the sputtering yield which, from equation 2, is directly proportional to the nuclear energy deposition near the surface, is reduced by channelling.

Figure 1 shows the energy dependence of the sputtering yield in the case of silicon bombarded with various primary ions at normal incidence (Wittmaack 1981a, 1985, 1991, Wittmaack and Poker 1990). Silicon is known to be rendered amorphous at bombardment fluences equivalent to the removal of less than a monolayer (Poate and Williams 1984) and therefore any effect of the sample crystallinity is suppressed. The threshold energy for sputtering of silicon ranges from about 30 to 100 eV, depending on the mass of the primary ion. At higher energies the sputtering yield increases, passes through a maximum and then decreases again. Above 1 keV the yield increases with increasing primary ion mass. For inert-gas ion bombardment this energy and mass dependence agrees with the predictions of Equation 2 (Wittmaack 1992a, Andersen and Bay 1981, Sigmund 1969 and 1981). With oxygen bombardment, however, the data deviate from a monotonic projectile-mass dependence; the yield is a factor of about 2.6–3.3 lower than expected from interpolation of the results for inert gases, see data in Figure 1, top left (Wittmaack and Poker 1990). This deviation results from oxygen bombardment at near normal incidence causing formation of SiO_2 (Reuter 1986, Dowsett 1991) for a depth that roughly equals the ion range (Dowsett 1991, Vancauwenberghe *et al.* 1992). As soon as sufficient oxygen has been implanted to complete the oxidation, the ratio of oxygen to silicon atoms removed by sputtering becomes 2:1. However, standard high-fluence methods for determining the sputtering yield record essentially only the removal of silicon atoms. If we ignore differences in the mean depth of origin and the surface binding energy of silicon and oxygen, as well as changes in the nuclear stopping power, one would expect a reduction in sputtering yield that is directly proportional to the change in silicon concentration, (*i.e.* a drop in yield by a factor of 3). This is in good agreement with the factor of 2.8 obtained at 3 keV (Figure 1).

The case of oxygen bombardment deserves particular attention because, as we shall see in Section 4, primary oxygen ions are commonly used to generate a high degree of ionisation in positive SIMS measurements. For negative ions, on the other hand, to achieve a high yield one uses caesium primary ions. The sputtering yield of silicon is not significantly different from that observed with xenon ions of almost the same mass as caesium. The 20% lower yield using caesium ions compared with xenon (Figure 1) may be due to the effect of the stationary caesium coverage of the bombarded surface.

Examples of the angle-of-incidence dependence of the silicon sputtering yield are shown in Figure 2a (Wittmaack 1985, Morgan *et al.* 1981, Wittmaack 1983a, Homma and Maruo 1989). We note that at impact angles below about 50° the measured yields using inert-gas and a caesium ion bombardment show dependences that agree reasonably well with Equation 4. At larger angles the yields pass through maxima and then drop to zero near 90° (not shown), because the amount of energy deposited in the sample decreases rapidly towards grazing incidence.

The results obtained for oxygen again deviate from the trend observed with inert-gas or caesium ion bombardment. For oxygen ions the change in sputtering yield with impact angle θ is exceptionally large because the amount of oxygen retained in the sample decreases rapidly as θ exceeds 27°±2°. At this angle the (partial) sputtering yield of silicon is ~ 0.5 atoms per oxygen atom, which means that the ratio of the oxygen-to-silicon removal rates, $R_{O,Si} = 1/Y_{Si}$, is 2:1, *i.e.* the same as the concentration ratio in SiO_2. For $\theta < 27°$ the results of Figure 2a yield $R_{O,Si} > 2$. In order to satisfy conservation of mass, we have to assume that the oxygen released in excess of $R_{O,Si} = 2$ must leave the sample by out-diffusion (Littlewood and Kilner 1988). For $\theta > 27°$ the (partial) sputtering yield of silicon becomes so high that $R_{O,Si} < 2$ and, consequently SiO_2 can no longer be formed. For more oblique oxygen-ion incidence, the angular dependence of the silicon sputtering-yield approaches that observed with inert gas ions.

It should be noted that silicon is not typical of the many kinds of samples encountered in SIMS and SNMS measurements. Most other materials retain only a small fraction of the injected oxygen. This is reflected in the angular dependence of the sputtering yield of oxygen-bombarded Ge (Wittmaack 1983a), InGaAs and InP (Homma

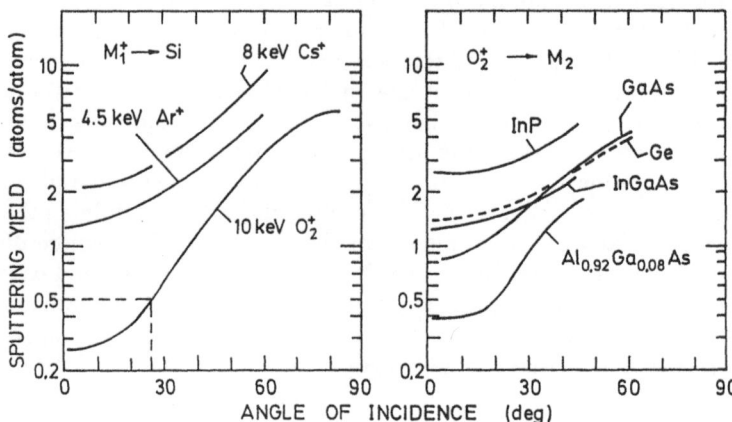

Figure 2. *Angular dependence of sputtering yields. (a) Silicon bombarded with various ions. (b) Different targets bombarded with oxygen ions.*

and Wittmaack 1989) shown in Figure 2b. With these target materials a yield depression at near normal incidence is not evident; see data for Ar and Cs bombardment in Figure 2a. On the other hand the angular dependence of the sputtering yield of GaAs (Wittmaack 1983a, Homma and Wittmaack 1989) and Al GaAs (Homma and Wittmaack 1989) suggests that moderate (GaAs) or even rather efficient beam induced oxidation (AlGaAs) can be achieved at near normal oxygen incidence. In the latter case this can be attributed to the fact that aluminium has a strong affinity to oxygen.

The fraction of oxygen retained in the sample after injection is determined by at least two factors: the sputtering yield, and the affinity of the matrix atoms to oxygen. The heat of formation of the oxide appears to be an important parameter (Homma and Wittmaack 1990). Recent experiments by Vancauwenberghe *et al.* (1991) have emphasised that low sputtering yields are a necessary prerequisite for achieving beam induced oxidation. Even with GaAs, for which only partial oxidation is observed at energies commonly employed in SIMS (typically 10 keV), complete oxidation of Ga can be achieved by lowering the oxygen energy to 500 eV (Vancauwenberghe *et al.* 1991). However, our understanding of beam-induced oxidation phenomena is still far from complete.

It is important to note that the sputtering yield corresponds to an integration over the *energy* and *angular distribution* of all atoms and molecules sputter-ejected from the sample (Hofer 1991). The differential quantities are very important and must be carefully considered in instrument design for SIMS and SNMS.

Figure 3 depicts the energy spectra of atomic and molecular Al_n^+ ions ($n = 1, 2$, and 6) emitted from sputter-cleaned aluminium under impact at normal incidence of

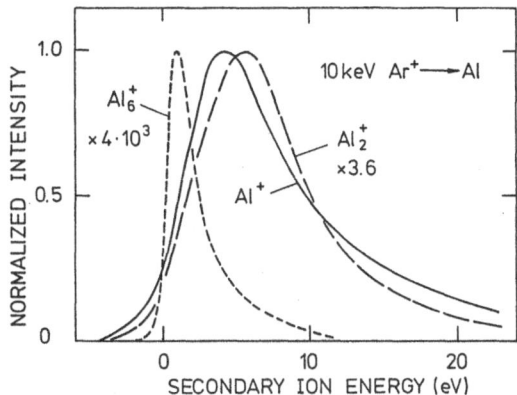

Figure 3. *Energy spectra for atomic and molecular secondary-ions sputtered from aluminium (Wittmaack 1975).*

10 keV Ar^+ ions (Wittmaack 1975). At very low energies the measured secondary ion yield increases rapidly, passes through a well defined peak and then decreases towards higher energies. For large ion clusters ($n \geq 3$) the peak moves towards low energies with increasing n. Moreover, the fall-off on the high-energy side of the peak is steeper for larger n.

The spectral features seen in Figure 3 are observed generally in SIMS, not only for homonuclear molecular ions but also for heteronuclear ions. Such 'cluster' ions often give rise to undesireable mass interferences with atomic ions of interest; one well-known example is the interference of $^{29}Si^{30}Si^{16}O^+$ with $^{75}As^+$ (Wittmaack 1977). As the results of Figure 3 suggest, this kind of interference can be successfully overcome, albeit at the expense of sensitivity, by setting the window of the secondary-ion energy analyser to sufficiently high energy so that only monomer (and possibly dimer) ions are transmitted to the detector. Even dimers can be removed completely by operating at pass energies above 500 eV (Schauer and Williams 1990).

The angular distribution of sputtered particles is another parameter that determines the measured signal intensity. In general one attempts to align the spectrometer such that the entrance aperture intersects the sputtered flux in the direction of maximum emission. At normal incidence the angular distribution is described reasonably well by a cosine power law,

$$\frac{dY}{d\Psi} = Y(0) \cos^m \Psi \tag{5}$$

where Ψ is the polar angle of emission and $1 \leq m \leq 2$ (Hofer 1991). At oblique incidence the axis of the emission cone tends to be tilted away from the surface normal. With single-crystal targets, preferential emission is observed, specifically along low-index directions (Hofer 1991).

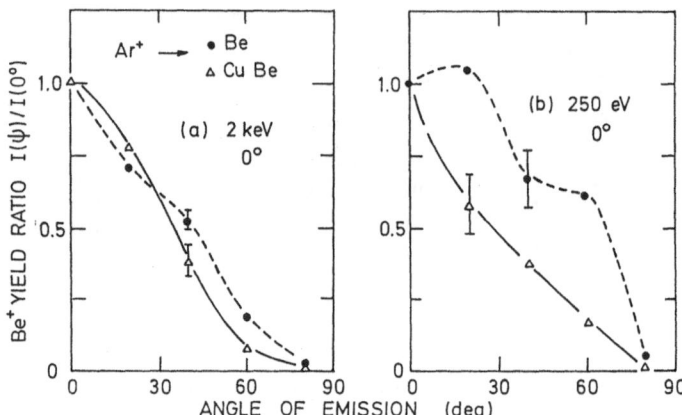

Figure 4. *Comparisons of the angular distributions of Be⁺ sputtered by Ar⁺ ions from two targets of different composition, using two different bombardment energies (Wucher and Reuter 1988).*

For the purpose of quantification, using sensitivity factors derived from standards, it is important to note that, under fixed bombardment conditions, the angular distribution for a certain element may change with sample composition. An important example of this is shown in Figure 4. Wucher and Reuter (1988) used electron impact SNMS to measure the angular distribution of Be sputtered from both an elemental beryllium target and from a CuBe alloy by Ar⁺ at normal incidence. The Be distributions observed using a bombardment energy of 2 keV, Figure 4a, are somewhat different, but in rough accordance with Equation 5 for $m=3$. However, at 250 eV the distributions for these two targets are dramatically different; see Figure 4b. This means that we must add an angular dependence factor in Equation 1 which is a function of the sample composition. In other words, a sensitivity factor determined for a given element in one type of sample cannot be used to quantify the measured signal for the same element in another sample. Note that this is presumably only the effect of sample composition on the angular distribution of sputtered atoms, and not necessarily an ionisation problem. The effect appears to be less dramatic at bombardment energies of a few keV, but the accuracy of a quantification procedure based upon the use of standards needs to be evaluated carefully before a definitive statement can be made.

3 Instrumentation for SIMS and SNMS

The instrumentation needed to perform SIMS or SNMS analyses is basically simple. The essentials are a vacuum chamber, an ion gun, a sample manipulator and a mass

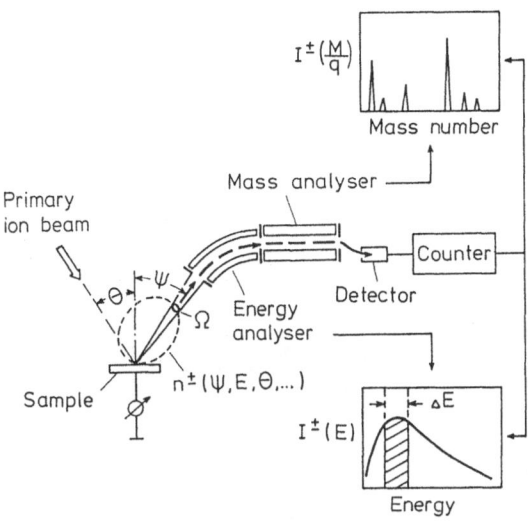

Figure 5. *Schematic of a quadrupole based* SIMS *instrument.*

spectrometer. The latter can be a magnetic sector, quadrupole filter or time-of-flight system. For analysis of the surface composition by low-fluence ('static') SIMS, time-of-flight instruments are usually employed, while for high fluence ('dynamic') SIMS either of the first two types of mass analysers are used. The design aspects that need to be considered in a quadrupole based SIMS instrument are reviewed elsewhere (Wittmaack 1982). A typical system is shown in Figure 5.

Most modern SIMS instruments are equipped with two ion guns, one as a source of oxygen ions and the other for caesium ions. Because of the large effect of the impact angle θ on the steady-state concentration of the implanted primary ions at the surface, or in the topmost layers of the sample (see Figure 2, and Section 4), it is important to allow for a free choice of θ, preferably from 0° to at least 80°. Rotatable sample stages can be easily fitted to quadrupole based systems because the low extraction field above the sample has a minor effect on the trajectories of the primary ions; care has to be taken to see that the acceptance of the secondary-ion energy analyser does not change too much with the tilt angle of the sample. With sector-field instruments the sample must not be deliberately tilted because of the high extraction field needed for efficient secondary-ion extraction. Sample tilt angles different from the nominal design orientation can severely distort the secondary-ion trajectories with a corresponding loss in transmission.

Quadrupole based SNMS and SIMS instruments are very similar and, in principle, a quadrupole SNMS may also be used for SIMS studies. However, optimisation for SNMS prevents high performance in the SIMS mode. There are two main differences between the two modes. First, in SNMS instruments only a single ion gun is needed, and the

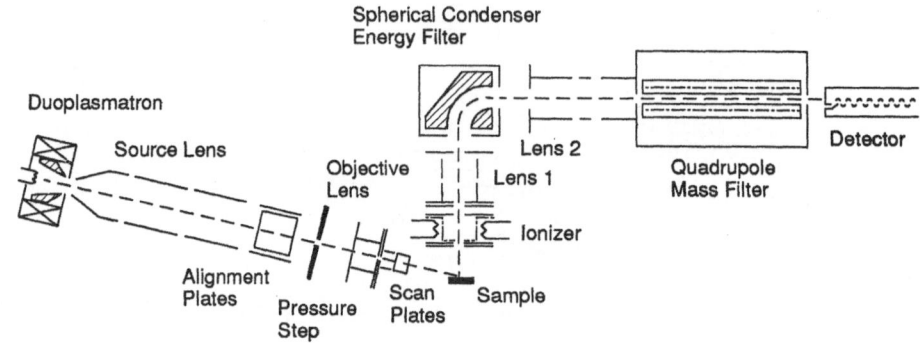

Figure 6. *Layout of an* SNMS *system with electron beam ionisation (Gnaser et al. 1985).*

primary ions should preferably be of an inert gas. Second, some means of post-ionising the sputtered neutrals in the space between the sample and the energy filter is needed. For post-ionisation by electron impact, two methods have been developed to date; the electron-beam and the electron-gas techniques.

Historically, the first approach used was to direct an electron beam at the cloud of sputtered particles (Honig 1958). More recent instruments use modified ionisation cells of the type commonly employed in residual gas analysers (Lipinsky *et al.* 1985, Gnaser *et al.* 1985, Jede *et al.* 1992). An example is depicted in Figure 6. To provide discrimination against secondary ions and ionised residual gases, simple suppressor plates (Gnaser *et al.* 1985) or more sophisticated ion-optical arrangements (Lipinsky *et al.* 1985, Jede *et al.* 1992) are added to the ioniser.

A different approach, for electron-beam SNMS, has been explored recently by Gersch and Wittmaack (1993). The idea is to bombard the sputtered neutrals with a focused electron beam (Figure 7) at a distance of 100 μm or less from the sample surface, rather than at the distance of some 20 mm used in standard electron beam SNMS systems (as, for example, in Figure 6). With the newer method the fractional ion yield can be improved by about an order of magnitude. On the other hand, discrimination between secondary ions and singly-charged post-ionised sputtered neutrals is much less efficient using focused e-beam SNMS. However, because of the relatively high electron energy (300 eV) employed in the prototype focused e-beam SNMS system, doubly (and even multiply) charged ions are produced with remarkable yields (Figure 8). Since secondary ions are usually only singly charged (with the exception of those of the light elements) doubly charged post-ionised neutrals can be used for quantitative analysis.

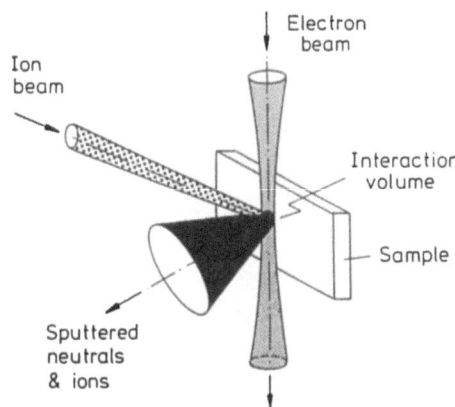

Figure 7. *Geometry employed in focused electron-beam (e-beam)* SNMS *(Gersch and Wittmaack 1993).*

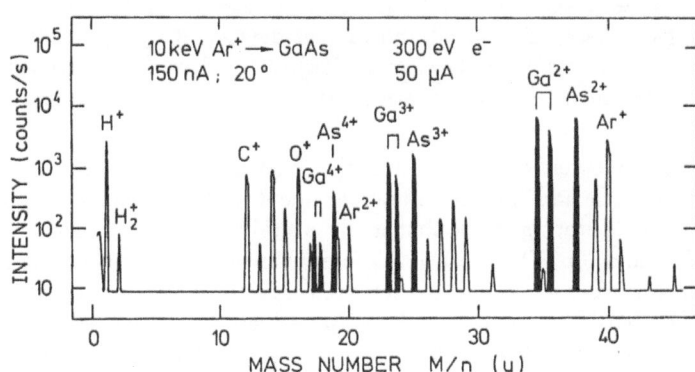

Figure 8. *Mass spectrum of doubly and multiply charged Ga and As ions produced when focused e-beam* SNMS *is combined with sputtering of GaAs by Ar⁺ bombardment (full mass lines) (Gersch and Wittmaack 1993). Note that the doubly-charged ion species of Ga and As exhibit almost the same intensity, as for singly-charged ions in standard* SNMS *(Gnaser et al. 1985).*

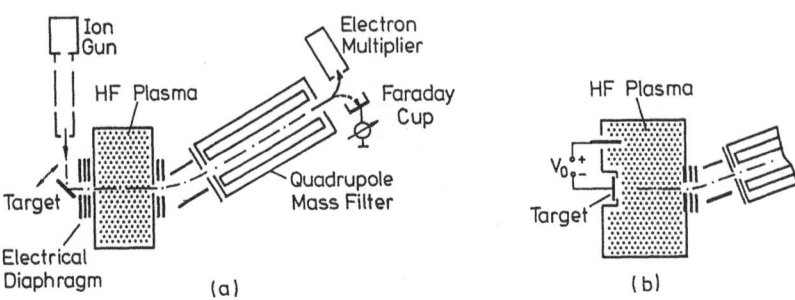

Figure 9. *Schematic of two alternative experimental arrangements for electron-gas* SNMS; *(a) external-bombardment mode; (b) direct-bombardment mode.*

As an alternative to ionising sputtered neutrals by electrons from a heated filament, a plasma may also be used as the source of electrons. This is the approach used in 'electron-gas' SNMS, introduced and further developed by Oechsner and co-workers (Oechsner and Gerhard 1972, Oechsner 1984). The electrons are generated in an inductively excited high-frequency plasma, with electron temperatures that correspond to energies of $\sim 10 - 15$ eV.

Two alternative versions of electron-gas SNMS can be distinguished, the external-bombardment mode and the direct-bombardment mode (Figure 9). The experimental arrangement for operating in the first mode is similar to standard electron-beam SNMS; *c.f.* Figures 6 and 9b. The essential difference is in the lay-out of the ionisation cell. Whereas very little work has been reported using the external-bombardment mode, the direct-bombardment mode has been investigated in some detail (Oechsner 1984, Wittmaack 1991, Jede *et al.* 1992). The advantage of the direct mode is that sputtering can be done routinely with bombardment energies as low as 200 eV, so that beam-induced broadening effects in depth profiling can be minimised. On the other hand, the operating conditions required for uniform erosion over the large area analysed (typically 4 mm diameter) are not straightforward, and the optimum combination of target position and target bias usually has to be found by trial and error. It should also to be mentioned, with reference to Figure 4, that quantification in low-energy SNMS is extremely difficult because of the change in angular distribution of sputtered neutrals with composition. To obtain quantitatively meaningful results, measurements at a few keV are preferable (Wucher and Reuter 1988).

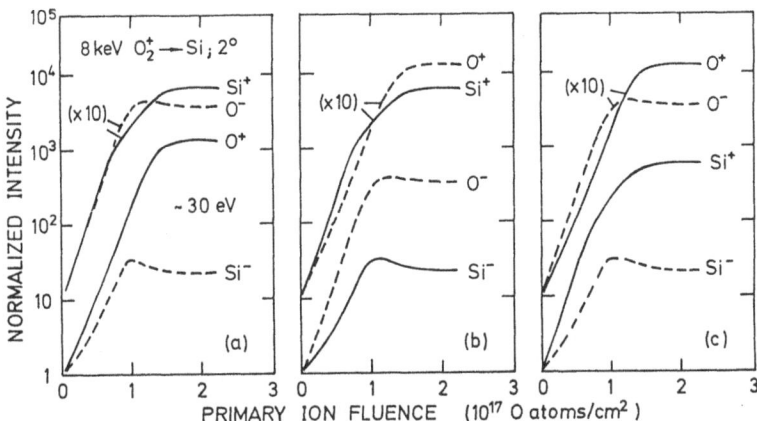

Figure 10. *Evolution of the signals of positively and negatively charged secondary ions of Si and O emitted from silicon bombarded with oxygen at near-normal incidence. The curves are normalised to the secondary ion intensities measured from clean silicon at the onset of oxygen bombardment. Panels a–c show identical data with different relative scaling factors.*

4   Achievement of efficient ionisation in SIMS

The common method for achieving efficient ionisation in SIMS is to bombard the sample with ions of either oxygen or caesium. An enhancement with ion fluence occurs using oxygen primary ions, as depicted in Figure 10 where the relative changes in the positive and negative secondary-ion yields are shown for a silicon sample progressively implanted with oxygen at near-normal incidence (Wittmaack, unpublished work).

A detailed interpretation of the observed enhancement is difficult. To illustrate this, the data in Figure 10 are presented in three different comparisons of positive and negative ion yields, using different combinations of relative scaling factors. Whereas the results of Figure 10a suggest a correlation between the emission of Si^+ and O^-, no such correlation is evident for Si^- and O^+. The correlation of Si^+ with O^- becomes more striking if we take into account the change in slope of the Si^+ build-up curve, observed at a fluence of about 0.7×10^{17} O atoms/cm². This is probably due to an oxygen-induced change in the partial sputtering yield of silicon, as discussed in Section 1 with reference to Figures 1 and 2. There appears also (Figure 10b) to be a correlation between the positive secondary-ion yields of Si^+ and O^+, whose transitions to steady-state intensities occur at the same fluence, and between Si^- and O^-, which exhibit similar local maxima features before attaining steady-state levels. However, hardly any correlation appears to exist (Figure 10c) between positive and negative secondary ions of the same element. In particular, for the matrix species, the positive ions exhibit a much stronger yield enhancement than the negative-ion species. For this reason oxygen bombardment is

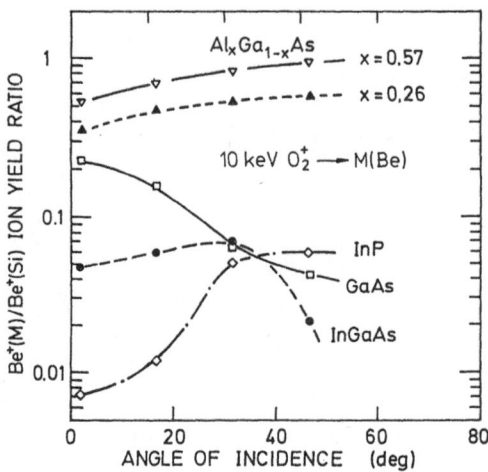

Figure 11. *Ratio of the fractional ion-yield of Be⁺, sputtered from different Be-implanted matrices, to the yield from Be-implanted Si (Homma and Wittmaack 1989).*

normally used only for positive secondary-ion analyses.

The most important feature of Figure 10 is the pronounced enhancement of positive secondary-ion yields, caused by beam-induced oxidation of silicon. If the retained oxygen is the cause of the strong enhancement, we could conclude, on the basis of the sputtering yield data (Figures 1 and 2), that a much less pronounced gain may be achieved with samples that show no oxygen effect in the angular dependence of their sputtering yields. Moreover the enhancement for silicon would then disappear at oblique beam incidence. This has in fact been observed (Wittmaack 1983a). Here, we shall therefore consider both these aspects in combination.

The common approach for studying the matrix dependence of the secondary ion yield is to implant a known fluence of a given dopant into a variety of matrices and to measure the fractional ion yield (Deline *et al.* 1978b, Williams 1979). This method is meaningful because it can be shown that, for oxygen and caesium ions the induced yield enhancements are almost the same for both matrix and dopant elements (Deline *et al.* 1978a, Wittmaack 1981b). In Figure 11 the matrix dependencies of the fractional ion-yield of Be⁺ for several matrices are shown as a ratio to the Be⁺ yield from silicon, for varying angle of incidence of the oxygen beam (Homma and Wittmaack 1989). Two findings are evident: first, the Be⁺ yield depends strongly (by a factor ∼100 or more) on the chemical identity of the matrix from which it is emitted; and second, the ratio of the Be⁺ yield for two matrices depends on the impact angle of the oxygen beam. Such strong matrix effects are a key issue in SIMS analyses. They are particularly severe for oxygen bombardment, and the physics of these yield enhancements is still not well understood. Reasonably successful models invoke the breaking of quasi-ionic bonds formed in an oxidising environment for the escaping metal atom (Yu and Mann 1986,

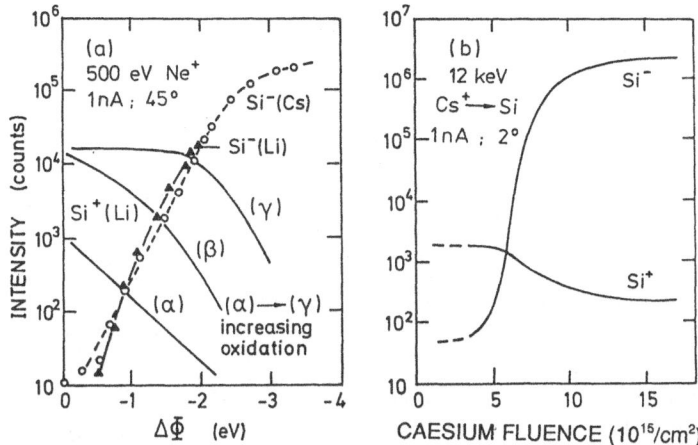

Figure 12. *(a) Intensity of Si⁻ and Si⁺ secondary ions sputtered by a low fluence of primary neon ions, from silicon, versus the change in sample work function (produced by vapour deposition of increasing amounts of Cs or Li) (Yu 1982, 1983). The Si⁺ data, α, β and γ, were obtained for slightly (to moderately) oxidised samples. (b) Si⁺ and Si⁻ sputtered from silicon by bombardment with Cs⁺ ions at near normal incidence (Wittmaack 1983b).*

Yu 1991). However, an analytically useful model is not yet available.

Turning to negative secondary-ion detection, even more pronounced matrix effects have been reported using caesium bombardment and negative secondary ion detection (Deline *et al.* 1978b). Here, however, a comprehensive theoretical model for the ionisation process is available (Lang 1983, Yu and Lang 1983, Yu 1991). This is the so-called electron-tunneling (or work function) model, which has been described in great detail in a recent review by Yu (1991). A test of the model is reasonably simple. The work function of a sample can be changed by depositing sub-monolayer quantities of alkali metals on the surface. For example Figure 12a shows the strong increase of the Si⁻ signal with decreasing work function. For sputtering, the sample was bombarded with low fluences of 500 eV Ne⁺ ions ($\sim 10^{12}$ Ne⁺ ions/cm² per data point), so that the change in alkali coverage during data acqustion was negligible. Compared with clean silicon, the degree of ionisation of Si⁻ can be increased by as much as a factor of 2x10⁴ by caesium deposition, an even larger enhancement than can be achieved in the positive SIMS mode by beam-induced oxidation (see Figure 10).

To first order, the dependence of the ionisation probability P^- on the work function Φ can be expressed (Lang 1983, Yu 1991) as

$$P^- \approx \exp[-(\Phi - E_A)/\varepsilon_n] \qquad (6)$$

where E_A is the electron affinity of the sputtered species, and ε_n is a characteristic energy parameter. The Si⁻(Cs) signal (Figure 12a) is in good agreement with the predicted exponential dependence of Equation 6 over several orders of magnitude. Note that the same work function dependence is observed for two quite different alkali metals, namely Li and Cs (Yu 1982).

The ionisation probability for positive secondary ions, P^+, from samples containing submonolayer surface coverages of alkali metals, can also be described by the electron tunneling model (Yu 1991, Yu and Lang 1983), *i.e.*

$$P^+ \approx \exp[-(E_I - \Phi)/\varepsilon_p] \tag{7}$$

where E_I is the ionisation potential of the sputtered atom. A dependence of this form has been observed for slightly oxidised silicon, curve (α) in Figure 12a (Yu 1983). With more completely oxidised samples the predicted exponential fall-off is shifted towards lower work functions, *i.e.* towards high Li coverage, curves (β) and (γ). This has been attributed by Yu to the formation of an electron band-gap in the oxidised sample (Yu 1983).

Essentially the same relative increase in the Si$^-$ yield as for vapour deposition of caesium (Figure 12a), can be achieved by bombarding a silicon sample with Cs$^+$ ions, (Figure 12b) (Wittmaack 1983b). Because of the different method of generating a surface layer of caesium, the fluence-dependence looks quite different from the work function dependence. Comparing the Si$^-$ and Si$^+$ signals, we observe the same 'anti-correlation' as in Figure 12a, *i.e.* the Si$^+$ yield decreases while the Si$^-$ yield increases. However, the reduction in the Si$^+$ yield produced by implantation of caesium is smaller than by vapour deposition (although the Si$^-$ yield enhancement is the same, as already noted). The reason for this is not yet clear. One difference is that the changes in Si$^+$ yield observed as a result of lithium deposition were obtained for samples of oxidised silicon, whereas the implantation relates to conditions of beam-induced sputter cleaning. It is possible also that the large difference in sputter-beam energy (0.5 keV for neon, 12 keV for caesium) plays a role; experiments with inert-gas primary ions show that the Si$^+$ yield increases more strongly with increasing impact energy than does the Si$^-$ yield (Wittmaack 1984).

In the absence of significant diffusion of implanted primary ions, the surface concentration of caesium during bombardment may be estimated from a simple ion-retention model (Schulz and Wittmaack 1976). Attempts to measure the surface caesium concentration by (AES) or (RBS) reveal a strong matrix dependence (Chelgren *et al.* 1979) and a pronounced dependence on the oxygen partial pressure during analysis (Menzel and Wittmaack 1981). More recently medium-energy ion scattering (MEIS) measurements, with improved vacuum conditions, have shown that caesium atoms are highly mobile even under bombardment with a 'small' primary ion like H$^+$ (at 52 keV); and also that caesium atoms residing at the surface can be removed efficiently by H$^+$ impact (Valizadeh *et al.* 1992).

Because of the difficulty in measuring the surface caesium coverage directly by ion-scattering, it is tempting to use the results obtained by Yu (1982, 1983) to derive the information from direct-implantation SIMS data (Figure 12b). Two important conclusions emerge from such a comparison. First, the steady-state caesium coverage of silicon during implantation at normal incidence is $\sim$ 20% of a monolayer. Secondly, since the Si$^-$ yield increases and the (Si$^+$) decreases with decreasing impact energy (Wittmaack 1983b), the caesium surface coverage apparently increases as the energy is reduced.

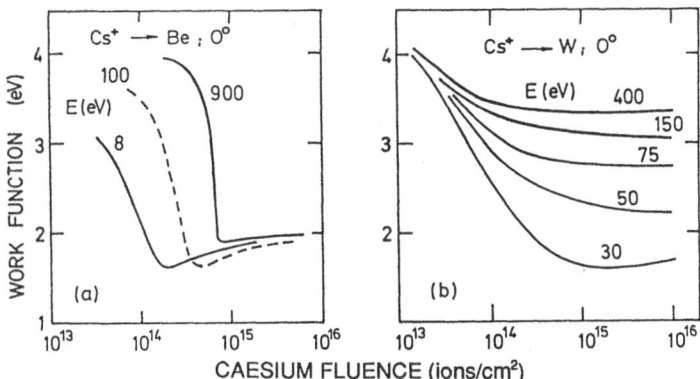

CAESIUM FLUENCE (ions/cm²)

Figure 13. *Evolution of the work function for (a) beryllium and (b) tungsten during implantation of low-energy Cs⁺ ions. (Tampa et al. 1988).*

This can be attributed to a decrease of the Si sputtering yield with decreasing energy (see Figure 1). Sputtering also appears to be the key factor responsible for the pronounced matrix effect observed in negative SIMS studies of silicon and metal silicides (Deline *et al.* 1978b, Chelgren *et al.* 1979). However the ejection characteristics of caesium and its diffusivity must be taken into account in order fully to understand such matrix effects.

It could be of relevance to perform a Cs-implantation experiment and measure positive and negative secondary-ion yields in combination with the work function changes. To date a detailed study has not been made, although a few results on work function changes during ion-implantation have been reported (Tampa *et al.* 1988). The original motivation for such measurements was not related to SIMS, but to H⁻-production in ion sources for fusion research (Tampa *et al.* 1986a,b). Figure 13a shows results obtained for low-energy Cs⁺ implantation in beryllium and tungsten (Tampa *et al.* 1988). In the case of beryllium the work function passes through a minimum before arriving at the steady-state level which is almost independent of energy (and equal to the work function of caesium (Tampa *et al.* 1988)). By contrast, the change in work function achieved under prolonged bombardment of tungsten becomes smaller with increasing energy. This may again be associated with a difference in matrix sputtering efficiency, but yield data (Andersen and Bay 1981) reveal only a small difference between Be and W for inert-gas ion bombardment at the same energy. With the low energies considered here, a dependence of primary-ion backscattering on the mass of the target atom, as well as the ion ranges (Tampa *et al.* 1988) and the diffusion characteristics must be taken into account when interpreting the results of Figure 13.

According to Figure 12a lowering of the sample work function has the effect of reducing the ionisation probability of positive secondary ions. This is observed not only with ions representative of the target material (Yu 1983, 1981), but also with

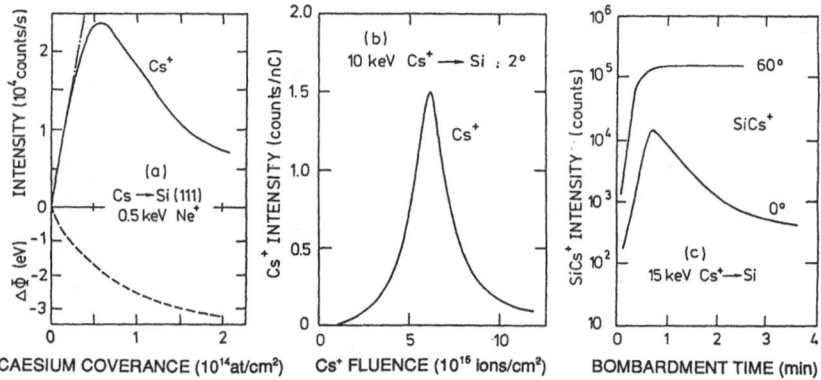

Figure 14. *(a) Cs$^+$ yield and work function change of silicon vs. the caesium coverage produced by vapour deposition (Yu 1984). (b) Cs$^+$ yield as a function of the fluence of Cs$^+$ implanted in silicon (Wittmaack 1983b). (c) Variation of the SiCs$^+$ yield as a function of time during bombardment of silicon with Cs$^+$ ions (Wittmaack 1992b).*

ions of the deposited (Yu 1984) or implanted (Wittmaack 1983b) alkali metal. Results illustrating the variation of the Cs$^+$ yield with the Cs coverage, or fluence, are compiled in Figures 14a and 14b, respectively. Also shown in Figure 14a is the change in work function with coverage (Yu 1984). The two Cs$^+$ curves in Figures 14a and 14b are similar; more specifically, we note that both yield curves increase from a very low initial level to a pronounced maximum, and then decrease to a steady-state level. The most important difference appears to be in the relative height of the peaks compared to the steady-state level, being opposite to that of Si$^+$ (Figure 12). Again the origin of this difference is unclear.

Whereas the results for Cs$^+$ in Figure 14a,b are not directly of interest analytically, the species MCs$^+$, which shows a fluence (or time dependence) similar to Cs$^+$, has become increasingly important recently for analysis. This dimer (or cluster) ion, MCs$^+$, is composed of a major or minor constituent M of the target and an ejected (resputtered) species of previously implanted Cs. For some elements such species can appear with remarkably high yields in the SIMS spectrum (Storms *et al.* 1977), but analytical applications have remained the exception until recently (Gauneau *et al.* 1988). Several papers over the past few years suggest that almost quantitative SIMS analyses of an element M can be achieved using MCs$^+$ ions (Gao 1988, Magee *et al.* 1990, Gnaser and Oechsner 1991). Hardly any matrix effect is observed, for example, with III–V semiconductor targets (Gao 1988, Magee *et al.* 1990). However, as observed in Figure 14c, the evolution of the SiCs$^+$ signal during the initial phase of Cs implantation depends strongly on the impact angle. At normal incidence the MCs$^+$ yield exhibits the same

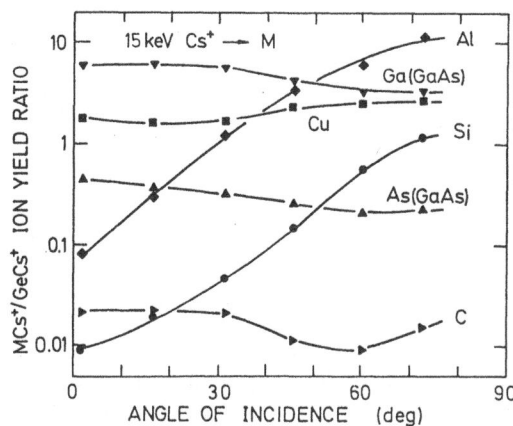

Figure 15. *Ratio of the MCs+ yield from the matrix M to the yield from Ge vs the impact angle of the Cs beam.*

kind of variation as the Cs+ yield [note the linear scale in Figure 14b as opposed to the logarithmic scale in Figure 14c]. At 60° the MCs+ yield increases monotonically with increasing time of bombardment (or fluence) and a local peak in the build-up curve is no longer observed (Wittmaack 1992b).

On the basis of these results we may draw the following conclusions: the steady-state caesium coverage of silicon with a Cs+ beam at normal incidence is high so that the ionisation probability of positive secondary ions, notably of Cs+ and MCs+, is reduced to below the maximum value (probably near to unity for Cs+). Since silicon has a comparatively low sputtering yield we might expect with other high yield target materials under Cs+ bombardment that the steady-state caesium coverage would be lower. Accordingly, the lowering of the positive secondary-ion yields should be less than with silicon. As a result of the matrix-dependent reduction in ion yield at normal incidence, we would then find a further matrix effect when using MCs+ secondary ions for quantitative analysis. The results in Figure 14c suggest, however, that we might be able to overcome this matrix effect by performing the analysis with the beam at glancing angle of incidence.

This hypothesis is supported by the results of detailed investigations into the angular dependence of MCs+ emission from a variety of target materials (Wittmaack 1992b), which show that for targets of medium atomic mass (Cu, Ge, GaAs) the ion-yields are roughly proportional to the product of the sputtering yield and the (estimated) instrument transmission; *i.e.* the ionisation and formation probability of the MCs+ ions is nearly independent of the impact angle θ. Accordingly, the yield-ratio MCs+/GeCs+ is almost independent of θ, as illustrated in Figure 15. (The small residual θ-dependence is probably due to differences in angular dependence of the sputtering yield). By contrast, the yield-ratio MCs+/GeCs+ for matrices M of Al or Si increases monotonically with increasing θ, by as much as two orders of magnitude, and only for angles above $\sim 70°$ are there indications of the yield-ratio saturating. It is evident that with samples like Al and Si the emission of MCs^+ ions suffers from a severe matrix effect, for impact angles $< 65°$. In this case the ionisation probability of the MCs^+ dimers falls well below the maximum, which can only be achieved when the bombardment-induced reduction of the sample work function is small (< 1 eV). This implies that the steady-state caesium coverage of the sample must stay below a certain critical value and one way of achieving this is to increase the matrix sputtering yield. The critical sputtering yield may be assessed as follows. For an approximately constant $SiCs^+/GeCs^+$ ion-yield ratio (Figure 15), the impact angle should be at least 70°. From Figure 2a the sputtering yield with this angle of incidence is 10–12 Si atoms per Cs^+ ion. This may be considered as the critical yield. (The differences in sputtering yields at 8 and 10 keV can be ignored; see Figure 1).

Investigations into the use of MCs^+ secondary ions for quantitative analysis are still at an early stage. It is possible that the MCs^+ mode might provide SIMS compositional analysis at the quantitative level presently achievable with AES. To this end, efforts to improve our understanding of the physics of MCs^+ emission are well motivated (see, for example, Fabian, this Proceedings). It should be noted that the elemental variation in sensitivity is also much smaller than in 'standard' SIMS using positive or negative secondary atomic ions. However, traditional SIMS will remain important because of the excellent sensitivity it achieves with most elements.

5   Electron-impact ionisation of sputtered neutrals

Compared with SIMS a discussion of ionisation phenomena in electron-impact SNMS is straightforward. The main features of the ionisation process are well understood, and the cross-sections are known for many elements. What we examine here are the experimental parameters that can be varied to maximise the ionisation probability and to minimise its variation from element to element.

In electron-impact SNMS the ionisation probability P^+ may be written (Gersch and Wittmaack 1993)

$$P^+ = \frac{F\sigma I_e}{ev_0 w} \tag{8}$$

where F is a factor describing the overlap between the electron beam and the cloud of sputtered atoms ($F < 1$), σ is the cross-section for electron-impact ionisation, I_e the electron current, e the elementary charge, v_0 the velocity of the neutrals normal to the electron beam axis, and w the width of the electron beam normal to the direction of propagation of electrons and sputtered neutrals. To maximise F, the electron beam should have a well-defined shape and be aligned carefully with respect to the cloud of sputtered neutrals. Moreover, the current per unit length I_e/w should be made as large as possible, the upper limit being set by the electron source design, and ultimately by space charge effects. For very high current densities it is necessary also to consider the potential trough generated by the electron beam. The depth of this trough can be easily a few volts (negative), so that post-ionised sputtered neutrals may not be able to escape from the ionisation volume (Gersch and Wittmaack 1993). In that sense, v_0 and

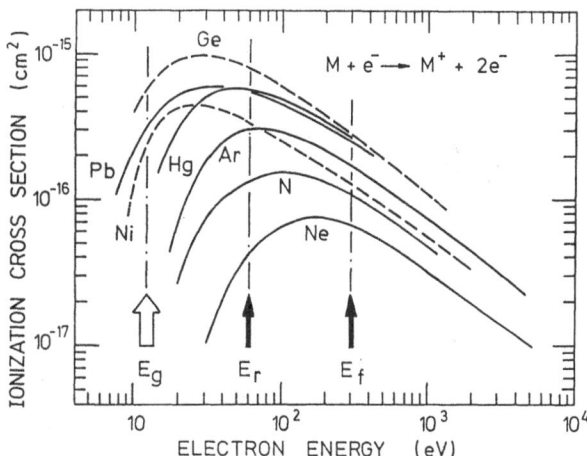

Figure 16. *Electron-impact ionisation cross-sections vs electron energy. Full lines: smoothed experimental data, dashed lines: theoretical results (Tawara and Kato 1987). The arrows mark the (mean) energies typically employed; in e-gas* SNMS *(E_g), remote e-beam* SNMS *(E_r), and focused e-beam* SNMS *(E_f).*

I_e/w are not really independent of each other. In order to optimise I_e/v_0w we have to take the energy spectrum of the sputtered neutrals into account (Sigmund 1981).

We now consider some aspects related to the choice of the electron energy. Figure 16 shows electron-impact ionisation cross-sections for a number of elements, taken from a recent compilation (Tawara and Kato 1987). All curves are of similar shape, but the low-energy threshold (*i.e.* the ionisation potential) and the position of the maximum are element specific. To maximise the cross-section for most elements, except for those featuring a high ionisation potential, we would choose an electron energy between 40 and 70 eV, *i.e.* the same energy as in residual gas analysis. Standard (or 'remote') e-beam SNMS systems are operated at this energy (marked E_r in Figure 16); the energy E_f used in focused (or 'matched') e-beam SNMS is 300 eV (Gersch and Wittmaack 1993). At both energies the cross-sections can be derived with reasonable accuracy from smoothed experimental data, as in Figure 16 (full lines).

The number of elements for which experimental data are available is limited (Tawara and Kato 1987), but often theoretical cross-sections provide an alternative (as for the dashed curves in Figure 16). The case of e-gas SNMS is quite different because ionisation is due to impact of low-energy plasma electrons with an assumed Maxwell-Boltzmann velocity distribution, and a characteristic temperature T_e corresponding then to an energy E_g of $\sim$8–13 eV (Jede *et al.* 1992, Oechsner 1984), which depends on the plasma conditions – notably the pressure (Jede *et al.* 1988). For most elements E_g is around the ionisation threshold, and in several cases below it. Ionisation can only be achieved by impact of electrons originating from the high-energy tail of the velocity distribution.

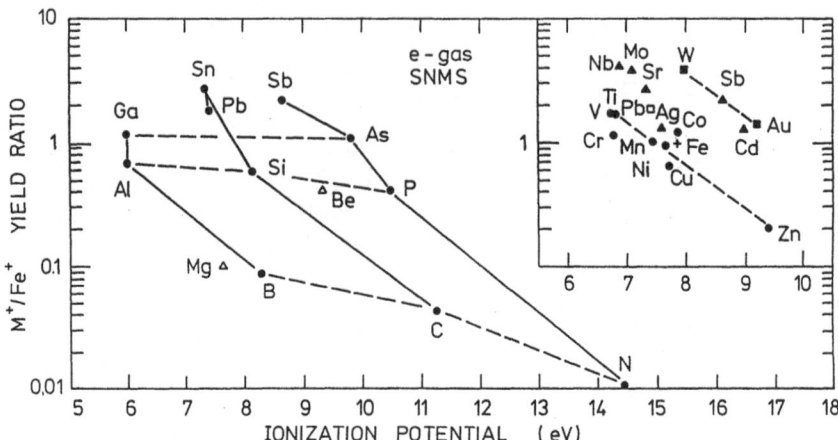

Figure 17. M^+/Fe^+ *yield ratio observed in electron-gas* SNMS *vs the ionisation potential of element M (Wucher et al. 1988).*

Accordingly, the ionisation efficiency is relatively poor for elements with a high ionisation potential. In e-gas SNMS we thus expect a much more pronounced variation in the elemental sensitivities then in e-beam SNMS.

This conclusion is supported by the experimental results reported by Wucher *et al.* (1988) who measured relative sensitivity factors for e-gas SNMS. For comparison with electron-impact cross-sections we show their data in Figure 17, in the form of M^+/Fe^+ yield-ratios, and as a function of the ionisation potential of the element studied. The results for elements of groups 3–5 of the periodic table are shown in this figure, with elements of the same group (row) connected by full (dashed) lines. A fairly smooth variation with ionisation potential within each group is evident. However, the difference in yield-ratio between elements is much larger than we expect from Figure 16 at electron energies above ∼40 eV. We compare, for example, the results for lead and nitrogen. The yield-ratio for Pb^+/N^+ in e-gas SNMS is 160. By contrast, the corresponding ratio of cross-sections, at 40 eV, is 6 and at 300 eV is 4. More generally, we find from such comparisons that in e-gas SNMS there is a discrimination in elemental sensitivity which increases with increasing ionisation potential. Thus the sensitivity achieved with this technique is reasonable for metals but poor for permanent gases and halides.

6   Conclusion

This introduction to SIMS and SNMS covers some aspects of quantitative analysis based on a combination of sputtering with mass spectrometry. The main problem is the difficulty of achieving a high degree of ionisation independent of the matrix from which a given element is sputtered. Exceptionally high sensitivities can be obtained for many elements using oxygen or caesium bombardment. However, SIMS suffers not only from a pronounced variation in elemental sensitivities (due to the frequently observed exponential dependence on ionisation potential or electron affinity) but sometimes also from severe matrix effects. There is evidence, however, that using the MCs$^+$ mode, the matrix effect can be reduced to a low or negligible level, provided favourable bombardment conditions (glancing angle of beam-incidence) can be used. Moreover, the variation in elemental sensitivities is then much smaller than in traditional or 'standard' SIMS.

With electron-impact SNMS, the achievable ionisation efficiencies are relatively low compared with SIMS, but the absence of matrix effects makes this a useful technique in many applications. Quantification is more difficult in e-gas SNMS because the ionising electrons are characterised by a broad velocity distribution, centred at energies near the ionisation threshold. Moreover, matrix effects may be encountered at low ion-energies, because the angular distribution of sputtered atoms changes with sample composition.

References

Andersen H H and Bay H L, 1981, in *Sputtering by Particle Bombardment I* p 145, ed BehrischR (Springer-Verlag, Berlin/Heidelberg)

Chelgren J E, Katz W, Deline V R, Evans, C A Jr, Blattner R J and Williams P, 1979, *J Vac Sci Technol* **16** 324

Deline V R, Evans C A Jr and Williams P, 1978a, *Appl Phys Lett* **33** 578

Deline V R, Katz W, Evans C A Jr and Williams P, 1978b, *Appl Phys Lett* **33** 832

Dowsett M G, 1991, *Fresenius J Anal Chem* **341** 224

Fine J, 1980, in *The Physics of Ionised Gases* p 379, ed Matic M (Boris Kidric Institute, Beograd)

Gao Y, 1988, *J Appl Phys* **64** 3760

Gauneau M, Chaplain R, Regreny A, Salvi M, Guillemot C, Azoulay R and Duhamel N, 1988, *Surface and Interface Analysis* **11** 545

Gersch H-U and Wittmaack K, 1993, *J Vac Sci Technol A* **11**

Gnaser H and Oechsner H, 1991, *Fresenius J Anal Chem* **341** 54

Gnaser H, Fleischhauer J and Hofer W O, 1985, *Appl Phys A* **37** 211

Hofer W O, 1991, in *Sputtering by Particle Bombardment III* p 15, eds Behrisch R and Wittmaack K (Springer-Verlag, Berlin/Heidelberg)

Homma Y and Maruo T, 1989, *Surface and Interface Analysis* **14** 725

Homma Y and Wittmaack K, 1989, *J Appl Phys* **65** 5061

Homma Y and Wittmaack K, 1990, *Appl Phys* **A50** 417

Honig R E, 1958, *J Appl Phys* **29** 549

Jede R, Peters H, Dünnebier G, Ganschow O, Kaiser U and Seifert K, 1988, *J Vac Sci Technol A* **6** 2271

Jede R, Ganschow O and Kaiser U, 1992, in *Practical Surface Analysis* (Second Edition) **2** 425, eds Briggs D and Seah M P (John Wiley and Sons, Chichester/New York)

Lang N D, 1983, *Phys Rev B* **27** 2019

Lipinsky D, Jede R, Ganschow O and Benninghoven A, 1985, *J Vac Sci Technol A* **3** 2007

Littlewood S D and Kilner J A, 1988, *J Appl Phys* **63** 2173

Magee C W, Harrington W L and Botnick E M, 1990, *Int J Mass Spectrom Ion Proc* **103** 45

Menzel N and Wittmaack K, 1981, *Nucl Instrum Meth* **191** 235

Morgan A E, de Grefte, H A M, Warmoltz N, Werner H W, 1981, *Appl Surface Sci* **9** 372

Oechsner H and Gerhard W, 1972, *Phys Lett A* **40** 211

Oechsner H, 1984, in *Thin Film and Profile Analysis* p 63, ed Oechsner H (Springer-Verlag, Berlin/Heidelberg)

Poate J M and Williams J S, 1984, in *Ion Implantation and Beam Processing* p13, eds Williams J S and Poate J M (Academic Press, Sydney)

Reuter W, 1986, *Nucl Instrum Meth B* **15** 173

Roosendaal H E, 1981, in *Sputtering by Particle Bombardment I* p 219, ed Behrisch R (Springer-Verlag, Berlin/Heidelberg)

Schauer S N and Williams P, 1990, in *Secondary Ion Mass Spectrometry SIMS VII* p 827, eds Benninghoven A, Evans C A, McKeegan K D, Storms H A and Werner H W (John Wiley and Sons, Chichester/New York)

Schulz F and Wittmaack K, 1976, *Rad Effects* **29** 31

Sigmund P, 1969, *Phys Rev* **184** 383

Sigmund P, 1981, in *Sputtering by Particle Bombardment I* p 9, ed Behrisch R (Springer-Verlag, Berlin/Heidelberg)

Storms H A, Brown K F and Stein J D, 1977, *Anal Chem* **49** 2023

Tampa G S, Carr W E and Seidl M, 1986a, *Appl Phys Lett* **48** 1048

Tampa G S, Carr W E and Seidl M, 1986b, *Appl Phys Lett* **49** 1511

Tampa G S, Carr W E and Seidl M, 1988, *Surf Sci* **198** 431

Tawara H and Kato T, 1987, *Atomic Data Nucl Data Tables* **36** 167

Valizadeh R, van den Berg J A, Badheka R, Al Bayati A, Armour D G and Sykes D, 1992, *Nucl Instrum Meth B* **64** 609

Vancauwenberghe O, Herbots N, Manoharan H and Ahrens M, 1991, *J Vac Sci Technol A* **9** 1035

Vancauwenberghe O, Herbots N and Hellman O C, 1992, *J Vac Sci Technol A* **10** 713

Williams P, 1979, *IEEE Trans Nucl Sci* **NS-26** 1807

Wittmaack K, 1975, *Surf Sci* **53** 626

Wittmaack K, 1976, *Appl Phys Lett* **29** 552

Wittmaack K, 1981a, *Appl Surface Sci* **9** 315

Wittmaack K, 1981b, *J Appl Phys* **52** 527

Wittmaack K, 1982, *Vacuum* **32** 65

Wittmaack K, 1983a, *Nucl Instrum Meth* **218** 307

Wittmaack K, 1983b, *Surface Sci* **126** 573

Wittmaack K, 1984, *Nucl Instrum Meth B* **2** 674

Wittmaack K, 1985, *J Vac Sci Technol A* **3** 1350

Wittmaack K, 1991, in *Sputtering by Particle Bombardment III* p 161, eds Behrisch R and Wittmaack K (Springer-Verlag, Berlin/Heidelberg)

Wittmaack K, 1992a, in *Practical Surface Analysis* (Second Edition) **2** 105, eds Briggs D and Seah M P (John Wiley andSons, Chichester/New York) pp 105-175

Wittmaack K, 1992b, *Nucl Instrum Meth B* **64** 621

Wittmaack K and Poker D B, 1990, *Nucl Instrum Meth B* **47** 224

Wucher A and Reuter W, 1988, *J Vac Sci Technol A* **6** 2316

Wucher A, Novak F and Reuter W, 1988, *J Vac Sci Technol A* **6** 2265

Yu M L, 1981, *Phys Rev B* **24** 1147

Yu M L, 1982, *Phys Rev B* **26** 4731

Yu M L, 1983, *Physica Scripta* **T6** 67

Yu M L, 1984, *Phys Rev B* **29** 2311

Yu M L, 1991, in *Sputtering by Particle Bombardment III* p 91 eds Behrisch R and Wittmaack K (Springer-Verlag, Berlin/Heidelberg)

Yu M L and Lang N D, 1983, *Phys Rev Lett* **50** 127

Yu M L and Mann K, 1986, *Phys Rev Lett* **57** 1476

Static SIMS

D Briggs

ICI Wilton Research Centre
Middlesbrough, UK

1 Quantification of static SIMS

1.1 Principles

In any SIMS experiment a sample in UHV is bombarded with a primary ion beam and material is sputtered from the surface. A few percent of the ejected atoms and groups of atoms (clusters) are charged positively or negatively; these are extracted into a mass spectrometer and mass analysed to provide, in separate experiments, positive and negative secondary-ion mass spectra. Thus, by its nature, SIMS is a destructive technique and dynamic SIMS obtains concentration depth profiles by erosion in depth—see Wittmaack in these proceedings. By contrast static SIMS (SSIMS) seeks to probe surface chemistry by minimising the ion dose accumulated during spectral acquisition. In this way 'virgin' material is analysed and the full power of analytical mass spectrometry is harnessed for surface chemical analysis.

Although applicable to all aspects of surface analysis, SSIMS has been applied mainly in three areas: (a) adsorption of small molecules onto metal surfaces (as in classical heterogeneous catalysis); (b) adsorption onto polymer surfaces and the related plastics technology; (c) sensitive analysis of large molecules (particularly biomolecules and drugs) in which SSIMS is used essentially as an alternative 'soft-ionisation' source for organic mass spectrometry. Only the second of these areas is addressed in this chapter.

1.2 Instrumentation

All the early developments in SSIMS of polymers utilised quadrupole mass spectrometry (QMS) and the principal technical development was the achievement of charge neutralisation, using simultaneous electron flooding without causing electron stimulated desorption (this was particularly important for imaging experiments, see later). High sensitivity QMS with a mass range of about 1000 was found to have sufficient sensitivity to record spectra from polymers under static conditions. Time-dependence measurements showed that the damage (*i.e.* the SSIMS) 'threshold' was typically a primary ion dose of 10^{13} ions cm^{-2} (*e.g.* using 2–4 keV Ar$^+$ or Xe$^+$, Briggs and Hearn 1986). Both positive and negative-ion spectra could be recorded with half the ion dose. The large acceptance area of the QMS means that stable current densities of less than 1 nA cm^{-2} can be achieved by irradiating an area of several mm^2.

The introduction of time-of-flight (TOF) SIMS provided higher sensitivity via increased transmission (relative to QMS), elimination of mass discrimination and parallel detection allowing spectra to be acquired for a dose of less than 10^9 ions. This led to the development of high-spatial-resolution imaging, as discussed later. TOF SIMS also offers a greatly increased mass range; at least to a mass-to-charge ratio, m/z of 10,000. The most recent analysers, (TOF reflectron design), also provide high mass resolution (upto 10,000 at $m/z = 300$). Details of instrumentation and experimental procedures are given,respectively, by Jede *et al.* (1992) and Briggs (1992).

1.3   Spectral interpretation

Systematic studies of homopolymers

The basis for the development of any new spectroscopic analytical technique is the creation of a spectral database from pure materials. This provides, progressively, for a base level of spectral interpretation through the matching of actual spectra with 'fingerprints' from the database. It is in the nature of organic polymers that systematic variations in structure arise readily through a homologous series, *e.g.* involving alkyl side chains, as a result of differing lengths of hydrocarbon backbone between in-chain functional groups and through variation of functional groups within an otherwise identical structure. This makes systematic interpretations of fragmentation patterns relatively straightforward for any such class of polymers. As an example consider Figure 1 which shows the negative-ion spectra for poly(methyl methacrylate) of PMMA (Hearn and Briggs 1988). Three sets of peaks clearly repeat at intervals of m/z of 100 (the mass of the MMA monomer).

The spectra for higher members of the series, *i.e.* alkylmethacrylate polymers with larger alkyl groups in the ester side-chain -COOR, are shown in Figure 2 (Hearn and Briggs 1988). By looking for consistent patterns in these fragmentation spectra we were able to assign the following structures, with masses given for PMMA (R = CH$_3$).

3M-15, 285   2M+55, 255   2M+41, 241   2M-15,185   M+55,155

For the polymers of Figure 2, R=C$_2$H$_5$ (PEMA), n=C$_3$H$_7$ (P^{n}PMA), s-C$_4$H$_9$ (P^{s}BMA)

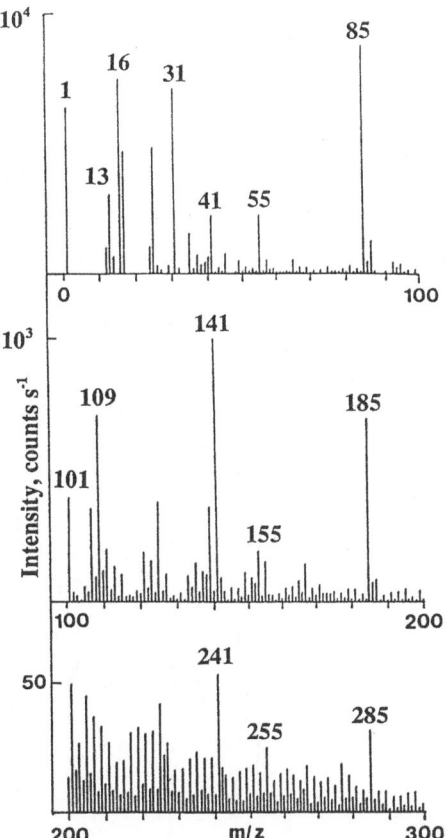

Figure 1. *Negative-ion spectrum of* PMMA. *Region m/z 200–300 uses three times the current density of region m/z 0–200.* QMS, *4keV Xe+ at 1nA cm⁻² (0–200).*

and cyclo-C_6H_{11} (P^{c}HMA). (2M–15) ions of mass are seen at m/z 213, 241, 269 and 321, and (M+55) ions appear at m/z 155, 169, 183 and 209. Furthermore, common ions throughout the series appear at m/z = 85 and 139, which means that these ions do not involve the R group. Poly(methacrylic anhydride) contains the proposed ring

structure in its repeat unit and the negative-ion spectrum is dominated by m/z 139 (Hearn and Briggs 1988).

Using this approach many other types of polymer spectra have been interpreted. These include poly(alkylacrylates) Hearn and Briggs (1988), polyamides (nylons),

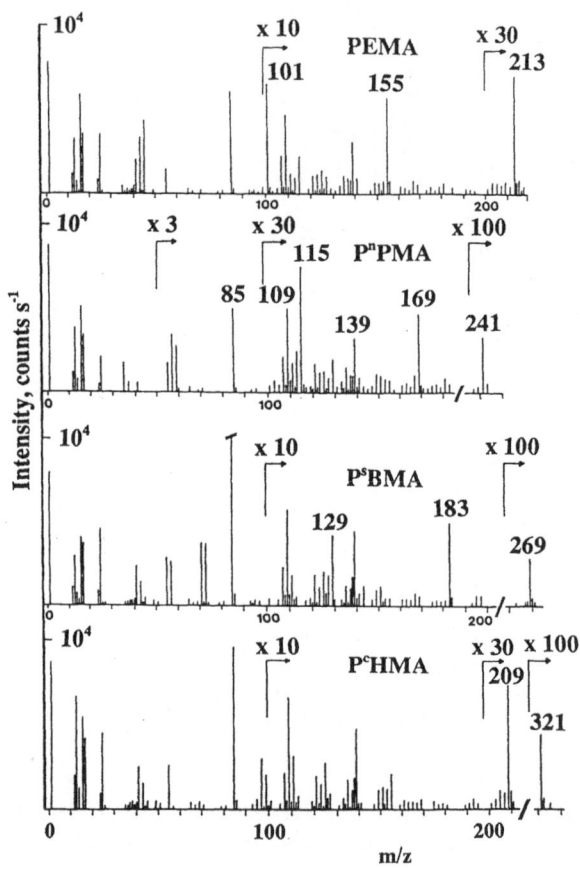

Figure 2. *Negative-ion spectra of poly(alkyl methacrylates). Experimental conditions as for Figure 1 (Hearn and Briggs 1988).*

Briggs (1987), aliphatic hydrocarbon polymers (van Oijj and Brinkhaus 1988, Briggs 1990), polyglycols (Hearn *et al.* 1988), segmented polyurethanes (Hearn *et al.* 1987, 1988), polycarbonates (Lub *et al.* 1988), polyesters (Davies *et al.* 1989), polyorthoesters (Davies *et al.* 1991a), polyanhydrides (Davies *et al.* 1991b) and a series of oxygen functionalised polymers including vinyl ether, ketone and carboxylate types (Chilkoti *et al.* 1992a). A database of fingerprint spectra, with assignments of principal peaks, for many common polymers and organic molecules commonly found on polymer surfaces, has also been compiled (Briggs *et al.* 1989).

In the course of such spectral interpretations most authors have compared and contrasted polymer fragmentation patterns with those of molecular analogues produced

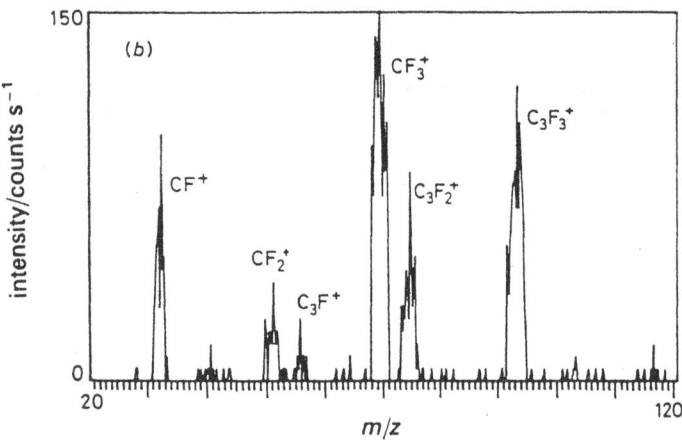

Figure 3. *Daughter-ion spectrum of $C_4F_5^+$ (m/z 143) from PTFE using a xenon target gas (Leggett et al. 1990b).*

by conventional mass spectrometry, particularly electron impact (EI) MS. No clear rules emerge, although stable positive ions, produced by any ionisation technique will obviously feature strongly and this suggests a correlation which may be more apparent than real. Short and Davies (1989) have, for instance, pointed out that whilst odd-electron ions dominate over even-electron ions in EIMS, the reverse is true in polymer SSIMS.

Tandem-SIMS studies

Tandem techniques (*i.e.* MS–MS) have proved most useful in the study of fragmentation pathways in conventional MS and they have recently been applied in this area with a view to elucidating the origin of the fragments observed in polymer SSIMS. These experiments, by Leggett *et al.* (1990a,b), utilise a triple quadrupole analyser fitted to a SSIMS instrument in place of the conventional single quadrupole analyser. If the triple quadrupole is Q_1–Q_2–Q_3, then a conventional SSIMS spectrum is recorded by scanning Q_1 with Q_2 and Q_3 in RF-only mode. In MS–MS (or daughter) mode, Q_1 is tuned to transmit an ion of interest, Q_2 is filled with a target gas at typically $\approx 10^{-5}$ mbar (the collision cell), and Q_3 is scanned. A daughter-ion spectrum is produced by collisionally activated dissociation (CAD) of the selected (parent) ion. If Q_1 and Q_3 are scanned together but with a fixed mass difference corresponding to the mass of a potential neutral loss fragment, then a spectrum of the ions resulting from such a neutral loss is obtained. By a combination of these two techniques a detailed picture of ion fragmentation pathways can be built up. PTFE has been examined in some detail, plus the effect of all important experimental variables — particularly target gas pressure and collision energy (Leggett *et al.* 1990b). Figure 3 shows a typical daughter-ion spectrum, whilst Table 1 illustrates a fragmentation pathway deduced for a characteristic PTFE fragment ion. The experiments can also specifically prove that a postulated fragmentation path-

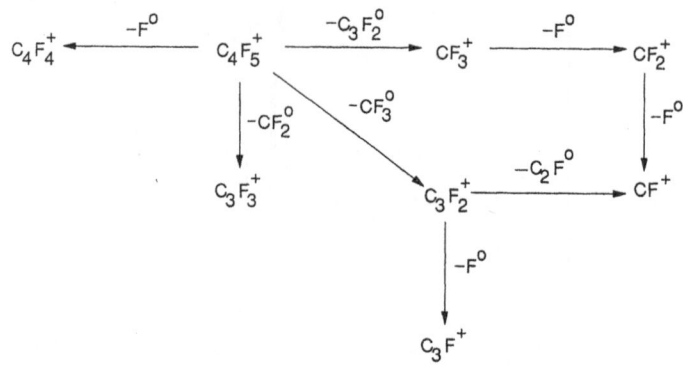

Table 1. *Fragmentation pathways for formation of the $C_4F_5^+$ ion from* PTFE.

way actually occurs. Thus, for example, the explanation offered by Briggs (1982) for the positive-ion spectrum of poly(ethylene-terephthalate), based on expectations from EI MS, has been completely vindicated (Leggett *et al.* 1990a). Many polymers give strong hydrocarbon-ion sequences in their positive-ion spectra, and polyethylene has been studied with this in mind (Leggett *et al.* 1991). From the data and from a study of PTFE it is clear that ions formed with high abundance in the CAD experiments are generally observed as strong peaks in the SSIMS spectrum. This supports the proposal that the probabilities of ion formation, during sputtering and during the collisionally activated dissociation of polymer fragments, are similar for given ions, *i.e.* the CAD process models sputtering processes.

Isotopic labelling

Comparisons of spectra for polymers in which some or all of the hydrogen atoms have been replaced by deuterium is one way of improving spectral interpretation, particularly in cases involving fragment ions of nominally the same mass which are therefore unresolved with conventional instruments; *e.g.* $C_xH_y^+$ or $C_xH_yO_z^+$ (Lub *et al.* 1989, Brinkhuis and van Ooij 1988). In aromatic systems, facile H–D exchange can, however, add its own complications (Chilkoti *et al.* 1991a). The scope for ^{13}C labelling was first outlined in a study of the gas-phase derivation of –OH groups using acetylchloride labelled at either C atom (Briggs and Munro 1987). Occhiello *et al.* (1990) have studied the oxygen plasma treatment of polypropylene and its surface ageing by using $^{16}O_2$ and $^{18}O_2$ for treatment and post-treatment exposures.

These techniques are being applied to SSIMS studies of plasma deposited films (PDF) of polymers formed from functional monomers, by incorporating stable isotopes into the original monomer (Chilkoti *et al.* 1991b, 1992b). For instance, Figure 4 shows positive-ion spectra for a isotope-labelled acetone derived PDF which showed that the majority of the m/z 43 peak in the labelled spectrum is due to CH_3CO^+ representative of the monomer functionality.

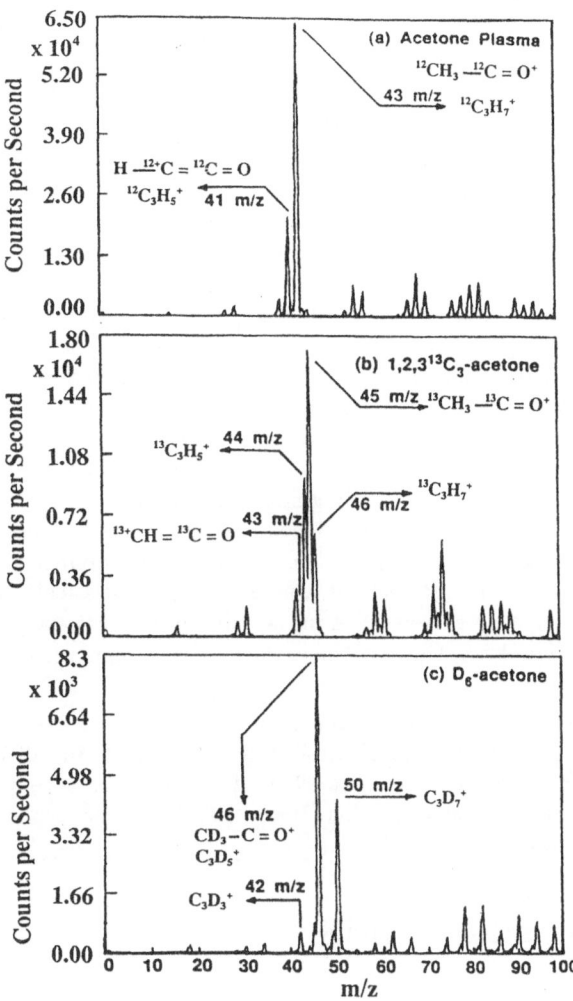

Figure 4. *Positive-ion spectra of acetone* PDFs *using labelled and unlabelled acetone monomer (Chilkoti et al. 1991b).*

Time of Flight SIMS (TOFSIMS) Studies

Reflectron-type TOF SIMS instruments are now becoming available from several manufacturers. They promise to simplify spectrum interpretation in two ways. Firstly, the mass resolution can be high enough to resolve overlapping peaks of the same nominal mass (see above) in the lower region of the spectrum; say m/z 0–250 (Briggs *et al.* 1984). Secondly, accurate mass determination may allow empirical formulae to be deduced for fragments (particularly for quasi-molecular ions).

1.4   Potential for quantification

Although the quantitative interpretation of polymer SSIMS data is not expected to be straightforward a number of studies have shown that the situation is far from hopeless. Relative intensities of atomic or quasi-atomic ions, *e.g.* $O^-:CH^-$, $F^-:CH^-$ show good correlations with quantitative x-ray photoelectron atomic concentrations when trends in composition are restricted to similar materials (Hearn *et al.* 1987, Chilkoti *et al.* 1990, 1992a) . Matrix effects are observed: for example, it has recently been shown that for three classes of oxygen-containing polymers the slopes of the correlations between $O^-:CH^-$ and bulk O:C are significantly different (Briggs 1992). Relative intensities of cluster ions have been shown to correlate with composition for random methacrylate copolymers (Briggs and Ratner 1988, Lub *et al.* 1989). Figure 5 shows relative intensity plots for a series of ethylmethacrylate: hydroxyethylmethacrylate (EMA:HEMA) random copolymers. Ion fragments at m/z 127 and 155 are representative of EMA and HEMA monomers respectively and relative intensity plots of the form $A/(A+B)$ give smooth trends (Figure 5)a. The intensities of two fragments believed to represent HEMA-EMA linked monomers are plotted in Figure 5b: in confirmation of these assignments both curves maximise in mid-range. The fragment representing EMA–EMA links (m/z 213) increases rapidly with EMA content (Figure 5c) as expected. The ratio $I(213)/(155)$ represents the ratio of diads to singles. For a random AB polymer for which diad (AA) signals come from sequences $B-(A)_n-B$, where $n > 2$, a statistical analysis predicts the ratio of diads to singles (*i.e.* isolated A units) will be proportional to the mole fraction of A, as seen in Figure 5c. Clearly, the surface of this copolymer is representative of the bulk composition.

Similar relative intensity plots have been used to follow surface segregation in block copolymers and blends (Hearn *et al.* 1987, 1988, Batia and Burrell 1990, Brinen *et al.* 1991, Michael *et al.* 1990, Niehaus *et al.* 1989). For hydrocarbon polymers, the $CH_2^-:C^-$ ratio is related to degree of unsaturation (Briggs 1990). Further aspects of quantification will be discussed later.

2   Imaging static SIMS

2.1   Principles

There are two approaches to secondary-ion imaging, leading respectively to microprobe and microscope instruments. In the former, a finely focused ion beam from a liquid metal ion source (LMIS) is raster-scanned over the surface. The intensity of a chosen secondary ion is recorded point-by-point to generate the image. Spatial resolution is essentially limited by the spot-size of the primary beam. In the imaging microscope, either the surface is flooded with primary ions or a fairly broad probe is scanned at TV rate over the surface. Imaging is stigmatic; that is the ion-optical system transports secondary ions through the mass filter to a two-dimensional detector whilst maintaining their spatial relationship at the sample. In this case spatial resolution is determined by the ion-optical system. The microscope instruments can also function in micro-probe mode, by employing a fine-focus beam, in which case the advantage of simulta-neous acquisition of the whole image is sacrificed. Microprobe SIMS instruments utilise

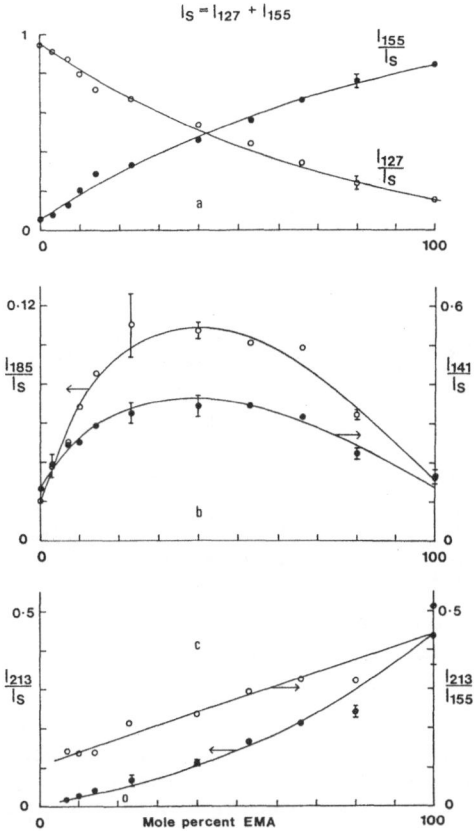

Figure 5. *Relative intensity plots for a series of ethylmethacrylate: hydroxyethyl-methacrylate (*EMA: HEMA*) random copolymers.*

quadrapole, magnetic sector or time-of-flight (TOF) mass spectrometers, while stigmatic ion microscope designs, involving magnetic sector and TOF mass spectrometers are also operational. For full details of the instrumental aspects of SIMS imaging, see for example Jede *et al.* (1992).

The earliest work on SSIMS imaging (Briggs 1983) employed a noble-gas ion source, producing a spot size of about 30μm, and a quadrupole mass spectrometer with image capture on a storage oscilloscope. For all-polymer heterogeneous surfaces, which require charge neutralisation, images were obtained using molecular fragments characteristic of the components. Spatial resolution of the order of the beam spot size was demonstrated. Using the same instrument, but employing a LMIS 10keV Ga^+ ion beam, Briggs (1989) encountered great difficulty in achieving the enhanced spatial resolution expected from the reduction of the probe size to about 0.2μm . This was due

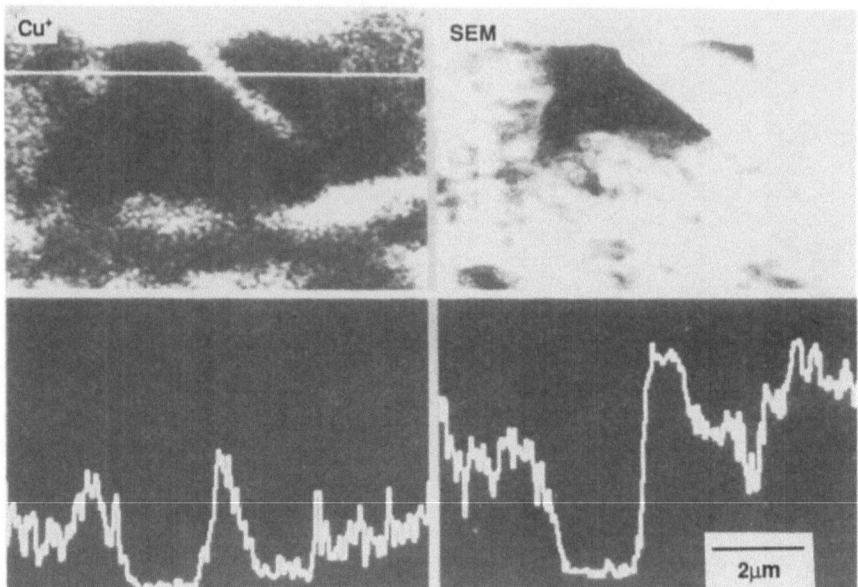

Figure 6. *Secondary electron and secondary-ion (Cu⁺) images of a feature of a copper grid (5µ nominal bar) with linescans measured at the same place in each image.*

to more severe charge neutralisation requirements and increased sample damage (the instantaneous beam current at any dwell point is much higher with the much smaller probe). The latter effect results essentially from the low transmission of a QMS; *i.e.* not enough signal can be detected from the area bombardeded by the microprobe before the static SIMS limit is exceeded. Until recently secondary-ion microscopes have been solely based on magnetic sector designs, optimised for high mass resolution dynamic SIMS and therefore inappropriate for static SIMS imaging.

The benefits in sensitivity of the TOF mass spectrometer over the QMS were demonstrated by Benninghoven (Steffans *et al.* 1984). These include greater transmission (> 10% TOF, about 0.1% QMS), parallel mass detection and zero mass discrimination, which can combine to give an improvement in sensitivity of about 10^4 for molecular-ion detection in imaging mode. An imaging TOF SIMS instrument was subsequently constructed in which a 30keV Ga⁺ beam of about 400Å diameter was employed (Waugh *et al.* 1988, Eccles and Vickerman 1989). Data from this instrument are discussed below.

2.2   Resolution limits

In static SIMS the aim is to obtain a spectrum or image in such a way that the collected secondary ions are ejected from a region of the surface that has not previously experienced primary-ion impact. When this is the case, the information obtained is representative of the 'virgin' surface composition. Therefore, the ultimate limit to spatial resolution is the size of the damage regime associated with primary-ion impact. Currently this is unknown. Monte Carlo simulations of collision cascades suggest that, within an area of radius about 20Å around the impact point, most chemical bonds will be broken and direct knock-on processes dominate leading to emission of atomic species and small fragments. It is believed that further away from the impact point, at say a distance of 30–40Å, low-energy recoils intersect the surface with enough energy to cause emission (desorption) of intact molecules (Magee 1983). However, there are several other possible mechanisms whereby energy from the primary impact zone may be converted into translation/rotational energy, such that molecules may be desorbed and ionised with insufficient vibrational energy to cause dissociation. The situation for large polymer molecules in a chain-entangled situation is unclear. However, it seems that a surface region of up to 100Å in diameter could easily be affected by a single high energy ($\sim$ 30keV) ion impact.

Pulsed liquid-metal ion guns are capable of depositing a small number of ions into a limited area, but this ultimate development is some way off. State-of-the-art probe diameters are about 200Å, from guns operated continuously. Images of a copper grid (with approximately 5μm bars and spaces) have been obtained using secondary-electron detection with continuous 30keV Ga$^+$ ion bombardment. At highest magnification each pixel corresponds to about 200Å, and clearly the spatial resolution (probe diameter) is within a factor of two of this dimension. However, for TOF SIMS the LMIS Ga$^+$ beam must be pulsed, which involves beam deflection and leads to some reduction in the beam spot size at the sample. This is shown in Figure 6 where a secondary-electron image described above and a Cu$^+$ secondary-ion image (pulsed beam) of the same feature are compared. The lower contrast in the secondary-ion image is evident, but the loss of spatial resolution is demonstrated by line scans across an edge. In this case the spot size degrades to about 2000Å (0.2μm). With more recent designs this degradation is reduced relative to the continuous beam spot size.

The simplest material for SSIMS imaging is a conductor with a 'patchy' inorganic overlayer of monolayer thickness. In this case the surface damage is simply the loss by sputtering of the overlayer. For SSIMS to be successful, a repeated imaging experiment of the same area must give the same result. Typically, an ion dose of 10^{14} ions cm^{-2} can be tolerated in such a case. An example is shown in Figure 7. This was an investigation of the poisoning of an unsupported silver catalyst, the topography of which is revealed by the secondary-electron 'SEM' image obtained after SIMS imaging; a continuous beam would give a relatively high ion dose. Secondary-ion images, obtained simultaneously, show the distribution of segregated or adsorbed calcium and iron. Contrast in the Ag$^+$ image reveals the polycrystallinity of the catalyst. Note the almost mutual exclusivity of Ca and Fe distributions, probably reflecting their preferential adsorption on different crystal faces. Sub-micron spatial resolution is demonstrated in these SIMS images.

A more difficult experiment is to image a delicate organic molecule on a conducting substrate. In Figure 8 images are shown for the small peptide 'gly–gly–gly' deposited on a deeply scratched aluminium foil substrate (The deep scratching generates shadowing effects, leading to image contrast) the quasi-molecular ion [M–H]$^-$ and the characteristic fragment ion CNO$^-$ are imaged using ion-doses above and below the static SIMS limit. Although the molecular-ion image is weak (on average less than 2 counts per pixel)

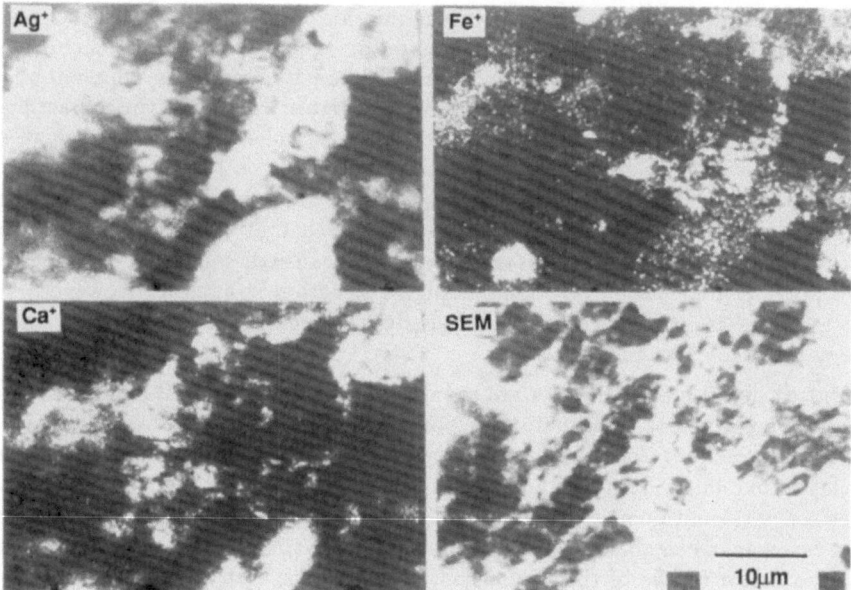

Figure 7. *Secondary-electron and secondary-ion (Ag⁺, Ca⁺, Fe⁺) images of a 'poisoned' silver catalyst.*

it correlates with that of the 'longer-lived' fragment. Summing the two CNO^- images gives the highest possible contrast. Figure 9 is a line-scan taken from a region of interest in the resulting image, showing sub-micron resolution (Briggs and Hearn 1988); note however that the initial image dose of 2×10^{13} ions cm^{-2} 'destroys' the peptide molecule. In this work Briggs and Hearn (1988) also imaged a large surfactant molecule, with an oligomeric distribution distributed on gold.

An ultimate level of difficulty is introduced by insulating materials, which require charge neutralisation. Low-energy electrons are flooded over the sample, between primary-ion pulses during the secondary-ion flight time (with the extraction voltage switched off). The species to be imaged tend to give low-intensity fragments, but often the parallel detection capability of the TOF instrument allows this to be overcome by image formation using the sum of several fragment-ions from a single molecular entity. In this way, the heterogeneous distribution of a silicone contaminant on polypropylene fibres has been demonstrated at about $1\mu m$ resolution (Briggs *et al.* 1990). Two-colour overlay images are required for visualisation (and do not reproduce well in black and white).

The limits of spatial resolution, in the most difficult case of imaging heterogeneous polymer systems, has been addressed for the particular instrument discussed here, by measuring the useful positive-ion yields (number of secondary ions detected per incident primary ion) for several polymers. Table 2 gives the useful ion-yields as the summation of given characteristic fragment ions, together with the minimum dimension of the

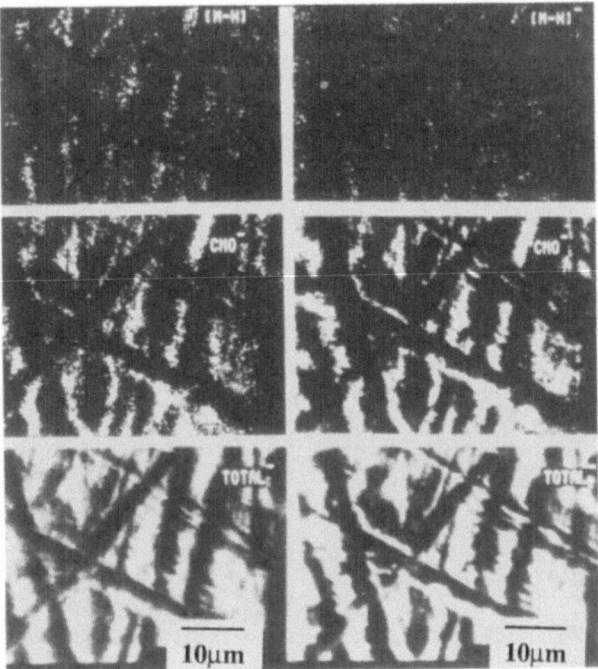

Figure 8. *Secondary-ion images for gly–gly–gly peptide on aluminium, using [M-H]⁻ (m/z 188), CNO⁻ (m/z 42), and total negative ions (m/z 0-200): Left : dose 6×10^8 ions. Right : additional dose of 5.4×10^9 ions to same 50μm $\times 50\mu$m area.*

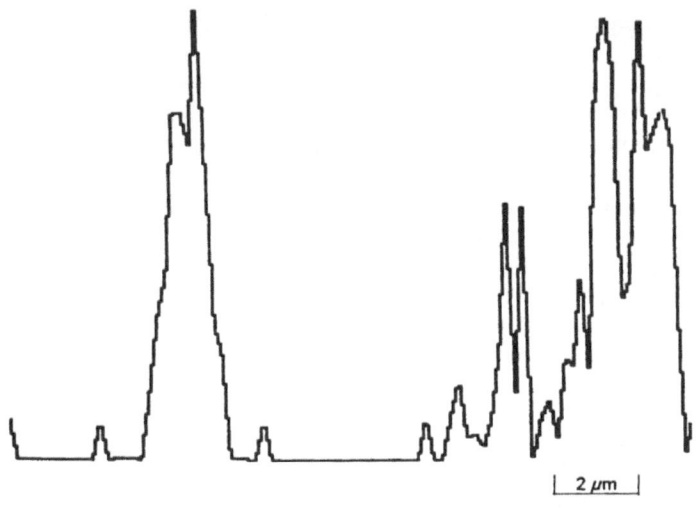

Figure 9. *Linescan across a small region of the resultant CNO⁻ image, obtained by adding the two images shown in Figure 8.*

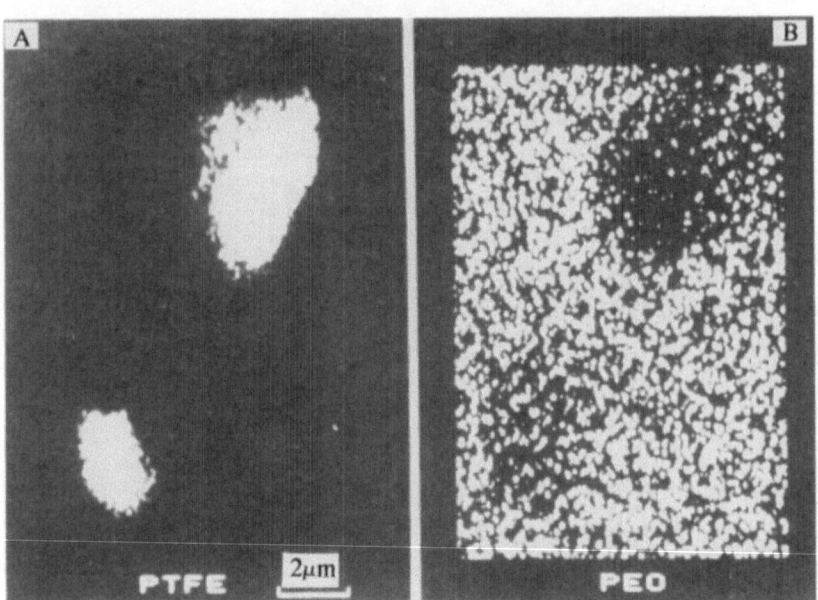

Figure 10. *Images of PTFE particles on PEG using the summation of positive ions listed in Table 1.*

(square) analysed area that can be imaged using a total ion dose of 10^{14} ions cm^{-2}. This is calculated for minimum numbers of counts per pixel (256×256 array) of 1 or 5. The result of the most favourable experiment, imaging PTFE particles on a PEG surface is shown in Figure 10. For a total dose of 10^{14} ion cm^{-2}, which is destructive, the contrast in the image is as expected from the data of Table 2 (Hearn 1992). The transmission of this (the first) imaging TOF SIMS instrument is not known, but is probably less than 10%. Improvements in transmission in current designs will lead to a proportional increase in detected signal and hence image contrast, and could easily lead to intensity increase by as much as an order of magnitude. Alternatively, the same contrast could be achieved for a static dose of 10^{13} ions cm^{-2}.

Polymer	Positive ions (m/z)	Useful yield	Width of analysis area (μm)	
			1 cpp	5 cpp
PFTE	31,69,93,100,131	6.3×10^{-3}	3.2	7.2
PEG	31,45,73,87,89,103,133	5.6×10^{-4}	10.8	24.3
PMMA	15,31,45,59,69,73,85,101	3.1×10^{-4}	14.5	32.3
PET	76,104,148,149,191,193	7.0×10^{-5}	30.6	68.4

Table 2. *Minimum analysis areas*

3 Static SIMS and its combination with XPS

3.1 Introduction

Continuing from Section 1, we discuss here the application of static SIMS in the field of polymer surfaces. Two comprehensive reviews have recently covered applications (including imaging) in the fields of inorganic and organic materials respectively (Reed and Vickerman 1992, Briggs 1992). It is particularly important to note that the detection of additive molecules, in the case of polymeric systems, and of molecular contaminants in general is a critical aspect of many applications. SSIMS and XPS are highly complementary and are increasingly used together in surface analysis : XPS detects all elements except hydrogen, is quantitative and gives a reasonable level of structural detail; SIMS detects hydrogen, but not all elements as directly, is not quantitative and provides a high level of structural detail; the two techniques have different but not mutually exclusive regimes of surface sensitivity and spatial resolution. These points are best illustrated by referring to two studies of biomedical polymer surfaces — a rapidly growing area of research, in which high resolution XPS and SSIMS are used in combination.

Segmented Polyurethane Surfaces

The synthesis of these materials (Hearn *et al.* 1988) is summarised in the scheme shown in Figure 11 The polyether unit, typically in the molecular weight range 400–2000, forms the 'soft' segment of the polyurethane, while the di-isocyanate plus chain extender (diamine or diol) forms the rigid 'hard' segment. These polymers have a tendency, dependent on the composition and processing history, to form discrete phases with domain sizes of 10–20 nm (in the bulk). A knowledge of surface structure, and of lateral and vertical heterogeneity, is of great importance for understanding many aspects of biocompatibility.

Although the pure polyethers can be distinguished by XPS, their identification in the polyurethane is difficult. However, their molecular fingerprint in SIMS is carried over into the polyurethane. Thus Figures 12–14 show SIMS spectra of poly(propylene glycol) or PPG, a hard segment model polymer (polyether replaced by butanediol), and a set of PPG/MDI/ED polyurethanes. As the molecular weight of the PPG 'soft' segment increases so the SIMS spectra become dominated by the 'soft' segment peaks. Partly this is due to the increasing fraction of PPG in the polymer, as shown in Figure 15a where peaks at $m/z = 59$ and 106 uniquely identify soft and hard segments respectively.

However, Figure 15b shows that increasing the bulk 'soft' segment to 'hard' segment ratio (SS/HS) leads to a dramatic increase in soft segment peak intensity, indicating surface segregation of PPG units. Detailed studies of the relative intensities of the PPG fragments in the pure polyether and in the polyurethane surface prove that this is not 'free' PPG, which is confirmed by studies of polyurethanes after solvent extraction or after deliberate doping with free PPG; *i.e.* the SIMS data is morphologically sensitive. On the other hand, SIMS cannot identify the chain-extender unit in the polyurethane while XPS can. In Figure 16 the low-intensity, high binding energy peak in (a) is due to carbamate groups (NH–CO–O) and in (b) to carbamate and urea groups (NH–CO–NH).

Polyetherurethane Synthesis

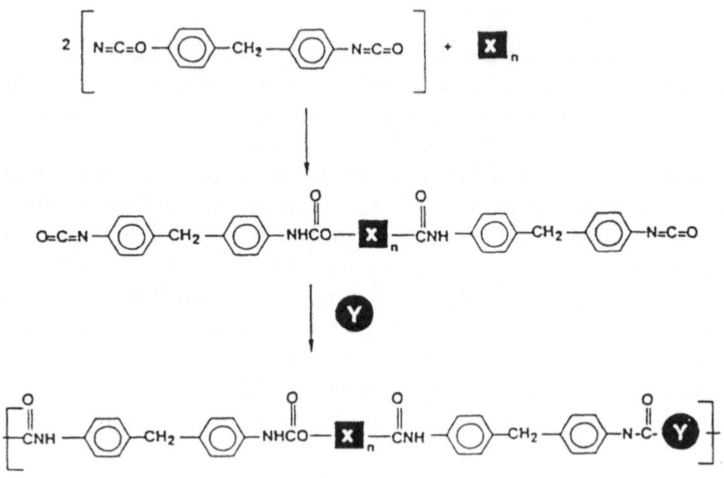

$$2 \left[N{=}C{=}O - \bigcirc - CH_2 - \bigcirc - N{=}C{=}O \right] + \boxed{X}_n$$

$$O{=}C{=}N - \bigcirc - CH_2 - \bigcirc - NHCO - \boxed{X}_n - CNH - \bigcirc - CH_2 - \bigcirc - N{=}C{=}O$$

$$\boxed{Y}$$

$$\left[CNH - \bigcirc - CH_2 - \bigcirc - NHCO - \boxed{X}_n - CNH - \bigcirc - CH_2 - \bigcirc - N\text{-}C\text{-} \boxed{Y} \right]$$

X	Abbreviation
-(-CH$_2$CH$_2$-O-)$_n$	PEG
$\underset{\text{-(-CH}_2\text{CH-O-)}_n}{\overset{\text{CH}_3}{\vert}}$	PPG
	DPG
	TPG
-(-CH$_2$CH$_2$CH$_2$CH$_2$O-)$_n$	PTMG
	BD unit

Y	Abbreviation
NH$_2$CH$_2$CH$_2$NH2	ED
OHCH$_2$CH$_2$CH$_2$CH$_2$OH	BD

Figure 11. *The synthesis of polymer surfaces*

These show slightly different chemical shifts, leading to a much broader peak in (b).

The surface segregation within the top monolayer or so (the SSIMS sampling depth) can in principle be quantitatively studied by angular resolved XPS. However, this relies on quantitative nitrogen analysis (specific to hard segment). For the PPG–PEU with the highest SS/HS ratio, the N content from the maximum XPS sampling depth (50–100Å) is less than 2 atomic %. Decreasing the sampling depth leads to decreasing N

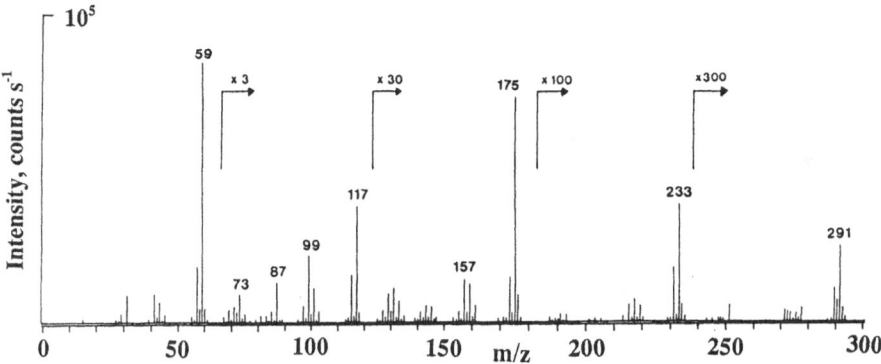

Figure 12. *Positive-ion spectrum of* PPG *(m. wt. = 425). (Hearn et al. 1988b)*

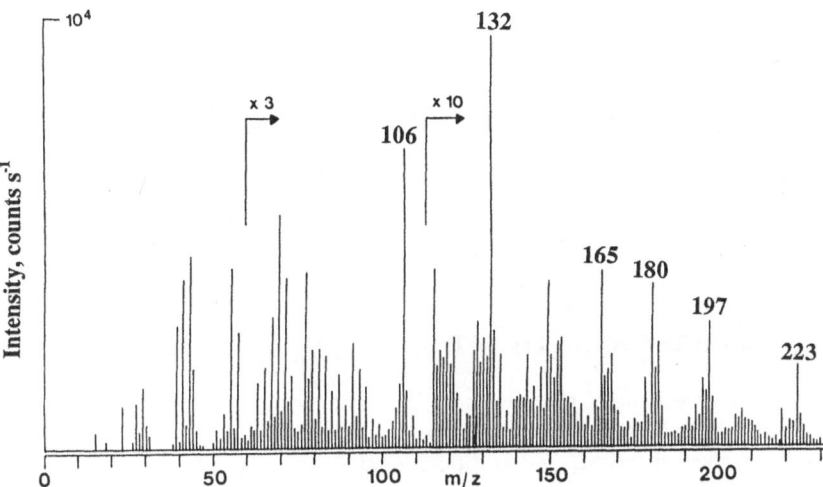

Figure 13. *Positive-ion spectrum of butanediol hard segment model polyurethane. (Hearn et al. 1988)*

concentration (as expected from the SIMS data) but also to a significant loss of signal-to-noise ratio with most XPS instruments. The detection limit for N is approximately 0.3 at %, so low take-off angle data can easily lead to the conclusion that there is no hard segment at the surface. The greater sensitivity of SIMS to the hard segment shows that this is never actually the case, which has important implications when constructing models of the polyurethane surface microstructure.

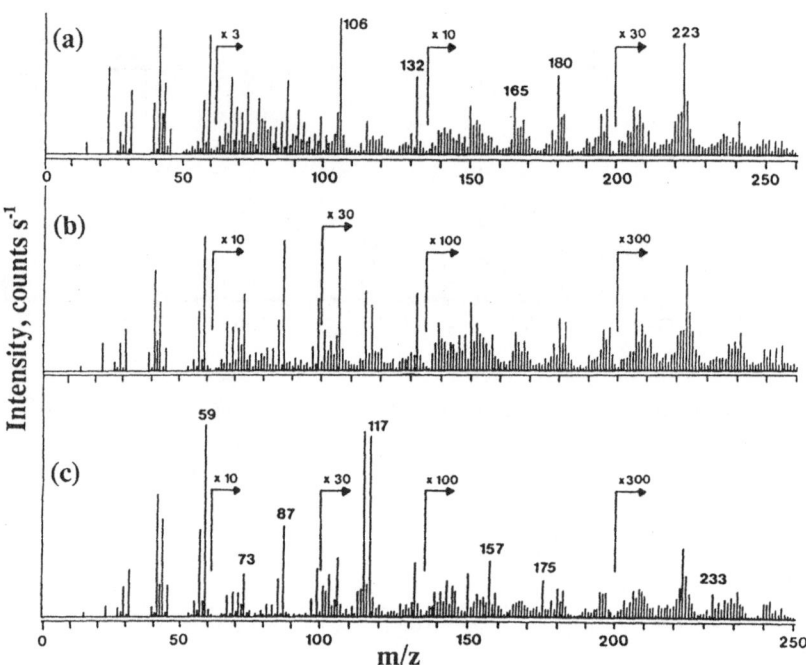

Figure 14. *Positive-ion spectra of* PPG/MDI/ED *polyurethanes prepared with (a)* PPG *425 (b)* PPG *775 (c)* PPG *2000. (Hearn et al. 1988b)*

Variation in surface chemistry of Biomer

Biomer is a polyurethane of the kind described above, but of commercially undisclosed composition. Variation in the properties of different batches of this medical grade product have been observed, and compositional variations have been detected by pyrolysis mass spectra. Although the bulk composition, measured by Fourier Transform infra-red spectroscopy FT-IR, of different batches is the same, the surface composition has recently been shown to vary greatly. Again, this knowledge will be very useful in understanding variations in biological response.

Two batches of Biomer, referred to as BSP and BSUA (for details see Tyler *et al.* 1992) are discussed. The x-ray photoelectron spectra shown in Figures 17 and 18 are clearly different. The BSP data are consistent with the expected polyurethane composition, based on PTMG, MDI and ED (from other analyses), as described above. The SIMS spectrum (Figure 19) confirms this, and also contains other fragment ions ($m/z = 147$, 161 and 177) typical of a hindered-phenol antioxidant. The SIMS spectra (Figure 20) for BSUA are very informative. None of the expected polyurethane peaks are present in the positive-ion spectrum, and the negative-ion spectrum (which is normally uninformative for polyurethanes of this kind) is also rich in information. Both spectra are consistent with the presence of a diisopropyl amino ethyl methacrylate (DPA-EMA), based on a sys-

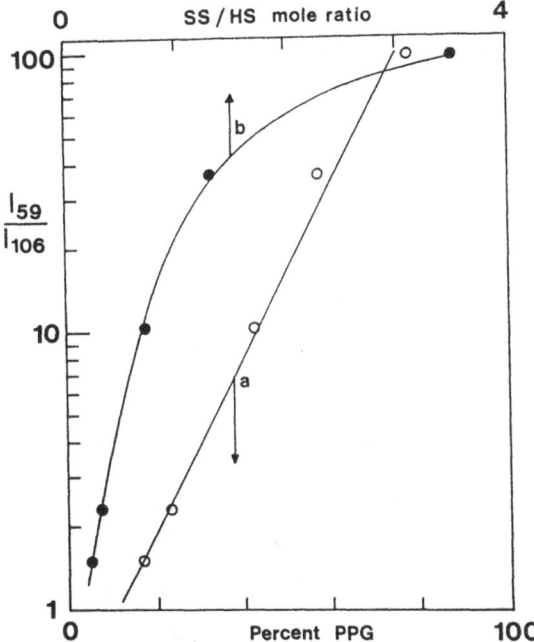

Figure 15. *Log(I_{59}/I_{106}) positive-ion peak intensity ratios plotted as a function of bulk concentration, expressed in terms of (a) % PPG content and (b) SS/HS ratio. (Hearn et al. 1988)*

tematic interpretation of the fragmentation patterns and on a knowledge of the spectra for a dimethyl amino ethyl methacrylate copolymer (Wilding *et al.* 1990). Exhaustive extraction experiments failed to convert the BSUA surface into a more BSP-like surface. Hence the controversy in the literature, relating to the biological interactions of Biomer, may be partly explained by these major differences in their surface chemistry, and the desirability of a full surface characterisation of such polymers prior to biological testing is indicated.

References

Batia Q S and Burrell M C, 1990, *Surface and Interface Analysis* **15** 388
Brinen J S, Greenhouse S and Jarrett P K, 1991, *Surface and Interface Analysis* **17** 259
Briggs D, 1982, *Surface and Interface Analysis* **4** 151
Briggs D, 1983, *Surface and Interface Analysis* **5** 113
Briggs D, 1987, *Org Mass Spec* **22** 91

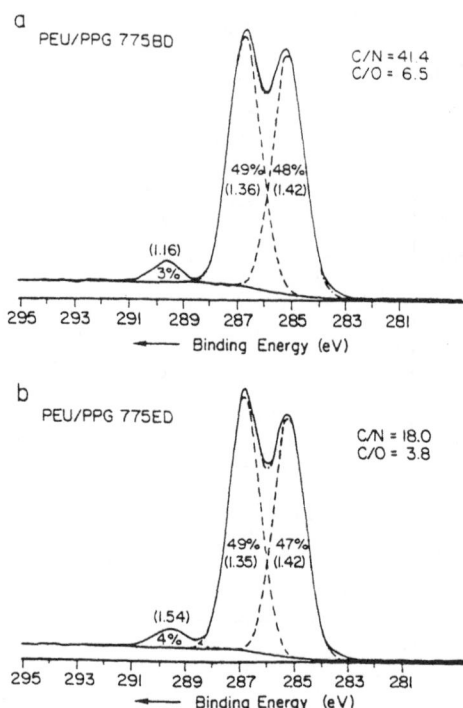

Figure 16. *C ls XP spectra of (a)* PPG 775/MDI/BD *and (b)* PPG 775/MDI/ED *polyurethanes. Peak full-width-half-maxima are indicated in parentheses. (Hearn et al. 1988)*

Briggs D, 1989, *in Fractography and Failure Mechanisms of Polymers and Composites* Ch 5, ed A C Roulin-Moloney, (Elsevier Applied Science, Barking)

Briggs D, 1990, *Surface and Interface Analysis* **15** 734

Briggs D, 1992, in *Practical Surface Analysis, Second Edition, Vol 2—Ion and Neutral Spectroscopy* (Ch 7), editors Briggs D and Seah M P (Wiley, Chichester)

Briggs D and Hearn M J, 1986, *Vacuum* **36** 1005

Briggs D and Munro H S, 1987, *Polym Commun* **28** 307

Briggs D and Ratner B D, 1988, *Polym Commun* **29** 6

Briggs D and Hearn M J, 1988, *Surface and Interface Analysis* **13** 181

Briggs D, Hearn M J and Ratner B D, 1984, *Surface and Interface Analysis* **6** 184

Briggs D, Brown A and Vickerman J C, 1989a, *Handbook of Static Secondary Ion Mass Spectrometry (SIMS)* (Wiley, Chichester)

Briggs D, Hearn M J, Fletcher J W, Waugh A R and McIntosh B J, 1990, *Surface and Interface Analysis* **15** 62

Brinkhuis P H G and van Ooij W J, 1988, *Surface and Interface Analysis* **11** 214

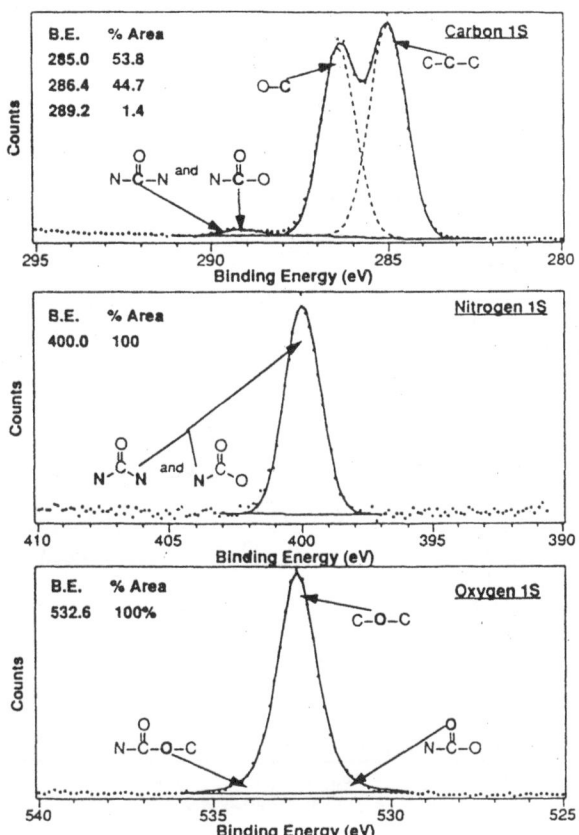

Figure 17. *Core level x-ray photoelectron spectra from batch* BSP. *In the oxygen 1s spectrum contributions from amide oxygens to high and low binding energy of the ether oxygen are too small to be accurately fitted. (Tyler et al. 1992)*

Chilkoti A, Castner D G, Ratner B D and Briggs D, 1990, *J Vac Sci, Technol A* **8** 2274

Chilkoti A, Castner D G and Ratner B D, 1991a, *Appl Spectrosc* **45** 209

Chilkoti A, Ratner B D and Briggs D, 1991b, *Anal Chem* **63** 1612

Chilkoti A, Ratner B D and Briggs D, 1992a, *Surface and Interface Analysis* **18** 604

Chilkoti A, Ratner B D, Briggs D and Reich F, 1992b, *J Polym Sci, Polym Chem* **30** 1261

Davies M C, Short R D, Khan M A, Watts J F, Brown A, Eccles A J, Humphrey P, Vickerman
   J C and Vert M, 1989, *Surface and Interface Analysis* **14** 115

Davies M C, Lynn R A P, Watts J F, Paul A J, Vickerman J C and Heller J, 1991a,
   Biomaterials **12** 305

Davies M C, Khan M A, Domb A, Larger R, Watts J F and Paul A J, 1991b,
   J Appl Polym Sci **42** 1597

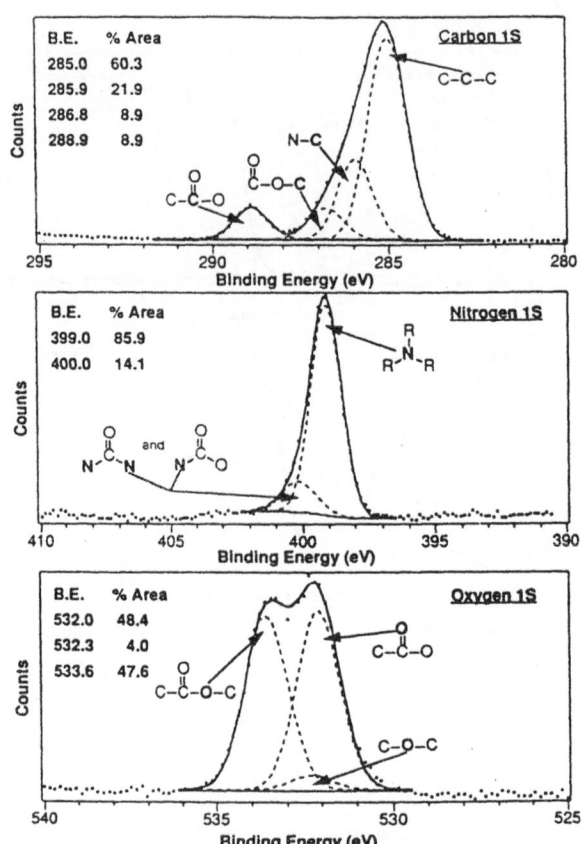

Figure 18. *Core level XP spectra from batch* BSUA. *(Tyler et al. 1992)*

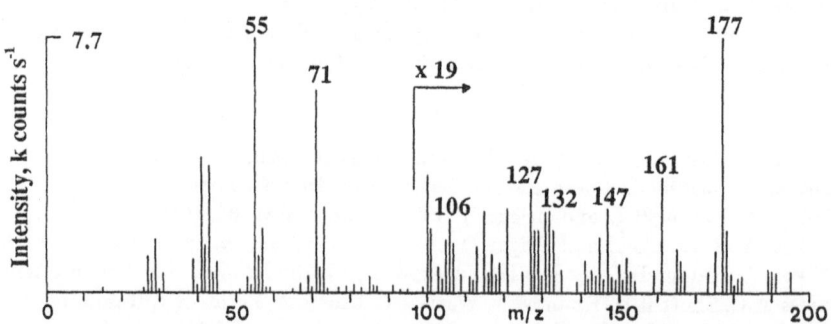

Figure 19. *Positive-ion spectrum from batch* BSP. *Peaks characteristic of* PTMG *(m/z 55, 71, 127), MDI (m/z 106, 132) and a hindered phenol antioxidant (m/z 147, 161, 177) are all clearly observed. (Tyler et al. 1992)*

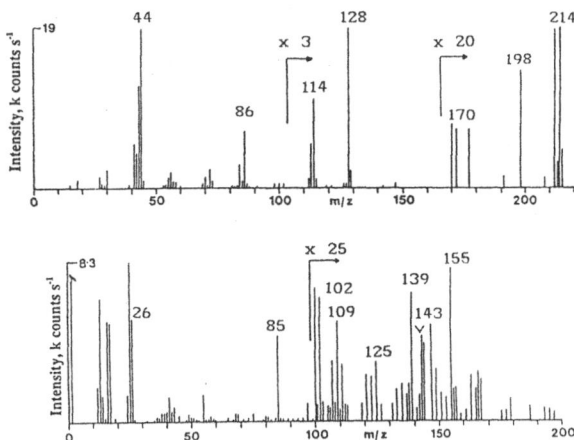

Figure 20. *Positive-ion (upper) and negative-ion (lower) spectra from batch* BSUA. *All the peaks are characteristic of* DPA–EMA. *(Tyler et al. 1992)*

Eccles J A and Vickermann J C, 1989, *J Vac Sci Technol A* **7** 234

Hearn M J and Briggs D, 1988, *Surface and Interface Analysis* **11** 198

Hearn M J, Briggs D, Yoon S C and Ratner B D, 1987, *Surface and Interface Analysis* **10** 384

Hearn M J, Ratner B D and Briggs D, 1988, *Macromol* **21** 2950

Hearn M J, 1992, *in SIMS VIII* eds Benninghoven A, Janssen K T F, Turnper J, and Werner H W (Wiley, Chichester)

Jede R, Ganschow O and Kaiser U, 1992, in *Practical Surface Analysis, Second Edn. Vol 2: Ion and Neutral Spectroscopy* Ch 2, eds Briggs D and Seah M P (Wiley, Chichester)

Leggett G J, Vickerman J C and Briggs D, 1990a, *Surface and Interface Analysis* **16** 3

Leggett G J, Vickerman J C and Briggs D, 1990b, *J Chem Soc, Faraday Trans* **86** 1863

Leggett G J, Briggs D and Vickerman J C, 1991, *Surface and Interface Analysis* **17** 737

Lub J, van Vroonhoven F C M B, van Lyen D and Benninghoven A, 1988, *Polymer* **29** 998

Lub J, van Vioonhoven F C B M, van Leyen D and Benninghoven A, 1989, *J Polym Sci, Polym Phys Ed* **27** 2071

Magee C W, 1983, *Int J Mass Spectrom Ion Phys* **49** 211

Michael R S, Katz W, Newman J and Moulder J, 1990, in *SIMS VII* p 773, eds Benninghoven A, Evans C A, McKeegan K D, Storms H A and Werner H W (Wiley, Chichester)

Niehuis E, van Veltzen P N T, Lub J, Heller T and Werner H W, 1989, *Surface and Interface Analysis* **14** 135

Occhiello E, Morra M, Garbassi F, Humphrey P and Vickerman J C, 1990, *SIMS VII* p 789, eds Benninghoven A, Evans C A, McKeegan K D, Storms H A and Werner H W (Wiley, Chichester)

Reed N M and Vickermann J C, 1992, in *Practical Surface Analysis, Second Edn, Vol 2: Ion and Neutral Spectroscopy* Ch 6, eds Briggs D and Seah M P (Wiley, Chichester)

Short R D and Davies M C, 1989, *Int J Ion Process and Format* **89** 149

Steffens P, Niehus E, Freise T, Griefendorf D and Benninghoven A, 1984, in *SIMS IV* p 404, eds Benninghoven A, Okano J, Shimizu R and Werner H W (Springer, Berlin)

Tyler B J, Ratner B D, Catner D G and Brigss D, 1992, *J Biomed Mater Res* **26** 273

van Ooij W J and Brinkhuis R H G, 1988, *Surface and Interface Analysis* **11** 430

Waugh A R, Kingham D R, Hearn M J and Briggs D, 1988, in *SIMS VI* p 231, edited
    by Benninghoven A, Huber A M and Werner H W (Wiley, Chichester)
Wilding J R, Melia C D, Short R D, Davies M C and Brown A, 1990,
    J Appl Polym Sci **39** 1827

Applications of Surface, Interface and Thin Film Analysis in an Industrial Research Laboratory

H W Werner

Technical University, Vienna & *formerly of*
Philips Research Laboratories, The Netherlands

1 Introduction

The ideal goals of an industrial enterprise are two-fold, to make money and thus to contribute to the general welfare of society. To achieve these goals products must be designed, developed and produced. The products themselves result either from a demand from the market (market pull) or are proposed by industry (technology push).

Instrumental analysis research is of paramount important right from the start of every production process. The industrial scientist provides the physical basis for the understanding of advanced production processes that incorporate new technology whilst the technological engineer develops processes based on these fundamental ideas. Finally at the production stage products are made, preferably cheaply and with high yield, by using such processes. The time delay between science and technology is often 5 to 10 years but in high technology there is no time delay between science and technology. The product must finally be sold in order to produce a cash-flow back into the company. The money earned is used to pay the salaries of the employees, to invest money in research and new technologies, and to satisfy the shareholders. In this way an industrial enterprise and hence its research organisation are embedded into society with technology as a tool. The approach in industry is pragmatic (one needs results, *i.e.* products, profit) and is problem oriented.

High technology is used more and more in industry in our 'silicon-age'. The larger the degree of sophistication of a technology (*e.g.* VLSI, submicron memories which have several million transistors on a chip a few mm^2), the more urgent the need for adequate analytical support of the technology from the outset (Grusserbauer and Werner 1991).

2 Analytical techniques for surface, interface and thin film analysis (SITA)

2.1 Techniques for element analysis

The diagnostic techniques used in integrated circuit (IC) analysis are complex and depend on different physical interactions. These techniques overlap in the analytical information that can be obtained (Werner 1989). In bulk analysis the average chemical composition across the whole sample is determined (up to several mm^3). When the sample itself has very small dimensions (typically μm^3) one speaks of bulk microanalysis. An example is chemical analysis of dust particles. In contrast to bulk analysis, distribution analysis gives information on the spatial distribution of the chemical composition across the sample. Depth analysis is the most frequently used mode: the sample surface is eroded continuously as a function of time by means, for example, of an ion beam to determine the concentration of a selected element as a function of depth.

2.2 Analytical features

To make a meaningful choice of diagnostic technique one must be aware of its analytical features. Consider for example the element range in chemical analysis. There are methods which can detect all the elements from hydrogen to uranium (*e.g.* SIMS), whilst with other methods it is difficult to detect low Z elements (*e.g.* XRF) or high Z elements (AES). The ability to separate one element from another is also important. For instance RBS is not capable of separating mercury from gold.

Another feature which must be considered is the limit of detection. This is the smallest detectable concentration of an element, Z, in a given matrix. The limit of detection is given in atoms per cubic centimetre (absolute detection limit) or in relative units (fractional detection limit). Some methods quote the relative concentration in ppmw (also denoted by $1\mu g/g$) and ppbw (also denoted by $1ng/g$). The term *detection power* is used to indicate the performance of an instrument. One should never speak of high sensitivity to indicate a high detection power.

The analysed area is also an important feature and the relevant size of the analysed area depends on the particular step in the fabrication process. Large area analysis is tolerable only before patterning has taken place; after patterning the necessity to analyse submicron features is evident.

In quantitative analysis (Werner 1980) using SIMS, XRF, *etc.*, the aim is to determine the concentration c of an element X from its measured signal I by a formula of the form $c = I/\rho$, where ρ is the sensitivity factor. Clearly the sensitivity ρ depends upon the instrument, the element and the matrix, but it may also depend on the concentration itself. Thus the relation between signal and concentration is of the form $I(c) = c\rho(c)$. The dependence of c on I of element X in the matrix M can be determined with the aid of calibration standards, *i.e.* samples of the matrix M containing element X in various known concentrations. The measured signals I from these calibration standards are plotted against the known concentrations to generate a calibration curve $I(c)$. Unknown concentrations can then be obtained from this calibration curve. The sensitivity factor

ρ (sometimes just referred to as 'sensitivity') is clearly the slope in the calibration curve if the function $\rho(c)$ is slowly varying. When ρ is constant, there is obviously a linear relationship between the measured signal and concentration. (Note that the term sensitivity is identical to the sensitivity factor as defined above; it is often erroneously used to indicate the limit of detection).

In analytical techniques we speak of a matrix effect when the sensitivity factor for an element depends on the matrix. In general any method can show a matrix effect, but in some methods (*e.g.* SIMS) it is quite pronounced. The SIMS sensitivity factor can change by a factor of 100 to 1000 when the medium changes from silicon to silicon dioxide. The effect however, may be strongly reduced when the silicon surface is saturated with oxygen, *i.e.* one always remains in a silicon dioxide-like matrix. Other methods show smaller matrix effects (up to about 50%). The matrix effect in any method can be corrected for by using calibration standards with similar composition.

A quantitative analysis must always be characterised by two figures of merit: accuracy and precision. *Precision* is a measure of the spread of the measured values around an average value; it is also a measure of the reproducibility of the measurement. A quantitative measure of precision is the standard deviation. The term random uncertainty (1 − precision) is also in use. *Accuracy* is a measure of the systematic deviation from a supposedly 'true' value. The term systematic uncertainty (1 − accuracy) is also in use. The influence of systematic and random uncertainties on a measurement is illustrated in Figure 1 (Bruninx and Bastings 1974). Systematic uncertainties caused by effects not yet identified may occur in any technique. The best way to reveal them is to use several methods simultaneously to solve the same problem. This technique also improves the reliability of an analytical result (synergetic effect of multi-method analysis).

The lateral resolution indicates the minimum dimension (preferably sub-micron) of features that can be observed separately on the sample. The lateral non-coherent resolution is defined as the distance R_1 between two features in the image (in the diffraction limited case the respective diffraction discs) at which the level of the valley between the two corresponding peaks is 75% of their maximum. The depth resolution of a depth profile on the other hand can be defined by using a step-function concentration profile. Ideally the measured profile would be abrupt, but in practice the concentration profile will appear to change over a finite depth due to the lack of infinite depth resolution. The depth resolution is then given by the measured depth across which the signal changes from 84% to 16% of its initial value. A second definition of depth resolution is given by the characteristic length λ (proportional to the slope of the concentration *vs.* depth curve) which gives the change in depth required to change the measured signal by one decade (Benninghoven *et al.* 1987).

All analytical techniques produce *some* destructive changes in an analysed material and this destructivity is an important feature of any analytical method. In the strictest sense, we define as destructive any permanent change induced in the state of a system (sample). In this sense any observation (analysis) based on the interaction between sample and observer (instrument) is destructive. Therefore, in practice, destruction is defined as any permanent change in the state of a system *above* a given level.

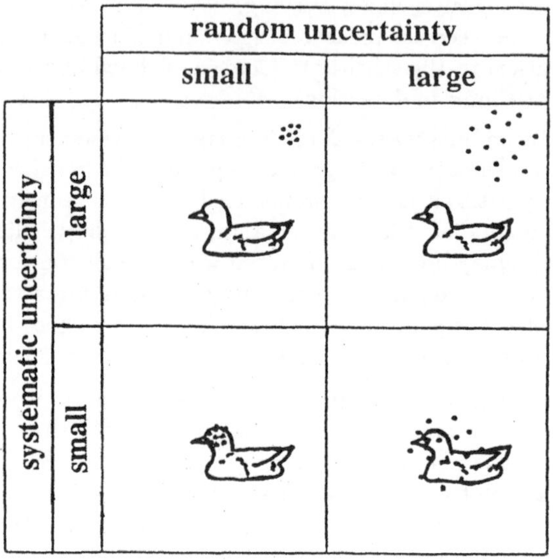

Figure 1. *The nature of random and systematic uncertainties that can occur in a measurement illustrated by shots fired at a target; every shot represents one measurement.*

2.3   Classification of analytical techniques

The principles of element distribution analysis are shown in Figure 2 (Werner 1984). The instrumental methods shown give information on the distribution of a large number of elements in the analysed medium. The methods by and large give a survey over a large number of elements (survey analysis).

Methods which can be used in element analysis are given in Table 1. Morphological methods are useful in telling us whether the sample is present as a single crystal; whether it is polycrystalline or amorphous and to what degree it is perfect (occurrence of defects, *e.g.*, dislocations *etc.*) (Table 2). The topography of a sample tells us such important things as surface roughness, layer thickness, line shapes, line widths *etc*. Appendix 1 lists the most important analytical techniques with their associated acronyms.

2.4   Comparison of methods

A detailed description of the analytical features of the most common techniques for interface and thin film analysis have been published elsewhere (Werner 1989). In Table 3 and 4 each technique for element distribution analysis and morphological analysis is given with its unique features and drawbacks.

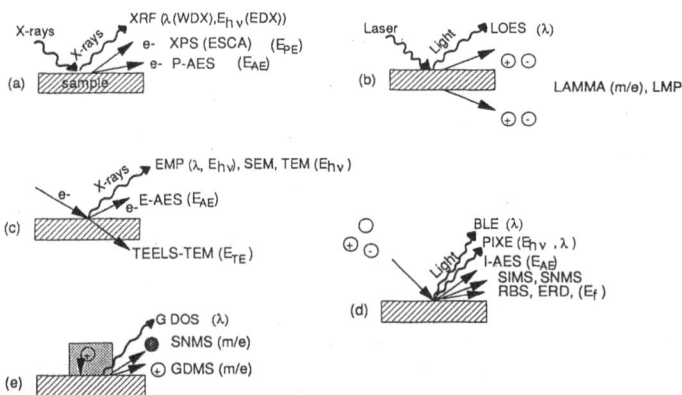

Figure 2. *Excitation and emission (schematic) in different methods for element analysis.*

Excitation	Photons ($h\nu$)	Electrons (e)	Ions (I)
(a) Photons	XRF	XPS (ESCA)	LMP (LAMMA)
	LOES		
(b) Electrons	EPMA (EMP)	AES, SAM	
	SEM-EDX	TEELS	
	TEM-EDX		
(c) Ions	PIXE		SIMS, SNMS
			RBS, ERD

Table 1. *Methods for element analysis (providing distribution or compound information)*

Excitation	Acoustic Waves	Photons	Electrons	Stylus
Acoustic Waves	SCAM			
Photons		Light Microscopy		
		ELL, XRD,		
Electrons			SEM, EBT	
			TEM, TED	
Stylus				STM, AFM

Table 2. *Methods providing morphological information (i.e. in general, no element information is obtained, except roughly through Z-contrast in the* TEM *and* SEM*).*

Method	Unique Feature	Drawback (Remedy)	Anal/yr
XRFA	High accuracy, fast, non-destructive	1mm^e	1,000
EPMA	High accuracy, fast, 1μm, diam, non-destructive	$Z \geq$ Na, ($\geq$ B)^b	1,000
ESCA	Compunds, insulators, non-destructive	2mm (2μm)^{e,c}	50
SAM	100Å(SAM) 100Å(SEM)	Insulators	500
TEM-EDX	100Å, $t = 500$Å	Sample preparation	100
TEM-EELS	Detection limit — 4 atoms with 1 nm probe size		
SEM-EDX	Elemental analysis + topography	$Z \geq$ Na, ($\geq$ B)^b 1μm	10,000
RBS	Quantitative, fast, non-destructive	1mm^e, (20μm)^e	2,000
ERD	Hydrogen detection	30 MeV	
SIMS	ppb^a, (500Å, LMI)^d	Matrix effect	500
SNMS	Quantitative	Useful yield	?

^{a}Thin film wafer analysis ^{b}Windowless detector ^{c}Small-spot ESCA
^{d}Not simultaneously with ^{a} ^{e}Diameter of analysed area

Table 3. *Comparison of some techniques for element distribution analysis in* VLSI *technology*

Method	Unique Feature	Analyses/yr
SEM	All-round 'workhorse'	10^4
SCOM	Simple, output in digital form (PIXELS	
SCAM	Depth analysis. Packaging inspection (non-consumptive)	
XRM^a	Non-consumptive. Packaging inspection	
TEM/STM	Atomic scale studies of growth processes Z-contrast element identification	(TEM) 100
X-TEM	Morphology in cross-section of ICs	1,000

^{a}x-ray radiography in microfocus mode = x-ray microscopy

Table 4. *Comparison of some techniques for morphological analysis in* VLSI *technology*

2.5 The use of analysis in wafer processing

The analyst in an industrial laboratory is constantly confronted with problems from technology. Problems from *research, development and production* (R & D and P) must be distinguished. In process research the analyst must contribute to the understanding of a physical phenomenon or chemical process that may be used during the production of a VLS-IC (very large scale, integrated circuit). In development, the need for analytical support is greatest when a new technology is just being introduced. At this stage the technological engineer needs analytical results to evaluate the degree to which he has succeeded in making a new device, or to determine why he has failed. Gradually, once the new process is under control, the number of samples to be analysed/day decreases. Finally, when a particular customer no longer returns for more analytical work, the analyst then has the satisfaction of knowing that he has helped to introduce a new technology.

In production the analyst's support consists of routine monitoring (off-line, on-line, or in-line) of a running process. The crisis which may arise when a process suddenly fails must be solved by the analyst, by pinpointing the reason for the sudden failure.

Due to this strong coupling between technology, physical background and analysis, it is very important that the analyst is thoroughly familiar with all of the stages in silicon wafer processing technology.

3 Atomic-structure analysis in process research (epitaxy, basis and application)

In the present context we shall use the term atomic structure in its broadest sense, and we will consider: (*i*) zones on an atomic scale *i.e.* the arrangement of atoms on the surface or at interfaces (detected by STM or TEM) or in the bulk (detected by TEM or XRD); (*ii*) the arrangement of atoms in rows (which is important for channeling effects, see below); (*iii*) the arrangement of atoms in monolayers, when we consider a monolayer thickness. Atomic structure is a key issue in *process* research and *process* development.

3.1 Process research

For a description of deposition techniques see for example, Brodie and Muray (1981), Pearce (1983), Muller and Kamins (1986) and Herman and Sitter (1988). These techniques include PVD (Physical Vapour Deposition), CVD (Chemical Vapour Deposition), VPE (Vapour Phase Epitaxy), SPE (Solid Phase Epitaxy *e.g.* $CoSi_2/Si$), MBE (Molecular Beam Epitaxy *e.g.* AlGaAs).

In PVD, CVD and VPE the deposited material is a product of a chemical reaction on the surface of the substrate or in the surroundings. In silicon technology most of the layer deposition processes are carried out in the vapour phase *i.e.* the material is transported in the gas phase.

The structure of deposited layers (crystalline, polycrystalline, amorphous) depends on the structure of the substrate onto which the layer is deposited (whether the substrate is amorphous or crystalline), the deposition conditions (temperature, deposition rate and gas pressure), and the reaction taking place during deposition by CVD. These reactions are promoted by heating the substrate in order to provide the necessary mobility of the adsorbed atoms. Energy can also be introduced into the system electrically, by generation of a plasma in the deposition chamber. The structure of the deposited layer can be single-crystalline, polycrystalline or amorphous.

SPE involves the epitaxial deposition (evaporation) of an element *e.g.* cobalt at room temperature followed by annealing (temperature treatment) whereas in MBE the material is deposited on the substrate by means of a molecular beam in UHV (10^{-10}torr). Very steep profiles can be made by means of MBE. In this way atomically sharp interfaces are made. The depth profile of the element then resembles a δ-function. SIMS-measurements give profiles $c(z)$ with a typical slope of 10–30Å per decade and FWHM 35Å.

3.2 Epitaxial growth

Epitaxy is the growth of single-crystalline layers on a crystalline substrate with a fixed crystallographic relation between the two (Matthews 1975). In order to have epitaxial layer growth one must consider among other factors the following parameters: surface reconstruction and relaxation, lattice mismatch, total energy of the system, crystal imperfections (crystallographic and chemical), material (different processes apply to homo- or hetero-epitaxy), crystal plane (hkl), impurities, arrival rate and mobility of adatoms (*i.e.* atoms which will form the future epilayer) and temperature (different phases exist only in a given temperature interval).

The surface of the substrate determines the growth and the chemical reaction which follows the arrival of atoms (including contamination and adsorption of unwanted atoms). A perfect lattice is a well-ordered structure such that the net force acting on each of the atoms is zero, whereas the structure of a surface is different from the bulk; due to the absence of neighbouring atoms on one side. The equilibrium position of the surface atoms is therefore different from those in the perfect (bulk) lattice positions and this results in a strained surface. In a relaxed surface the symmetry of the bulk in a plane parallel to the surface is retained, but the spacing and arrangement are different from the bulk in directions perpendicular to the surface. The strained surface will rearrange in such a way that the symmetry in the plane of the surface is different from an equivalent plane in the bulk. One speaks of ($m \times n$) reconstruction when the mesh (period) is m-times larger than the original undisturbed lattice parameter in one direction and n-times larger in a direction that is perpendicular to the first one. This of course assumes that the surface in its initial stage (prior to reconstruction) occurs in a completely perfect state. A real surface, however is not perfectly flat and may have steps on a atomic scale. These steps can be studied with the *scanning tunneling microscope* (STM).

3.3 Analytical support for process research

An example of analytical support for process research is the investigation of silicon prepared by MBE using a STM, in particular homoepitaxy of silicon on Si (001) surfaces. Van de Walle *et al.* (1991) and Hoeven *et al.* (1989) have shown with the STM, that epitaxial growth on Si (001) takes place on two types of steps (called A-type and B-type respectively). Epitaxial growth of silicon was found to take place preferentially on B-type steps (terraces) since the energy release, when a Si-dimer is attached to a B-step, is larger than when attached to an A-step.

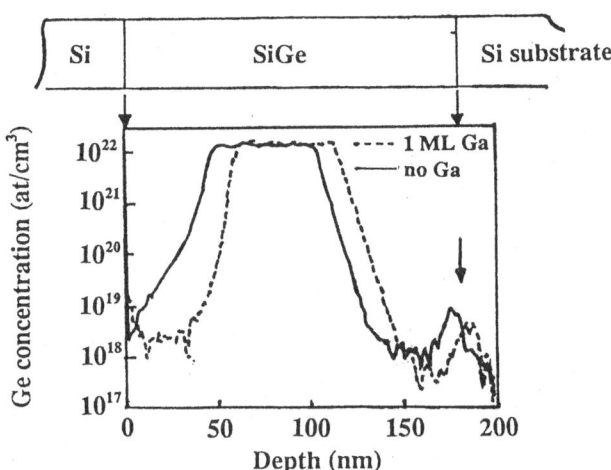

Figure 3. *Ge depth profiles of* MBE-*grown samples G0113 and G0114 as obtained with 3 keV O_2^+. The arrow denotes the interface with the substrate. (– –) G0114, grown at 835 K without precautions (- -) G0113, grown at 835 K after depositing one monolayer of gallium onto the cleaned substrate. (Zalm et al. 1991)*

Another example of analytical support for process research is in the assessment of growth processes with SIMS. For MBE the good depth resolution of SIMS can be used to check the sharpness of Ge-profiles in SiGe layers (Figure 3).

4 Analytical support to process development (diffusion and implantation)

4.1 Diffusion

Introductions to diffusion are given by Tsai (1983), Grove (1967), Le Claire (1949, 1953). Diffusion (and implantation) are used in semiconductor device processing in order to obtain n-type or p-type dopant profiles in semiconductors. The term diffusion refers to the random movement of particles in a medium due to their thermal energy. Diffusion of atomic particles through a solid takes place via interactions between these particles with the lattice atoms or lattice defects (point defects). During this diffusion the particles involved assume different positions in the lattice which are characterised by different energies. In order that diffusion can take place a certain energy barrier ΔG must be surmounted. The energy is delivered by the thermal energy of the diffusing particles and is therefore strongly temperature dependant.

The general diffusion equation can be written as:

$$\frac{\partial c}{\partial t} = \left(\frac{\partial D}{\partial x}\right)\left(\frac{\partial c}{\partial x}\right) + D\frac{\partial^2 c}{\partial^2 x} \tag{1}$$

where the diffusivity D may depend on both x and c, particularly for large implantation doses which cause local strains. (Emitter push). For low concentrations the dependence of D on both c and x may be ignored and we obtain the standard diffusion equation, or Fick's law:

$$\frac{\partial c}{\partial t} = D\frac{\partial^2 c}{\partial x^2}. \tag{2}$$

For large concentrations, where D will depend on the concentration, one needs insight into the atomic processes. However, as long as D is constant, Equation 2 is valid and can be used for prediction in technology.

The application of Fick's Law in technology can now be discussed in two regimes of the solid state diffusion used in semiconductor technology to form diffused layer dopants (impurities). The first regime is the *predeposition step* or *constant source diffusion* in which c_s, defined as $c(x = 0, t)$, is held constant. In the second regime, called *drive-in diffusion* the total amount of dopant is held constant. Predeposition and drive-in diffusion are used in IC technology to make an npn or pnp transistors.

4.1.1   Predeposition or constant source diffusion

The constant flow of gas onto the surface forms a thin solid film on the surface ($x = 0$) and the number of atoms of dopants in such a film is large enough so as not to be depleted in the course of diffusion. the solution of Equation 2 with $c_s = c(0, t)$ held constant is

$$c(x,t) = c_s \operatorname{erfc}\left\{\frac{x}{2(Dt)^{1/2}}\right\} \tag{3}$$

where $2(Dt)^{1/2}$ may be identified as the diffusion length. The total number of dopants (per cm^2) is given by

$$Q(t) = \int_0^\infty c(x,t)\, dx \tag{4}$$

Hence

$$Q(t) = \frac{2}{c_s(\pi Dt)^{1/2}} \tag{5}$$

and we see that $Q(t)$ increases as $\sqrt{t}$.

4.1.2   Drive-in diffusion

In order to achieve a constant value of Q, the surface must be sealed; for example this may be achieved by a layer of SiO_2) which is impermeable even at high temperatures. The solution of Equation 2 which gives Q constant is a Gaussian distribution function:

$$c(x,t) = \frac{Q}{(\pi Dt)^{1/2}} \exp\left\{\frac{-x^2}{4Dt}\right\} \tag{6}$$

We now see that where $c_s(t) = c(0, t) = Q(\pi D t)^{-1/2}$ and hence c_s decreases as $1/t^{1/2}$. It is straightforward to check that

$$\int_0^\infty c(x, t) dt = \int_0^\infty \frac{Q}{(\pi D t)^{1/2}} \exp\left\{\frac{-x^2}{4Dt}\right\} dt \tag{7}$$

is indeed equal to Q.

At this point the general formulation of the diffusion equation (Bockris and Reddy 1967) should be considered. Equation 2 above takes into account only x and t dependence of the concentration. In this case one can expect that the diffusion current, $j = -\partial c / \partial x$, flows down a concentration gradient However, there are cases in practice where the diffusion current may be reversed. This can be ascribed to a thermodynamic 'pseudo-force' F_D, defined as

$$F_D = \frac{dG}{dx} \tag{8}$$

where G is the Gibbs free energy or thermodynamic potential. This gives a diffusion current

$$j = -D\frac{\partial c}{\partial x} - B\frac{\partial G}{\partial x}.$$

For a chemical reaction to occur δG is always negative so if $B\,\partial G/\partial x$, is great enough the diffusion current may be reversed.

4.1.3 Intrinsic point defects and the diffusion mechanism

The simplest and most basic intrinsic point defects (intrinsic means belonging to the lattice itself, in this case silicon) in crystals are vacancies (V) which are unoccupied sites and self-interstitials (I) which are additional atoms of the same species as the crystal, squeezed in between the atoms on normal lattice sites. The symbol S (self-substitutional) is used to denote a silicon atom on its normal lattice site. Vacancies can be created when a self-substitutional atom (S) is pushed out of its place. A Schottky vacancy (surface atom, V) is formed if a pushed-out atom goes to the surface, leaving a vacancy V in the bulk. A Frenkel vacancy (I,V – pair), is formed if the pushed-out atom moves into an interstitial site I in the bulk leaving a (bulk) vacancy V.

4.1.4 Extrinsic point defects

Dopants in semiconductors (n-type or p-type) can be considered as extrinsic (not belonging to the original lattice), see Figure 4 (Frank *et al.* 1985). An example of an interstitial dopant A_i (where A = dopant atom) is Fe or Li in silicon (lithium has a small diameter and therefore fits well into the silicon lattice). Example of substitutional dopants A_s in silicon (group IV) are P, As, Sb (group V) or B, Al, Ga (group III). The substitutional elements are in a charged state, *i.e.* P^+, As^+, Sb^+ (donors in Si) and B^-, Al^-, Ga^- (acceptors in Si).

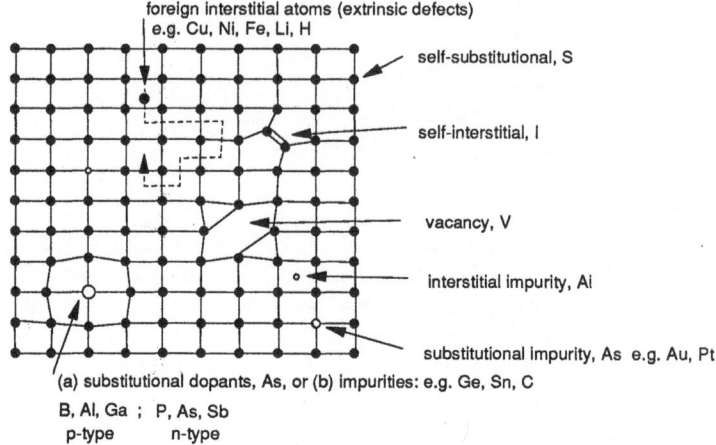

Figure 4. *Various types of point defect (extrinsic or X-type and intrinsic) in a Si-lattice shown schematically.*

4.1.5   Frank-Turnbull diffusion mechanisms

In direct interstitial diffusion dissolved atoms (Fe, Li in Si) diffuse by 'hopping' from interstitial site to interstitial site (Figure 4). No other defects are needed for this mechanism, therefore it is a fast diffusion and much faster than for substitutionally dissolved elements, which move via a vacancy mechanism (see below).

If E_a is the activation energy for diffusion in $D = D_0 \exp(-E_a/kT)$, then for interstitial diffusion E_a lies in the range 0.6–1.2 eV and for vacancy diffusion E_a lies in the range 3–4 eV.

In the vacancy diffusion mechanism substitutionally dissolved atoms (including self-substitutionals S) require intrinsic point defects (vacancies) for diffusion (Figure 5a). Here two particles are exchanged. In the Frank-Turnbull (Frank *et al.* 1985) mechanism by contrast, exchange of A_i, A_s, V, occurs (Figure 5b). Other possible mechanisms are the 'kick-out' mechanism (Figure 5c) and the 'interstitialcy' mechanism or 'kick-in/kick-out' mechanism. This is an indirect mechanism, over two processes (Figure 6).

$$A_s + I \rightarrow A_i + S$$

Anomalous diffusion processes have been studied by means of analytical techniques. For example the effectiveness of drive-in diffusion processes have been studied by SIMS and RBS (Josquin *et al.* 1983).

4.2   Diffusion in device fabrication

Polycrystalline silicon (also referred to as polysilicon or π-silicon) diffusion sources are used in IC technology to make transistors. Typically the substrate is covered with polysilicon layer (about 0.5 μm thick) which is implanted with a dopant. This layer then

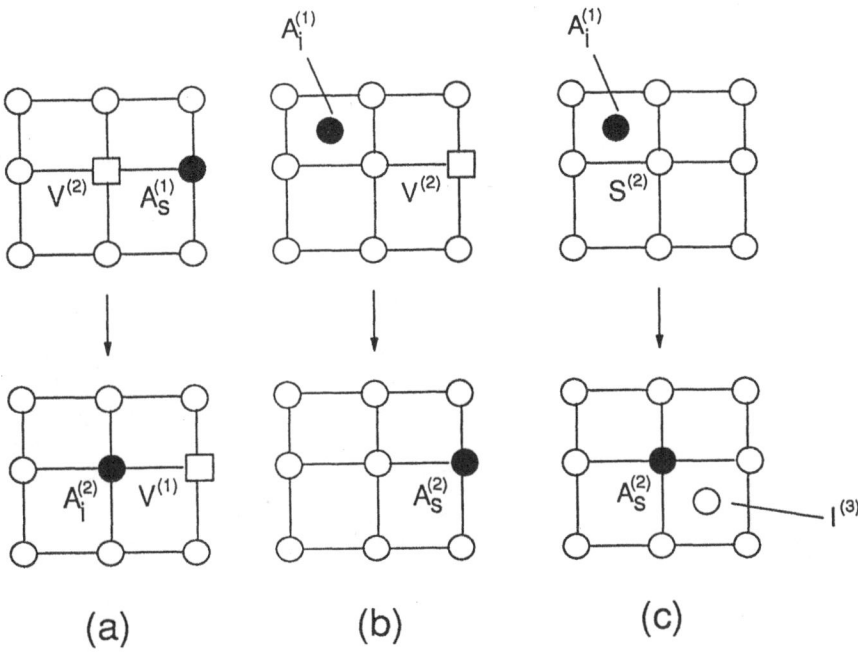

Figure 5. *(a) vacancy, (b) Frank-Turnbull and (c) 'kick-out' mechanisms. • dopant atom, ○ lattice atom, □ vacancy. Subscripts: S = substitutional, superscripts = position e.g. $A_s^{(2)}$: dopant atom (A), substitutional on position (2); Note: I = self interstitial, S = self substitutional = normal lattice position.*

acts as a drive-in source (Section 4.1) in a subsequent temperature cycle.). It is found that the presence of an oxide layer (native oxide) on the polysilicon/monocrystalline silicon interface gives better diffusion than the oxygen-free interface. The reason is the presence of the native oxide since it prevents realignment of the polylayer into monosilicon. Monosilicon has a smaller diffusivity (vacancy mechanism) than polysilicon. Grain boundary diffusion, involving an interstitial mechanism is much faster.

The emitter push effect is used in the fabrication of, for example, npn narrow-base bipolar transistors. Enhanced diffusion of B occurs, *i.e.* the boron atoms penetrate deeper than would happen without P doping (Nicholas 1966). This can be explained by observing that the diffusivity in Equation 1 varies with a^2, where a is the lattice constant (strained).

Implantation (see Maydell, this Proceedings) can also be used to introduce dopants. An individual implanted ion undergoes scattering events with electrons and atoms in the target, reducing the energy of the ion until it comes to rest. Point defects and even small amorphous zones may result (damage). The total integrated path length of the ion, R, is called the range. The projection from the end point of the trajectory onto

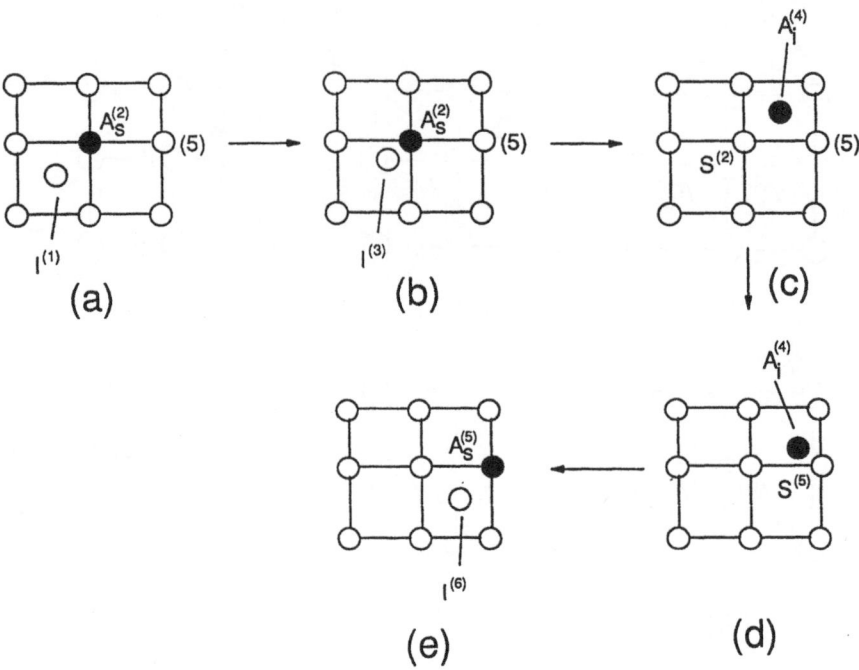

Figure 6. *The interstitialcy mechanism, labelling as in Figure 5.*

the original direction of the ion is R_p, called the projected range. The projection of R_p in turn onto the direction normal to the surface (z-direction) is called the penetration depth. Only for normal incidence ($\theta = 0$), is R_p equivalent to the penetration depth. The range along the path R is not easily accessible so one prefers to use R_p.

Some ions are statistically 'lucky'; they encounter fewer scattering events in a given distance in the target and come to rest beyond the average projected range. Other ions are 'unlucky'; they have more than the average number of scattering events, and come to rest between the surface and the average projected range. The fluctuation or straggle in the projected range is ΔR_p. There is also a fluctuation in the position of the final ion perpendicular to the direction of the incident ion, called the lateral straggle, $\Delta R_\perp$. The depth distribution of stopped ions implanted in a random direction can be approximated by a symmetric Gaussian distribution function ($\Delta R_p = \sigma$).

The implanted concentration $n_I(R_p)$ in the top of the profile has been approximated by (Seidel 1983).

$$n_I(R_p) \simeq \frac{0.4\Phi}{\Delta R_p} \tag{9}$$

Setting $\Delta R_p \simeq 0.4 R_p$, (see Zalm 1992) we obtain

$$n_I(R_p) = \Phi/R_p \tag{10}$$

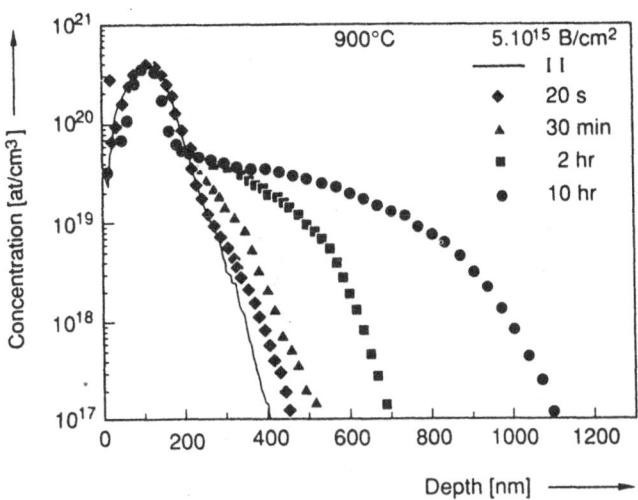

Figure 7. *Time evolution of atomic profiles after diffusion at 900°C, for implantation doses of (a) $5 \times 10^{15} cm^{-2}$ and (b) $3 \times 10^{14} cm^{-2}$.*

In crystalline silicon we must distinguish between implantation profiles, in crystallographically open directions (channeling) and in random directions.

Microbeam analytical techniques have a role in the study of implantation profiles. For example, SIMS concentration profiles can be obtained of implants in both amorphous (a) and polycrystalline (π) silicon. Damage can be determined by RBS (channeling) and TEM. The interaction of ions with target atoms via nuclear stopping results in displacement of lattice atoms (generation of interstitials and vacancies which is equivalent to damage in the crystal lattice).

After every ion implantation, annealing must be carried out. The purpose of annealing is to remove damage generated by the implantation, to place the implanted atoms on substitutional (electrically active) lattice sites and achieve the desired depth distribution. There is a fluence dependence of annealing behavior. Hofker (1975) has shown with SIMS, that boron implanted in silicon precipitates (fixed fraction), when the solid solubility limit of 2×10^{20} is exceeded.

For boron implanted in silicon, in the low concentration region $c_B < 10^{18}$ cm^3, transient (fast) diffusion occurs. We speak of transient diffusion because it only occurs temporarily (mobile fraction). In the high concentration region there is a peak which does not change with anneal time—the static fraction. (See Cowern *et al.* (1990)and also Figure 7).

Experimental results show that there is an electrically inactive static peak (high boron concentration) between $z = 0$ and 200 nm. The measurements of Cowern (1990) have shown that this peak consists of the electrically non-active *i.e.* non-substitutional boron atoms. These observations can be explained by clustering of boron atoms. The

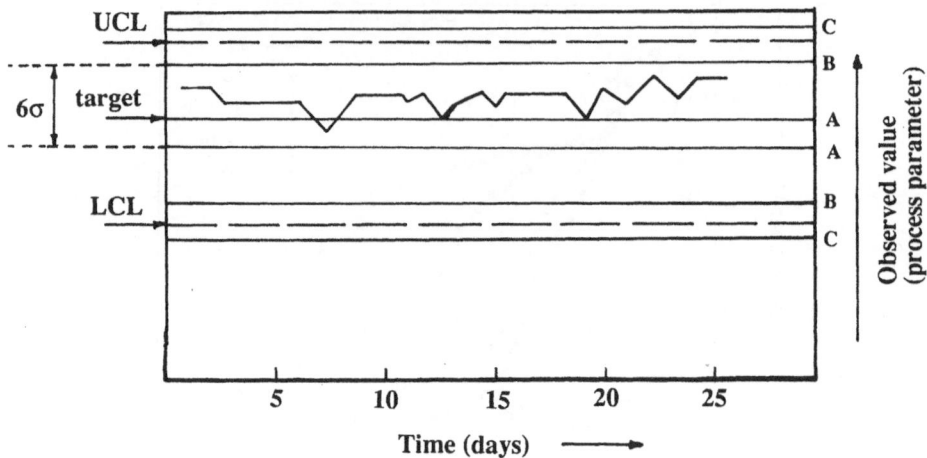

Figure 8. *Process control chart (observed values of the process parameters, e.g. layer thickness, focus value in lithography, etc., versus time, usually in days (de Vries 1990).*

cluster intensity is proportional to the boron concentration. Trapping of boron occurs via clustering. The maximum of this peak therefore occurs at the maximum of the as-implanted profile, and is not related to the damage profile (as in older models) where precipitation or trapping at fixed defects was postulated. It should be noted that here clustering occurs at a concentration smaller than the solubility limit of boron $(2 \times 10^{20}/cm^3)$.

There is also an electrically active profile due to a low concentration of substitutional boron (B_s) ions. The broadening of this profile is ascribed to fast interstitial diffusion of the boron. The boron impurities are generated by the ion implantation.

5 Process control by analytical techniques

5.1 Steps to process control

The steps to process control involve investigation of the mechanism, process development, prototype design, processing and testing followed by yield/failure and reliability analysis of the process. Finally statistical process control is carried out by means of a process control chart (Figure 8).

In 'Failure and Yield' analysis (Van der Wijk 1990) the yield (Y) is defined as

$$Y = \frac{\text{amount of functional products}}{\text{total amount of products}} \tag{11}$$

If the process is functional this means the total process yield is cumulative and depends on the number, n of steps and the yield per step, Y_s. In VLSI there are typically 400 steps with $Y_s=99\%$. The yield after 400 steps is then $Y_{400}=2\%$

Failure, defined by

$$\text{Failure} = 1 - \text{Yield} \tag{12}$$

refers to device properties, *i.e.* it refers to devices that do not meet the target specifications, whereas yield refers to the process. There are several categories of failure, for example, hard permanent failure and soft failure.

Product failure analysis deals with fully processed products (with or without encapsulation) that have failed to perform according to specifications. To discover the cause and mechanism of the failure is the principle aim of the analysis. Many analytical techniques have to be at hand in order to cover the broad range of possible failures.

Reliability is the probability that the product performs according to specifications during a specified period of life. The projected lifetime of an electronic device such as an IC is commonly specified at 25 years. The rate of failures during this time can be described by the 'bathtub' curve (Van der Wijk 1990) which illustrates that initially the 'infant mortality' plays a dominant role followed by a period of steady-state failure. The end of the curve is formed by the wear-out failures.

Obviously without accelerating the wear-out mechanism it is impossible to verify whether or not the expected lifetime is actually achieved. Therefore, reliability tests are performed under accelerated conditions (accelerated stress testing), such as increased temperature, on a large set of devices to ensure statistically significant results. For many failure mechanisms, the Arrhenius equation adequately describes the relation between the temperature T and the reaction rate R

$$R(T) = R_\infty \exp\left[1 - (E_a/kT)\right] \tag{13}$$

where E_a is the activation energy for a certain wear-out mechanism and k is Boltzmann's constant (8.6×10^{-5} eVK^{-1}). Average activation energies are of the order of a few tenths of an electron-volt. The test temperature is chosen to be so high that within a reasonable span of time (days rather than weeks) the devices under each test reach the end of the bathtub curve. For example, if the normal operating temperature of a device is 60°C, an increase to 160°C accelerates a failure mechanism by a factor of ~200 if the activation energy is 0.65 eV. That means that effectively 100 days is equivalent to 25 years.

5.2 Process monitoring using statistical methods

A product is the result of a process (an ordered sequence of events or steps). The product is characterised by specific physical properties or chemical properties called output parameters. Examples are layer thickness, junction depth, sheet resistance, threshold voltage, current amplification, and leakage current.

Product output parameters (OP) are stochastic variables i.e they have an average value ⟨OP⟩ and fluctuations Δ(OP) around this average value. The input parameters, I, for a process are temperature, gas pressure, annealing time, annealing temperature, implantation fluence and energy, equipment factors, and even operator performance ('blue Monday' syndrome).

Even if one keeps all input parameters constant there will still be small fluctuations in the process. This is modified by means of a (statistical) process-monitored control chart, a plot of the output parameter as a function of time or sampling sequence. (See Figure 8 and De Vries 1990). The aim of process-monitoring is to find out if fluctuations in a given process are within the required limits and are according to statistics or are systematic deviations.

We impose UCL (upper control level), LCL (lower control level) limits. These limits may not be transgressed. If we define UCL$-$LCL $= 6\sigma$ (where σ is the standard deviation of a Gaussian or normal distribution, then 99.73% of all values must lie between $x-3\sigma$ and $x+3\sigma$ then approximately one in one thousand will be outside the (UCL$-$LCL) interval. In Figure 8 requirements for the process (*i.e.* the chosen values for (UCL and LCL) are less stringent than the statistical fluctuations, given by 6σ, *i.e.* UCL$-$LCL $> 6\sigma$. The process is said to be under control. Instructions can be displayed to the production operator which are based on these parameters and indicate that production should be continued, stopped, or other action taken.

5.3   Process-monitoring using analytical techniques

SIMS has been successfully used to monitor the implantation flux in VLSI-processing. In a typical case systematic decrease of the implantation fluence was revealed from a process control chart. The points in the control chart (channeling and non-channeling implantations) were derived from broadening (channeling) and non-broadening (non-channeling) of SIMS in the implantation depth profile.

Every product which is delivered from the n'th step of a process will have been preceded by $(n-1)$ steps. Recent studies at Philips Research Laboratories have shown that it is possible to analyse the production in an early process step i ($i \ll n$) and hence predict performance from the quality of the product at step n. This procedure saves valuable redesign time and speeds up production. For example, surface roughness on active layers of a CD can be determined by high resolution x-ray diffraction (XRD). This saves unnecessary redesign of the following process steps (Van der Sluis 1991). Another example is the use of the SEM to measure line width and window dimensions— see Troost 1991. A linear relation has been found between the SEM width and the electrical line width. One can therefore predict early on at the 'front-end' of the process if the linewidth will pass the electrical quality test at the 'back-end' of the process. A substantial amount of time can be saved in this way.

References

Benninghoven A, Ruedenauer F G and Werner H W, 1987, *Secondary Ion Mass Spectrometry* (J Wiley and Sons, New York)

Bockris J O'M and Reddy A K N, 1967, *Modern Electrochemistry*, p 289 (Plenum Press)

Brodie I and Muray J, 1981a, *The Physics of Microfabrication*, p 227 (Plenum Press)

Brodie I and Muray J, 1981b, *The Physics of Microfabrication*, p 349 (Plenum Press)

Bruninx E and Bastings L C, 1974, *Philips Tech Review* **34** *350*

Cowern N E B, Janssen K T F and Jos H F F, 1990, *J Phys A: Math Gen* **68** 6191

De Vries C, 1990, private communication

Frank W, Goesele U M, Mehrer H and Seeger A, 1985, in *Diffusion in Crystalline Solids*,
  p 63, ed D Khang (Academic Press, New York)
Grusserbauer M and Werner H W, 1991, *Analysis of Microelectronic Materials and Devices*
  (John Wiley and Sons, Chichester/New York)
Grove A S, 1967, *Physics and Technology of Semiconductor Devices*, p 35
  (J Wiley, New York)
Herman M A and Sitter H, 1988, *Molecular Beam Epitaxy* (Springer, Berlin/Heidelberg)
Hoeven A J, Lenssinck J M, Dijkkamp D, van Loenen E J and Dieleman J, 1989,
  Phys Rev Lett **63** 1830
Hofker W, 1975, *Implantation of Boron in Silicon, Ph.D Thesis, Univ. of Amsterdam* p79
Josquin W J M J, Boudewijn P R and Tamminga Y, 1983, *Appl Phys Lett* **43** 960
Le Claire A D, 1949, in *Progress in Metal Physics* **1** 306, ed B Chalmers (Pergammon)
Le Claire A D, 1953, in *Progress in Metal Physics* **4** 295, ed B Chalmers (Pergammon)
Matthews J W, 1975, *Epitaxial Growth* Part B (Academic Press, New York)
Muller R S and Kamins T I, 1986, *Device Electronics for Integrated Circuits*, p 94
  (J Wiley, New York/London)
Nicholas K H, 1966, *Solid State Electronics* **9** 35
Pearce C W, 1983, in *VLSI Technology*, p 51, ed S M Sze (McGraw-Hill, New York)
Seidel T E, 1983, in *VLSI Technology*, p 219, ed S M Sze (McGraw-Hill, New York)
Troost K, 1991, private communication
Tsai J C C, 1983, in *VLSI Technology* p 169, ed S M Sze (McGraw-Hill, New York)
Van de Walle G F A, van Loenen E J, Elswijk H B and Hoeven A J, 1991, in
  Analysis of Micrelectronic Materials and Devices, p 657 (J Wiley, Chichester/New York)
Van der Sluis, 1991, private communication
Van der Wijk A and Werner H W, 1990, *Surface and Interface Anal* **16** 253
Werner H W, 1980, *Surf Interface Anal* **2** 56
Werner H W, 1984, in *Thin Film and Depth Profiles Analysis*, p 5 ed H Oechsner
  (Springer Verlag, Berlin/New York)
Werner H W, 1989, in *Microelectronic Materials and Processes*, NATO ASI Series **164** 845
  ed R A Levy (Kluwer Academic Publishers, Dordrecht/London)
Zalm P, Vriezema C J, Gravesteijn D J, van de Walle G F A and de Boer W B, 1991, *Surface
  and Interface Analysis* **17** 556
Zalm P, 1992, private communication

Ion-Induced Auger Electron Emission From Solids

Derek J Fabian

University of Strathclyde
Glasgow, Scotland

1 Introduction

Ionisation of the surface atoms of a solid by bombardment with charged particles is an important field of study from the viewpoint of the fundamental physics it invokes. The subject has also several aspects of relevance to ion-beam methods and to surface analytical physics generally. The information we can extract from the ionisation processes caused by ion bombardment varies according to the mass, energy and charge of the incident particle. Ionisation may also not be the only process involved.

Light high-energy ions (H^+ or He^+ in the MeV range) lose their energy to surface atoms in collision processes and the energy spectrum of those that are backscattered and emerge from the solid (Rutherford back-scattering, or RBS; see Maydell in this Proceedings) can provide information on the surface and subsurface chemical composition.

Ions of relatively low energy (10–40eV), on the other hand, lose their charge by electron tunnelling of a valence-band electron from the solid. The neutralisation occurs very near to the surface and the process is radiationless when the energy released is taken up by the promotion of another valence-band electron to above the vacuum level; *i.e.* when it involves an Auger or 'electron pair' transition (Figure 1). If directed at a suitable angle to the surface, the second or 'Auger' electron escapes from the solid and is emitted with a kinetic energy determined by the initial energies of the two valence-band electrons. The probability of such an Auger transition occurring depends, among other factors, on the densities of occupied and unoccupied electron states in the valence and conduction bands — giving rise to an early probe of electronic structure known as 'ion neutralisation spectroscopy' (see, for example, Fabian 1968).

Ions of intermediate energy (2–300 keV) cause atoms and ions to be sputtered from a solid surface, and mass spectrometry of the secondary ions (*i.e.* SIMS; see chapters

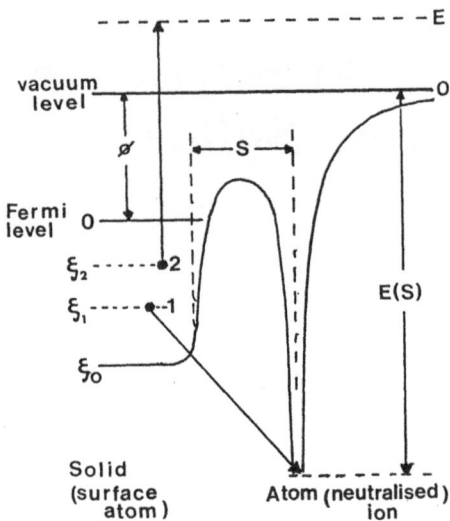

Figure 1. *Energy-level diagram depicting the ion-neutralisation process; the electron-pair transition of valence-band electrons 1 and 2, with initial energies ξ_1 and ξ_2, respectively neutralise the ion and promote an Auger electron to above the vacuum level.*

by Wittmaack and Briggs, this Proceedings) is used extensively to study surface and interface chemistry. Bombardment of solids with ions of energies typically used in SIMS also excites secondary electrons and this 'ion-induced' electron emission is employed in a modern SIMS instrument to provide an 'SEM' image of the surface analysed — in the same manner as in a scanning electron microscope. However, ions in this energy range also cause electron emission arising from Auger decay in excited atoms of the bulk material and in sputtered electronically-excited neutrals and ions outside (as they leave) the solid. Auger transitions again play a part in the ion neutralisation process but, with more energy being released in the primary and subsequent collisions, the two electrons involved can come from inner-shell levels of the surface atoms. In the case of light elements, magnesium, aluminium and silicon, where atoms are frequently excited in the 2p shell by ion impact, the energy-spectrum of emitted electrons contains sharp atomic-like Auger peaks characteristic of the target element.

This review predominantly discusses the ion-induced Auger emission from sputtered excited particles. While the atomic-like spectra are distinctly (and quantitatively) characteristic of the chemical element, restriction of the method to magnesium, aluminium and silicon, and to their compounds or alloys, makes it one of limited application in surface analysis. However, an understanding of the mechanisms of 2p-electron excitation in atoms of these elements, under ion-bombardment, forms an important part in our over-all picture of secondary-ion formation and of ion-beam methods in general.

2 Mechanisms of excitation by ion bombardment

The yield of sputtered ions from a solid, produced by ion-impact, is determined by factors that include sputter rate, surface composition and probability of ionisation of a given species. The first and last of these depend on the chemical nature of the species and matrix material. For metals and semiconductors, the ionisation involves electron tunnelling and models for explaining the ion yield assume ionisation (or neutralisation) of excited species as they leave the surface. For ionic solids, a 'bond breaking' mechanism predominates; excited 'ionic' fragments leave the surface, and in some cases their neutralisation is then followed by auto-ionisation of the excited neutrals formed.

It is well established that the characteristic atomic-like peaks in the ion-induced Auger spectra derive from electron-pair transitions involving 2p holes in excited target atoms (Legg *et al.* 1980), and it is generally accepted that the excitation of the 2p vacancies arise during energetic collisions initiated by the impact of projectile ions with target atoms (Fano and Lichten 1965, Barat and Lichten 1972). Controversy continues, however, concerning the relative importance, for production of 2p-excited target atoms, of the 'primary' projectile/target-atom collision and the resulting target-atom/target-atom collisions that occur in the subsequent collision cascade. In other words, the question of whether asymmetric or symmetric collisions predominate in the promotion of a 2p electron of a target atom has still to be unambiguously answered. Baragiola *et al.* (1982 1991), as well as other authors (*e.g.* Bonanno *et al.* 1992) have addressed the question in detail; *e.g.* Baragiola recently found evidence to show that, with noble-gas projectile ions, Si–Si collisions predominate in the formation of 2p holes in silicon target atoms at projectile energies $\lesssim$ 3.8 keV.

The 2p promotion occurs during the lifetime of the transient 'quasi-molecule' formed from the energetically colliding particles, by the process known as Pauli excitation, first described by Fano and Lichten (1965) for asymmetric collisions and resulting from the crossing of the molecular orbitals created in the collision molecule from the atomic orbitals of the separate atoms when they momentarily combine. Molecular orbital correlation diagrams for asymmetric collisions, depicting the relationship between the atomic orbitals of the separated atoms and those of the single 'united' atom formed if they were made so to unite, are well described for various combinations of projectile ion and target atoms by Barat and Lichten (1972), together with the so-called rules for predicting when molecular orbital crossing and 'promotion' will occur.

The molecular orbital energy levels can be viewed as a slowly time-varying function of the separation of the collision particles. Inner-shell levels of the separated atoms (2p orbitals for aluminium, magnesium or silicon) evolve into molecular orbitals of the combined atoms and, if the distance of closest approach is small enough, can be promoted into unoccupied orbitals of the valence band, where they may remain when the particles separate. Figure 2 illustrates this mechanism for Al + Al collisions. For example, electron promotion by crossing from a $4f\sigma$ molecular orbital of the collision molecule to a $3p\pi$ can occur at internuclear separations $\lesssim$ 0.5A, and interaction of the $4f\sigma$ orbital with the Al valence band produces a 'virtual bound' state which can result in promotion of a 2p electron to a valence-shell 3p level in one of the collision partners.

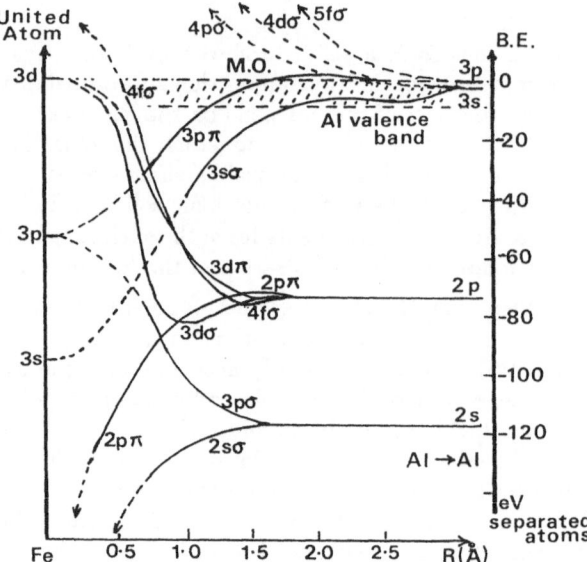

Figure 2. *Molecular orbital correlation and 'evolution' diagram for Al + Al isolated atoms forming a united (Fe) atom. The valence band formed from (3s3p) orbitals plus some 3d is for solid aluminium (and does not directly correlate with the valence band created for the solid formed by the 'united' atoms). Electron promotion occurs by crossing of 4fσ and valence band 3pπ molecular orbitals, and interaction of the 4fσ with the Al valence band.*

In the case of asymmetric collisions specific promotions are possible from either atom. The correlation diagram for Ar + Al collisions, extending the scheme described by Barat and Lichten (1972), is given in Figure 3. On separation the promoted electron may remain with either the lighter collision partner (Barat and Lichten 1972) or alternatively, for L-shell excitations, the collision partner with the lower 2p binding energy (Powell 1978).

3   Instrumentation

The minimum equipment needed for measurements of ion-induced Auger electron emission is illustrated in Figure 4, while a more detailed schematic of a suitable instrument (a SIMS-Auger instrument used in our experimental work at Strathclyde, Maydell *et al.* 1992) is shown in vertical section in Figure 5. Basic measurements require a suitable ion gun, a vacuum chamber with sample loading facility, and some form of electron-energy analyser. The incident ions employed are most commonly those of noble gases (*e.g.* Ar⁺, Ne⁺) though caesium or gallium and even oxygen or nitrogen ion-beams can be

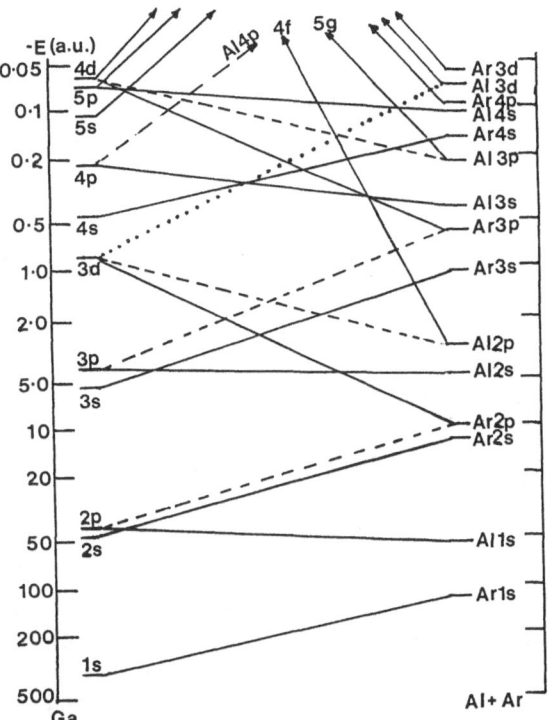

Figure 3. *Molecular orbital correlation diagram for Ar+Al isolated atoms, forming a united (Ga) atom. Electron promotion in Ar atoms can occur by crossing of the Ar 2pσ molecular orbital with the Ar 3pπ.*

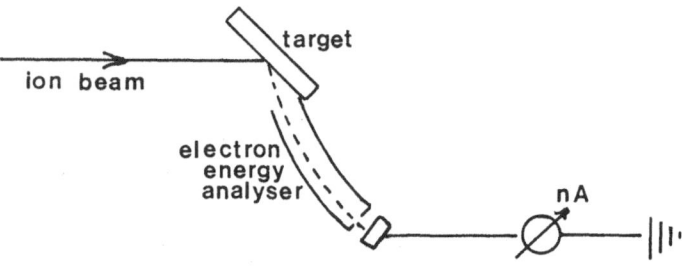

Figure 4. *Basic equipment for measurement of ion-induced Auger electron emission. Count times are arbitrary, to show atomic-like peak intensities (with respect to the background intensities) highest for magnesium, lowest for silicon.*

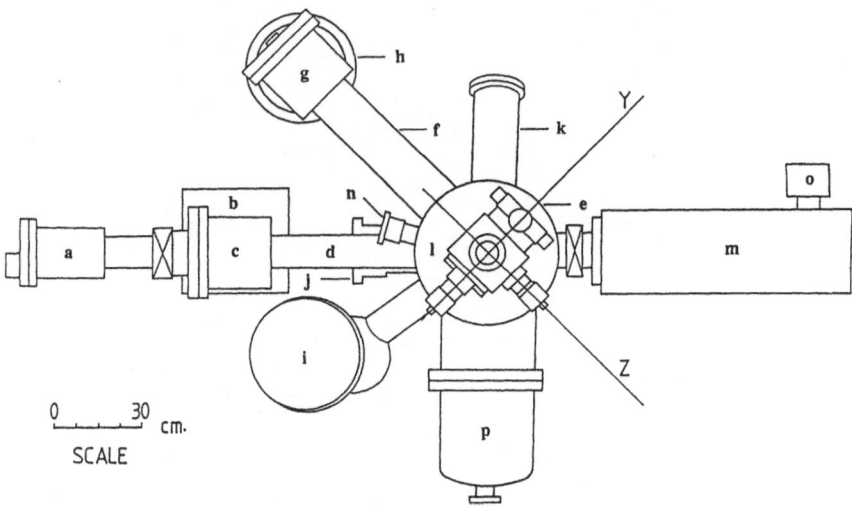

Figure 5. *Vertical schematic elevation of a* SIMS-*Auger instrument used for ion-induced Auger electron emission studies: (a) Duoplasmatron ion gun* (Ar^+, O_2^+); *(b) Thermal ionisation* Cs^+ *ion source; (c) Wien filter (for selecting ions from the duoplasmatron source; (d) Final stage of 'dual' (DP/Cs) ion-beam column; (e) Sample analyser chamber; (f) Secondary ion extraction column; (g) Secondary-ion energy filter; (h) Quadrupole mass spectrometer; (i) Electron-energy analyser (180°-hemispherical); (j) 10 keV electron gun (for electron-induced* AES*). (Reproduced with permission of IOP Publishing from Maydell et al. 1992)*

used. The electron spectrometer can be of any standard design (hemispherical, cylindrical mirror, electrostatic or magnetic sector), with energy resolution 0.5 eV or better if atomic-like Auger peaks are to be resolved.

Argon ions are generated in a duoplasmatron source which operates on a mixture of argon and oxygen gases. Ar^+ ions are separated out from the duoplasmatron beam with a Wien filter, accelerated to 10 keV and focused to a beam diameter of 5 μm. Caesium ions are generated in a thermal ionisation source and are steered, also by Wien filter, so that they share the same final stage (horizontal; labelled 'b' in Figure 5) of the primary-ion column. Secondary electrons are collected in a detection solid angle of 5 msr, at an angle of 36° to the ion beam. The angle of ion-beam incidence on the target can be varied (through 180°) by rotation of the sample, with the angle of electron detection varying at the same time — being fixed (at 36°) with respect to the ion-beam column.

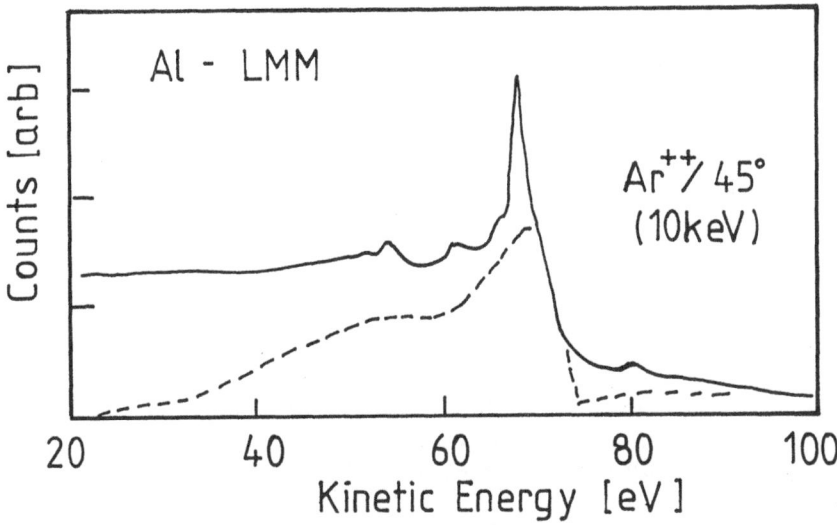

Figure 6. *Ion-induced* LMM *Auger emission from aluminium (solid curve) measured using 10 keV Ar⁺⁺ ions incident at 45° to the sample surface; superimposed on the electron-induced bulk solid valence-band* LVV *Auger emission (dashed curve) measured using a 3 keV electron beam.*

4 Ion-induced Auger electron spectra

A typical ion-induced Auger spectrum for aluminium, measured using 10 keV Ar⁺⁺ ion bombardment ($\equiv$ 20 keV Ar⁺ ions), is illustrated in Figure 6. Characteristic atomic-like Auger peaks arising from sputtered excited aluminium atoms (both ions and neutrals) are observed superimposed on the background valence-band Auger emission.

The spectrum in Figure 6 was recorded with an argon ion-beam incident on the sample at 45°. A sample manipulator (seen depicted in the schematic elevation, Figure 5) permits 360° rotation of the sample with respect to the ion beam; *i.e.* about an axis normal to the Z and Y directions, along which the sample can be translated for accurate positioning of the point of impact of the ions on the sample surface. With the instrument illustrated the angle of collection of electrons is fixed at 36° to the ion beam.

Typical caesium-ion induced Auger spectra for magnesium, aluminium and silicon, observed at 45° ion-beam incidence, are shown in Figure 7. For assignments of atomic-like peaks to Auger decay in neutral or excited ion species see Maydell and Fabian (1992). The intensity of atomic-like Auger electron emission from magnesium, using a 10 keV (15 nA) caesium-ion beam, is of the order of twice that from aluminium and ten times that from silicon. A similar order-of-magnitude higher intensity of atomic-like Auger emission from aluminium compared with silicon is observed with argon-

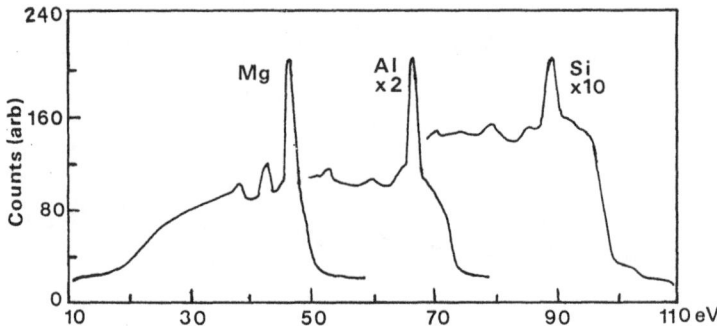

Figure 7. *Auger electron spectra for magnesium, aluminium and silicon, induced by bombardment with a 10 keV (15 nA) caesium-ion beam at 45° incidence. (Reproduced with permission of IOP Publishing from Maydell et al. 1992)*

ion bombardment. Relative atomic-like peak intensities (with respect to background intensity) can give an almost direct quantitative measure of the relative concentration for these elements in a sample (*e.g.* Mg or Al in a MgAl alloy; or of Si in SiC).

Note that when the ion-beam impinges the aluminium sample at normal incidence the relative intensity of the atomic-like Auger peaks observed, with respect to the background bulk solid (ion-induced) emission, is much lower than in Figure 6. This is also true for magnesium and silicon; and the same holds when caesium primary-ions are used. Indeed with silicon, when a Cs^+ primary-ion beam impinges the sample at normal incidence, we find that no atomic-like Auger peaks can be observed; whereas, with a Cs^+ ion beam at 45° to the surface normal (and thus only a 9° angle of detection of the electrons), the spectrum shown in Figure 8 is measured, while with ion-beam incidence 55° and glancing-angle electron detection we observe a dramatic increase in relative intensity of the atomic-like peaks (Figure 9).

5 Discussion

The dramatic dependence of the relative intensities of atomic-like peaks, to the background silicon (bulk solid) valence-band Auger emission, is consistent with the interpretation that these peaks arise from excited ions and neutrals outside the solid. Similar angle-of-incidence effects are observed for aluminium and magnesium, though in those cases some measurable intensity of the atomic-like emission is obtained with the primary-ion beam at normal incidence. An increase in sputtering rate is expected with increasing angles of ion-beam incidence, while glancing-angle electron detection will also suppress the bulk solid background emission (as a consequence of the finite depth of escape of electrons). However, the absolute increase in the intensity of atomic-like peaks with angle of ion-beam incidence is found to exceed by far that expected from the

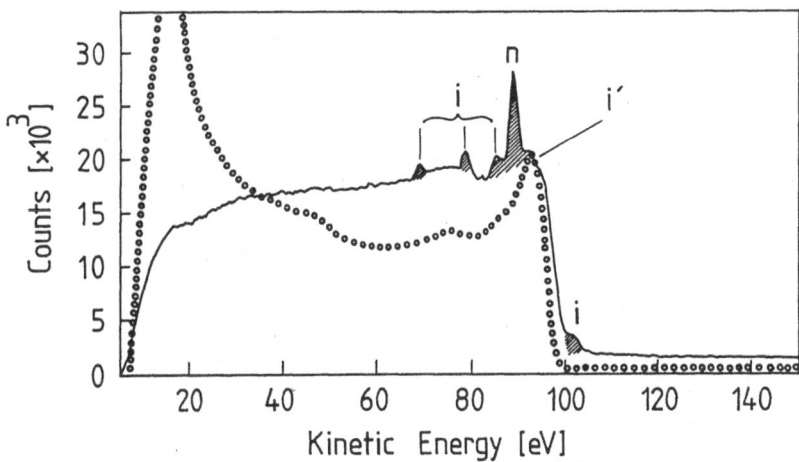

Figure 8. *Ion-induced* LMM *Auger emission from silicon, using 10 keV Cs⁺ ions incident at 45°, with electrons collected at a 9° detection angle to the sample surface. Dotted curve: silicon electron-induced* LVV *valence band Auger emission measured with 2 keV electrons at 45° incidence. (Reproduced with permission of IOP Publishing from Maydell et al. 1992)*

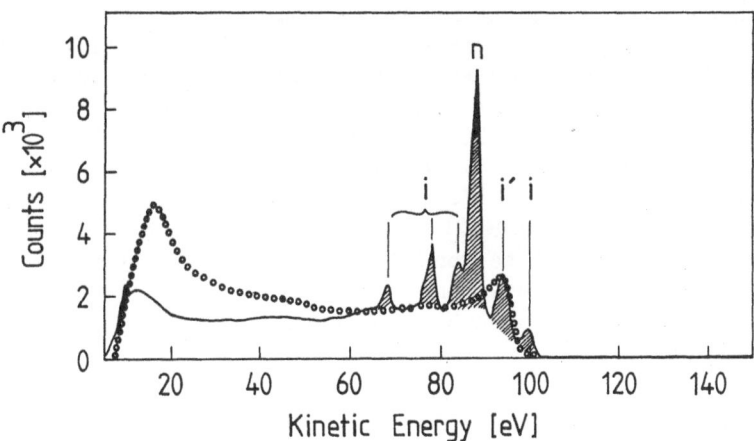

Figure 9. *Ion-induced* LMM *Auger emission from silicon measured with 10 keV Cs⁺ ions at 55° incidence and electron detection at glancing angle. The electron-induced spectrum is shown dotted. (Reproduced with permission of IOP Publishing from Maydell et al. 1992)*

known increase in rate of sputtering. We find this to be especially so with caesium-ion bombardment and mechanisms for formation of the 2p-excited neutral and ionic species outside the solid surface have to be examined.

Bombardment of silicon and aluminium with 10 keV caesium ions produces emission of SiCs$^+$ and AlCs$^+$ molecular ions respectively (Maydell 1992a, Maydell and Fabian 1992). Bombardment with argon ions, on the other hand, produces very little if any emission of molecular ions containing argon. We note also, with caesium-ion bombardment, an order of magnitude higher intensity of atomic-like Auger emission from magnesium and aluminium, by comparison with silicon. Detailed studies of x-ray emission from the sputtered excited particles (such as in earlier work by Cairns 1971) may help to throw further light on these and related problems.

Auger peaks characteristic of the projectile atom have been frequently observed in ion-induced electron emission spectra of transition-metal targets, *e.g.* argon LMM-Auger peaks in the electron emission from chromium, manganese, iron and cobalt induced by argon-ion bombardment (Legg *et al.* 1980), but not in the Auger spectra induced by argon-ion bombardment of aluminium (*e.g.* Baragiola 1982, Whaley and Thomas 1984, Maydell 1992a). However, Legg *et al.* (1980) also reported no argon LMM emission from silicon on bombardment with argon ions, whereas we have observed (Maydell 1992b) argon Auger emission in the atomic-like Auger spectra from both silicon and graphite, induced by 10 keV argon-ion bombardment. This might be attributable to accumulation of argon by ion implantation, and thus to symmetric collisions between the argon-ion projectiles and argon atoms in the surface layers of the target. Surface and subsurface accumulation of argon is known to occur in both silicon and carbon when argon ion-beams are used for cleaning these target materials (Coburn 1976, Webb *et al.* 1985), and has recently been observed in molybdenum and nickel irradiated with argon (Kosyachkov and Cherepin 1989, Filius *et al.* 1989).

It could be the case also that the Ar LMM peaks observed on argon-ion bombardment of transition metals arise from Ar–Ar (symmetric) collisions; Legg and coworkers did not address this possibility, and thus no evidence was reported in their work to exclude accumulation of subsurface argon in transition metals using 20–100 keV argon ions. They discuss their observations solely in terms of excitation of a 2p vacancy in an Ar$^+$ projectile asymmetric collision with a target atom. In our work using caesium-ion bombardment (Maydell 1992a, Maydell and Fabian 1992), with a SIMS-Auger instrument (Figure 5; Maydell *et al.* 1992), some surface accumulation of caesium ions occurs, resulting in energy shifts of the atomic-like Auger peaks from changes in target material workfunction. If symmetric primary collisions occur, excited primary ions leaving the surface can be neutralised by the same Auger process, involving surface atoms, as depicted in Figure 1.

Surface and sub-surface compositional effects therefore undoubtedly complicate the ionisation mechanisms and probabilities of ion-induced Auger emission. Surface oxidation (*e.g.* Koichiro and Tanaka 1984) and more importantly bulk oxidation, using an oxygen primary-ion beam to induce the Auger emission (Maydell 1992b), have been shown to have significant effects on the ion-induced Auger spectra.

Acknowledgements

It is a pleasure to acknowledge the careful experimental work of Dr E A Maydell on which this study is based.

References

Baragiola R A, Alonso E V and Raiti H J L, 1982, *Phys Rev A* **25** 1969

Baragiola R A, Nair L and Madey T E, 1991, *Nucl Instrum Methods in Phys Res* **B58** 322

Barat M and Lichten W, 1972, *Phys Rev A* **6** 211

Bonanno A, Zoccali P, Mandarino N, Oliva A and Xu F, 1992, in *Ionisation of Solids by Heavy Particles*, ed Baragiola R A (NATO ASI Series, Proc NATO ARW) in press

Cairns J A, 1971, *Nucl Instrum Methods* **92** 507

Coburn J W, 1976, *J Vac Sci Technol* **13** 1037

Fabian D J, 1968, in *Soft X-Ray Band Spectra and Electronic Structure of Metals and Materials* p 215, ed Fabian D J (Academic Press, London/NY).

Fano U and Lichten W, 1965, *Phys Rev Lett* **14** 627

Filius H A, Van Veen A, Bijkerk K R and Evans J H, 1989, *Radiation Effects and Defects in Solids* **108** (1) 1

Koichara S and Tanaka S, 1984, *Jap J Appl Phys* **21** L529

Kosyachkov A A and Cherepin V T, 1989 *Sol State Commun* **69** 659

Legg K O, Metz W A and Thomas E W, 1980, *J Appl Phys* **51** 4437

Maydell E A 1992a, *Surf Interf Anal* **19** 65

Maydell E A, 1992b, in *Ionisation of Solids by Heavy Particles* ed Baragiola R A (NATO ASI Series, Proc NATO ARW) in press

Maydell E A and Fabian D J, 1992, *Nucl Instrum Methods in Phys Res* **B67** 610

Maydell E A, Bolouri H and Fabian D J, 1992, *Meas Sci Technol* **3** 1087

Powell R A, 1978, *J Vac Sci Technol* **15** 1797

Webb A P, El-Hossary F M, Fabian D J and Maydell-Ondrusz E A, 1985, *Thin Solid Films* **129** 281

Whaley R and Thomas E W, 1984, *J Appl Phys* **56** 1505

Resonance Ionisation Mass Spectrometry (RIMS)

K W D Ledingham

University of Glasgow
Scotland

1 Introduction

1.1 RIMS technology

Resonance Ionisation Mass Spectroscopy(RIMS) is a relatively new analytical technique which has only been developed in the last ten years or so. The physics on which it is based is however much older. Goppert-Mayer (1931) derived the basic expression for the two photon transition rate (the simplest RIMS process) from second order perturbation theory as long ago as 1931. A sensitive laser spectroscopic technique was proposed by Ambartzumian and Letokhov (1972) twenty years ago, but it is only in the last decade that the pioneering work of Letokhov (1987) and Hurst and Payne (1988) has come to fruition.

RIMS is a particular case of the more general analytical technique of laser mass spectrometry. There are many arrangements of lasers and mass spectrometers. The lasers are used for two purposes. Firstly, they gasify or atomise (desorption/ablation) solid samples and secondly they ionise resonantly or non-resonantly the emitted neutral atoms or molecules. The mass spectrometers can be magnetic, quadrupole or much more commonly time of flight instruments.

Acronyms abound in the field of mass spectrometry and with the aid of Figure 1 a number of these will be explained. When an ion beam hits a target and sputters off ions the procedure is called secondary ionisation mass spectrometry (SIMS). If a pulsed laser beam strikes the target and desorbs/ablates positive or negative ions the process is called laser-induced mass analysis (LIMA). In the sputtering or ablation process, the yield of ions is several orders of magnitude lower than the production of neutral atoms or molecules. When the neutral plume is post-ionised by a second laser system

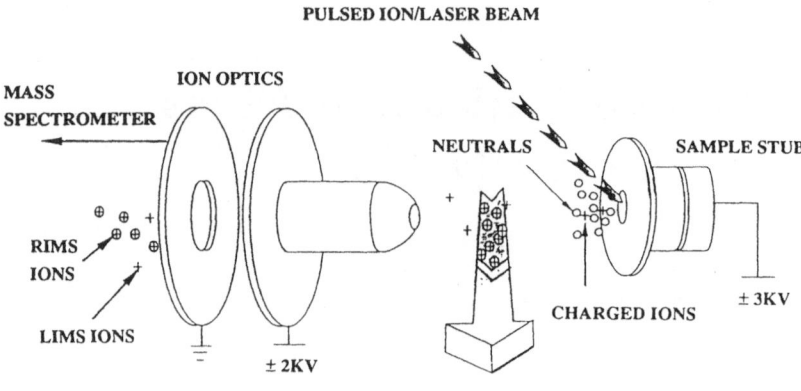

Figure 1. *Ions and neutral atoms are emitted from a surface after bombardment with an ion or laser beam.*

the process is called surface analysis by laser ionisation (SALI) if the laser wavelength is off resonance, and resonance ionisation mass spectrometry (RIMS) if the lasers are resonant. If the plume consists of molecules rather than atoms the process is usually called resonance enhanced multiphoton ionisation (REMPI). In principle, laser post-ionisation of neutrals should always be more sensitive than LIMA or SIMS.

The reasoning that leads to the development of RIMS include:

1. The sensitivity is likely to be increased over LIMA and SIMS since the number of neutrals is between two and three orders greater than the number of ions.

2. The matrix problem associated with RIMS is greatly reduced. Matrix problems in SIMS can be at the level of some orders of magnitude while for RIMS it has been shown to be about a factor of two.

1.2   Excitation Schemes in RIMS

Atoms of each element have a unique set of excited states which may be reached by the absorption of one or more photons at the correct frequency providing certain optical selection rules are satisfied. An excited atom would normally decay back to the ground state with a characteristic time of about 10ns. However before decaying, the atom can absorb another photon which may take the atom to a higher excited state or cause ionisation. The cross-section for resonant absorption of one photon is typically of the order of 10^{-11} to 10^{-12} cm^2 and a laser fluence of about 1μJ cm^{-2} is sufficient to cause saturation excitation of the atoms in the laser beam. The cross-sections for photo-ionisation are normally 10^6 times smaller and laser fluences of about 100 mJcm^{-2} are required for saturation. Most modern tunable lasers require moderate focusing to reach

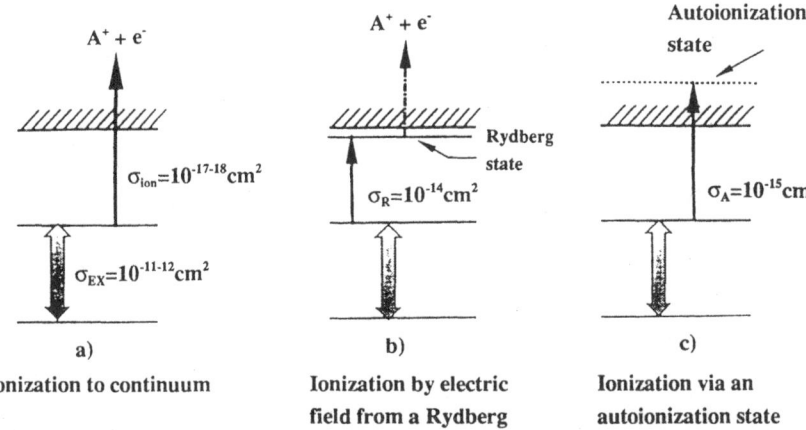

Figure 2. *a) An electron in its ground state absorbs a photon and is raised to an excited state. Ionisation by absorption of a second photon has a small cross section. b) The atom is excited to a Rydberg state and is finally ionised by a pulsed electric field with high efficiency. c) The final step is to an autoionisation state with a large cross section.*

these values. Two other ionisation procedures have been employed to alleviate this problem (Bekov and Letokhov, 1983). These are shown in Figure 2b and c. In b) the atom is excited to a Rydberg state and is finally ionised by a pulsed electric field with high efficiency. The other method c) is to ionise the atom via autoionisation states. The rate limiting step in b), c) has a cross-section some two orders of magnitude larger than in a) and requires much lower fluences to reach saturation.

The laser linewidth typically used in RIMS measurements is between 0.1 and 1.0 cm^{-1} and hence all the isotopes of an element are ionised simultaneously. Separation of the various isotopes is achieved in the mass spectrometer.

All atomic transitions between the ground state and some excited states, with the exception of He and Ne, may be resonantly excited with commercially available tunable dye lasers. Hurst and Payne (1988) have proposed five basic ionisation schemes according to the relative energy positions of the intermediate states to the continuum. These are shown on Figure 3. In the first scheme a level exists at an energy that is more than half the value required to ionise the atom. Hence the atom can be ionised by two photons from the same dye laser, with the first photon being resonant. In the second scheme the output of the tunable dye laser must be frequency doubled to excite the atom resonantly. The atom is then ionised by a photon from the more intense fundamental beam.

Figure 4 shows the whole periodic table with one of the schemes being ascribed for each element. It can be seen from Figure 4 that many elements can be ionised using a single dye laser that has a frequency-doubling capacity. For complete elemental coverage, one requires a large pump laser, an excimer or Nd:YAG laser and two tunable lasers, one which is frequency doubled. In fact with a single dye, 39 elements can

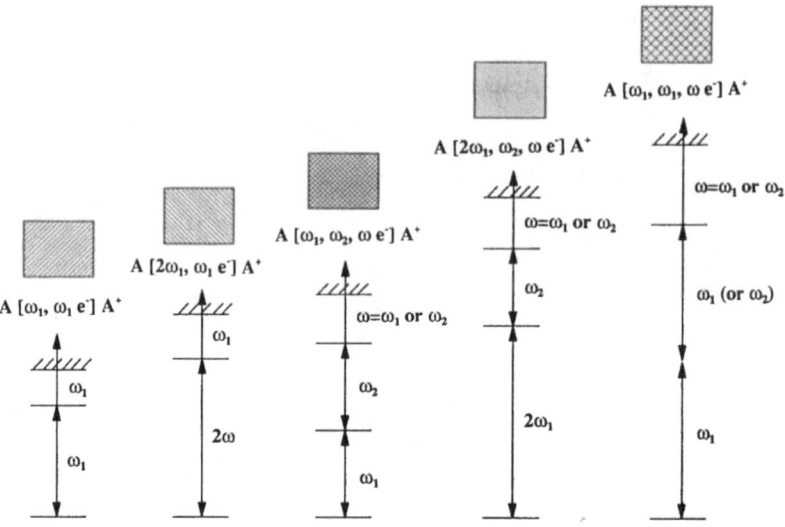

Figure 3. *The five ionisation schemes which can ionise every element in the periodic table except helium and neon.*

be ionised, enabling in principle, computer-controlled element changes in a matter of seconds (Thonnard *et al.* 1989).

RIMS is the acronym used to label a technique primarily associated with the detection of atoms, which usually have a simple set of energy levels. The same techniques can be used with molecules, but molecular structure is considerably more complex, since each electronic level has an associated set of vibrational and rotational levels. For molecules the bound-bound and the bound-continuum transitions are about the same order of magnitude at about 10^{-18} cm^2.

Figure 5 shows several different molecular ionisation schemes. It should be remembered that for each vibrational level shown in Figure 5 there is a manifold of rotational levels which often makes the number of available excitational levels Figure 4 continuous to the ionisation level. Because of this complexity the absorption features of some molecules at room temperature are in general broad and structureless with consequent loss in selectivity. To improve selectivity these can be cooled down to a few sharp peaks by supersonic or jet cooling. This is carried out by seeding the molecules into a light carrier gas (*e.g.* He or Ar at a few atmospheres pressure) and expanding into vacuum through a narrow orifice.

1.3 Sputtering and desorption/ablation

Sputtering in SIMS is normally carried out using either inert gas or oxygen beams. The most important aspect of ion sputtering is to maximise the sputtering yield, *i.e.*

I	II	III	IV	V	VI	VII	VIII			O
I H										
3 Li	4 Be	5 B	6 C	7 N	8 O	9 F				
11 Na	12 Mg	13 Al	14 Si	15 P	16 S	17 Cl				
19 K	20 Ca	21 Sc	22 Ti	23 V	24 Cr	25 Mn	26 Fe	27 Co	28 Ni	
29 Cu	30 Zn	31 Ga	32 Ge	33 As	34 Se	35 Br				36 Kr
37 Rb	38 Sr	39 Y	40 Zr	41 Nb	42 Mo	43 Tc	44 Ru	45 Rh	46 Pd	
47 Ag	48 Cd	49 In	50 Sn	51 Sb	52 Te	53 I				54 Xe
55 Cs	56 Ba		72 Hf	73 Ta	74 W	75 Re	76 Os	77 Ir	78 Pt	
79 Au	80 Hg	81 Tl	82 Pb	83 Bi	84 Po					
87 Fr	88 Ra									

57La	58Ce	59Pr	60Nd	61Pm	62Sm	63Eu	64Gd	65Tb	66Dy	67Ho	68Er	69Tm	70Yb	71Lu
			92U			95Am				90Es				

Figure 4. *Periodic table with an appropriate ionisation scheme for each element.*

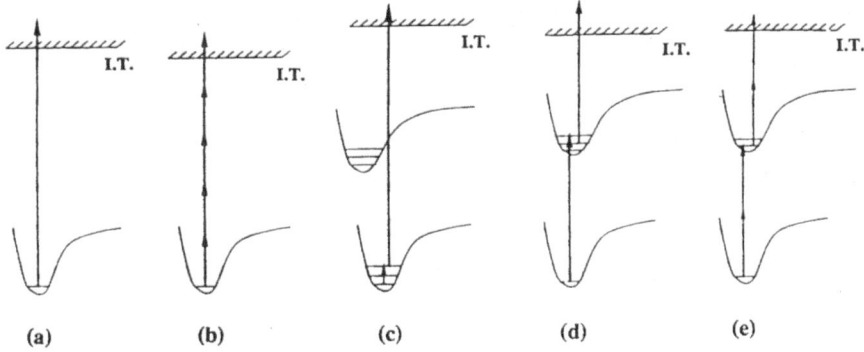

(a)   (b)   (c)   (d)   (e)

Figure 5. *Several different molecular ionisation schemes: i) single photon ionisation; ii) non-resonant multiphoton ionisation (MPI); iii) resonant 2-photon ionisation via ground state rovibrational level; iv) same via excited state rovibrational levels; v) resonance enhanced MPI (2 photons excite and 2 photons ionise).*

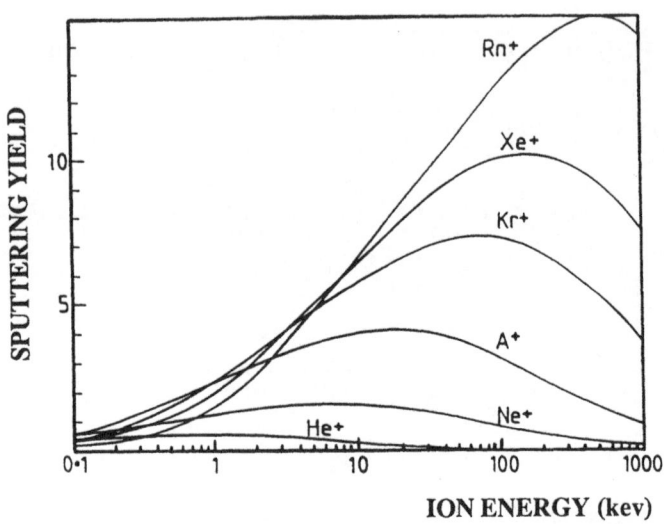

Figure 6. *Calculated sputtering yields for aluminium as a function of ion energy and ion mass (Townsend et al. 1976). (Reproduced by permission of the authors and Academic Press.)*

the number of neutral atoms created from the sample per incident ion. The yield is a complicated function of many parameters among which are the incident ion mass and energy as well as the incident angle, *i.e.* the angle between the beam and normal to the sample. Figures 6 and 7 show how the sputtering yield varies as a function of the mass, energy and angle of incidence of the incident ion(Townsend *et al.* 1976, Oechsner 1973). Clearly an argon ion gun with 5-10 keV energy fitted at an angle of incidence of about 60 − 70° to the normal is a sensible choice for sputtering and maximises the sputtering yield at about 5. Typically, pulsed ion guns have currents of about a few microamps in pulses of about 1 μsec duration, and hence the number of sputtered neutral atoms per microamp is about 3×10^7. Sputtering is a very well characterised procedure and is essential for depth profiling. When a laser beam strikes the surface of a sample, vaporisation is referred to as desorption if the photon flux is less than 10^8 Wcm^{-2} or ablation if the flux exceeds this value. Above flux levels of 10^8 Wcm^{-2}, microscopic craters and measurable material removal exist. This is the region where LIMA operates. Very little work has been done at flux levels below damage threshold although recently Borthwick *et al.* (1992a,b) have shown that at 10^8 Wcm^{-2}, the number of desorbed aluminium atoms is between 10^9 and 10^{10} *i.e.* about 10^{-2} of a monolayer per shot. The desorbing laser is normally focused and studies have shown that there is little effect on the temporal distribution of neutral atoms from an aluminium sample when the wavelength is changed between 266, 355, 532 and 1064nm, as shown in Figure 8. Much work must still be done on laser desorption/ablation to put

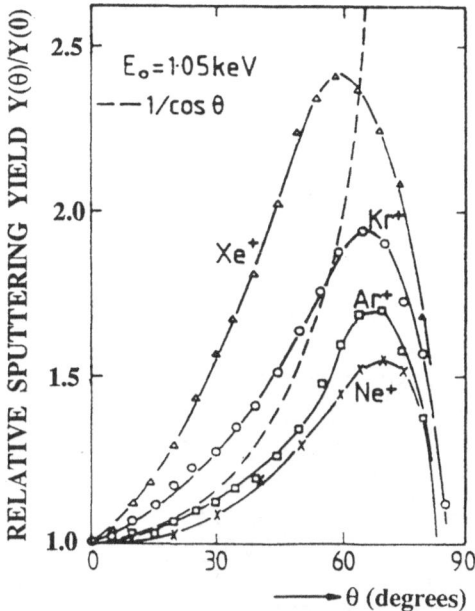

Figure 7. *Relative sputtering yields ($\theta°/0°$). For Ne^+, Ar^+, Kr^+ and Xe^+ as a function of angle of incidence (θ) to the normal (Oechsner 1973). (Reproduced by permission of the authors and Springer-Verlag)*

it on the same experimental and theoretical basis as sputtering. In addition, because of the poorly understood spatial and temporal stability of present lasers, the inherent shot to shot reproducibility of laser desorption is inferior by far to that of ion sputtering.

2 Experimental

2.1 Typical arrangement for analysis

There are many arrangements for RIMS analysis using different types of mass spectrometers — magnetic sector, quadrupole, and time-of-flight of both linear and reflectron types. There are many types of lasers used, both of cw and pulsed designs. By far the most common configuration is a pulsed laser arrangement coupled to a time of flight mass spectrometer.

The Glasgow University design (Towrie *et al.* 1990), described here, is a typical RIMS system now available commercially. Figure 9 shows the system in detail. The spherical sample chamber is 30cm in diameter with as many ports as possible facing the centre of the chamber, the point at which the sample stub is positioned. The sample is mounted on an $xyz\theta$ manipulator and can be inserted and withdrawn from the sample chamber using a rapid transfer probe. Fast sample exchanges (typically 5 minutes) can be made without disruption of the main chamber vacuum. The sample

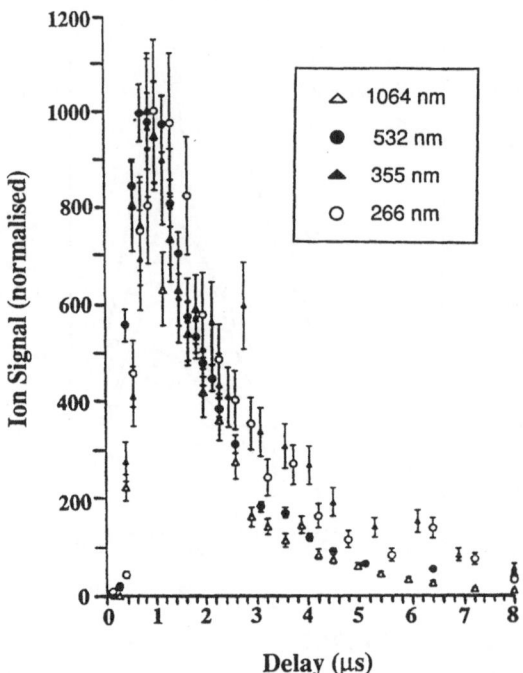

Figure 8. *The effect of changing the ablation laser wavelength on the temporal distribution of neutral atoms from an aluminium sample*

chamber is pumped by an oil diffusion pump fitted with a cold trap and a titanium sublimation pump, having a base pumping speed greater than 800 ls^{-1} and capable of maintaining a base pressure of less than 10^{-7} Pa in the chamber. The ion extract optics and the reflectron time-of-flight TOF spectrometer are shown in Figure 9b. The sample is maintained at a voltage of about 2000V with the first extraction electrode at 1400V. The reflectron TOF system has an overall drift length of 3m. The principal factor limiting the resolution of a conventional TOF is the spread of the initial ion energies in the ablation process. This spread of energies can be compensated using a reflectron TOF mass spectrometer in which the high energy ions penetrate deeper into an electrostatic ion reflector and hence experience a longer flight time than ions of lower energy. The FWHM of the present system is about 1000 for ions of about 40amu. A thin wire, 0.005 cm in diameter, follows the ion path through the flight tube, providing an electrostatic guide for the ions, increasing the transmission of the mass spectrometer. The wire operates at about −10V.

Ablation is carried out using a Quantel Nd:YAG laser operated either at 1064nm or one of its harmonics, although normally 532nm is the preferred wavelength. If sputtering is used the system is fitted with a Kratos Penning ion gun. Both inert gases and oxygen ions can be used with energies up to 15keV and with beam currents up to 5μA.

The ionisation laser system consists of a Spectron Nd:YAG laser powering two dye

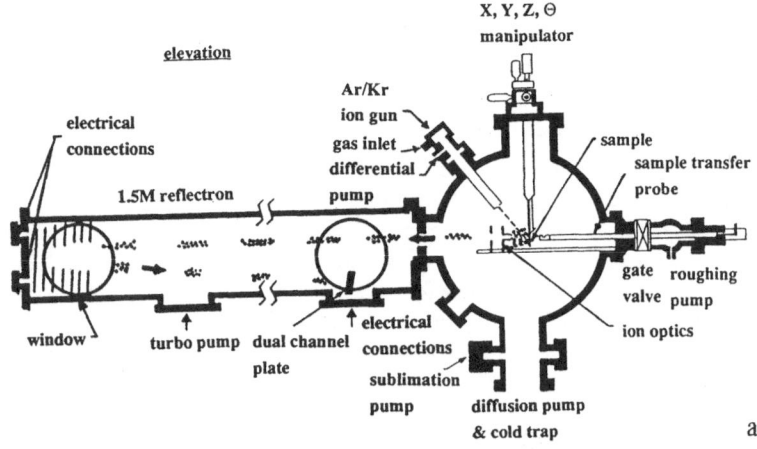

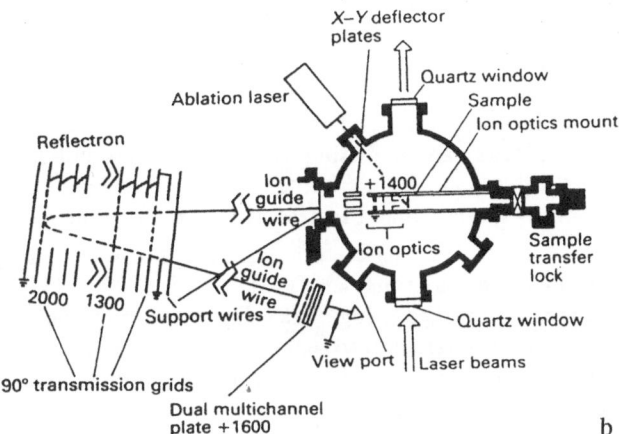

Figure 9. *(a) The Glasgow Resonant Ionisation Mass Spectrometer and (b) details of the electrostatic reflector*

lasers one of which can be frequency doubled. The lasers operate at repetition rates of 10Hz and with pulse durations of about 10ns. The tunable laser pulse energies depend on the dyes but are normally greater than 1mJ. The time between the ablation/sputter and ionisation system can be varied between 0.1 and 10μs, although this typically is 1μs. The ionising laser is introduced into the sample chamber parallel and as close to the sample stub as possible to maximise the overlap with the ablation plume. The ions are detected in a dual channel plate detector and passed to a data acquisition which measures and stores mass spectra and laser pulse energies on a pulse-to-pulse basis. A Lecroy 2261 transient recorder coupled to a COMPAQ 386/25 PC forms the basis of the system. Ion signals from the multichannel plate detector are digitised by the transient recorder which provides 640 time channels (11 bit resolution) each of 10–100ns width.

The RIMS instrument described above is being routinely used for trace analysis and for studies of the characteristics of laser ablation from solid samples. Figure 10a shows a RIMS signal for gold in copper at 10ppm level, accumulated over 10^4 shots. Figure 10b indicates how linear the technique is over a wide concentration range from majors to minors for gold in copper (McCombes *et al.* 1991).

RIMS trace detection down to parts per billion are now common in semiconductors and indeed sensitivities down to parts in 10^{12} have been reported for specific elements (Bekov *et al.* 1985). Atom Sciences Inc can now make noble gas detection measurements to levels below 100 atoms in five minutes which makes solar neutrino, small meteorite and ground water dating experiments feasible (Thonnard *et al.* 1992). Attogram detection limits and isotopic selectivity greater than 10^{10} has been demonstrated for rare radionuclides using cw laser RIMS (Bushaw 1992).

2.2   General advantages of RIMS over SIMS

SIMS is a very mature analytical technique which is widely used for characterising materials. It has however two limitations which are not shared by RIMS. Firstly is the very strong matrix effect, the often unknown relationship between the chemical composition of a solid and the secondary ion yield of the analyte in the solid. Another difficulty with SIMS is isobaric interferences. Isobars are nuclei which have the same A value but different Z values. SIMS cannot distinguish easily isobars unless the resolution of the mass spectrometer is extremely high. RIMS on the other hand easily distinguishes isobars since the ionising laser is tuned to transitions in a specific element.

Figure 11 shows a RIMS analysis on a series of samples with silicon in different matrices *e.g.* steel, niobium and tungsten (Parks *et al.* 1986). This is a remarkable result which could not be obtained using SIMS. The graph is reasonably linear and the lack of matrix dependence is attributed to separation in time and space of the sputtering and the ionisation processes. In the SIMS procedure these processes are coupled.

Figure 12 shows RIMS and SIMS depth profiles which are used to locate a buried monolayer of Co. In the SIMS spectrum the isobars ^{59}Co and the silicon dimer $^{29}Si^{30}Si$ cannot be distinguished whereas in the RIMS depth profile the ionisation lasers are tuned to ionise cobalt alone (Downey and Hosack 1990).

2.3   Quantification in RIMS

A number of laboratories have demonstrated the extreme sensitivity and selectivity of RIMS but less effort has been devoted to the questions of accuracy and precision which are normally as important to analysts. RIMS of sputtered atoms (SIRIS) has the potential of being a quantitative ultra-trace technique. The reason for this is that the sputtering of clean metals by noble gas ions produces a plume dominated by ground state neutral atoms. In addition there is large data base of sputtering rates. If the sputtering rate, the transmission of the mass spectrometer and the degree of laser ionisation are all known then it is possible to draw quantitative conclusions regarding impurity concentrations.

Recently Argonne National Laboratory using an ion sputtering plus RIMS instru-

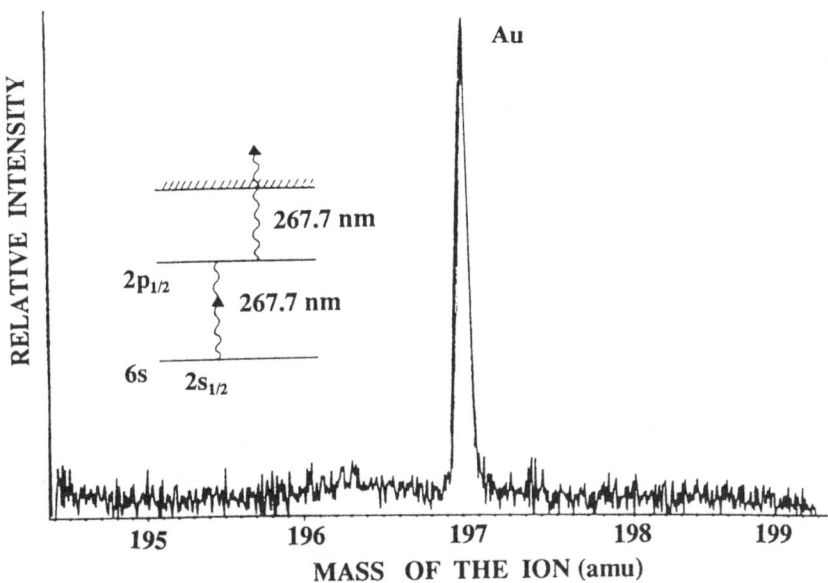

Figure 10. *a) RIMS signal for gold in copper at 10ppm, accumulated over 10^4 shots.*

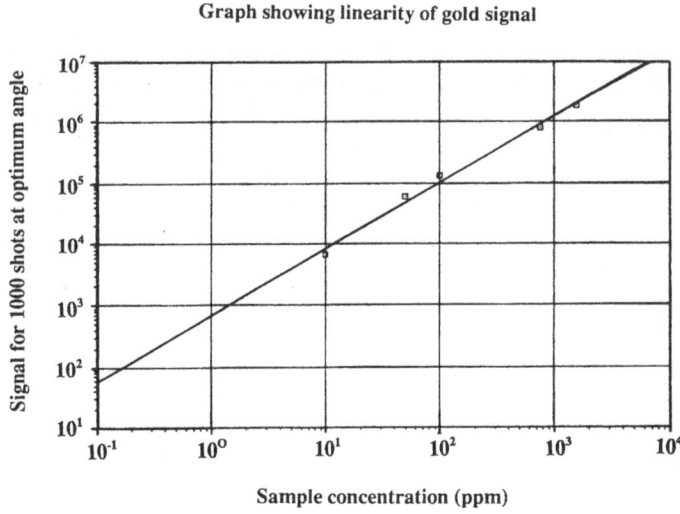

Figure 10. *b) RIMS signal versus gold in copper concentration illustrating the linearity of the process.*

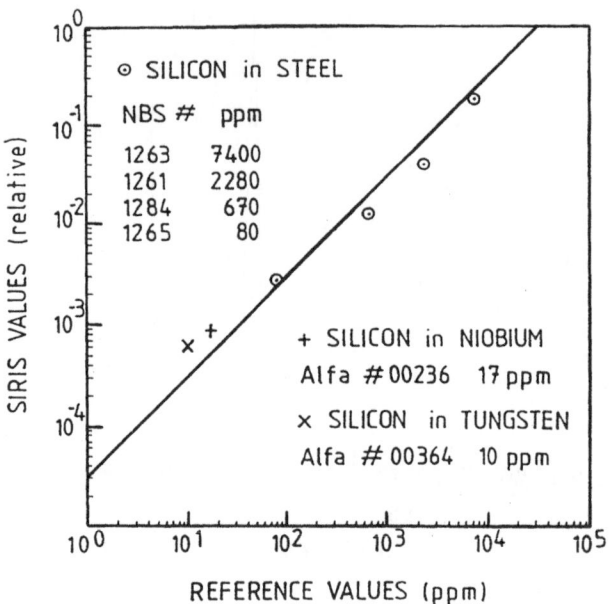

Figure 11. *Comparison of SIRIS measurements using certified reference values of silicon in different matrices (Parks et al. 1986). (Reproduced by permission of the authors and the Institute of Physics)*

ment, have analysed a number of well characterised metallic samples containing iron, copper, molybdenum, indium and lead. In addition Si samples contaminated with Cr, Cu, Ni, Fe and also with implanted Mo were analysed by RIMS and compared with TXRF and known implantation rates. The significant finding is that trace impurity measurements down to ppb levels in metal matrices can be made quantitative by employing pure metal targets as calibration standards. This discovery substantially reduces the effort required for quantitative analysis, since a single standard can be used for determining concentrations spanning nine orders of magnitude. A comparison of the different techniques for the contaminated silicon produced agreement, to within a factor of three, down to parts per billion sensitivity (Calaway et al. 1992). It would appear that RIMS has the potential to become an ultra-trace quantitative analytical technique.

3   Some important applications of RIMS

3.1   Semiconductor materials and structures

Depth-resolved information about dopant and impurity distributions in semiconductors like Si and GaAs is of great importance in the electronics industry. Techniques such as

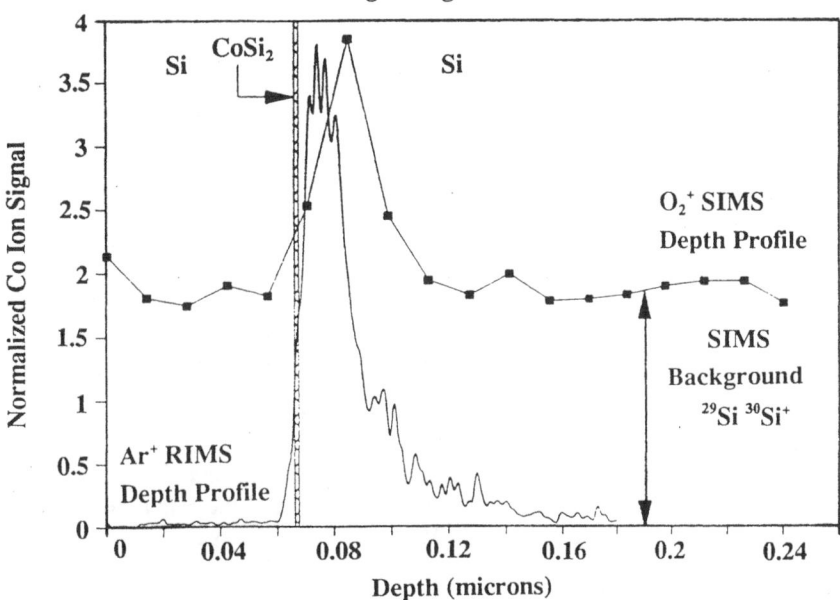

Figure 12. *Comparison of RIMS and SIMS depth profiles to detect a buried monolayer of Co (Downey and Hosack 1990). (Reproduced by permission of the authors and the American Vacuum Society)*

SIMS are very mature and are used routinely to characterise such materials. However for quantitative results, SIMS requires calibration with standards for every element in each matrix encountered. Alternatively for RIMS, assuming that the photoionisation is saturated, the signals have been demonstrated to be almost element independent. It has been shown that equal amounts of impurities such as Be, Al and Co in Si give equivalent signals. This is what is meant by matrix independent. Figure 13 demonstrates this elemental independence (Downey and Emerson 1991).

Some of the most difficult materials to characterise are compound semiconductors such as GaAs, AlAs, AlGaAs, InGaAs, and InAlAs because such devices use thin layers of these materials. Figure 14 shows a comparison between a SIMS and RIMS depth profile to detect beryllium in a layered structure of GaAs and AlAs. Alternative layers *e.g.* GaAs and AlAs, 100nm thick, were grown on a GaAs substrate at 400°C and doped with Be to nominal uniformity at approximately 2×10^{19} cm^{-3}. It is known that under these growth conditions Be is ten times more soluble in GaAs than in AlAs. It can be seen that the RIMS profile represents the layered structure better than SIMS. Moreover, the Be spikes known to accumulate at the interfaces, are of equal widths in the RIMS profile whereas in the SIMS data the widths depend on the ordering of the layers (Downey *et al.* 1990).

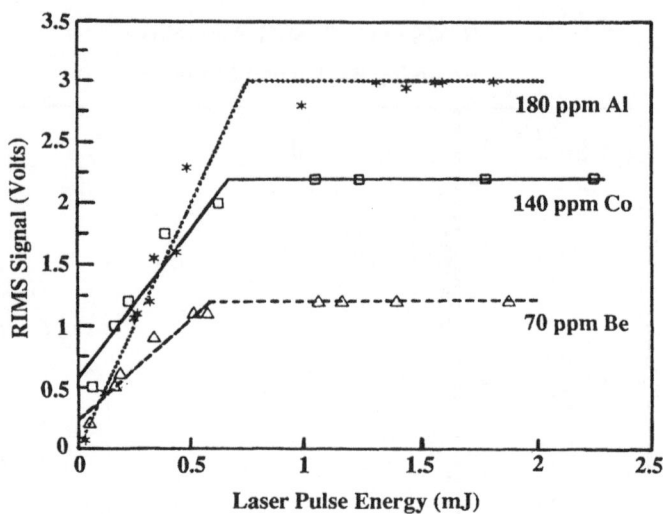

Figure 13. *Demonstration of RIMS saturation for Al, Be and Co implanted into Si. Above 1 mJ per pulse each element responds similarly and the signals are proportional to concentration (Downey and Emerson 1991)*

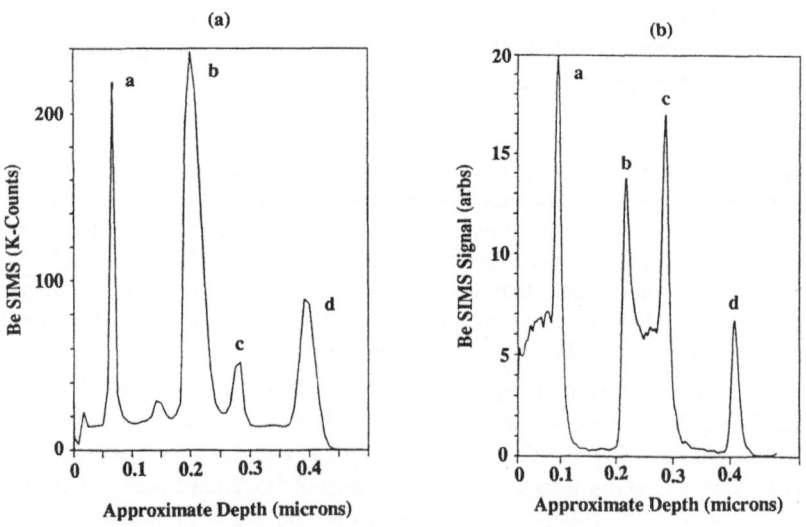

Figure 14. *Depth profiles for a Be doped GaAs/AlAs sample. The top layer and that between b and c are GaAs and a–d mark the interfaces (Downey et al. 1990). (Reproduced by permission of the authors and John Wiley and Sons.)*

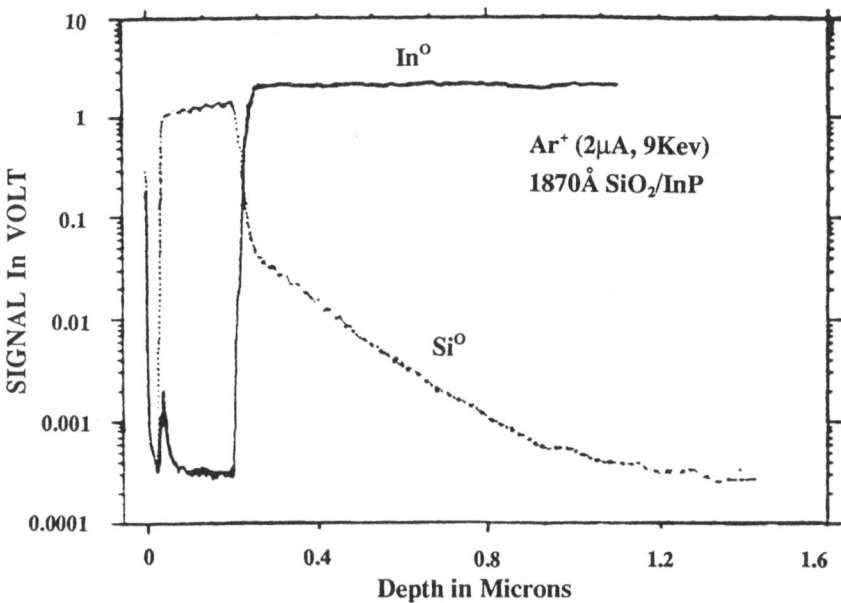

Figure 15. *RIMS of sputtered neutral Si and In atoms in a depth profile through an Au/SiO₂/InP test sample. The signal has not been normalised: note the correct Si and In signal levels (Arlinghaus et al. 1990). (Reproduced by permission of the authors and the American Vacuum Society.)*

Figure 15 shows a SIRIS Si/In depth profile through a Au-SiO₂-InP layered structure. The uncorrected SIRIS signal for Si and In have the approximately correct 1.0 to 1.5 ratio, indicating the essentially identical response to Si and In by the RIMS process (Arlinghaus *et al.* 1990).

It must be emphasised that RIMS is no way a competitor to SIMS. SIMS has been established for more than 25 years and is accepted as a very important technique for characterising materials and the procedure at present is much simpler than RIMS. It is hoped that RIMS can be established as a complementary technique which can be used as a powerful aid for quantification of SIMS.

3.2 Laser ablation and resonant ionisation of neutrals

As has been demonstrated, LIMA is a well established laser microprobe trace analytical technique in which a focused laser is directed at the surface of a sample. The ions created are analysed normally by a reflectron time of flight mass spectrometer to ppm sensitivities or lower. Its speed and versatility make it a very useful instrument although it is essentially a qualitative technique. The reason for this is that the ablation and the ionisation processes are coupled, which leads to matrix dependent ionisation, making quantification difficult.

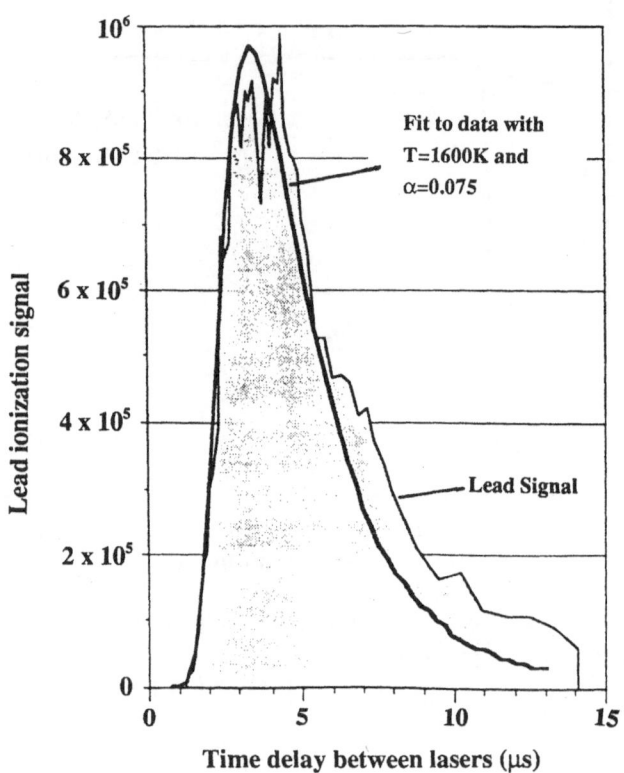

Figure 16. *Lead sample ablated with radiation of 532nm wavelength. Post ionisation is carried out non-resonantly. The fitted curve is a single Maxwell-Boltzman distribution at temperature* $T = 1600K$ *and* $\alpha = 0.075$.

An alternative approach is to separate the ablation and ionisation processes in time and space. This process has been called laser ablation resonant ionisation spectroscopy (LARIS). LARIS is not nearly so well characterised as SIRIS but it is of great interest for analysts involved in the characterisation of insulators (these charge up when sputtered by ion beams), biomolecules and for studying the plasma processes involved in the laser-solid interaction. The normal LARIS procedure is to ablate different samples with various wavelengths and fluences and then ionise the emitted neutrals by either resonant or non-resonant multiphoton processes (Borthwick *et al.* 1992a). The energies of the neutrals are determined by varying the delay time between the ablation and ionising lasers and plotting the signal intensity of a specific mass as a function of delay time. Figure 16 shows the spectrum of lead neutral atoms emitted by ablating at 532nm at a power level less than 10^8 Wcm^{-2} and ionising non-resonantly at 266nm. The curve has been fitted to a Maxwell-Boltzmann distribution at $T = 1600K$ (α is the semi-angle of the expanding plume).

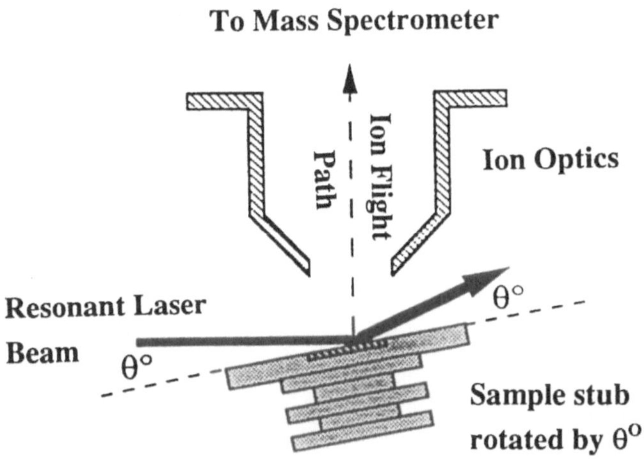

To Mass Spectrometer

Ion Flight Path

Ion Flight Path

Ion Optics

Resonant Laser

Beam $\theta°$

$\theta°$

Sample stub rotated by $\theta°$

Figure 17. *The geometric arrangement for RLA measurements with the laser beam at grazing incidence (θ) to the sample stub.*

In a series of experiments carried out at Glasgow University it has been shown that both ablation and resonant ionisation may be performed with a single laser pulse to provide enhancements of several hundred. This procedure has been called resonant laser ablation (RLA). A pulsed laser beam from a tunable dye laser is directed at grazing incidence (5°) to a series of samples — GaAs, AlGaAs, and NIST steels with traces of aluminium as shown in Figure 17. The dye laser output was frequency doubled and both the fundamental and the doubled beams were moderately focused with a 30cm lens to a diameter of 0.5mm resulting in an irradiation area on the sample surface of about $2.5 \times 10^{-3} cm^2$. The ions created were analysed in time-of-flight mass spectrometers of either linear or reflectron type. The ion signal for Al from AlGaAs as a function of the ablation laser wavelength is shown in Figure 18. The resonance effects are clearly seen with a signal enhancement of greater than two orders of magnitude. Careful calibration of the laser wavelength indicated that the resonances correspond to known atomic transitions and are sharp, indicating that the neutral atoms were first desorbed from the surface of the sample, resonantly excited and then subsequently ionised in the gas phase by the same laser pulse. This is therefore a two step process similar to LARIS and completely different from LIMA. One of its advantages is that it requires a simpler experimental arrangement than LARIS. Recent work has suggested that the width of the peaks, increasing with both increasing laser fluence and angle of incidence, is at least in part caused by atomic collisions within the plume (Wang *et al.* 1992 and all the references therein).

All of the initial work on RLA was carried out on majors. A steel sample (NIST SRM 1263A) containing 5000 ppm Al was chosen to demonstrate that the technique was also sensitive to minor levels. Mass spectra were obtained on and off resonance

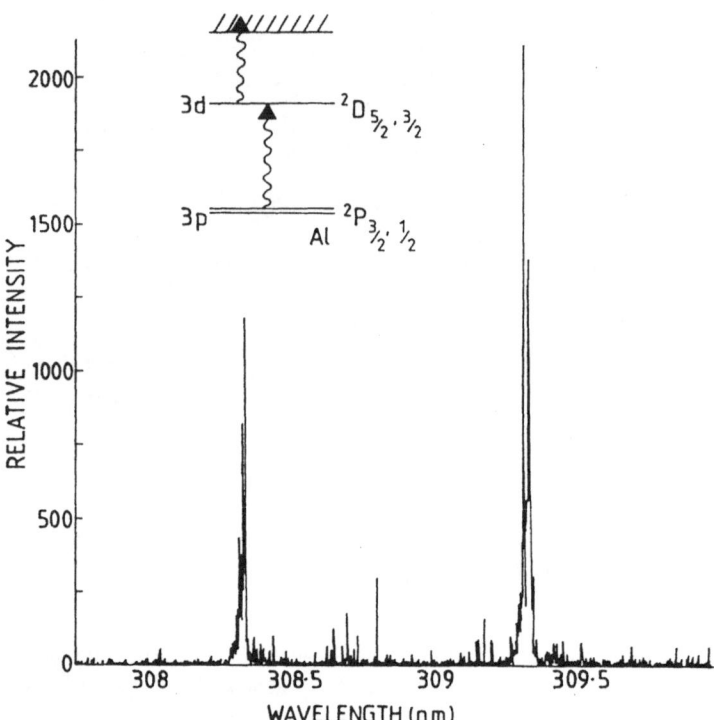

Figure 18. *The ion signal for Al as a function of the ablation laser wavelength. The enhanced yield at wavelengths corresponding to the excitation of the 3d level from the ground state doublet is clearly observed.*

and are shown in Figure 19. It can be seen that the yield of Al ions is highly enhanced at resonance and a sensitivity of a few ppm has been estimated from the signal to noise ratio from this data. A second steel sample (NIST SRM 1261A) containing 500 ppm Al was also analysed which produced an aluminium signal five times smaller than the 1263A sample. The agreement with the nominal ratio (10:1) is considered to be reasonable at this stage of the quantitative development of RLA (Borthwick *et al.* 1992b).

RLA has been demonstrated for calcium, aluminium, gallium and iron from a variety of surfaces, shiny to rough, and from metals, semiconductors and non-conductors. It is felt that the importance of RLA may be greater in connection with molecular desorption and detection from metal and non-metal surfaces.

It should be emphasised that both LARIS and RLA are essentially surface analytical techniques.

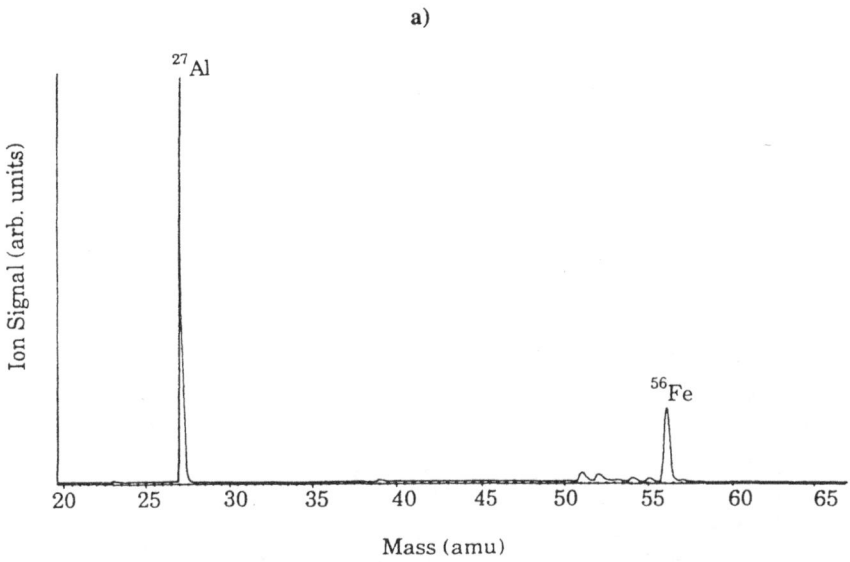

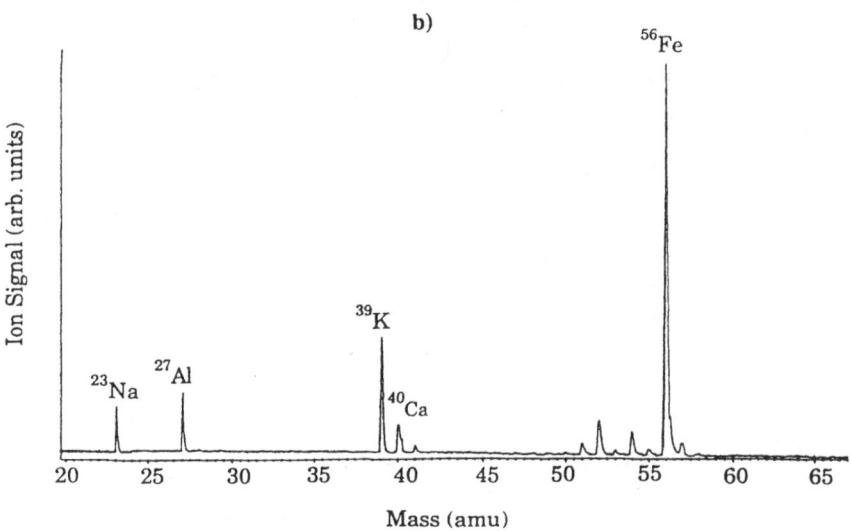

Figure 19. *a) Spectrum of NIST SRM 1263A with the laser tuned to the Al resonance at 308.3nm for 500 shots. b) Off-resonant spectrum at 307.5 nm. Scale approx. ×10 compared to a).*

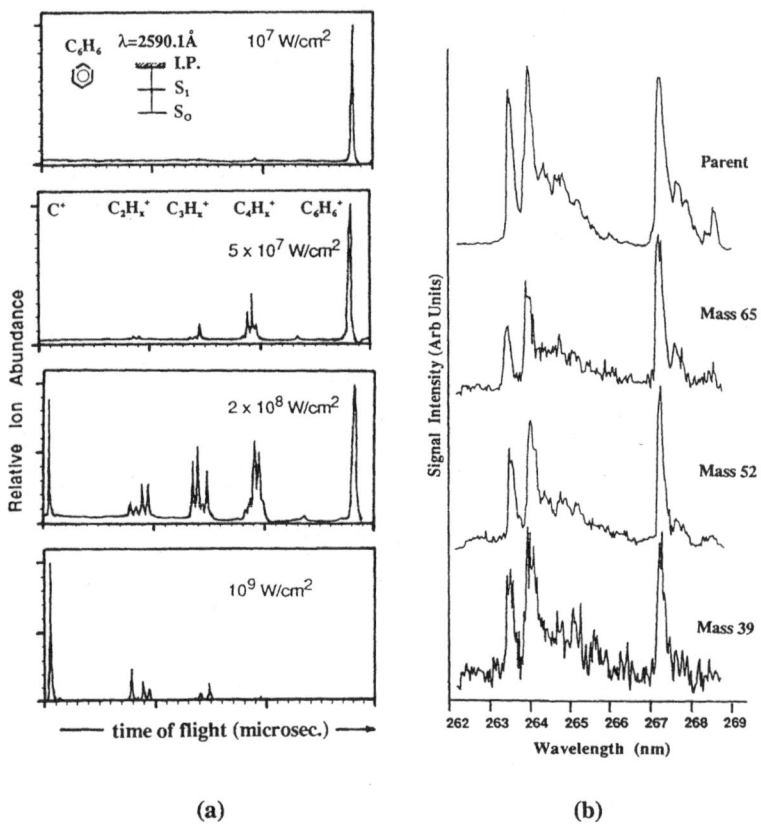

Figure 20. *a) Fragmentation pattern of benzene as a function of laser fluence. As the fluence increases the fragmentation becomes greater. (Reproduced by permission of the authors and Marcel Dekker Inc.); b) Wavelength dependence of the toluene parent and fragment ions. They are identical.*

3.3   The detection of molecules by RIMS

The most common RIMS scheme for analytical purposes using molecules is resonant two-photon ionisation (R2PI) in which one photon excites the molecule to an electronic state and a second photon causes molecular ionisation. Since molecules normally have ionisation potentials between 7 and 13 eV, R2PI require UV photons. Tunable radiation of dye lasers with frequency doubling can provide UV radiation down to 210nm. Excimer lasers are powerful sources of UV and VUV *e.g.* XeCl (308nm), KrF (248nm), ArF (193nm) and F_2 (155nm) and are particularly useful for the photoionisation step in R2PI.

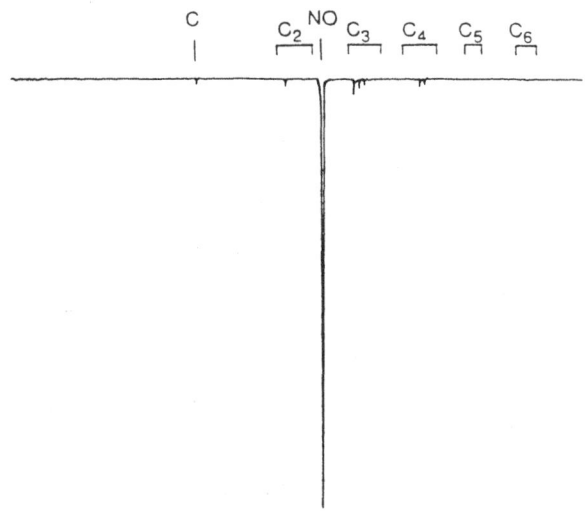

Figure 21. *The fragmentation of nitrobenzene at 226nm. The* NO^+ *is by far the largest feature and can be used as an indicator of the presence of this group of molecules.*

In contrast to atoms, molecules in excited states can undergo different photophysical and photochemical transformations. Photodissociation can compete with photoexcitation and photoionisation processes. For larger laser fluences, the fragmentation of the molecule is more extensive. Figure 20a shows the fragmentation of benzene following an R2PI process, indicating that at low fluence only the parent peak is visible and at high fluence carbon ions are the most abundant fragments (Boesl *et al.* 1987). The fragmentation pattern gives important information regarding molecular structure. In addition the wavelength dependence of the parent or fragment ions can also be used as a fingerprint for trace detection of molecules. Figure 20b gives a comparison of the wavelength dependence of the toluene parent ion with that of some of the more prominent fragment ions. From the great similarity of all the fragment spectra it can be concluded that fragmentation is initiated from the excited states of the parent (Marshall *et al.* 1991).

In a series of recent experiments, REMPI has been used to detect and distinguish nitroaromatic molecules (these include many of the explosive materials). A tunable laser has been used to irradiate nitrobenzene and nitrotoluene in gas phase and it has been observed that at specific wavelengths the NO^+ ion is by far the largest signal (Figure 21). In addition the fragmentation pattern of the two molecules is completely different as can be seen from Figure 22 and moreover the wavelength dependence of the fragments can also be distinguished. It is believed that when these molecules are irradiated with UV then the molecule first photodissociates to a NO_2 molecule and a

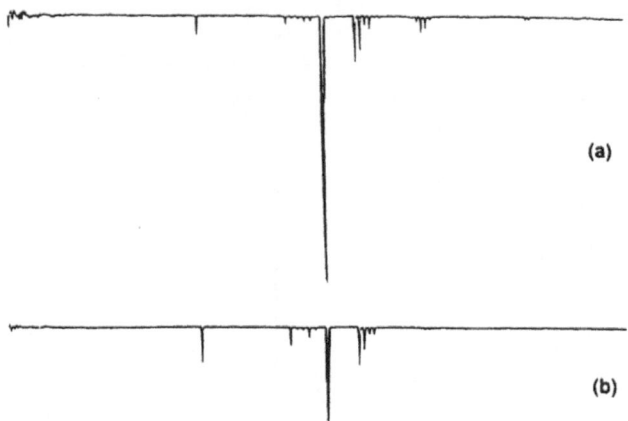

Figure 22. *a) The TOF mass spectrum of nitrobenzene recorded at a wavelength of 245.4nm and a fluence of 8.5mJ/mm² b) The TOF mass spectrum of o-nitrotoluene vapour at a laser wavelength of 246.2nm and a fluence of 20mJ/mm². The fragmentation pattern of the two molecules can be distinguished.*

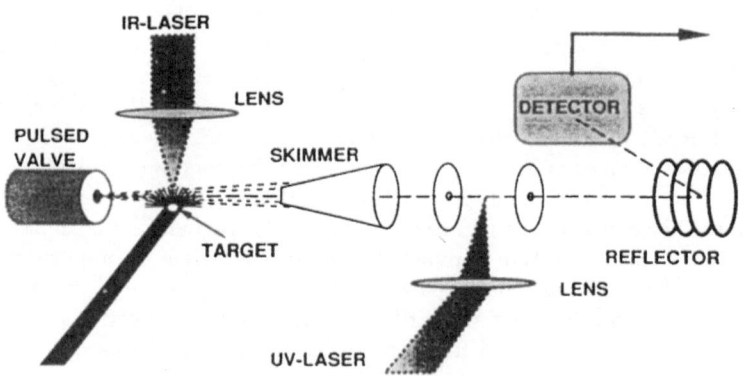

Figure 23. *The sample is irradiated by an IR laser to produce molecules which are seeded into a supersonic jet. The cooled molecules are post ionised non-resonantly at 266nm and then analysed by a reflectron TOF mass spectrometer (Grotemeyer and Schlag 1989).*

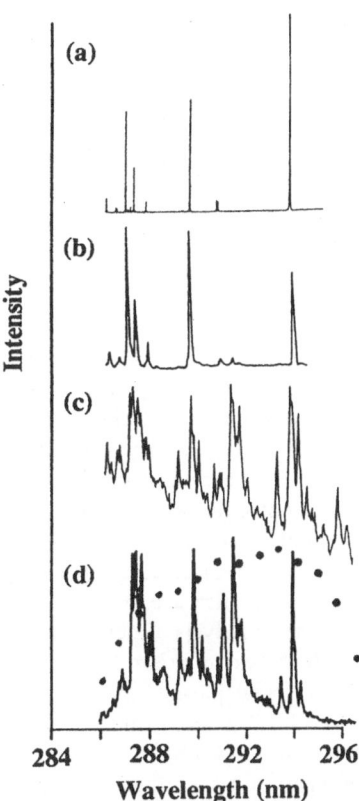

Figure 24. *Aniline spectrum recorded by (a) jet-cooled laser induced fluorescence, (b) jet-cooled R4PI, (c) single-photon UV absorption and (d) R2PI. The laser power profile is also recorded. It is clear that the cooling considerably simplifies the spectrum.*

C_6H_5 or C_7H_7 fragment which are then both ionised by resonant multiphoton processes. With the nitroaromatics studied to date, a large NO^+ is always present at 226nm and can be used as an indicator of this group of molecules. Sensitivity levels lower than 10^5 molecules have been demonstrated. The different types of nitroaromatics can be distinguished by studying the fragmentation patterns of the hydrocarbon ions or the wavelength dependence of these ions (Marshall *et al.* 1992 and references therein). Other groups of molecules can be detected in a similar manner — *e.g.* the PO radical can be used to identify the presence of organophosphonates, of which some are nerve gas agents.

To improve molecular selectivity, albeit at the expense of sensitivity, the molecular rotational degrees of freedom can be reduced to a few sharp atomic-like peaks by introducing the molecules into supersonic beams. Using the Joule-Kelvin effect, rapid cooling is realised by converting the energy of the internal rotational degrees of freedom into translational energy via gas collisions. A system using jet cooling in conjunction with laser desorption and postionisation is shown in Figure 23 (Grotemeyer and Schlag 1989). An interesting comparison of the different techniques of exciting and ionising a molecule is provided in Figure 24 where aniline spectra recorded under various conditions are presented. Jet cooling sharpens the resonance features considerably ((Marshall *et al.* 1991). With conventional analytical techniques molecular isomers are extremely difficult to distinguish from each other. Although they have the same molecular weight, they have different absorption spectra and their ionisation potentials may also be different. Supersonic jet-cooling is essential to reduce the rotational-vibrational structure, so that the distinguishing sharp absorption peaks result. Figure 25 shows R2PI spectra for four isomers of dichlorotoluene. The wavelength spectra were obtained by expanding several ppm of each compound in a 1-atm reservoir of argon into vacuum at 10^{-6} torr (Tembruell *et al.* 1985). If the molecules are not in the gas phase then a variation of laser postionisation of neutral molecules has been developed which involves the desorption of molecules from surfaces essentially intact using an infrared (CO_2) laser, and then after a suitable delay the ionisation of the neutral molecules produced using a quadrupled Nd:YAG laser. The ions are subsequently detected using a TOF. The appropriate delay ($70-90\mu s$) between the desorbing and ionising laser is chosen so that the ionising laser intercepts as many molecules as possible. Zare has called this procedure REMPI although it should be pointed out that the single 266nm ionising wavelength cannot distinguish the different molecules and this is done by the mass spectrometer. If the ionising laser fluence is low, the spectra are dominated by parent peak production. REMPI is especially important for the analysis of non-volatile compounds and those that decompose on heating. Figure 26 shows an example of a cocktail of equimolar amounts of a number of PTH amino acids. The parent peaks labelled a-e are almost equal in height (Engelke *et al.* 1987). In addition REMPI has been shown to be extremely sensitive. By way of example, Hahn *et al.* (1987) have demonstrated how the signal height of the parent peak, for a similar series of substances, can vary with sample concentration and have found a linear dependence ranging over five orders of magnitude. The amount of adsorbate desorbed per laser shot can range from nanomoles (100 monolayers) to femtomoles (10^{-4} monolayers).

4   Conclusions

RIMS has been shown to be a trace analytical technique of considerable potential, qualitatively and quantitatively, for both atoms and molecules. Already several commercial RIMS instruments of various designs are available, while in addition, laser microprobes can be fitted with post-ablation ionisation capacity, both resonant and nonresonant. One of the difficulties that manufacturers identify with RIMS technology is the use of dye lasers, with their inherent bulkiness and operating difficulties. With the development of solid state tunable lasers (Ti-sapphire and optical parametric oscillators), as well as semiconductor diode lasers, it is likely that RIMS will soon reach its full potential and be accorded similar acceptance to SIMS (Ledingham and Singhal 1991).

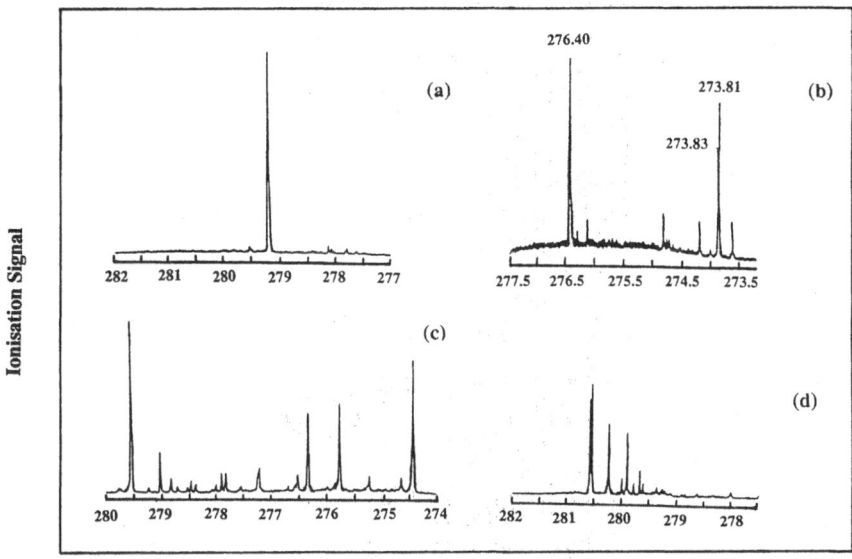

Figure 25. *Resonant two-photon ionisation spectrum taken in an expansion of 1 atm back-pressure of Ar of: a) 2,4 dichlorotoluene b) 2,6 dichlorotoluene c) 2,5 dichlorotoluene d) 3,4 dichlorotoluene. Only the parent molecular ion is monitored in these spectra using a TOF mass spectrometer. (Tembruell et al. 1985) (Reproduced by permission of the authors and the American Chemical Society)*

Acknowledgements

It is a pleasure to acknowledge the staff and research students of the Glasgow University LIS group without whose hard work little of this would have been possible.

References

Ambartzumian R V and Letokhov V S, 1972 *Appl Optics* **11** 354

Arlinghaus H F, Spaar M T and Thonnard N, 1990 *J Vac Sci Technol* **A8** 2318

Bekov G I and Letokhov V S, 1983 *Appl Phys* **B30** 161

Bekov G I, Letokhov V S and Radaev V N, 1985 *J Opt Soc Am* **B2** 1554

Boesl U, Grotemeyer J Walter K, and Schlag E W 1987 *Analytical Instrumentation* **16** 151

Borthwick I S, Ledingham K W D, Sander J and Singhal R P, 1992a to be published

Borthwick I S, Ledingham K W D and Singhal R P, 1992b *Spectrochimica Acta B* to be published

Bushaw B A *Resonance Ionisation Spectroscopy 92, Santa Fe, USA*, 1992 (Inst Phys) to be published

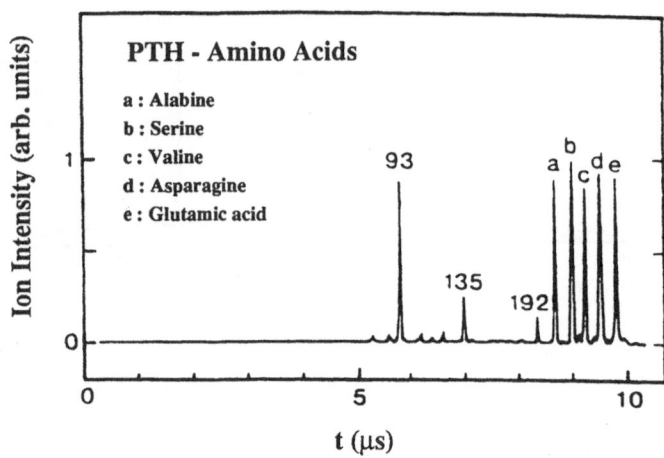

Figure 26. *Laser desorption-multiphoton ionisation TOF mass spectra of an equimolar mixture of five PTH-amino acids (Engelke et al. 1987). (Reproduced by permission of the authors and the American Chemical Society.)*

Calaway W F, Coon S R, Pellin M J, Young C E, Whitten J E, Wiens R C, Gruen D M, Stingeder G, Penka V, Grasserbauer M, and Burnett D S, 1992 *Resonance Ionisation Spectroscopy 92 Santa Fe, USA*, 1992 to be published

Downey S W, Emerson A B, Kopf R F and Kuo J M, 1990 *Surf Int Anal* **15** 781

Downey S W and Hosack R S, 1990 *J Vac Sci Technol* **A8** 791

Downey S W and Emerson A B, 1991 *Anal Chem* **63** 916

Engelke F, Hahn J H, Henke W and Zare R N, 1987 *Anal Chem* **59** 909

Goppert-Mayer M, 1931 *Ann Phys, Leipzig* **9** 273

Grotemeyer J and Schlag E W, 1989 *Acc Chem Res* **22** 399

Hahn J H, Zenobi R and Zare R N, 1987 *J Am Chem Soc* **109** 2842

Hurst G S and Payne M G, 1988 *Principles and Applications of Resonance Ionisation Spectroscopy* (Adam Hilger, Bristol and Philadelphia: 1988)

Ledingham K W D and Singhal R P, 1991 *J Anal Atom Spectrom* **6** 73

Letokhov V S, 1987 *Laser Photoionisation Spectroscopy* (Academic Press, London, 1987)

Marshall A, Clark A, Jennings R, Ledingham K W D, and Singhal R P, 1991 *Meas Sci Technol* **2** 1078

Marshall A, Clark A, Jennings R, Ledingham K W D, and Singhal, 1992 *Int J Mass Spectrom Ion Processes* **112** 273

McCombes P T, Borthwick I S, Jennings R, Ledingham K W D and Singhal R P, 1991 *Optical Spectroscopy* eds R S Stewart and J E Lawler (Inst Phys Conf Ser **113**) 163

Oechsner H, 1973 *Z Phys* **261** 37

Parks J E, Beekman D W, Moore L J, Schmitt H W, Spaar M T Taylor E H, Hutchinson J M R and Fairbank W M Jr, 1986 *Resonance Ionisation Spectroscopy 86* (Inst Phys Conf Ser 84) 157

Tembruell R, Sin C H, Li P, Pang H M and Lubman D M, 1985 *Anal Chem* 57 1186

Thonnard N, Parks J E, Willis R D, Moore L J and Arlinghaus H F, 1989 *Surface and Interface Analysis* **14** 751

Thonnard N, Wright M C, Davis W A and Willis R D, 1992 *Resonance Ionisation Spectroscopy 92 Santa Fe USA* 1992 (Inst Phys) to be published

Townsend P D, Kelly J C, Hartley N E W, 1976 *Ion Implantation, Sputtering and their Applications* (Academic Press, London) p122

Towrie M, Drysdale S L T, Jennings R, Land A P, Ledingham K W D, McCombes P T, Singhal R P, Smyth M H C and McLean C J, 1990 *Int J of Mass Spec and Ion Proc* **26** 309

Wang L, Ledingham K W D, McLean C J and Singhal R P, 1992 *Appl Phys* **B54** 71

APPENDIX 1

List of Acronyms

The following acronyms are used in the papers presented in this volume:

AEM	Analytical Electron Microscopy
AES	Auger Electron Spectroscopy
AFM	Atomic Force Microscope
BLE	Bombardment Induced Light Emission
COAM	Computer Assisted Microscopy
EBIC	Electron Beam Induced Current
EBT	Electron Beam Testing in SEM
EDS	Energy Dispersive Spectroscopy (= EDX)
EDX	Energy Dispersive x-ray Analysis (= EDS)
EELS	Electron Energy-Loss Spectroscopy
ELL	Ellipsometry
EPMA	Electron Probe Micro Analysis (= EMP)
ERD	Elastic Recoil Detection
ESCA	Electron Spectroscopy for Chemical Analysis
EXAFS	Extended x-ray Absorption Fine Structure
GDMS	Glow Discharge Mass Spectrometry
GDOS	Glow Discharge Optical Spectrometry
HEED	High Energy Electron Diffraction
HEIS	High Energy Ion Scattering (= RBS)
HR-XRD	High Resolution Electron Diffraction
HTSC	High Temperature Superconductors
ISS	Ion Scattering Spectroscopy (RBS + LESI)
LAMMA	Laser Microprobe Mass Analysis = LMP
LEED	Low Energy Electron Diffraction
LEIS	Low Energy Ion Scattering
LIMA	Laser Induced Micro Analysis (= LMP)
LMP	Laser Micro Probe (= LIMA)
LMS	Laser Mass Spectrometry
LOES	Laser Optical Emission Spectroscopy
NAA	Neutron Activation Analysis
NRA	Nuclear Reaction Analysis
OBIC	Optical Beam Induced Conductivity
OM	Optical Microscopy
PAM	Photo Acoustic Microscopy
PES	Photo Electron Spectroscopy

/continued

PIXE	Proton Induced x-ray Emission
PRA	Prompt Radiation Analysis
RBS	Rutherford Backscattering Spectrometry (= HEIS)
REM	Reflection Electron Spectrometry
RHEED	Reflection High Energy Electron Diffraction
SAM	Scanning Auger Microprobe
SCAM	Scanning Acoustic Microscopy
SCOM	Scanning Optical Microscopy
SEM	Scanning Electron Microscopy
SIMS	Secondary Ion Mass Spectrometry
SNMS	Secondary Neutrals Mass Spectrometry
STM	Scanning Tunneling Microscopy
TED	Transmitted Electron Diffraction
TEELS	Transmission Electron Energy-Loss Spectrometry
TEM	Transmission Electron Microscopy
TR-XRFA	Total Reflection—x-ray Fluorescence Analysis
UHV-TEM	Ultra High Vacuum–TEM
UHR-TEM	Ultra High Resolution–TEM
UPS	Ultraviolet Photoelectron Spectroscopy
WDS	Wavelength Dispersive Spectroscopy (= WDX)
WDX	Wavelength Dispersive x-ray (detection)
XPS	x-ray Photoelectron Spectroscopy (= ESCA)
XRD	x-ray Diffraction
XRF	x-ray Fluorescence
XRE	x-ray Emission Spectroscopy
XRM	x-ray Microscopy
XRT	x-ray Topography

Contributed Papers

The titles of papers presented by participants at the School, either orally or as posters, are listed below.

Aloupi, E	Investigation of ancient paint layers using SEM/EDX and PIXE microprobe
Aloupogiannis, P	Rare earth silicide thin films study by RBS and NRA methods
Baker, M A	Expert Systems in Electron Spectroscopy
Belu, A M	Polymer characterization using time of flight SIMS
Berger, A	Elemental mapping: image formation and resolution limits
Borthwick, I	Resonant ionisation mass spectrometry for trace analysis of metals
Coufal, J C	The detection of electrons with semiconductor detectors
da Silva, P L N B	Metal-ceramic interfaces for biomaterials
de Jesus, J C	Surface analysis applied to catalysis
Fakkeldij, E J M	Initial states of oxidation of $\gamma' - Fe_4N_{1-x}$ layers studied with microbeam analysis techniques
Farnworth, M A	Surface analysis using a combined LIMA/FAB-SIMS instrument
Faryna, M	An XPS/SAM/SEM/SIMS/EDS multi-technique system – possibilities and limitations
Cao, L L	Chemical structure characterisation of lubrication films using XPS and scanning AES microprobe techniques
Horvath, G	Surface segregation in $SrTiO_3$
Jenkins, S N	Transmission X-ray photoelectron spectroscopy: XPS with single micrometer lateral resolution
Jimenez, C	RBS and NRA as quantitative techniques applied to TiN_x films
Johannson, E L-S	Surface studies on TiO_2 pigments
Kooi, B J	Stages of oxidation of iron-nitride surface layers studied by microbeam analysis
Lekki, J	Tribological properties of implanted silicon crystals
Mercier, F	Application of microanalysis techniques to oilfield rocks
Michaud, V	Applications of microanalysis techniques in earth sciences and environmental areas
Olsson, C	FeCl3 testing of stainless steel evaluated by AESAND XPS
Perez-Campos, R	Characterization of quasicrystalline phases in ternary and quaternary alloys
Plesanovas, A	Photoelectron spectroscopy of poly-vanadium transition metal acids
Prieto, P	XPS and AES study of transition metal nitrides

/continued

Participants

●Miss E Aloupi
Laboratory of Archaeometry
Institute of Materials Science
NCSR Demokritos
15310 Ag. Paraskevi Attiki
Greece

●Dr P Aloupogannis
Demokritos
Nat. Res. Centre for Phys. Sciences
153-0 Ag Paraskevi
PO Box 60228, Attikis
Greece

●Dr M A Baker
Dept. of Materials Science
University of Surrey
Guilford
Surrey GU2 5XH, UK

●Mrs T Bartlett
Materials Divn.
Herschel Building
University of Newcastle
Newcastle upon Tyne NE1 7RU
UK

●Dr S Baunack
Inst. für Festkörper und
Werkstofforschung/ISF
Helmholtzstr. 20
D-O-8027 Dresden
Germany

●Dr P Beccat
I-F-P
1 & 4 Avenue de Bois Preau
92506 Rueil
Malamison Cedex
France

●Ms A M Belu
Dept. of Chemistry, CB #3290
University of North Carolina
Chapel Hill
NC 27599-3290, USA

●Mr R L L Berbara
Dept. of Biological Sciences
University of Dundee
Dundee DD1 4HN, UK

●Mr A Berger
Inst. fur Angewandte Physik
Licht- und Teilchenoptik
6100 Darmstadt
Hochschulstr. 6
Germany

●Dr C L Bianchi
Dept. of Physical Chemistry
University of Milan
Via Golfi 19
20133 Milan
Italy

●Mr I Borthwick
Department of Physics & Astron.
University of Glasgow
Glasgow G12 8QQ, UK

●Prof D Briggs
Wilton Materials Res. Centre
ICI
P.O. Box No.90
Wilton, Middlesbrough
Cleveland TS6 8JE, UK

●Mr M Brogan
Dept of APEME
University of Dundee
Dundee DD1 4HN
Scotland, UK

●Mr M Bronold
Hahn-Meitner-Institut
Bereich S Glienickes Str. 100
W-1000 Berlin 39
Germany

●Prof J Campbell
College of Phys. & Eng. Science
University of Guelph
Ontario
Canada

●Prof L L Cao
Dept of Chemistry
Tsinghua University
Beijing 100084
China

•Miss M J Caturla
Jaune I, 13, 3th
Departament de Fisica Aplicada
Universitat d'Alacant
99 Alacant
Spain

•Prof J Cazaux
L.A.S.S.I.
Universite de Reims
B.P.347 Faculte des Sciences
51062 Reims Cedex
France

•Miss H Chen
Boeretang 200
Vito
B-2400 Mol
Belgium

•Miss A Cook
Department of APEME
University of Dundee
Dundee DD1 4HN
Scotland, UK

•Mr J C de Jesus
INTEVEP S.A.
Los Teques
P.O. Box 76343
Venezuela

•Dr M T Domenech Carbo
Dept. de Conserv. y Rest.
Facultad de Bellas Artes
Universitat de Valencia
Camino de Vera 14
46022-Valencia
Spain

•Miss J Dumville
Dept. of Mat. Science & Metall.
University of Cambridge
Pembroke Street
Cambridge, UK

•Prof T N Durlu
Department of Physics
Ankara University
Besevler
06100 Ankara
Turkey

•Prof R F Egerton
Dept of Physics
University of Alberta
Edmonton, Alberta
Canada T6G 2J1

•Dr D Fabian
Dept. of Physics & Appl. Phys.
University of Strathclyde
107 Rottenrow
Glasgow G4 0NG, UK

•Miss E J M Fakkeldij
Lab. of Metallurgy
Delft University of Technology
Rotterdamseweg 137
2628 AL Delft
The Netherlands

•Dr C Fanizza
c/o ISPESL
Monteporzio Catone (RM) 00040
Italy

•Mr M A Farnworth
Pilkington Technology Centre
Hall Lane
Lathom, nr. Ormskirk
Lancashire L40 5UF, UK

•Dr Marek Faryna
Jagiellonian University
Slafibs
ul. Karasia 3
30-060 Krackow
Poland

•Dr D Finlayson
Department of Physics and Astronomy
University of St Andrews
St Andrews, Fife
Scotland, UK

•Prof A G Fitzgerald
Dept of APEME
University of Dundee
Dundee DD1 4HN
Scotland, UK

•Mr H Fonseca
Dept. of Biological Sciences
University of Dundee
Dundee DD1 4HN
Scotland, UK

•Dr J V Gimeno Adelantado
Dept. de Quimica Analitica
Facultad de Quimica
Universitat de Valencia
c/ Doctor Moliner 50
46100-Burjassot (Valencia)
Spain

•Mr J Goward
Dept of APEME
University of Dundee
Dundee, DD1 4HN
Scotland, UK

•Dr R Gurler
Anadolu University
Metalurji Enstitusu Yunusemre
Kampusu
Eskisehir 26470
Turkey

•Dr H Hammer
Inst. für Angewandte Physik
Heinrich Heine Univ. Düsseldorf
Universtätsstrasse 1
D 4000 Düsseldorf
Germany

•Mr G Horvath
Dept. of Atomic Physics
Technical University of Budapest
1111 Budapest
Budafoki ut 8
Hungary

•Mr S N Jenkins
Department of Material Science
University of Surrey
Guildford
Surrey GU2 5XH, UK

•Mrs L-S Johansson
Materials Research
University of Turku & Abo Akedemi
ElectroCity, Tykistokatu 2 D
SF-20520 Turku
Finland

•Dr P John
Dept. of Chemistry
Heriot-Watt University
Riccarton
Edinburgh EH14 4AS
Scotland, UK

•Prof D Joy
EM Facility
University of Tennessee
F239 Life Sciences Building
Knoxville
Tennessee 37996-0810, USA

•Mr R Keatch
Dept of APEME
University of Dundee
Dundee, DD1 4HN
Scotland, UK

•Ir. B J Kooi
Laboratory of Metallurgy
Delft University of Technology
Rotterdamesweg 137
NL-2628 AL Delft
The Netherlands

•Prof P Kruit
Faculty of Applied Physics
Delft University of Technology
P.O. Box 5046
2600 GA Delft
The Netherlands

•Dr S Kuypers
Vito (Materials Section)
Boeretang 200
B-2400 Mol
Belgium

•Dr C Lamberto
c/o ISPESL
00040 Monteporzio Catone (RM)
Italy

•Ms L Lanier
ENS des Mines de Nancy
Lab. de Science et Genie d. Surface
Parc de Saurupt
54042 Nancy Cedex
France

•Dr K W D Ledingham
Dept. of Physics & Astronomy
University of Glasgow
Glasgow G12 8QQ, UK

•Miss E R Leitao
Dept of Metallurgy
Engineering Faculty
Oporto University
Rua dos Bragas
4099 Porto Codex
Portugal

•Mr J Lekki
Dept. of Applied Spectroscopy
Institute of Nuclear Physics
ul. Radzikowskiego 152
31-342 Cracow
Poland

•Mr S Li
Department of APEME
University of Dundee
Dundee DD1 4HN
Scotland, UK

•Miss R Maldonado
Institut für Metallforsch. Münster
Wilhelm-Klemm Strasse 10
D-4400 Munster
Germany

•Dr E Maydell
Dept. of Metallurgy & Eng. Mat.
University of Strathclyde
Colville Building
48 North Portland Street
Glasgow G1 1XN
Scotland, UK

•Miss F Mercier
Laboratoire 'Pierre Sue'
Groupe Physico-Chimie
Bat. 637 - CE Saclay
91191 Gif-sur-Yvette Cedex
France

•Dr V Michaud
Laboratoire 'Pierre Sue'
Groupe Environnement Continental
CE - Saclay
Bat. 637 - 91191 Gif-sur-Yvette
France

•Ms S S Montoro
Department of Chemistry
Univ. Instelling Antwerpen
B-2610 Antwerpen-Wilrijk
Belgium

•Mr C O Olsson
Materials Centre
P.O. Box 764
S-78127 Borlange
Sweden

•Dr E Pavlidou
Physics Dept. - Solid State
Aristotle University of Thesaloniki
54006 Thesaloniki
Greece

•Miss L C Pereira
Dept. of Chemistry
University of Lisbon
LNETI/ICEN
P.2685
Portugal

•Mrs T P Pereira da Silva
Centro de Cristalografia e Minerl.
Al. D. Afonso Henriques
41 ESQ.
1000 Lisboa
Portugal

•Dr R Perez-Campos
Lab. de Cuernavaca Unam
P.O. Box 139-B
62191 Cuernavaca Mor.
Mexico

•Mr A Plesanovas
Semiconductor Physics Institute
Lithuanian Academy of Science
A. Gostauto 11
Vilnius 232600
Lithuania

•Mr A Ploessl
Department of APEME
University of Dundee
Dundee DD1 4HN
Scotland, UK

•Prof M Prutton
Department of Physics
University of York
Heslington
York, UK

•Ing M P Raanes
Dept. of Metallurgy
NTH
N-7034 Trondheim
Norway

•Dr V Radmilovic
c/o Dr W A Soffa
Dept. of Materials Sci. and Eng.
University of Pittsburgh
848 Benedum Hall
Pittsburgh PA 15261, USA

•Mr E Redmard
Dept of Electrical Engineering-
Physical Electronics
Tel-Aviv University
Tel-Aviv 69978
Israel

•Miss S Rio
Laboratoire 'Pierre Sue'
Centre d'Etudes Nucl, de Saclay
91191 Gij Sur Yvette Cedex
France

•Miss A G Rolo
Departmento de Fisica
Universidade do Minho
Lago do Paco
4700 Braga
Portugal

•Mr M Rose
Dept of APEME
University of Dundee
Dundee, DD1 4HN
Scotland, UK

•Miss A T C Santos
Departamento de Eng. Quimica
FEUP
Universidade do Porto
Rua dos Bragas
4099 Porto Codex
Portugal

•Dr A Santucci
SNAMPROGETTI Research Labs.
Via Maritano
n. 26 - I-20097 S. Donato Milanese
Milano
Italy

•Mr C Scott
Dept. of Physics and Astronomy
University of Glasgow
Glasgow
Scotland, UK

•Dr M P Seah
Division of Materials Metrology
National Physical Laboratory
Teddington
Middlesex, UK

•Mr L J Seijbel
Faculty of Applied Physics
Delft University of Technology
Lorentzweg 1
2628 CJ Delft
The Netherlands

•Mr D M Shirokov
Lab of Neutron Physics
Joint Inst for Nuclear Research
141980 Dubna,Moscow Region
Russia

•Ms P L N Silva
Dept. Eng. Metalurgica
R. dos Bragas
4099 Porto Codex
Portugal

•Miss J M Smith
Dept. of Physics and Astronomy
University of Glasgow
Glasgow G12 8QQ
Scotland, UK

•Prof J K Solberg
Dept. of Metallurgy
N-7034 Trondheim - NTH
Norway

•Mr B Stenbom
Department of Physics
Chalmers University of Technology
S-412 96 Goteburg
Sweden

•Miss C Steukers
Rue de College 9
5000 Namur
Belgium

•Dr B E Storey
Dept of APEME
University of Dundee
Dundee DD1 4HN
Scotland, UK

•Dr J M Titchmarsh
Materials & Chemistry Division
Harwell Laboratory
AEA Technology
Oxfordshire OX11 0RA, UK

•Prof P Trebbia
L.A.S.S.I.
Univ. de Reims Champagne-Ardenne
U.F.R. Sciences
B.P. 347 F51062 Reims Cedex
France

•Mr J F Trigo
Departmento Fisica Aplicada
C-XII Univ. Autonoma de Madrid
28049 Madrid
Spain

•Dr M L Trudeau
Hydro-Quebec Research Institute
1800 Montee Ste-Julie
Varennes
Quebec J3X 1S1
Canada

•Dr A O Tooke
Dept of APEME
University of Dundee
Dundee DD1 4HN
Scotland, UK

•Mr N M Uzunov
Department of Applied Physics
University of Shumen
"K Preslavsky"
Shumen 9700
Bulgaria

•Prof G Van der Laan
Daresbury Laboratory
Warrington
Cheshire WA44 4AD, UK

•Dr P Van Espen
Department of Chemistry
University of Antwerp
Universiteitsplein 1
B2610 Antwerp-Wilrijk
Belgium

•Dr H Viefhaus
Max-Planck-Institut für Eisenf.
Max-Planck-Str. 1
D4000 Dusseldorf
Germany

•Mr J M Walton
Department of Physics
University of York
Heslington
York YO1 5DD, UK

•Dr H Watton
Department of APEME
University of Dundee
Dundee DD1 4HN
Scotland, UK

•Mr R Watts
Department of Physics
University of York
Heslington
York YO1 5DD, UK

•Prof H Werner
Philips Research Laboratories
P.O. Box 80 000
5600 JA Eindhoven
The Netherlands

•Prof D B Williams
Dept. of Materials Science and Eng.
Lehigh University
Whitaker Laboratory # 5
Bethlehem PA 18015-3195, USA

•Dr K Wittmaack
ATOMIKA Analysetechnik
Bruckamnnring 6
D-8042 Oberschliesshelm
Germany

•Mr R A Wood
321 Stardust Lane
Columbia, MO 65201, USA

•Mr F Yubero
Departamento Fisica Aplicada
C-XII Univ. Autonoma Madrid
28049 Madrid
Spain

•Mr S M Zemyan
Dept. of Mat. Sci. and Eng.
Lehigh University
Whitaker Laboratory, Lab. #5
Bethlehem
Pennsylvania 18015, USA

•Dr I Zotov
Institute of Metal Physics
Vernadsky str 36
Kiev
Ukraine

Index